Graduate Texts in Physics

Graduate Texts in Physics publishes core learning/teaching material for graduate- and advanced-level undergraduate courses on topics of current and emerging fields within physics, both pure and applied. These textbooks serve students at the MS- or PhD-level and their instructors as comprehensive sources of principles, definitions, derivations, experiments and applications (as relevant) for their mastery and teaching, respectively. International in scope and relevance, the textbooks correspond to course syllabi sufficiently to serve as required reading. Their didactic style, comprehensiveness and coverage of fundamental material also make them suitable as introductions or references for scientists entering, or requiring timely knowledge of, a research field.

Michele Fabrizio

A Course in Quantum Many-Body Theory

From Conventional Fermi Liquids to Strongly Correlated Systems

 Springer

Michele Fabrizio
Department of Physics
International School for Advanced Studies
Trieste, Italy

ISSN 1868-4513   ISSN 1868-4521 (electronic)
Graduate Texts in Physics
ISBN 978-3-031-16307-4   ISBN 978-3-031-16305-0 (eBook)
https://doi.org/10.1007/978-3-031-16305-0

Cover image: Olga_Kostrova

This Springer imprint is published by the registered company Springer Nature Switzerland AG
The registered company address is: Gewerbestrasse 11, 6330 Cham, Switzerland

To my family, especially my wife Laura

Preface

This book collects the Lecture Notes of the Ph.D. course on Many-Body Theory that I have been teaching for many years now. As such, the book is not intended to be a compendium of all possible topics in Many-Body Theory, but just aims to provide Ph.D. students with concepts and tools that I personally consider indispensable to tackle modern issues of Many-Body Theory, most notably strongly correlated electron systems. For that same reason, the book unavoidably reflects my own limited research experience.

Besides a core part, Chaps. 1–4, where basic tools are presented in detail, there are frequent excursions in topics of strongly correlated electron systems, among which the two final chapters are on Luttinger Liquids and the Kondo Effect. In addition, the microscopic derivation of Landau's Fermi Liquid Theory in Chap. 5 includes conventional Fermi Liquids as a special case, but also less conventional states, like, for instance, the Mott insulators of Sect. 5.6 that display the same spin and thermal properties of normal metals, which might be relevant to strongly correlated materials.

All chapters contain very few simple problems, whose solution is not even shown. In addition, most chapters include sections with applications that are simply more involved problems whose solution is explicitly derived.

Trieste, Italy   Michele Fabrizio

Acknowledgements I am grateful to Prof. Claudio Castellani, whose master's course on Quantum Many-Body Theory first sparked my interest in the field, and to Profs. Philippe Noziéres and Erio Tosatti, who nourished and mentored that interest.

Contents

Second Quantization

1

The first difficulty encountered in solving a many-body problem is how to deal with a many-body wavefunction. The reason is that a many-body wavefunction has to take into account both the indistinguishability of the particles as well as their statistics, whereas in first quantisation any operator, including the Hamiltonian, does not have those properties. Therefore, it is desirable to have at disposal an alternative scheme where the indistinguishability principle as well as the statistics of the particles are already built in the expression of operators. That is actually the scope of second quantisation.

1.1   Fock States and Space

Let us take a system of N particles, either fermions or bosons. The Hilbert space spans a basis of N-body orthonormal wavefunctions which should satisfy both the indistinguishability principle as well as the appropriate statistics of the particles. The simplest way to construct this space is as follows.

We start by choosing an orthonormal basis of single-particle wavefunctions:

$$\phi_a(x), \qquad a = 1, 2, \ldots.$$

Here x is a generalised coordinate that includes both the space coordinate $\mathbf{r}$ as well as, e.g., the z-component, σ, of the spin, which is half-integer for fermions and integer for bosons. The suffix a is a quantum label and, by definition,

$$\int dx \, \phi_a^*(x)\, \phi_b(x) = \delta_{ab}, \quad \sum_a \phi_a^*(x)\, \phi_a(y) = \delta(x - y), \qquad (1.1)$$

© The Author(s), under exclusive license to Springer Nature Switzerland AG 2022
M. Fabrizio, *A Course in Quantum Many-Body Theory*, Graduate Texts in Physics,
https://doi.org/10.1007/978-3-031-16305-0_1

where $\int dx \ldots$ means $\sum_\sigma \int d\mathbf{r} \ldots$, while $\delta(x - y) \equiv \delta(\mathbf{r} - \mathbf{r}')\delta_{\sigma\sigma'}$ with $x = (\mathbf{r}, \sigma)$ and $y = (\mathbf{r}', \sigma')$. A generic N-body wavefunction with the appropriate symmetry properties can be constructed through the above single-particle states. Since the particles are not distinguishable, we do not need to know which particle occupies a specific state. Instead, what we need to know are just the occupation numbers n_a's, i.e., the number of particles occupying each single-particle state ϕ_a. That number is either $n_a = 0, 1$ for fermions, because of the Pauli principle, or an arbitrary integer $n_a \geq 0$ for bosons. Apart from that, the occupation numbers should satisfy the trivial particle-conservation constraint

$$\sum_a n_a = N.$$

Since the occupation numbers are the only ingredients we need in order to build up the N-body wavefunction, we can formally denote the latter as the *ket*

$$|n_1, n_2, \ldots\rangle, \tag{1.2}$$

which is called a *Fock state*, while the space spanned by the Fock states is called *Fock space*. Within the Fock space, the state with no particles, the vacuum, will be denoted by $|0\rangle$.

For instance, if the N fermions with coordinates x_i, $i = 1, \ldots, N$, occupy the states a_j, $j = 1, \ldots, N$, with $a_1 < a_2 < \cdots < a_N$, namely $n_a = 1$ for $a \in \{a_j\}$, otherwise $n_a = 0$, then the appropriate wavefunction is the Slater determinant

$$\Psi_{\{n_a\}}(x_1, \ldots, x_N) = \sqrt{\frac{1}{N!}} \begin{vmatrix} \phi_{a_1}(x_1) & \cdots & \phi_{a_1}(x_N) \\ \phi_{a_2}(x_1) & \cdots & \phi_{a_2}(x_N) \\ \vdots & \vdots\vdots\vdots & \vdots \\ \phi_{a_N}(x_1) & \cdots & \phi_{a_N}(x_N) \end{vmatrix}, \tag{1.3}$$

which is therefore the first-quantisation expression of the Fock state $|\{n_a\}\rangle$ with the same occupation numbers. The above wavefunction satisfies the condition of being antisymmetric if two coordinates are interchanged, namely two columns in the determinant. Analogously, it is antisymmetric by interchanging two rows, i.e., two quantum labels.

On the contrary, if we have N bosons with coordinates x_i, $i = 1, \ldots, N$, which occupy the states a_j, $j = 1, \ldots, M$, $M \leq N$ and $a_1 < a_2 < \cdots < a_M$, with occupation numbers n_{a_j}, then the appropriate wavefunction is the permanent

$$\Phi_{\{n_a\}}(x_1, \ldots, x_N) = \sqrt{\frac{\prod_j n_j!}{N!}} \sum_P \phi_{a_1}(x_{p_1}) \ldots \phi_{a_1}(x_{p_{n_{a_1}}})$$

$$\phi_{a_2}(x_{p_{n_{a_1}+1}}) \ldots \phi_{a_2}(x_{p_{n_{a_1}+n_{a_2}}}) \tag{1.4}$$

$$\ldots \phi_{a_M}(x_{p_{N-n_{a_M}+1}}) \ldots \phi_{a_M}(x_{p_N}),$$

where the sum is over all non-equivalent permutations p's of the N coordinates.[1] Indeed, the wavefunction is even by interchanging two coordinates or two quantum labels.

In conclusion, the space spanned by all possible Slater determinants built with the same basis set of single-particle wavefunctions constitutes an appropriate Hilbert (Fock) space for many-body fermionic wavefunctions. Analogously, the space spanned by all possible permanents is an appropriate Hilbert space for many-body bosonic wavefunctions.

In the following, we will introduce operators acting in the Fock space. We will consider separately the fermionic and bosonic cases.

1.2 Fermionic Operators

Let us introduce the creation, $c_a^\dagger$, and annihilation, c_a, operators which add or remove, respectively, one fermion in state a. The operator $c_a^\dagger c_a$ first annihilates then creates a particle in a, which can be done as many times as many particles occupy that state. Therefore,

$$c_a^\dagger c_a |n_a\rangle = n_a |n_a\rangle, \tag{1.5}$$

so it acts like the occupation number operator $c_a^\dagger c_a \equiv \hat{n}_a$. Since by the Pauli principle $n_a = 0, 1$, then

$$c_a^\dagger c_a | 0\rangle = 0, \quad c_a^\dagger c_a | n_a = 1\rangle = | n_a = 1\rangle. \tag{1.6}$$

Analogously, the operator $c_a c_a^\dagger$ first creates then destroys a fermion in state a, which it cannot do if a is occupied because of the Pauli principle, while it can if it is empty. Therefore,

$$c_a c_a^\dagger | 0\rangle = | 0\rangle, \quad c_a c_a^\dagger | n_a = 1\rangle = 0. \tag{1.7}$$

Thus, either a is empty or occupied, the following equation holds:

$$\left(c_a c_a^\dagger + c_a^\dagger c_a\right) | n_a\rangle = | n_a\rangle, \quad n_a = 0, 1, \tag{1.8}$$

which leads to the operator identity

$$\left(c_a c_a^\dagger + c_a^\dagger c_a\right) = \{c_a, c_a^\dagger\} = 1, \tag{1.9}$$

where the symbol $\{\ldots\}$ means the anti-commutator. Moreover, since we cannot create nor destroy two fermions in the same state, it also holds that

$$\{c_a, c_a\} = \{c_a^\dagger, c_a^\dagger\} = 0. \tag{1.10}$$

---

[1] Non-equivalent means for instance that $\phi_i(x)\phi_i(y)$ is equivalent to $\phi_i(y)\phi_i(x)$.

Equations (1.9) and (1.10) are the anti-commutation relations satisfied by the fermion operators with the same quantum label. Going back to (1.6) and (1.7), we readily see that all are satisfied if

$$c_a^\dagger \,|\, 0\rangle = |\, n_a = 1\rangle, \quad c_a^\dagger \,|\, n_a = 1\rangle = 0, \quad c_a \,|\, 0\rangle = 0, \quad c_a \,|\, n_a = 1\rangle = |\, 0\rangle, \tag{1.11}$$

also showing that $c_a^\dagger$ is the Hermitian conjugate of c_a.

Let us now consider the action of the above operators on a Fock state. First, we need to provide a prescription to build a Fock state by means of the creation operators. We shall assume that, if

$$|\, n_a = 1\rangle = c_a^\dagger \,|\, 0\rangle,$$

then by definition and for $b \neq a$

$$c_b^\dagger \,|\, n_a = 1\rangle = c_b^\dagger c_a^\dagger \,|\, 0\rangle \equiv |\, n_b = 1, n_a = 1\rangle. \tag{1.12}$$

Since the Slater determinant, hence the corresponding Fock state, is odd by interchanging two rows, then

$$|\, n_b = 1, n_a = 1\rangle = -\,|\, n_a = 1, n_b = 1\rangle \equiv -c_a^\dagger c_b^\dagger \,|\, 0\rangle. \tag{1.13}$$

Comparing (1.13) with (1.12), we conclude that

$$c_a^\dagger c_b^\dagger = -c_b^\dagger c_a^\dagger \;\Rightarrow\; \left\{ c_a^\dagger, c_b^\dagger \right\} = 0. \tag{1.14}$$

The Hermitian conjugate thus implies that also

$$\left\{ c_a, c_b \right\} = 0. \tag{1.15}$$

We note that because of (1.10), both (1.14) and (1.15) remain valid even if $a = b$.

Finally, we need to extract the reciprocal properties of $c_b^\dagger$ and c_a for $a \neq b$. We assume the following result:

$$c_a \,|\, n_a = 1, n_b = 1\rangle = c_a c_a^\dagger c_b^\dagger \,|\, 0\rangle \equiv |\, n_b = 1\rangle = c_b^\dagger \,|\, 0\rangle. \tag{1.16}$$

Since $|\, 0\rangle = c_a c_a^\dagger \,|\, 0\rangle$, and $c_a^\dagger c_b^\dagger = -c_b^\dagger c_a^\dagger$, it follows that

$$-c_a c_b^\dagger c_a^\dagger \,|\, 0\rangle = c_b^\dagger c_a c_a^\dagger \,|\, 0\rangle,$$

namely that, for $a \neq b$,

$$\left\{ c_a, c_b^\dagger \right\} = 0. \tag{1.17}$$

All the above anti-commutation relations can be cast in the simple formulas

$$\boxed{\left\{c_a, c_b^\dagger\right\} = \delta_{ab}, \quad \left\{c_a^\dagger, c_b^\dagger\right\} = 0, \quad \left\{c_a, c_b\right\} = 0, \quad \forall\, a, b}. \qquad (1.18)$$

If we extend our prescription (1.12) to more than two electrons, we find that any Fock state has the simple expression

$$|n_1, n_2, \dots\rangle = \prod_{i \geq 1} \left(c_i^\dagger\right)^{n_i} |0\rangle, \qquad (1.19)$$

where the occupation numbers $n_i = 0, 1$.

1.2.1 Fermi Fields

Till now we have defined fermionic operators for a given set of single-particle wavefunctions. Let us now introduce new operators which are independent of that choice. We define annihilation and creation Fermi fields by

$$\Psi_\sigma(\mathbf{r}) \equiv \Psi(x) = \sum_i \phi_i(x)\, c_i \,, \quad \Psi(x)^\dagger = \sum_i \phi_i(x)^*\, c_i^\dagger \,. \qquad (1.20)$$

They have the following anti-commutation relations:

$$\begin{aligned}
\left\{\Psi(x), \Psi(y)^\dagger\right\} &= \sum_{ij} \phi_i(x)\, \phi_j(y)^* \left\{c_i, c_j^\dagger\right\} \\
&= \sum_i \phi_i(x)\, \phi_i(y)^* = \delta(x - y),
\end{aligned} \qquad (1.21)$$

as well as

$$\left\{\Psi(x), \Psi(y)\right\} = \left\{\Psi(x)^\dagger, \Psi(y)^\dagger\right\} = 0, \qquad (1.22)$$

which are indeed independent of the basis. If we change the basis via the unitary transformation $\hat{U}$ acting on the basis set

$$\phi_i(x) = \sum_\alpha U_{i,\alpha}\, \phi_\alpha(x),$$

with unitary $\hat{U}$, then

$$\Psi(x) = \sum_i \phi_i(x)\, c_i = \sum_{i,\alpha} \phi_\alpha(x)\, U_{i,\alpha}\, c_i = \sum_\alpha \phi_\alpha(x)\, c_\alpha, \qquad (1.23)$$

showing that the proper transformation of the fermionic operators is

$$c_\alpha = \sum_i U_{i,\alpha}\, c_i\,.$$ (1.24)

Through the Fermi fields we can give a representation of the N-fermion Fock space not through a basis of single-particle wavefunctions but directly in the space of the coordinates. Specifically, such Fock space is spanned by the states

$$|\,x_1,\dots,x_N\rangle \equiv \frac{1}{\sqrt{N!}}\,\Psi^\dagger(x_1)\dots\Psi^\dagger(x_N)\,|\,0\rangle\,,$$ (1.25)

which are indeed antisymmetric upon exchanging two coordinates. We note that, through (1.1) and (1.20),

$$\int dx_1\dots dx_N\ |\,x_1,\dots,x_N\rangle\langle x_1,\dots,x_N\,|$$

$$= \frac{1}{N!}\int dx_1\dots dx_N\,\Psi^\dagger(x_1)\dots\Psi^\dagger(x_N)\,|\,0\rangle\langle 0\,|\,\Psi(x_N)\dots\Psi(x_1)$$

$$= \frac{1}{N!}\sum_{a_1,\dots,a_N} c^\dagger_{a_1}\dots c^\dagger_{a_N}\,|\,0\rangle\langle 0\,|\,c_{a_N}\dots c_{a_1}$$ (1.26)

$$= \sum_{a_1<a_2\cdots<a_N} c^\dagger_{a_1}\dots c^\dagger_{a_N}\,|\,0\rangle\langle 0\,|\,c_{a_N}\dots c_{a_1} \equiv \mathbb{I}_N\,,$$

where $\mathbb{I}_N$ is the identity in the N-fermion Hilbert space. In addition, it is straightforward to show that the Slater determinant identified by the occupation numbers $\{n_a\} \equiv (n_1, n_2 \dots)$ can be simply written as

$$\Psi_{\{n_a\}}(x_1,\dots,x_N) = \langle x_1,\dots,x_N\,|\,n_1,n_2,\dots\rangle$$

$$= \frac{1}{\sqrt{N!}}\,\langle 0\,|\,\Psi(x_N)\dots\Psi(x_1)\,|\,n_1,n_2,\dots\rangle\,.$$ (1.27)

1.2.2 Second Quantisation of Multiparticle Operators

Let us consider a generic single-particle operator in the first quantisation

$$\hat{V} = \sum_{i=1}^{N} V(x_i)\,,$$ (1.28)

where the sum runs over all N particles and $V(x_i)$ is an operator acting both on space coordinates and spins, namely

$$V(x) \equiv V_{\sigma\sigma'}(\mathbf{r}, \mathbf{p})\,,$$ (1.29)

where $\mathbf{p} = -i\hbar\nabla$ is the conjugate momentum of $\mathbf{r}$.

Using (1.27) and the indistinguishability of the fermions, the matrix element between two different Slater determinants can be shown to be

$$\langle n_1', n_2', \ldots \mid \hat{V} \mid n_1, n_2, \ldots \rangle$$

$$\equiv \sum_{i=1}^{N} \int \prod_{i=1}^{N} dx_i \, \Psi_{\{n_a'\}}^*(x_1, \ldots, x_N) \, V(x_i) \, \Psi_{\{n_a\}}(x_1, \ldots, x_N)$$

$$= N \int \prod_{i=1}^{N} dx_i \, \Psi_{\{n_a'\}}^*(x_1, \ldots, x_N) \, V(x_1) \, \Psi_{\{n_a\}}(x_1, \ldots, x_N) \qquad (1.30)$$

$$= \frac{N}{N!} \int \prod_{i=1}^{N} dx_i \, V(x_1) \, \langle n_1', n_2', \ldots \mid \Psi^\dagger(x_1) \Psi^\dagger(x_2) \ldots \Psi^\dagger(x_N) \mid 0 \rangle$$

$$\langle 0 \mid \Psi(x_N) \ldots \Psi(x_2) \Psi(x_1) \mid n_1, n_2, \ldots \rangle \, .$$

We note that the integral over $x_2, \ldots, x_N$ can be readily performed through (1.26)

$$\frac{N}{N!} \int dx_2 \ldots dx_N \, \Psi^\dagger(x_2) \ldots \Psi^\dagger(x_N) \mid 0 \rangle \langle 0 \mid \Psi(x_N) \ldots \Psi(x_2) = \mathbb{I}_{N-1} \, , \quad (1.31)$$

so that, dropping the identity operator $\mathbb{I}_{N-1}$,

$$\langle n_1', n_2', \ldots \mid \hat{V} \mid n_1, n_2, \ldots \rangle = \langle n_1', n_2', \ldots \mid \int dx_1 \, \Psi^\dagger(x_1) \, V(x_1) \, \Psi(x_1) \mid n_1, n_2, \ldots \rangle \, .$$
$$(1.32)$$

It thus follows that the second quantised expression of $\hat{V}$ is simply

$$\hat{V} = \int dx \, \Psi^\dagger(x) \, V(x) \, \Psi(x) \quad = \sum_{ab} V_{ab} \, c_a^\dagger c_b \, , \qquad (1.33)$$

where

$$V_{ab} \equiv \int dx \, \phi_a^*(x) \, V(x) \, \phi_b(x) = \sum_{\sigma\sigma'} \int d\mathbf{r} \, \phi_a^*(\mathbf{r}, \sigma) \, V_{\sigma\sigma'}(\mathbf{r}, \mathbf{p}) \, \phi_b(\mathbf{r}, \sigma').$$
$$(1.34)$$

As an example, let us consider the density operator that, in first quantisation, reads

$$\hat{\rho}_\sigma(\mathbf{r}) = \sum_{i=1}^{N} \delta(\mathbf{r} - \mathbf{r}_i) \, \delta_{\sigma,\sigma_i} \, ,$$

while in second quantisation is

$$\hat{\rho}_\sigma(\mathbf{r}) = \sum_{\sigma'} \int d\mathbf{r}' \, \Psi_{\sigma'}(\mathbf{r}')^\dagger \, \delta(\mathbf{r} - \mathbf{r}') \, \delta_{\sigma,\sigma'} \, \Psi_{\sigma'}(\mathbf{r}')$$
$$(1.35)$$

$$= \Psi_\sigma(\mathbf{r})^\dagger \Psi_\sigma(\mathbf{r}) = \Psi(x)^\dagger \Psi(x),$$

which also shows that $\Psi(x)^\dagger$ is nothing but the operator which creates a particle at x, while $\Psi(x)$ destroys it.

Let us continue and consider a two-particle operator

$$\hat{U} = \frac{1}{2} \sum_{i \neq j} U(x_i, x_j). \tag{1.36}$$

We can simply repeat the previous steps and find that

$$\langle n_1', n_2', \ldots \mid \hat{U} \mid n_1, n_2, \ldots \rangle$$

$$= \frac{1}{2} \sum_{i \neq j} \int \prod_{i=1}^{N} dx_i \, U(x_i, x_j) \, \Psi_{\{n_a'\}}^*(x_1, \ldots, x_N) \, \Psi_{\{n_a\}}(x_1, \ldots, x_N)$$

$$= \frac{N(N-1)}{2} \int \prod_{i=1}^{N} dx_i \, \Psi_{\{n_a'\}}^*(x_1, \ldots, x_N) \, U(x_1, x_2) \, \Psi_{\{n_a\}}(x_1, \ldots, x_N)$$

$$= \frac{N(N-1)}{N!} \int \prod_{i=1}^{N} dx_i \, U(x_1, x_2)$$

$$\langle n_1', n_2', \ldots \mid \Psi^\dagger(x_1) \, \Psi^\dagger(x_2) \, \Psi^\dagger(x_3) \ldots \Psi^\dagger(x_N) \mid 0 \rangle$$

$$\langle 0 \mid \Psi(x_N) \ldots \Psi(x_3) \, \Psi(x_2) \, \Psi(x_1) \mid n_1, n_2, \ldots \rangle ,$$

which, since,

$$\frac{N(N-1)}{N!} \int dx_3 \ldots dx_N \, \Psi^\dagger(x_3) \ldots \Psi^\dagger(x_N) \mid 0 \rangle \langle 0 \mid \Psi(x_N) \ldots \Psi(x_3) = \mathbb{I}_{N-2} ,$$

leads to the second quantised expression

$$\hat{U} = \frac{1}{2} \int dx_1 \, dx_2 \, \Psi^\dagger(x_1) \, \Psi^\dagger(x_2) \, U(x_1, x_2) \, \Psi(x_2) \, \Psi(x_1)$$

$$= \frac{1}{2} \sum_{abcd} U_{abcd} \, c_a^\dagger \, c_b^\dagger \, c_c \, c_d , \tag{1.37}$$

where

$$U_{abcd} = \int dx \, dy \, \phi_a(x)^* \, \phi_b(y)^* \, U(x, y) \, \phi_c(y) \, \phi_d(x) .$$

Analogously, a generic m-particle operator

$$\hat{U}_m = \frac{1}{m!} \sum_{i_1 \neq i_2 \neq \cdots \neq i_m} U(x_1, x_2, \ldots, x_m)$$

translates in second quantisation into

$$\hat{U}_m = \frac{1}{m!} \int dx_1 \ldots dx_m \, \Psi^\dagger(x_1) \ldots \Psi^\dagger(x_m) \, U(x_1, \ldots, x_m) \, \Psi(x_m) \ldots \Psi(x_1) \,.$$

$$(1.38)$$

We conclude by emphasising the advantages of second quantisation with respect to first quantisation. In the former, any multiparticle operator depends explicitly on the particle coordinates. It is only the wavefunction that contains information about the indistinguishability of the particles as well as their statistics. On the contrary, in second quantisation, those properties are hidden in the definition of creation and annihilation operators; hence, the multiparticle operators do not depend anymore on the particle coordinates. Moreover, in second quantisation, we can also introduce operators which have no first-quantisation counterpart. For instance, we can define particle-non-conserving operators which connect subspaces with different numbers of particles. For instance,

$$\sum_{ab} \left(\Delta_{ab} \, c_a^\dagger c_b^\dagger + \Delta_{ab}^* \, c_b \, c_a \right)$$

is a Hermitian operator that connects Fock states with particle numbers differing by two. Those operators are useful when discussing superconductivity.

1.3 Bosonic Operators

As previously done for the fermionic case, we introduce the creation, $d_a^\dagger$, and its Hermitian conjugate, the annihilation d_a, operators which, respectively, create and destroy a boson in state a. Again the operator $d_a^\dagger d_a$ counts how many times we can destroy and create back a boson in state a; hence, it is just the occupation number n_a. However, since the Pauli principle does not hold for bosons, the operator $d_a \, d_a^\dagger$ first adds one boson in state a, next increases $n_a \to n_a + 1$, and finally destroys one boson in that same state. This latter process can be done $n_a + 1$ times, being $n_a + 1$ the actual occupation number once one more boson has been added. Therefore,

$$d_a^\dagger d_a | n_a \rangle = n_a \, | n_a \rangle, \quad d_a \, d_a^\dagger \, | n_a \rangle = \left(1 + n_a \right) | n_a \rangle \,; \qquad (1.39)$$

Hence, the following commutation relation holds:

$$[d_a, d_a^\dagger] = d_a d_a^\dagger - d_a^\dagger d_a = 1, \qquad (1.40)$$

where $[\ldots]$ denotes the commutator. Equation (1.39) is satisfied, e.g., by

$$d_a \, | n_a \rangle = \sqrt{n_a} \, | n_a - 1 \rangle, \quad d_a^\dagger \, | n_a \rangle = \sqrt{n_a + 1} \, | n_a + 1 \rangle \,, \qquad (1.41)$$

which we assume hereafter. The permanent, contrary to the Slater determinant, is invariant upon interchanging two quantum labels. As a consequence, bosonic operators corresponding to different states commute instead of anti-commuting as the fermionic ones. Namely, for $a \neq b$,

$$\left[d_a^\dagger, d_b\right] = \left[d_a^\dagger, d_b^\dagger\right] = 0.$$

Therefore, in general,

$$\boxed{\left[d_a, d_b^\dagger\right] = \delta_{ab}, \qquad \left[d_a^\dagger, d_b^\dagger\right] = \left[d_a, d_b\right] = 0.} \tag{1.42}$$

Moreover, through (1.41) and (1.42), we can write a generic Fock state as

$$|n_1, n_2, \dots\rangle = \prod_i \frac{(d_i^\dagger)^{n_i}}{\sqrt{n_i!}} \, |0\rangle. \tag{1.43}$$

1.3.1   Bose Fields and Multiparticle Operators

The analogous role of the Fermi fields is now played by the Bose fields defined through

$$\Phi(x) = \sum_a \phi_a(x) \, d_a, \quad \Phi(x)^\dagger = \sum_a \phi_a(x)^* \, d_a^\dagger, \tag{1.44}$$

which satisfy the commutation relations

$$\left[\Phi(x), \Phi(y)^\dagger\right] = \delta(x - y), \quad \left[\Phi(x), \Phi(y)\right] = \left[\Phi(x)^\dagger, \Phi(y)^\dagger\right] = 0. \tag{1.45}$$

Exactly like in the fermionic case, if we define

$$|x_1, \dots, x_N\rangle \equiv \frac{1}{\sqrt{N!}} \, \Phi^\dagger(x_1) \dots \Phi^\dagger(x_N) \, |0\rangle, \tag{1.46}$$

we find that

$$\int dx_1 \dots dx_N \, |x_1, \dots, x_N\rangle\langle x_1, \dots, x_N|$$

$$= \frac{1}{N!} \int dx_1 \dots dx_N \, \Phi^\dagger(x_1) \dots \Phi^\dagger(x_N) \, |0\rangle\langle 0| \, \Phi(x_N) \dots \Phi(x_1) \tag{1.47}$$

$$= \frac{1}{N!} \sum_{a_1, \dots, a_N} d_{a_1}^\dagger \dots d_{a_N}^\dagger \, |0\rangle\langle 0| \, d_{a_N} \dots d_{a_1} \equiv \mathbb{I}_N,$$

as well as that the permanent with occupation numbers $\{n_a\} \equiv (n_1, n_2 \ldots)$ is just

$$
\begin{aligned}
\Phi_{\{n_a\}}(x_1, \ldots, x_N) &= \langle x_1, \ldots, x_N \mid n_1, n_2, \ldots \rangle \\
&= \frac{1}{\sqrt{N!}} \, \langle 0 \mid \Phi(x_N) \ldots \Phi(x_1) \mid n_1, n_2, \ldots \rangle .
\end{aligned}
\tag{1.48}
$$

These two equations imply that the second quantised expressions of the operators are simply the same as in the fermionic case; for instance, a single-particle operator is

$$
\hat{V} = \int dx \, \Phi(x)^\dagger \, V(x) \, \Phi(x) ,
\tag{1.49}
$$

a two-particle one

$$
\hat{U} = \frac{1}{2} \int dx \, dy \, \Phi(x)^\dagger \Phi(y)^\dagger \, U(x, y) \, \Phi(y) \Phi(x) ,
\tag{1.50}
$$

and an m-particle operator

$$
\hat{U}_m = \frac{1}{m!} \int dx_1 \ldots dx_m \, \Phi(x_1)^\dagger \ldots \Phi(x_m)^\dagger \, U(x_1, \ldots, x_m) \, \Phi(x_m) \ldots \Phi(x_1) .
\tag{1.51}
$$

1.4 Canonical Transformations

In general, an interacting Hamiltonian that contains besides bilinear also quartic and higher order terms in creation and annihilation operators cannot be diagonalised. On the contrary, a bilinear Hamiltonian is diagonalisable by a canonical transformation that preserves the commutation/anti-commutation properties of the operators.

Since the interaction is commonly analysed perturbatively starting from an appropriate non-interacting theory, it is useful to begin with bilinear Hamiltonians and introduce the canonical transformations that diagonalise them.

The simplest bilinear Hamiltonian is the second quantised expression of a non-interacting first-quantisation Hamiltonian, which has the general form

$$
H = \sum_{ab} t_{ab} \, c_a^\dagger c_b ,
\tag{1.52}
$$

both for fermions and bosons. Since H is Hermitian, then

$$
t_{ab} = t_{ba}^* .
$$

If we define a column vector c, i.e., a spinor, with components c_a, its Hermitian conjugate $c^\dagger$, a row vector, and the Hermitian matrix $\hat{t}$ with elements t_{ab}, the Hamiltonian can be shortly written as

$$
H = c^\dagger \, \hat{t} \, c .
\tag{1.53}
$$

Since $\hat{t}$ is Hermitian, there exists a unitary transformation $\hat{U}$, i.e., satisfying $\hat{U}^\dagger \hat{U} = \hat{U}\hat{U}^\dagger = \hat{I}$, with $\hat{I}$ the identity matrix, such that

$$\hat{t} = \hat{U}^\dagger \hat{E} \hat{U} \;\Rightarrow\; \hat{E} = \hat{U}\hat{t}\hat{U}^\dagger, \tag{1.54}$$

where $\hat{E}$ is diagonal with real elements ϵ_a. Therefore, if we define a new spinor operator $\boldsymbol{d}$ through

$$\boldsymbol{d} \equiv \hat{U}\boldsymbol{c}, \tag{1.55}$$

with elements

$$d_a = \sum_b U_{ab}\, c_b, \tag{1.56}$$

then

$$H = \boldsymbol{d}^\dagger \hat{E} \boldsymbol{d} = \sum_\alpha \epsilon_a\, d_a^\dagger d_a \tag{1.57}$$

is diagonal. We have now to check whether the above is a canonical transformation. If the c_a's are fermionic/bosonic operators, then

$$\left\{d_a, d_b^\dagger\right\}_\pm = \sum_{nm} U_{an} U_{mb}^\dagger \left\{c_n, c_m^\dagger\right\}_\pm = \sum_n U_{an} U_{nb}^\dagger = \delta_{ab},$$

where $\{\dots\}_\pm$ stands for the anti-commutator (+) and commutator (-), respectively. Therefore, also the d_a's are fermions/bosons; hence, $\hat{U}$ is indeed canonical.

Once the Hamiltonian has been transformed into the diagonal form (1.57), the problem is solved. Indeed any Fock state constructed through the new basis set with operators d_a, namely a wavefunction $|\{n_a\}\rangle$ where each state a is occupied by n_a d_a-particles, is an eigenstate of the Hamiltonian

$$H\; |\{n_a\}\rangle = \left(\sum_a \epsilon_a n_a\right) |\{n_a\}\rangle. \tag{1.58}$$

It is important to remark that in the case of bosons the eigenvalues ϵ_a must all be greater or equal to zero; otherwise, the Hamiltonian would be ill-defined since the ground state would correspond to putting an infinite number of bosons in the most negative energy state, thus with infinitely negative total energy. The condition $\epsilon_a \geq 0$ implies that $\hat{t}$ must be semi positive definite.

An equivalent way to implement the canonical transformation (1.55) is through a unitary operator

$$U = e^{i\,\boldsymbol{c}^\dagger \hat{\phi} \boldsymbol{c}} \equiv e^{i\varphi}, \tag{1.59}$$

where $\hat{\phi} = \hat{\phi}^\dagger$ is Hermitian and thus $U^\dagger U = 1$ unitary. We recall that, given two operators A and B, then

$$e^{-iA} B e^{iA} = \sum_{n \geq 0} \frac{(-i)^n}{n!} C_n, \quad C_0 = B, \quad C_n = \left[A, C_{n-1}\right], \tag{1.60}$$

which can be readily demonstrated by expanding the exponentials, so that, in the specific case of $U^\dagger c U$

$$C_0 = c, \quad C_1 = \left[\varphi, c\right] = -\hat{\phi} c, \quad C_n = (-1)^n \hat{\phi}^n c, \tag{1.61}$$

namely

$$U^\dagger c U = \sum_{n \geq 0} \frac{i^n}{n!} \hat{\phi}^n c = e^{i\hat{\phi}} c, \quad U^\dagger c^\dagger U = c^\dagger e^{-i\hat{\phi}}, \tag{1.62}$$

which coincides with (1.55) if $e^{i\hat{\phi}} = \hat{U}$, and thus $e^{-i\hat{\phi}} = \hat{U}^\dagger$. In this case

$$U^\dagger c U = \hat{U} c, \quad U^\dagger c^\dagger U = c^\dagger \hat{U}^\dagger, \tag{1.63}$$

which also imply

$$U c U^\dagger = \hat{U}^\dagger c, \quad U c^\dagger U^\dagger = c^\dagger \hat{U}. \tag{1.64}$$

In this formulation, the diagonalization of the Hamiltonian means the following. Since the eigenstates and eigenvalues of a Hermitian operator do not change under a unitary transformation, then we are allowed to study, instead of H, the transformed Hamiltonian $\tilde{H} \equiv U H U^\dagger$, which, through (1.64) and noting that $\hat{U} \hat{\imath} \hat{U}^\dagger = \hat{E}$, can be written as

$$\tilde{H} \equiv U H U^\dagger = U c^\dagger \hat{\imath} c U^\dagger = U c^\dagger U^\dagger \hat{\imath} U c U^\dagger = c^\dagger \hat{U} \hat{\imath} \hat{U}^\dagger c = c^\dagger \hat{E} c, \tag{1.65}$$

and therefore, it is diagonal.

Conversely, $H = U^\dagger \tilde{H} U$, so that its eigenstates are simply $U^\dagger \mid \{n_a\}\rangle$, where $\mid \{n_a\}\rangle$ is a Fock state. Indeed

$$H U^\dagger \mid \{n_a\}\rangle = U^\dagger \tilde{H} \mid \{n_a\}\rangle = \left(\sum_a \epsilon_a n_a\right) U^\dagger \mid \{n_a\}\rangle, \tag{1.66}$$

which coincides with (1.58). In this language, the partition function

$$\begin{aligned} Z &\equiv \mathrm{Tr}\left(e^{-\beta H}\right) = \mathrm{Tr}\left(U^\dagger U e^{-\beta H}\right) = \mathrm{Tr}\left(U e^{-\beta H} U^\dagger\right) \\ &= \mathrm{Tr}\left(e^{-\beta U H U^\dagger}\right) = \mathrm{Tr}\left(e^{-\beta \tilde{H}}\right) \equiv \tilde{Z} \end{aligned} \tag{1.67}$$

is evidently invariant under the unitary transformation, and the thermal average of any operator A can be simply calculated through the diagonal Hamiltonian $\tilde{H}$ with A transformed into $U A U^\dagger$,

$$\langle A \rangle = \frac{1}{Z} \text{Tr}\left(e^{-\beta H} A\right) = \frac{1}{Z} \text{Tr}\left(U^\dagger U e^{-\beta H} A\right) = \frac{1}{Z} \text{Tr}\left(e^{-\beta \tilde{H}} U A U^\dagger\right). \tag{1.68}$$

1.4.1 Canonical Transformations with Charge Non-conserving Hamiltonians

The above results show that a unitary transformation of the fermionic/bosonic operators c_a preserves the anti-commutation/commutation relations. Indeed, if

$$c_a \to U^\dagger c_a U , \quad c_a^\dagger \to U^\dagger c_a^\dagger U , \tag{1.69}$$

with any unitary operator U, i.e., such that $U^\dagger U = U U^\dagger = 1$, then

$$\left\{U^\dagger c_a U , U^\dagger c_b^\dagger U\right\}_\pm = U^\dagger \left\{c_a, c_b^\dagger\right\}_\pm U = \delta_{ab} ,$$
$$\left\{U^\dagger c_a U , U^\dagger c_b U\right\}_\pm = U^\dagger \left\{c_a, c_b\right\}_\pm U = 0 . \tag{1.70}$$

In second quantisation, we have the opportunity to introduce bilinear Hamiltonians that does not conserve the number of particles:

$$H = \sum_{ab} \left(t_{ab} c_a^\dagger c_b + \Delta_{ab} c_a c_b + \Delta_{ab}^* c_b^\dagger c_a^\dagger\right) = \boldsymbol{c}^\dagger \hat{t} \, \boldsymbol{c} + \left(\boldsymbol{c}^T \hat{\Delta} \, \boldsymbol{c} + H.c.\right) \tag{1.71}$$

and are relevant for a wide class of physical problems. Again $\hat{t} = \hat{t}^\dagger$, while $\hat{\Delta}$ is antisymmetric for fermions and symmetric for bosons. Since H is bilinear in single-particle operators, it can always be diagonalised by a proper unitary transformation (1.69). However, if we write as before $U = e^{i\varphi}$, now the Hermitian operator φ must be of the form

$$\varphi = \boldsymbol{c}^\dagger \hat{\phi} \, \boldsymbol{c} + \left(\boldsymbol{c}^T \hat{\psi} \, \boldsymbol{c} + H.c.\right), \tag{1.72}$$

with $\hat{\phi} = \hat{\phi}^\dagger$ and $\hat{\psi}$ antisymmetric/symmetric for fermions/bosons. In other words, φ is still bilinear but does not conserve the number of particles. It is always possible to find $\hat{\phi}$ and $\hat{\psi}$ such that

$$U^\dagger H U = \boldsymbol{c}^\dagger \hat{E} \, \boldsymbol{c} \equiv \tilde{H} , \tag{1.73}$$

with $\hat{E}$ diagonal with real elements ϵ_a. It follows that, given a Fock state $\mid \{n_a\}\rangle$, then $U \mid \{n_a\}\rangle$ is eigenstate of H with eigenvalue $\sum_a \epsilon_a n_a$. As previously discussed, in the case of bosons $\epsilon_a \geq 0$, which poses constraints on $\hat{t}$ and $\hat{\Delta}$.

We just note that, while $| \{n_a\}\rangle$ is eigenstate of the number of particles with eigenvalue $N = \sum_a n_a$, the eigenstate $U | \{n_a\}\rangle = e^{i\varphi} | \{n_a\}\rangle$ of the Hamiltonian is not, since φ does not commute with the number of particle operator $\sum_a c_a^\dagger c_a$.

In the case of fermions, the diagonalization of the Hamiltonian (1.71) can still be recast to that of a Hermitian matrix. Indeed, if we introduce the new spinors

$$\psi \equiv \begin{pmatrix} c \\ c^{\dagger T} \end{pmatrix}, \quad \psi^\dagger = \begin{pmatrix} c^\dagger, & c^T \end{pmatrix},$$

the Hamiltonian (1.71) can be rewritten, dropping constant terms, as

$$H = \psi^\dagger \begin{pmatrix} \hat{t}/2 & \hat{\Delta}^\dagger \\ \hat{\Delta} & -\hat{t}^T/2 \end{pmatrix} \psi = \psi^\dagger \hat{H} \psi,$$

where now $\hat{H}$ is Hermitian and thus has real eigenvalues. We further note that, since $\hat{t}$ is Hermitian, then $\hat{t}^T = \hat{t}^*$, while, since $\hat{\Delta}$ is antisymmetric, then $\hat{\Delta}^\dagger = -\hat{\Delta}^*$, so that

$$\hat{H} = \begin{pmatrix} \hat{t}/2 & \hat{\Delta}^\dagger \\ \hat{\Delta} & -\hat{t}^T/2 \end{pmatrix} = \begin{pmatrix} \hat{t}/2 & -\hat{\Delta}^* \\ \hat{\Delta} & -\hat{t}^*/2 \end{pmatrix} = -\begin{pmatrix} 0 & \hat{I} \\ \hat{I} & 0 \end{pmatrix} \hat{H}^* \begin{pmatrix} 0 & \hat{I} \\ \hat{I} & 0 \end{pmatrix} \equiv -\hat{\sigma}_1 \hat{H}^* \hat{\sigma}_1,$$

where $\hat{\sigma}_1$ is a generalised first Pauli matrix, thus $\hat{\sigma}_1 \hat{H} \hat{\sigma}_1 = -\hat{H}^*$. Therefore, if

$$\chi_\lambda = \begin{pmatrix} u_\lambda \\ v_\lambda \end{pmatrix} : \quad \hat{H} \chi_\lambda = \lambda \chi_\lambda, \quad \lambda \in \mathbb{R},$$

so that $\hat{H}^* \chi_\lambda^* = \lambda \chi_\lambda^*$, then

$$\hat{H} \hat{\sigma}_1 \chi_\lambda^* = \hat{\sigma}_1 \hat{\sigma}_1 \hat{H} \hat{\sigma}_1 \chi_\lambda^* = -\hat{\sigma}_1 \hat{H}^* \chi_\lambda^* = -\lambda \hat{\sigma}_1 \chi_\lambda^*.$$

In other words,

$$\hat{\sigma}_1 \chi_\lambda^* = \begin{pmatrix} v_\lambda^* \\ u_\lambda^* \end{pmatrix} \equiv \chi_{-\lambda}$$

is the eigenstate with opposite eigenvalue. Therefore, the unitary matrix

$$\hat{U} = \begin{pmatrix} u_{\lambda_1} \\ v_{\lambda_1} \end{pmatrix} \begin{pmatrix} u_{\lambda_2} \\ v_{\lambda_2} \end{pmatrix} \cdots \begin{pmatrix} v_{\lambda_1}^* \\ u_{\lambda_1}^* \end{pmatrix} \begin{pmatrix} v_{\lambda_2}^* \\ u_{\lambda_2}^* \end{pmatrix} \cdots$$

brings the Hamiltonian into the diagonal form

$$\hat{U}^\dagger \hat{H} \hat{U} = \begin{pmatrix} \hat{\lambda} & 0 \\ 0 & -\hat{\lambda} \end{pmatrix},$$

where $\hat{\lambda}$ is diagonal with components λ_i, $i = 1, 2, \ldots$, which also implies that the canonical transformation

$$\psi = \hat{U} \begin{pmatrix} d \\ d^{\dagger T} \end{pmatrix}$$

allows to rewrite H in (1.71) as

$$H = \psi^{\dagger} \hat{H} \psi = \left(d^{\dagger}, d \right) \begin{pmatrix} \hat{\lambda} & 0 \\ 0 & -\hat{\lambda} \end{pmatrix} \begin{pmatrix} d \\ d^{\dagger T} \end{pmatrix} = \sum_i \lambda_i \left(2 d_i^{\dagger} d_i - 1 \right),$$

namely in a diagonal form.

1.4.2   Harmonic Oscillators

When dealing with bosons, bilinear Hamiltonians very often reduce to coupled harmonic oscillators, like, e.g., in the case of lattice vibrations in the harmonic approximation. In those cases, it is more convenient to apply canonical transformations directly to the conjugate variables rather than to the bosonic creation and annihilation operators. For that, let us consider the prototypical example of a single harmonic oscillator with Hamiltonian

$$H = \frac{\hbar^2}{2m} p^2 + \frac{g}{2} x^2 , \tag{1.74}$$

with x and p conjugate variables, i.e., $[x, p] = i$. We consider the transformation

$$x = \sqrt{K} \, X , \quad p = \frac{1}{\sqrt{K}} \, P , \tag{1.75}$$

which evidently preserves the commutation relations and therefore is canonical. In terms of the new variables

$$H = \frac{\hbar^2}{2mK} P^2 + \frac{Kg}{2} X^2 . \tag{1.76}$$

If we fix K such that

$$\frac{\hbar^2}{mK} = K g \;\Rightarrow\; K^2 = \frac{\hbar^2}{mg} , \tag{1.77}$$

we readily find that

$$H = \frac{\hbar}{2} \sqrt{\frac{g}{m}} \left(P^2 + X^2 \right) \equiv \frac{\hbar \omega}{2} \left(P^2 + X^2 \right) . \tag{1.78}$$

Hereafter, we denote the above expression as the 'canonical form' of the Hamiltonian. Indeed, if we now introduce bosonic creation and annihilation operators through

$$X = \frac{1}{\sqrt{2}}\left(a + a^\dagger\right), \quad P = -\frac{i}{\sqrt{2}}\left(a - a^\dagger\right), \tag{1.79}$$

the Hamiltonian becomes

$$H = \hbar\omega\left(a^\dagger a + \frac{1}{2}\right), \tag{1.80}$$

i.e., the standard expression of a harmonic oscillator Hamiltonian with frequency ω. We now consider the unitary operator

$$U = \exp\left[-i\frac{\alpha}{2}\left(x\,p + p\,x\right)\right],$$

whose action on the conjugate variables can be readily found to be

$$U^\dagger\, x\, U = e^{\alpha}\, x, \quad U^\dagger\, x\, U = e^{-\alpha}\, p,$$

which yields the canonical transformation (1.75) if $e^{2\alpha} = K$. In other words,

$$U^\dagger\, H\, U = \frac{\hbar\omega}{2}\left(p^2 + x^2\right)$$

directly brings the Hamiltonian in the canonical form. The eigenstates are therefore $U\,|\,n\rangle$, where $|\,n\rangle$ is the Fock state with n bosons.

1.5 Application: Electrons in a Box

Let us consider electrons in a square box of linear length L with periodic boundary conditions. As a basis of single-particle wavefunctions, we use simple plane waves and spin, namely the quantum label $a = (\mathbf{k}, \sigma)$ with

$$\mathbf{k} = \frac{2\pi}{L} \left(n_x, n_y, n_z \right) ,$$

being n_i's integers, and $\sigma = \uparrow, \downarrow$. Hence, the single-particle wavefunctions of the basis set are

$$\phi_a(x) = \frac{1}{\sqrt{L^3}} e^{i\mathbf{k}\cdot\mathbf{r}} \chi_\sigma .$$

The annihilation and creation operators are defined as $c_{\mathbf{k}\sigma}$ and $c^\dagger_{\mathbf{k}\sigma}$, respectively; hence, the Fermi fields are

$$\Psi_\sigma(\mathbf{r}) = \frac{1}{\sqrt{L^3}} \sum_{\mathbf{k}} e^{i\mathbf{k}\cdot\mathbf{r}} c_{\mathbf{k}\sigma} , \quad \Psi^\dagger_\sigma(\mathbf{r}) = \frac{1}{\sqrt{L^3}} \sum_{\mathbf{k}} e^{-i\mathbf{k}\cdot\mathbf{r}} c^\dagger_{\mathbf{k}\sigma} . \tag{1.81}$$

The kinetic energy in the first quantisation reads

$$H_{kin} = \sum_{i=1}^{N} -\frac{\hbar^2}{2m} \nabla_i^2$$

and is diagonal in the spin. Since

$$\frac{1}{L^3} \int d\mathbf{r}\, e^{-i\mathbf{k}\cdot\mathbf{r}} \left(-\frac{\hbar^2}{2m} \nabla^2 \right) e^{i\mathbf{p}\cdot\mathbf{r}} = \frac{\hbar^2 k^2}{2m} \delta_{\mathbf{k}\mathbf{p}} ,$$

then, on the plane-wave basis, the kinetic energy in the second quantisation becomes

$$H_{kin} = \sum_{\mathbf{k}\sigma} \frac{\hbar^2 k^2}{2m} c^\dagger_{\mathbf{k}\sigma} c_{\mathbf{k}\sigma} \equiv \sum_{\mathbf{k}\sigma} \epsilon_{\mathbf{k}}\, c^\dagger_{\mathbf{k}\sigma} c_{\mathbf{k}\sigma} . \tag{1.82}$$

The ground state with N, assumed to be even, electrons is the Fermi sea $|FS\rangle$ obtained by filling with spin-up and spin-down electrons the momentum states from $\mathbf{k} = 0$ up to the Fermi momentum k_F defined through

$$N = 2 \sum_{|\mathbf{k}| \le k_F} ,$$

namely

$$|FS\rangle = \prod_{\mathbf{k}:\, |\mathbf{k}| \le k_F} c^\dagger_{\mathbf{k}\uparrow} c^\dagger_{\mathbf{k}\downarrow} |0\rangle ; \tag{1.83}$$

hence, the occupation number in momentum space is

$$n_{\mathbf{k}\uparrow} = n_{\mathbf{k}\downarrow} = \theta \left(k_F - |\mathbf{k}| \right) ,$$

where the θ-function is defined through $\theta(x) = 1$ if $x \ge 0$, otherwise $\theta(x) = 0$.

Let us add an electron-electron interaction of the general form

$$H_{int} = \frac{1}{2} \sum_{i \neq j} U(\mathbf{r}_i - \mathbf{r}_j).$$

In second quantisation, it becomes

$$H_{int} = \frac{1}{2} \sum_{\sigma\sigma'} \int d\mathbf{x}\, d\mathbf{y}\, \Psi_\sigma^\dagger(\mathbf{x}) \Psi_{\sigma'}^\dagger(\mathbf{y})\, U(\mathbf{x} - \mathbf{y})\, \Psi_{\sigma'}(\mathbf{y}) \Psi_\sigma(\mathbf{x})$$

$$= \frac{1}{2L^3} \sum_{\sigma\sigma'} \sum_{\mathbf{k}_i, i=1,\dots,4} c_{\mathbf{k}_1\sigma}^\dagger c_{\mathbf{k}_2\sigma'}^\dagger c_{\mathbf{k}_3\sigma'} c_{\mathbf{k}_4\sigma}$$

$$\frac{1}{L^3} \int d\mathbf{x}\, d\mathbf{y}\, e^{-i k_1 \cdot \mathbf{x}} e^{-i k_2 \cdot \mathbf{y}} e^{i k_3 \cdot \mathbf{y}} e^{i k_4 \cdot \mathbf{x}}\, U(\mathbf{x} - \mathbf{y}).$$

We define the Fourier transform as

$$U(\mathbf{q}) = \int d\mathbf{r}\, e^{-i\mathbf{q}\cdot\mathbf{r}}\, U(\mathbf{r}),$$

so that

$$U(\mathbf{x} - \mathbf{y}) = \frac{1}{L^3} \sum_{\mathbf{q}} e^{i\mathbf{q}\cdot(\mathbf{x}-\mathbf{y})}\, U(\mathbf{q}),$$

and the interaction is finally found to be

$$H_{int} = \frac{1}{2L^3} \sum_{\sigma\sigma'} \sum_{\mathbf{k}\mathbf{p}\mathbf{q}} U(\mathbf{q})\, c_{\mathbf{k}\sigma}^\dagger c_{\mathbf{p}+\mathbf{q}\sigma'}^\dagger c_{\mathbf{p}\sigma'} c_{\mathbf{k}+\mathbf{q}\sigma}. \tag{1.84}$$

We showed that the electron density for spin σ in second quantization is

$$\rho_\sigma(\mathbf{r}) = \Psi_\sigma^\dagger(\mathbf{r})\, \Psi_\sigma(\mathbf{r}).$$

Its Fourier transform is

$$\rho_\sigma(\mathbf{q}) = \int d\mathbf{r}\, e^{-i\mathbf{q}\cdot\mathbf{r}}\, \Psi_\sigma^\dagger(\mathbf{r})\, \Psi_\sigma(\mathbf{r}) = \sum_{\mathbf{k}} c_{\mathbf{k}\sigma}^\dagger c_{\mathbf{k}+\mathbf{q}\sigma}.$$

Let us consider interacting electrons also in the presence of a single-particle potential, provided, e.g., by the ions,

$$H_{pot} = \sum_{\sigma} \int d\mathbf{x}\, V(\mathbf{x})\, \rho_\sigma(\mathbf{x}),$$

so that the total Hamiltonian reads

$$H = H_{kin} + H_{int} + H_{pot}.$$

Since the Hamiltonian conserves the total number of electrons, the electron density summed over the spins, i.e., $\rho = \rho_\uparrow + \rho_\downarrow$ must satisfy a continuity equation. Let us define the Heisenberg evolution of the density through

$$\rho(\mathbf{r}, t) = e^{iHt} \rho(\mathbf{r}) e^{-iHt}.$$

Since the integral of the density over the whole volume is the total number of electrons, N, which is conserved, it follows that

$$\int d\mathbf{r} \frac{\partial \rho(\mathbf{r}, t)}{\partial t} = \frac{\partial}{\partial t} \int d\mathbf{r}\, \rho(\mathbf{r}, t) = \frac{\partial N}{\partial t} \equiv 0.$$

This condition is satisfied if

$$\frac{\partial \rho(\mathbf{r}, t)}{\partial t} = -\nabla \mathbf{J}(\mathbf{r}, t),$$

where $\mathbf{J}(\mathbf{r}, t)$ is the current density operator. In fact, the integral over the volume of the left-hand side is also equal to minus the flux of the current out of the surface of the sample. If the number of electrons is conserved, it means that the flux of the current through the surface vanishes, which is the desired result. The equation

$$\frac{\partial \rho(\mathbf{r}, t)}{\partial t} + \nabla \mathbf{J}(\mathbf{r}, t) = 0, \tag{1.85}$$

is the *continuity equation* associated to the number of particles, which is a conserved quantity. Similar equations can be derived for any conserved quantity.

1.6   Application: Electron Lattice Models and Emergence of Magnetism

Let us consider the electron Hamiltonian in presence of the periodic potential provided by the ions in a lattice:

$$H = \sum_{i=1}^{N} \frac{\mathbf{p}_i^2}{2m} + \sum_i \sum_{\mathbf{R}} V(\mathbf{r}_i - \mathbf{R}) + \frac{1}{2} \sum_{i \neq j} U(\mathbf{r}_i - \mathbf{r}_j) \tag{1.86}$$

$$= H_{kin} + H_{el-ion} + H_{el-el} = H_0 + H_{el-el},$$

where $\mathbf{R}$ are lattice vectors. We start by rewriting the Hamiltonian in the second quantisation. For that purpose, we need to introduce a basis set of single-particle wavefunctions. Since the Hamiltonian is spin independent, it is convenient to work with factorised single-particle wavefunctions: $\phi(\mathbf{r}, \sigma) = \phi(\mathbf{r}) \chi_\sigma$. As a basis for the space-dependent $\phi(\mathbf{r})$, we use Wannier orbitals $\phi_{n,\mathbf{R}}(\mathbf{r})$ satisfying

$$\int d\mathbf{r} \, \phi_{n,\mathbf{R}_1}(\mathbf{r})^* \phi_{m,\mathbf{R}_2}(\mathbf{r}) = \delta_{nm} \, \delta_{\mathbf{R}_1\mathbf{R}_2} \,, \quad \sum_{n,\mathbf{R}} \phi_{n,\mathbf{R}}(\mathbf{r})^* \phi_{n,\mathbf{R}}(\mathbf{r}') = \delta(\mathbf{r} - \mathbf{r}') \,,$$

as well as

$$\phi_{n,\mathbf{R}+\mathbf{R}_0}(\mathbf{r}) = \phi_{n,\mathbf{R}}(\mathbf{r} - \mathbf{R}_0) \,. \tag{1.87}$$

Consequently, we associate to any wavefunction $\phi_{n,\mathbf{R}}(\mathbf{r}) \chi_\sigma$ creation, $c^\dagger_{n,\mathbf{R},\sigma}$, and annihilation, $c_{n,\mathbf{R},\sigma}$, operators and introduce the Fermi fields

$$\Psi_\sigma(\mathbf{r}) = \sum_{n,\mathbf{R}} \phi_{n,\mathbf{R}}(\mathbf{r}) \, c_{n,\mathbf{R},\sigma} \,, \quad \Psi_\sigma^\dagger(\mathbf{r}) = \left(\Psi_\sigma(\mathbf{r}) \right)^\dagger \,.$$

Let us start by second quantization of the non-interacting part of the Hamiltonian, H_0 in (1.86):

$$H_0 = \sum_\sigma \int d\mathbf{r} \, \Psi_\sigma^\dagger(\mathbf{r}) \left[-\frac{\hbar^2 \nabla^2}{2m} + \sum_\mathbf{R} V(\mathbf{r} - \mathbf{R}) \right] \Psi_\sigma(\mathbf{r})$$

$$= \sum_\sigma \sum_{nm} \sum_{\mathbf{R}_1,\mathbf{R}_2} t^{nm}_{\mathbf{R}_1,\mathbf{R}_2} \, c^\dagger_{n,\mathbf{R}_1,\sigma} \, c_{m,\mathbf{R}_2,\sigma} \,.$$

The matrix elements are

$$t^{nm}_{\mathbf{R}_1,\mathbf{R}_2} = \int d\mathbf{r} \, \phi_{n,\mathbf{R}_1}(\mathbf{r})^* \left[-\frac{\hbar^2 \nabla^2}{2m} + \sum_\mathbf{R} V(\mathbf{r} - \mathbf{R}) \right] \phi_{m,\mathbf{R}_2}(\mathbf{r}) \tag{1.88}$$

and satisfy

$$t^{nm}_{\mathbf{R}_1,\mathbf{R}_2} = \left(t^{mn}_{\mathbf{R}_2,\mathbf{R}_1} \right)^* \,. \tag{1.89}$$

By the property (1.87), it follows that

$$t^{nm}_{\mathbf{R}_1+\mathbf{R}_0,\mathbf{R}_2+\mathbf{R}_0} = \int d\mathbf{r} \, \phi_{n,\mathbf{R}_1+\mathbf{R}_0}(\mathbf{r})^* \left[-\frac{\hbar^2 \nabla^2}{2m} + \sum_\mathbf{R} V(\mathbf{r} - \mathbf{R}) \right] \phi_{m,\mathbf{R}_2+\mathbf{R}_0}(\mathbf{r})$$

$$= \int d\mathbf{r} \, \phi_{n,\mathbf{R}_1}(\mathbf{r} - \mathbf{R}_0)^* \left[-\frac{\hbar^2 \nabla^2}{2m} + \sum_\mathbf{R} V(\mathbf{r} - \mathbf{R}) \right] \phi_{m,\mathbf{R}_2}(\mathbf{r} - \mathbf{R}_0)$$

$$= \int d\mathbf{r} \, \phi_{n,\mathbf{R}_1}(\mathbf{r})^* \left[-\frac{\hbar^2 \nabla^2}{2m} + \sum_\mathbf{R} V(\mathbf{r} + \mathbf{R}_0 - \mathbf{R}) \right] \phi_{m,\mathbf{R}_2}(\mathbf{r}) = t^{nm}_{\mathbf{R}_1,\mathbf{R}_2} \,,$$

since

$$\sum_{\mathbf{R}} V(\mathbf{r} + \mathbf{R}_0 - \mathbf{R}) = \sum_{\mathbf{R}} V(\mathbf{r} - \mathbf{R}).$$

Let us now introduce the operators in the reciprocal lattice through the canonical transformation

$$c_{n,\mathbf{k},\sigma} = \frac{1}{\sqrt{V}} \sum_{\mathbf{R}} e^{-i\mathbf{k}\cdot\mathbf{R}} c_{n,\mathbf{R},\sigma}, \tag{1.90}$$

and its inverse

$$c_{n,\mathbf{R},\sigma} = \frac{1}{\sqrt{V}} \sum_{\mathbf{k}} e^{i\mathbf{k}\cdot\mathbf{R}} c_{n,\mathbf{k},\sigma}, \tag{1.91}$$

where V is the number of lattice sites and $\mathbf{k}$ belongs to the reciprocal lattice, namely

$$\sum_{\mathbf{R}} e^{i\mathbf{k}\cdot\mathbf{R}} = V \, \delta_{\mathbf{k}0}, \quad \sum_{\mathbf{k}} e^{-i\mathbf{k}\cdot\mathbf{R}} = V \, \delta_{\mathbf{R}0}.$$

One can check that the above transformation is indeed canonical:

$$\left\{ c_{n,\mathbf{k}_1,\sigma_1}, c^{\dagger}_{m,\mathbf{k}_2,\sigma_2} \right\} = \frac{1}{V} \sum_{\mathbf{R}_1,\mathbf{R}_2} e^{-i\mathbf{k}_1\cdot\mathbf{R}_1} \, e^{i\mathbf{k}_2\cdot\mathbf{R}_2} \left\{ c_{n,\mathbf{R}_1,\sigma_1}, c^{\dagger}_{m,\mathbf{R}_2,\sigma_2} \right\}$$

$$= \delta_{nm} \, \delta_{\sigma_1\sigma_2} \frac{1}{V} \sum_{\mathbf{R}_1} e^{-i(\mathbf{k}_1-\mathbf{k}_2)\cdot\mathbf{R}_1} = \delta_{nm} \, \delta_{\sigma_1\sigma_2} \, \delta_{\mathbf{k}_1\mathbf{k}_2}.$$

Let us substitute (1.91) into (1.88):

$$H_0 = \frac{1}{V} \sum_{\sigma} \sum_{nm} \sum_{\mathbf{R}_1,\mathbf{R}_2} \sum_{\mathbf{k}_1,\mathbf{k}_2} e^{-i\mathbf{k}_1\cdot\mathbf{R}_1} e^{i\mathbf{k}_2\cdot\mathbf{R}_2} \, t^{nm}_{\mathbf{R}_1,\mathbf{R}_2} \, c^{\dagger}_{n,\mathbf{k}_1,\sigma} c_{m,\mathbf{k}_2,\sigma}.$$

We observe that

$$\sum_{\mathbf{R}_1,\mathbf{R}_2} e^{-i\mathbf{k}_1\cdot\mathbf{R}_1} \, e^{i\mathbf{k}_2\cdot\mathbf{R}_2} \, t^{nm}_{\mathbf{R}_1,\mathbf{R}_2} = \frac{1}{V} \sum_{\mathbf{R}_1,\mathbf{R}_2,\mathbf{R}_0} e^{-i\mathbf{k}_1\cdot\mathbf{R}_1} \, e^{i\mathbf{k}_2\cdot\mathbf{R}_2} \, t^{nm}_{\mathbf{R}_1-\mathbf{R}_0,\mathbf{R}_2-\mathbf{R}_0}$$

$$= \frac{1}{V} \sum_{\mathbf{R}_1,\mathbf{R}_2,\mathbf{R}_0} e^{-i\mathbf{k}_1\cdot(\mathbf{R}_1+\mathbf{R}_0)} \, e^{i\mathbf{k}_2\cdot(\mathbf{R}_2+\mathbf{R}_0)} \, t^{nm}_{\mathbf{R}_1,\mathbf{R}_2}$$

$$= \sum_{\mathbf{R}_1,\mathbf{R}_2} e^{-i\mathbf{k}_1\cdot\mathbf{R}_1} \, e^{i\mathbf{k}_2\cdot\mathbf{R}_2} \, t^{nm}_{\mathbf{R}_1,\mathbf{R}_2} \, \frac{1}{V} \sum_{\mathbf{R}_0} e^{-i(\mathbf{k}_1-\mathbf{k}_2)\cdot\mathbf{R}_0}$$

$$= \delta_{\mathbf{k}_1\mathbf{k}_2} \sum_{\mathbf{R}_1,\mathbf{R}_2} e^{-i\mathbf{k}_1\cdot(\mathbf{R}_1-\mathbf{R}_2)} \, t^{nm}_{\mathbf{R}_1,\mathbf{R}_2} = V \, \delta_{\mathbf{k}_1\mathbf{k}_2} \, t^{nm}_{\mathbf{k}_1},$$

where we define

$$t_{\mathbf{k}}^{nm} \equiv \frac{1}{V} \sum_{\mathbf{R}_1, \mathbf{R}_2} e^{-i\mathbf{k}\cdot(\mathbf{R}_1 - \mathbf{R}_2)} \, t_{\mathbf{R}_1, \mathbf{R}_2}^{nm} = \sum_{\mathbf{R}} e^{-i\mathbf{k}\cdot\mathbf{R}} \, t_{\mathbf{R}, 0}^{nm} \,. \tag{1.92}$$

Since (1.89) holds, it also follows that

$$
\begin{aligned}
\left(t_{\mathbf{k}}^{nm} \right)^* &= \sum_{\mathbf{R}} e^{i\mathbf{k}\cdot\mathbf{R}} \left(t_{\mathbf{R},0}^{nm} \right)^* = \sum_{\mathbf{R}} e^{i\mathbf{k}\cdot\mathbf{R}} \, t_{0,\mathbf{R}}^{mn} = \sum_{\mathbf{R}} e^{i\mathbf{k}\cdot\mathbf{R}} \, t_{-\mathbf{R},0}^{mn} \\
&= \sum_{\mathbf{R}} e^{-i\mathbf{k}\cdot\mathbf{R}} \, t_{\mathbf{R},0}^{mn} = t_{\mathbf{k}}^{mn} \,,
\end{aligned}
\tag{1.93}
$$

namely the matrix $\hat{t}_{\mathbf{k}}$, with elements $t_{\mathbf{k}}^{mn}$, is Hermitian. The Hamiltonian is therefore

$$H_0 = \sum_{\sigma} \sum_{nm} \sum_{\mathbf{k}} t_{\mathbf{k}}^{nm} \, c_{n,\mathbf{k},\sigma}^{\dagger} c_{m,\mathbf{k},\sigma} \,. \tag{1.94}$$

Since $\hat{t}_{\mathbf{k}}$ is Hermitian, we can write

$$\hat{t}_{\mathbf{k}} = \hat{U}(\mathbf{k})^{\dagger} \, \hat{\epsilon}_{\mathbf{k}} \, \hat{U}(\mathbf{k}) \quad \rightarrow \quad t_{\mathbf{k}}^{nm} = U^{\dagger}(\mathbf{k})_{ni} \, \epsilon_{i,\mathbf{k}} \, U(\mathbf{k})_{im} \,,$$

with $\hat{U}(\mathbf{k})^{\dagger} \, \hat{U}(\mathbf{k}) = \hat{I}$. Therefore, upon applying the canonical transformation

$$c_{i,\mathbf{k},\sigma} = \sum_{n} U(\mathbf{k})_{in} \, c_{n,\mathbf{k},\sigma} \,,$$

the non-interacting Hamiltonian acquires a diagonal form

$$H_0 = \sum_{\sigma} \sum_{i} \sum_{\mathbf{k}} \epsilon_{i,\mathbf{k}} \, c_{i,\mathbf{k},\sigma}^{\dagger} c_{i,\mathbf{k},\sigma} \,. \tag{1.95}$$

The index i identifies the band, and $\epsilon_{i,\mathbf{k}}$ is the energy dispersion in the reciprocal lattice: we have thus obtained the band structure.

We can formally write

$$c_{i,\mathbf{k},\sigma} = \frac{1}{\sqrt{V}} \sum_{\mathbf{R}} e^{-i\mathbf{R}\cdot\mathbf{k}} \, c_{i,\mathbf{R},\sigma} \,,$$

thus introducing a new basis of Wannier functions. Since the Fermi field is invariant upon the basis choice, then

$$
\begin{aligned}
\Psi_{\sigma}(\mathbf{r}) &= \sum_{\mathbf{R},n} \phi_{n,\mathbf{R}}(\mathbf{r}) \, \chi_{\sigma} \, c_{n,\mathbf{R},\sigma} = \frac{1}{\sqrt{V}} \sum_{\mathbf{R},n,\mathbf{k}} \phi_{n,\mathbf{R}}(\mathbf{r}) \, \chi_{\sigma} \, e^{i\mathbf{k}\cdot\mathbf{R}} \, c_{n,\mathbf{k},\sigma} \\
&= \frac{1}{\sqrt{V}} \sum_{\mathbf{R},n,i,\mathbf{k}} \phi_{n,\mathbf{R}}(\mathbf{r}) \, \chi_{\sigma} \, e^{i\mathbf{k}\cdot\mathbf{R}} \, U^{\dagger}(\mathbf{k})_{ni} \, c_{i,\mathbf{k},\sigma} \\
&= \frac{1}{V} \sum_{\mathbf{R},\mathbf{R}',n,i,\mathbf{k}} \phi_{n,\mathbf{R}}(\mathbf{r}) \, \chi_{\sigma} \, e^{i\mathbf{k}\cdot(\mathbf{R}-\mathbf{R}')} \, U^{\dagger}(\mathbf{k})_{ni} \, c_{i,\mathbf{R}',\sigma} \equiv \sum_{\mathbf{R}'} \phi_{i,\mathbf{R}'}(\mathbf{r}) \, \chi_{\sigma} \, c_{i,\mathbf{R}',\sigma} \,,
\end{aligned}
$$

thus implying the following expression of the new Wannier functions:

$$\phi_{i,\mathbf{R}'}(\mathbf{r}) = \frac{1}{V} \sum_{n,\mathbf{R},\mathbf{k}} e^{i\mathbf{k}\cdot(\mathbf{R}-\mathbf{R}')} U^\dagger(\mathbf{k})_{ni} \, \phi_{n,\mathbf{R}}(\mathbf{r}) \,. \tag{1.96}$$

Going back to the Hamiltonian in the diagonal basis, we can also rewrite it as

$$
\begin{aligned}
H_0 &= \sum_\sigma \sum_i \sum_\mathbf{k} \epsilon_{i,\mathbf{k}} \, c^\dagger_{i,\mathbf{k},\sigma} c_{i,\mathbf{k},\sigma} \\
&= \frac{1}{V} \sum_\sigma \sum_i \sum_\mathbf{k} \sum_{\mathbf{R}_1,\mathbf{R}_2} \epsilon_{i,\mathbf{k}} \, e^{i\mathbf{k}\cdot(\mathbf{R}_1-\mathbf{R}_2)} c^\dagger_{i,\mathbf{R}_1,\sigma} c_{i,\mathbf{R}_2,\sigma} \\
&\equiv \sum_\sigma \sum_i \sum_{\mathbf{R}_1,\mathbf{R}_2} t^i_{\mathbf{R}_1,\mathbf{R}_2} c^\dagger_{i,\mathbf{R}_1,\sigma} c_{i,\mathbf{R}_2,\sigma} \,,
\end{aligned}
\tag{1.97}
$$

namely like a tight-binding Hamiltonian diagonal in the band index. Once we know the band structure, the ground state of the non-interacting Hamiltonian is simply obtained by filling all the lowest bands with the available electrons. If the highest occupied band is full, the model is a band insulator, otherwise is a metal. In particular, since each band can accommodate $2V$ electrons, V of spin up and V of spin down, a necessary condition for a band insulator is to have an even number of available electrons per unit cell. This is not sufficient since the bands may overlap.

1.6.1 Hubbard Models

Hereafter, we consider the case of a putative metal, in which the highest occupied bands are partly filled. In many physical situations, the Wannier orbitals $\phi_{i,\mathbf{R}}(\mathbf{r})$ of (1.96) for those bands, the conduction ones, are quite delocalised; hence, the lattice vector label $\mathbf{R}$ looses its physical meaning. In those cases, although formally exact, the tight-binding Hamiltonian (1.97) is of little use since the hopping matrix elements $t^i_{\mathbf{R}_1,\mathbf{R}_2}$ are very long ranged.

However, there is a wide class of materials, commonly called *strongly correlated* ones, where the tight-binding formalism is meaningful. There, the valence bands derive from d or f orbitals of transition metals, rare earth or actinides, and the Wannier orbitals keep noticeable atomic character, thus leading to short-range hopping elements $t^i_{\mathbf{R}_1,\mathbf{R}_2}$ in (1.97). Let us consider just the above circumstance and write down the left-over electron-electron interaction on the Wannier basis. We further assume that there is only one conduction band well separated from lower and higher ones, so that we can safely neglect interband transition processes due to interaction and just project the latter onto the conduction band. For that reason, we hereafter drop the band index i, so that the Wannier orbitals of the conduction band are simply denoted as $\phi_\mathbf{R}(\mathbf{r})$, and the tight-binding Hamiltonian projected on the same band reads

$$H_0 = \sum_\sigma \sum_{\mathbf{R}_1,\mathbf{R}_2} t_{\mathbf{R}_1,\mathbf{R}_2} c^\dagger_{\mathbf{R}_1\sigma} c_{\mathbf{R}_2\sigma} \,. \tag{1.98}$$

The interaction term is, correspondingly,

$$H_{int} = \frac{1}{2} \sum_{\sigma_1 \sigma_2} \sum_{\mathbf{R}_1, \mathbf{R}_2, \mathbf{R}_3, \mathbf{R}_4} U_{\mathbf{R}_1, \mathbf{R}_2; \mathbf{R}_3, \mathbf{R}_4} c_{\mathbf{R}_1 \sigma_1}^{\dagger} c_{\mathbf{R}_2 \sigma_2}^{\dagger} c_{\mathbf{R}_3 \sigma_2} c_{\mathbf{R}_4 \sigma_1} , \qquad (1.99)$$

where

$$U_{\mathbf{R}_1, \mathbf{R}_2; \mathbf{R}_3, \mathbf{R}_4} = \int d\mathbf{r} d\mathbf{r}' \, \phi_{\mathbf{R}_1}(\mathbf{r})^* \phi_{\mathbf{R}_2}(\mathbf{r}')^* \, U(\mathbf{r} - \mathbf{r}') \, \phi_{\mathbf{R}_3}(\mathbf{r}') \phi_{\mathbf{R}_4}(\mathbf{r}) . \qquad (1.100)$$

Let us make use of our assumption of well-localised Wannier orbitals, namely that $\phi_{\mathbf{R}}(\mathbf{r})$ decays sufficiently fast with $|\mathbf{r} - \mathbf{R}|$. Within this assumption, the leading matrix element is when all lattice sites are the same:

$$U_{\mathbf{R}, \mathbf{R}; \mathbf{R}, \mathbf{R}} \equiv U. \qquad (1.101)$$

This term gives rise to an interaction

$$H_U = \frac{U}{2} \sum_{\mathbf{R}} \sum_{\sigma \sigma'} c_{\mathbf{R}\sigma}^{\dagger} c_{\mathbf{R}\sigma'}^{\dagger} c_{\mathbf{R}\sigma'} c_{\mathbf{R}\sigma} = U \sum_{\mathbf{R}} n_{\mathbf{R}\uparrow} n_{\mathbf{R}\downarrow} , \qquad (1.102)$$

since $c_{\mathbf{R}\sigma} c_{\mathbf{R}\sigma} = 0$ by the Pauli principle, where the operator

$$n_{\mathbf{R}\sigma} = c_{\mathbf{R}\sigma}^{\dagger} c_{\mathbf{R}\sigma} ,$$

counts the number of spin-σ electrons at site $\mathbf{R}$.

The interaction term (1.102), which simply describes an on-site Coulomb repulsion, plus the hopping (1.98) yield the so-called *Hubbard model* [1,2], which is the prototype of strongly correlated lattice models.

Let us continue and consider in (1.100) two other cases: either (1) $\mathbf{R}_1 = \mathbf{R}_4$, $\mathbf{R}_2 = \mathbf{R}_3$ with $\mathbf{R}_2$ nearest neighbour of $\mathbf{R}_1$ or (2) $\mathbf{R}_1 = \mathbf{R}_3$, $\mathbf{R}_2 = \mathbf{R}_4$ still with $\mathbf{R}_2$ nearest neighbour of $\mathbf{R}_1$. In case (1), we obtain

$$U_1 \sum_{<\mathbf{R}\mathbf{R}'>} n_{\mathbf{R}} n_{\mathbf{R}'} ,$$

where $n_{\mathbf{R}} = \sum_{\sigma} n_{\mathbf{R}\sigma}$, $< \mathbf{R}\mathbf{R}' >$ stands for the sum over nearest neighbour sites, and $U_1 = U_{\mathbf{R}, \mathbf{R}'; \mathbf{R}', \mathbf{R}}$. In case (2), we instead find

$$U_2 \sum_{<\mathbf{R}\mathbf{R}'>} \sum_{\sigma \sigma'} c_{\mathbf{R}\sigma}^{\dagger} c_{\mathbf{R}'\sigma'}^{\dagger} c_{\mathbf{R}\sigma'} c_{\mathbf{R}'\sigma} = -U_2 \sum_{<\mathbf{R}\mathbf{R}'>} \sum_{\sigma \sigma'} c_{\mathbf{R}\sigma}^{\dagger} c_{\mathbf{R}\sigma'} c_{\mathbf{R}'\sigma'}^{\dagger} c_{\mathbf{R}'\sigma} ,$$

where $U_2 = U_{\mathbf{R}, \mathbf{R}'; \mathbf{R}, \mathbf{R}'}$. One can easily show that

$$\sum_{\sigma \sigma'} c_{\mathbf{R}\sigma}^{\dagger} c_{\mathbf{R}\sigma'} c_{\mathbf{R}'\sigma'}^{\dagger} c_{\mathbf{R}'\sigma} = \frac{1}{2} \left[n_{\mathbf{R}} n_{\mathbf{R}'} + 4 \mathbf{S}_{\mathbf{R}} \cdot \mathbf{S}_{\mathbf{R}'} \right], \qquad (1.103)$$

where the spin-density operator

$$\mathbf{S_R} = \frac{1}{2} \sum_{\alpha\beta} c^{\dagger}_{\mathbf{R}\alpha} \, \boldsymbol{\sigma}_{\alpha\beta} \, c_{\mathbf{R}\beta} \,,$$

being $\boldsymbol{\sigma} = (\sigma_1, \sigma_2, \sigma_3)$, with σ_i's the Pauli matrices. Therefore, upon defining $V = U_1 - U_2/2$ and $J_{ex} = -2U_2$, the two nearest neighbour interaction terms lead to

$$H_{n.n.} = V \sum_{<\mathbf{RR'}>} n_{\mathbf{R}} \, n_{\mathbf{R'}} + J_{ex} \sum_{<\mathbf{RR'}>} \mathbf{S_R} \cdot \mathbf{S_{R'}} \,. \qquad (1.104)$$

Since $U_1 > U_2 > 0$, the first term describes a nearest neighbour repulsion, while the second a spin exchange which tends to align the spin ferromagnetically since $J_{ex} < 0$, so called *direct exchange*. Therefore, although we have started from a spin-independent interaction, projecting onto the Wannier basis makes a spin interaction emerge, thus showing in a simple way how magnetism raises out of the charge Coulomb repulsion.

1.6.2   Mott Insulators and Heisenberg Models

Let us summarise the approximate Hubbard Hamiltonian which we have so far derived, by further assuming a nearest neighbour hopping:

$$\begin{aligned} H = -t \sum_{\sigma} \sum_{<\mathbf{RR'}>} \left(c^{\dagger}_{\mathbf{R}\sigma} c_{\mathbf{R'}\sigma} + H.c. \right) + U \sum_{\mathbf{R}} n_{\mathbf{R}\uparrow} n_{\mathbf{R}\downarrow} \\ + V \sum_{<\mathbf{RR'}>} n_{\mathbf{R}} \, n_{\mathbf{R'}} + J_{ex} \sum_{<\mathbf{RR'}>} \mathbf{S_R} \cdot \mathbf{S_{R'}} \equiv H_0 + H_{int}. \end{aligned} \qquad (1.105)$$

By construction $U > V > 0$ and $J_{ex} < 0$. We consider the case in which the number of conduction electrons is equal to the number of sites N, i.e., density equal to half-filling. In the absence of interaction, the hopping forms a band that can accommodate $2N$ electrons while there are just N of them: the band is therefore half-filled and the system metallic.

Let us analyse the opposite case of a very large $U \gg t, V, |J_{ex}|$. In this case, it is better to start from the configuration which minimises the Coulomb repulsion U and treat what is left in perturbation theory. That lowest energy electronic configuration is the one in which each site is singly occupied. Indeed, the energy cost in having just an empty site and a doubly occupied one instead of two singly occupied sites is given by

$$E(2) + E(0) - 2E(1) = U,$$

and it is much larger than the hopping energy t. In this situation, the model describes an insulator but of a particular kind. Namely, the insulating state is driven by the strong correlation, while the conventional counting argument instead predicts a metal.

This correlation-induced insulator is called a *Mott insulator* after Sir Nevil Mott [3]. Therefore, as the strength of U increases with respect to the bandwidth, which is proportional to t, an interaction-driven metal-to-insulator transition, commonly referred to as a *Mott transition*, must occur at some critical U_c.

However, the configuration with one electron per site is hugely degenerate, since the electron can have either spin up or down, thus a degeneracy 2^N that is going to be split by the other terms in the Hamiltonian. The nearest neighbour interaction V is not effective in splitting the degeneracy, contrary to the direct exchange that favours a ferromagnetic ordering.

However, this is not the only source of spin correlations. Indeed, also the hopping term is able to split the degeneracy within second order in perturbation theory. Specifically, if $\mathcal{P}$ is the projector onto the degenerate ground state manifold with energy $E_0 = 0$, at second order, the hopping Hamiltonian H_0 generates an effective spin exchange operator, called super-exchange.

$$H_{s-ex} = \sum_n \frac{1}{E_0 - E_n} \, \mathcal{P} \, H_0 \mid n \rangle \langle n \mid H_0 \, \mathcal{P} \,, \qquad (1.106)$$

where $\mid n \rangle$ is an intermediate excited state with energy E_n. Because of the projector, the only allowed hopping processes are those where one electron hops to a nearest neighbour site and then hops back to the initial site. It follows that the intermediate states allowed in the sum are just those where one site is doubly occupied and a nearest neighbour one empty, thus $E_n = U$ independent of n and

$$H_{s-ex} = -\frac{1}{U} \sum_n \mathcal{P} \, H_0 \mid n \rangle \langle n \mid H_0 \, \mathcal{P} = -\frac{1}{U} \, \mathcal{P} \, H_0 \, H_0 \, \mathcal{P}$$

$$= -\frac{t^2}{U} \sum_{\sigma\sigma'} \sum_{<\mathbf{RR'}>} \left(c_{\mathbf{R}\sigma}^\dagger \, c_{\mathbf{R'}\sigma} \, c_{\mathbf{R'}\sigma'}^\dagger \, c_{\mathbf{R}\sigma'} + c_{\mathbf{R'}\sigma}^\dagger \, c_{\mathbf{R}\sigma} \, c_{\mathbf{R}\sigma'}^\dagger \, c_{\mathbf{R'}\sigma'} \right) \qquad (1.107)$$

$$= -\frac{2t^2}{U} \sum_{\sigma\sigma'} \sum_{<\mathbf{RR'}>} c_{\mathbf{R}\sigma}^\dagger \, c_{\mathbf{R'}\sigma} \, c_{\mathbf{R'}\sigma'}^\dagger \, c_{\mathbf{R}\sigma'} \,,$$

which, through (1.103) and since $n_{\mathbf{R}} = 1$, is simply

$$H_{s-ex} = \frac{4t^2}{U} \sum_{<\mathbf{RR'}>} \left(\mathbf{S}_{\mathbf{R}} \cdot \mathbf{S}_{\mathbf{R'}} - \frac{1}{4} \right) \,. \qquad (1.108)$$

Therefore, the second-order perturbation theory in the hopping gives rise to an antiferromagnetic spin super-exchange. In conclusion, the effective Hamiltonian at large U describes localised spin-1/2 (each site occupied by one electron) coupled by the overall spin exchange (dropping constant terms)

$$H_{Heis} = J \sum_{<\mathbf{RR'}>} \mathbf{S}_{\mathbf{R}} \cdot \mathbf{S}_{\mathbf{R'}} \,, \qquad (1.109)$$

where

$$J = J_{ex} + \frac{4t^2}{U} = -2U_2 + \frac{4t^2}{U}.$$

The spin exchange may be either ferromagnetic or antiferromagnetic depending on the strength of the direct-exhange constant with respect to the super-exchange one. The effective Hamiltonian (1.109) thus corresponds just to a spin-1/2 Heisenberg model.

1.7 Application: Spin-Wave Theory

When the hopping involves further neighbour sites and more than a single band, as it is commonly the case for magnetic ions, the corresponding Heisenberg model has the general expression

$$H = \sum_{\mathbf{RR'}} J_{\mathbf{R},\mathbf{R'}} \, \mathbf{S_R} \cdot \mathbf{S_{R'}} \, . \tag{1.110}$$

The Hamiltonian to be Hermitian requires that $J_{\mathbf{R},\mathbf{R'}} = J_{\mathbf{R'},\mathbf{R}} \in \mathbb{R}$, while translational symmetry that

$$J_{\mathbf{R},\mathbf{R'}} = J_{\mathbf{R}-\mathbf{R_0},\mathbf{R'}-\mathbf{R_0}} \, , \quad \forall \, \mathbf{R_0} \, . \tag{1.111}$$

For instance, taking $\mathbf{R_0} = \mathbf{R'}$ or $\mathbf{R_0} = \mathbf{R}$, and using the property $J_{\mathbf{R},\mathbf{R'}} = J_{\mathbf{R'},\mathbf{R}}$,

$$J_{\mathbf{R},\mathbf{R'}} = J_{\mathbf{R}-\mathbf{R'},0} = J_{0,\mathbf{R'}-\mathbf{R}} = J_{\mathbf{R'}-\mathbf{R},0} \, . \tag{1.112}$$

We can introduce Fourier transforms defined through

$$S(\mathbf{q}) = \sum_{\mathbf{R}} e^{-i\mathbf{q}\cdot\mathbf{R}} \, \mathbf{S_R} \, , \quad \mathbf{S_R} = \frac{1}{N} \sum_{\mathbf{q}} e^{i\mathbf{q}\cdot\mathbf{R}} \, S(\mathbf{q}) \, , \tag{1.113}$$

where N is the number of lattice sites, and note that

$$
\begin{aligned}
J_{\mathbf{q},\mathbf{q'}} &= \frac{1}{N} \sum_{\mathbf{RR'}} e^{-i\mathbf{q}\cdot\mathbf{R}} \, e^{i\mathbf{q'}\cdot\mathbf{R'}} \, J_{\mathbf{R},\mathbf{R'}} = \frac{1}{N} \sum_{\mathbf{R_0}} \frac{1}{N} \sum_{\mathbf{RR'}} e^{-i\mathbf{q}\cdot\mathbf{R}} \, e^{i\mathbf{q'}\cdot\mathbf{R'}} \, J_{\mathbf{R},\mathbf{R'}} \\
&= \frac{1}{N} \sum_{\mathbf{R_0}} \frac{1}{N} \sum_{\mathbf{RR'}} e^{-i\mathbf{q}\cdot\mathbf{R}} \, e^{i\mathbf{q'}\cdot\mathbf{R'}} \, J_{\mathbf{R}-\mathbf{R_0},\mathbf{R'}-\mathbf{R_0}} \\
&= \frac{1}{N} \sum_{\mathbf{R_0}} e^{-i(\mathbf{q}-\mathbf{q'})\cdot\mathbf{R_0}} \frac{1}{N} \sum_{\mathbf{RR'}} e^{-i\mathbf{q}\cdot\mathbf{R}} \, e^{i\mathbf{q'}\cdot\mathbf{R'}} \, J_{\mathbf{R},\mathbf{R'}} \\
&= \delta_{\mathbf{q},\mathbf{q'}} \frac{1}{N} \sum_{\mathbf{RR'}} e^{-i\mathbf{q}\cdot(\mathbf{R}-\mathbf{R'})} \, J_{\mathbf{R},\mathbf{R'}} = \delta_{\mathbf{q},\mathbf{q'}} \sum_{\mathbf{R}} e^{-i\mathbf{q}\cdot\mathbf{R}} \, J_{\mathbf{R},0} \equiv \delta_{\mathbf{q},\mathbf{q'}} \, J(\mathbf{q}) \, .
\end{aligned}
\tag{1.114}
$$

It follows that the Hamiltonian in Fourier space is simply

$$H = \frac{1}{N} \sum_{\mathbf{q}} J(\mathbf{q}) \, S(-\mathbf{q}) \cdot S(\mathbf{q}) \,. \tag{1.115}$$

Moreover, since $S_{\mathbf{R}} = S_{\mathbf{R}}^{\dagger}$,

$$S(\mathbf{q})^{\dagger} = \sum_{\mathbf{R}} e^{i\mathbf{q}\cdot\mathbf{R}} \, S_{\mathbf{R}} = S(-\mathbf{q}) \,, \tag{1.116}$$

and, since $J_{\mathbf{R},\mathbf{R'}} \in \mathbb{R}$,

$$\begin{aligned} J(\mathbf{q})^{*} &= \sum_{\mathbf{R}} e^{i\mathbf{q}\cdot\mathbf{R}} \, J_{\mathbf{R},0} = J(-\mathbf{q}) = \sum_{\mathbf{R}} e^{i\mathbf{q}\cdot\mathbf{R}} \, J_{\mathbf{R}-\mathbf{R},0-\mathbf{R}} = \sum_{\mathbf{R}} e^{i\mathbf{q}\cdot\mathbf{R}} \, J_{0,-\mathbf{R}} \\ &= \sum_{\mathbf{R}} e^{i\mathbf{q}\cdot\mathbf{R}} \, J_{-R,0} = \sum_{\mathbf{R}} e^{-i\mathbf{q}\cdot\mathbf{R}} \, J_{R,0} = J(\mathbf{q}) \,, \end{aligned}$$

$$\tag{1.117}$$

namely $J(\mathbf{q}) = J(-\mathbf{q}) \in \mathbb{R}$. In conclusion,

$$H = \frac{1}{N} \sum_{\mathbf{q}} J(\mathbf{q}) \, S(\mathbf{q})^{\dagger} \cdot S(\mathbf{q}) \,. \tag{1.118}$$

1.7.1 Classical Ground State

Let us first assume that the spin $S_{\mathbf{R}}$ is a classical vector satisfying

$$S_{\mathbf{R}} \cdot S_{\mathbf{R}} = S^{2} \,, \quad \forall \, \mathbf{R} \,, \tag{1.119}$$

so that the Hamiltonian becomes

$$H = \frac{1}{N} \sum_{\mathbf{q}} J(\mathbf{q}) \, S(\mathbf{q})^{*} \cdot S(\mathbf{q}) \,. \tag{1.120}$$

The classical ground state is actually identified by a vector $\mathbf{Q}$ and a classical spin configuration

$$S(\mathbf{Q}) = S(-\mathbf{Q})^{*} \neq 0 \,, \quad S(\mathbf{q}) = 0 \,, \; \forall \mathbf{q} \neq (\mathbf{Q}, -\mathbf{Q}) \,. \tag{1.121}$$

To prove that, let us assume it is indeed the case, and first identify what $\mathbf{Q}$ is. Under that assumption, and if $\mathbf{Q} \neq -\mathbf{Q}$, where the equality holds apart from a reciprocal lattice vector G, then

$$S_{\mathbf{R}} = \frac{1}{N} \left(S(\mathbf{Q}) \, e^{i\mathbf{Q}\cdot\mathbf{R}} + S(\mathbf{Q})^{*} \, e^{-i\mathbf{Q}\cdot\mathbf{R}} \right) \,. \tag{1.122}$$

In order to verify (1.119), We have to impose that

$$S(\mathbf{Q}) \cdot S(\mathbf{Q}) = S(-\mathbf{Q}) \cdot S(-\mathbf{Q}) = 0\,,$$

$$S(\mathbf{Q})^* \cdot S(\mathbf{Q}) = S(-\mathbf{Q}) \cdot S(\mathbf{Q}) = \frac{N^2}{2}\, S^2\,, \qquad (1.123)$$

which can be satisfied by taking any two orthogonal real unit vectors, $\mathbf{u}_1$ and $\mathbf{u}_2$, thus $\mathbf{u}_i \cdot \mathbf{u}_j = \delta_{ij}$, and writing

$$S(\mathbf{Q}) = \frac{S N}{2}\,(\mathbf{u}_1 - i\,\mathbf{u}_2)\,, \quad S(-\mathbf{Q}) = S(\mathbf{Q})^* = \frac{S N}{2}\,(\mathbf{u}_1 + i\,\mathbf{u}_2)\,. \qquad (1.124)$$

With that choice

$$S_{\mathbf{R}} = S\,\mathbf{u}_1\,\cos\mathbf{Q}\cdot\mathbf{R} + S\,\mathbf{u}_2\,\sin\mathbf{Q}\cdot\mathbf{R}\,. \qquad (1.125)$$

If instead $\mathbf{Q} = -\mathbf{Q} + \mathbf{G}$, i.e., $2\mathbf{Q} = \mathbf{G}$, then $S(\mathbf{Q}) = S(-\mathbf{Q}) \in \mathbb{R}$, so that, since

$$\mathbf{Q}\cdot\mathbf{R} = \frac{1}{2}\,\mathbf{G}\cdot\mathbf{R} \equiv \pi\,n_{\mathbf{R}}(\mathbf{G})\,, \qquad (1.126)$$

with $n_{\mathbf{R}}(\mathbf{G})$ integer, then

$$S_{\mathbf{R}} = \frac{1}{N}\,S(\mathbf{Q})\,\cos\mathbf{Q}\cdot\mathbf{R} = \frac{1}{N}\,S(\mathbf{Q})\,(-1)^{n_{\mathbf{R}}(\mathbf{G})}\,. \qquad (1.127)$$

Equation (1.119) is now satisfied if

$$S(\mathbf{Q}) = N\,S\,\mathbf{u}\,, \quad \mathbf{u}\cdot\mathbf{u} = 1\,. \qquad (1.128)$$

In both cases, $2\mathbf{Q} \neq \mathbf{G}$ and $2\mathbf{Q} = \mathbf{G}$, the energy of the classical configuration is

$$E(\mathbf{Q}) = \frac{1}{N}\left(J(\mathbf{Q})\,S(\mathbf{Q})^* \cdot S(\mathbf{Q}) + J(-\mathbf{Q})\,S(-\mathbf{Q})^* \cdot S(-\mathbf{Q})\right) = N\,S^2\,J(\mathbf{Q})\,, \qquad (1.129)$$

which is minimised by $\mathbf{Q}$ such that

$$J(\mathbf{Q}) = \min_{\mathbf{q}} J(\mathbf{q})\,. \qquad (1.130)$$

We note that the energy does not depend on the choice of $\mathbf{u}_1$ and $\mathbf{u}_2$, if $2\mathbf{Q} \neq \mathbf{G}$, or $\mathbf{u}$ if $2\mathbf{Q} = \mathbf{G}$, which reflects the invariance of the Hamiltonian under spin O(3) rotations and, if $2\mathbf{Q} \neq \mathbf{G}$, an additional $U(1)$ symmetry related to $S(\mathbf{Q})$ being complex.

Now suppose we choose a combination of two $\mathbf{Q}$, $\mathbf{Q}_1$ and $\mathbf{Q}_2$, assuming, just for simplicity, that both satisfy $2\mathbf{Q}_i = \mathbf{G}_i$. In this case,

$$S_{\mathbf{R}} = \frac{1}{N}\,S(\mathbf{Q}_1)\,\cos\mathbf{Q}_1\cdot\mathbf{R} + \frac{1}{N}\,S(\mathbf{Q}_2)\,\cos\mathbf{Q}_2\cdot\mathbf{R}\,. \qquad (1.131)$$

The equation (1.119) is fulfilled if

$$S(\mathbf{Q}_1) = N S \cos\phi\, \mathbf{u}_1\,, \quad S(\mathbf{Q}_2) = N S \sin\phi\, \mathbf{u}_2\,, \quad \mathbf{u}_i \cdot \mathbf{u}_j = \delta_{ij}\,, \quad (1.132)$$

in which case the energy of such configuration reads

$$E(\mathbf{Q}_1, \mathbf{Q}_2, \phi) = N S^2 \left(\cos^2\phi\, J(\mathbf{Q}_1) + \sin^2\phi\, J(\mathbf{Q}_2) \right). \quad (1.133)$$

It is easy to realise that this energy is always higher than choosing the single $\mathbf{Q}$ that minimises $J(\mathbf{q})$, thus proving the claim (1.121).

We end by observing that the average over the Brillouin zone of $J(\mathbf{q})$, i.e.,

$$\frac{1}{N} \sum_{\mathbf{q}} J(\mathbf{q}) \equiv J_{\mathbf{R},\mathbf{R}} = 0\,, \quad (1.134)$$

is just the on-site exchange, which is zero. Therefore, if the average over the Brillouin zone vanishes, then $J(\mathbf{q})$ must have both positive and negative values; hence, the minimum, i.e., $J(\mathbf{Q}_1)$, is necessarily negative and so is the classical energy.

1.7.1.1 Symmetry Considerations

The lattice Hamiltonian is generally invariant under a space group S, i.e., under the symmetry transformations $g \in S$. It follows that, if $J(\mathbf{q})$ is minimised by $\mathbf{Q}_1$, then

$$J\big(g(\mathbf{Q}_1)\big) = J(\mathbf{Q}_1) \quad (1.135)$$

namely is minimum for the whole set of vectors obtained by applying any $g \in S$ to $\mathbf{Q}_1$. In general $g(\mathbf{Q}_1)$ for all $g \in S$ generates a finite number of inequivalent $\mathbf{Q}_i$, $i = 1, n_*$, called the star of $\mathbf{Q}_1$. We emphasise that the Hamiltonian (1.110) implicitly assumes, besides spin O(3), also inversion symmetry, in which absence additional terms would be present, as, e.g., $S_{\mathbf{R}} \wedge S_{\mathbf{R}'} \cdot S_{\mathbf{R}''}$. It follows that S at least includes inversion, $\mathbf{Q}_1 \to -\mathbf{Q}_1$, besides the identity.

If $n_* > 1$, the classical configuration choosing any of the $\mathbf{Q}_i$, e.g., $\mathbf{Q}_1$, breaks the space group symmetry. Specifically, such classical configuration will be invariant under a subgroup $S' \subseteq S$ that contains all $g' \in S'$ such that $g'(\mathbf{Q}_1) = \mathbf{Q}_1$, apart from a reciprocal lattice vector.

If the star of $\mathbf{Q}$ contains just $\mathbf{Q}$, which also implies that $-\mathbf{Q} \equiv \mathbf{Q}$ apart from a reciprocal lattice vector $\mathbf{G}$, the classical configuration still seems to break translation symmetry, which belongs to the space group, since

$$S_{\mathbf{R}} = N S u \cos\mathbf{Q} \cdot \mathbf{R} = N S u \cos\frac{\mathbf{G} \cdot \mathbf{R}}{2} \quad (1.136)$$

is not invariant under translation unless $\mathbf{G} = \mathbf{0}$, namely $\mathbf{Q} = \mathbf{0} = \mathbf{\Gamma}$ equal to the $\mathbf{\Gamma}$ point. However, if we combine the translation $T_{\mathbf{R}_0}$ such that

$$\cos\frac{\mathbf{G} \cdot (\mathbf{R} + \mathbf{R}_0)}{2} = -\cos\frac{\mathbf{G} \cdot \mathbf{R}}{2}\,, \quad (1.137)$$

with time reversal, $\mathcal{T}(S_{\mathbf{R}}) = -S_{\mathbf{R}}$, we do recover full translational symmetry.

In conclusion, any classical configuration that minimises the energy is unavoidably not invariant under spin O(3) symmetry, time reversal symmetry, and, eventually, under the space group symmetry of the Hamiltonian, which entails the existence of a whole manifold of different classical configurations with the same energy.

1.8   Beyond the Classical Limit: The Spin-Wave Approximation

In reality $S_{\mathbf{R}}$ is not a classical vector, but a quantum operator whose components satisfy the commutation relations

$$\left[S_{i\mathbf{R}} , S_{j\mathbf{R}'} \right] = i\, \delta_{\mathbf{R},\mathbf{R}'}\, \epsilon_{ijk}\, S_{k\mathbf{R}} , \qquad (1.138)$$

with ϵ_{ijk} the antisymmetric tensor and taking $\hbar = 1$, which translates, for the Fourier transformed operators, into

$$\left[S_i(\mathbf{q}) , S_j(\mathbf{q}') \right] = i\, \epsilon_{ijk}\, S_k(\mathbf{q} + \mathbf{q}') . \qquad (1.139)$$

Our aim here is to add the quantum mechanics implicit in the non-trivial commutation relations under the assumption that the quantum fluctuations do not alter completely the classical ground state. For that, we associate to the chosen classical configuration, assuming for simplicity that $J(\mathbf{q})$ is minimum for $\mathbf{q} = \mathbf{Q}$ such that $2\mathbf{Q} = \mathbf{G}$, the vacuum $| 0 \rangle$ of the quantum fluctuations. Specifically, if we choose the classical state such that

$$S_i^{\text{class}}(\mathbf{q}) = N\, S\, \delta_{i,z}\, \delta_{\mathbf{q},\mathbf{Q}} , \qquad (1.140)$$

namely taking $\boldsymbol{u} = (0, 0, 1)$, we define $| 0 \rangle$ through

$$\langle 0 | S_i(\mathbf{q}) | 0 \rangle = S_i^{\text{class}}(\mathbf{q}) = N\, S\, \delta_{i,z}\, \delta_{\mathbf{q},\mathbf{Q}} . \qquad (1.141)$$

We thus write

$$S_i(\mathbf{q}) = S_i^{\text{class}}(\mathbf{q}) + \delta S_i(\mathbf{q}) , \qquad (1.142)$$

where $\delta S_i(\mathbf{q})$ is an operator that describes the quantum fluctuation corrections, while $S_i^{\text{class}}(\mathbf{q})$ just a c-number, and satisfies

$$\langle 0 | \delta S_i(\mathbf{q}) | 0 \rangle = 0 . \qquad (1.143)$$

We adopt a semiclassical approach, and thus assume that $\delta S(\mathbf{q}) \sim O(\sqrt{N})$, unlike $S_z^{\text{class}}(\mathbf{Q}) \sim N$, and proceed consistently. For instance,

$$\begin{aligned}
\left[S_x(\mathbf{q}) , S_y(\mathbf{q}') \right] &= \left[\delta S_x(\mathbf{q}) , \delta S_y(\mathbf{q}') \right] \\
&= i\, S_z(\mathbf{q} + \mathbf{q}') = i\, \delta_{\mathbf{q}+\mathbf{q}',\mathbf{Q}}\, N\, S + \delta S_z(\mathbf{q} + \mathbf{q}') \simeq i\, \delta_{\mathbf{q}+\mathbf{q}',\mathbf{Q}}\, N\, S + \cdots ,
\end{aligned}$$
$$(1.144)$$

where ... mean corrections subleading in N. It follows that, at leading order in N, the above commutation relation is fulfilled if

$$\delta S_x(\mathbf{q}) \simeq \sqrt{NS}\, x_{\mathbf{q}}\,, \quad \delta S_y(\mathbf{q}) \simeq \sqrt{NS}\, p_{\mathbf{q}-\mathbf{Q}}\,, \tag{1.145}$$

where $x_{\mathbf{q}} = x^\dagger_{-\mathbf{q}}$ and $p_{\mathbf{q}} = p^\dagger_{-\mathbf{q}}$ are conjugate variables satisfying

$$\left[x_{\mathbf{q}}\,,\, p^\dagger_{\mathbf{q}'} \right] = i\,\delta_{\mathbf{q},\mathbf{q}'}\,, \tag{1.146}$$

and such that the expectation values on the vacuum,

$$\langle 0 \mid x_{\mathbf{q}} \mid 0 \rangle = \langle 0 \mid p_{\mathbf{q}} \mid 0 \rangle = 0\,. \tag{1.147}$$

It follows that, if we write these conjugate variables in terms of bosonic operators, i.e.,

$$x_{\mathbf{q}} = \frac{1}{\sqrt{2}} \left(b_{\mathbf{q}} + b^\dagger_{-\mathbf{q}} \right)\,, \quad p_{\mathbf{q}} = -\frac{i}{\sqrt{2}} \left(b_{\mathbf{q}} - b^\dagger_{-\mathbf{q}} \right)\,, \tag{1.148}$$

then $\mid 0 \rangle$ is their vacuum.

However, we have to enforce two more commutation relations, specifically

$$\left[S_x(\mathbf{q})\,,\, S_z(\mathbf{q}') \right] = -i\, S_y(\mathbf{q} + \mathbf{q}')\,, \quad \left[S_y(\mathbf{q})\,,\, S_z(\mathbf{q}') \right] = i\, S_x(\mathbf{q} + \mathbf{q}') \tag{1.149}$$

which imply that

$$\begin{aligned}
\left[S_x(\mathbf{q})\,,\, S_z(\mathbf{q}') \right] &= \left[\delta S_x(\mathbf{q})\,,\, \delta S_z(\mathbf{q}') \right] = \sqrt{NS} \left[x_{\mathbf{q}}\,,\, \delta S_z(\mathbf{q}') \right] \\
&= -i\,\delta S_y(\mathbf{q} + \mathbf{q}') = -i\sqrt{NS}\, p_{\mathbf{q}+\mathbf{q}'-\mathbf{Q}}\,, \\
\left[S_y(\mathbf{q})\,,\, S_z(\mathbf{q}') \right] &= \left[\delta S_y(\mathbf{q})\,,\, \delta S_z(\mathbf{q}') \right] = \sqrt{NS} \left[p_{\mathbf{q}-\mathbf{Q}}\,,\, \delta S_z(\mathbf{q}') \right] \\
&= i\,\delta S_x(\mathbf{q} + \mathbf{q}') = i\sqrt{NS}\, x_{\mathbf{q}+\mathbf{q}'}\,,
\end{aligned} \tag{1.150}$$

namely

$$\left[x_{\mathbf{q}}\,,\, \delta S_z(\mathbf{q}') \right] = -i\, p_{\mathbf{q}+\mathbf{q}'-\mathbf{Q}}\,, \quad \left[p_{\mathbf{q}-\mathbf{Q}}\,,\, \delta S_z(\mathbf{q}') \right] = i\, x_{\mathbf{q}+\mathbf{q}'}\,. \tag{1.151}$$

The solution for $\delta S_z(\mathbf{q})$ with the condition (1.143) can be readily shown to be

$$\delta S_z(\mathbf{q}) = -\frac{1}{2} \sum_{\mathbf{k}} \left(x^\dagger_{\mathbf{k}} x_{\mathbf{k}+\mathbf{q}-\mathbf{Q}} + p^\dagger_{\mathbf{k}} p_{\mathbf{k}+\mathbf{q}-\mathbf{Q}} - \delta_{\mathbf{q},\mathbf{Q}} \right)\,, \tag{1.152}$$

since

$$\langle 0 \mid x^\dagger_{\mathbf{q}} x_{\mathbf{q}'} \mid 0 \rangle = \langle 0 \mid p^\dagger_{\mathbf{q}} p_{\mathbf{q}'} \mid 0 \rangle = \frac{1}{2}\,\delta_{\mathbf{q},\mathbf{q}'}\,. \tag{1.153}$$

In conclusion, at leading order in an expansion for large N, we can represent the spin operators as

$$S_x(\mathbf{q}) = \sqrt{N\,S}\; x_{\mathbf{q}},$$
$$S_y(\mathbf{q}) = \sqrt{N\,S}\; p_{\mathbf{q}-\mathbf{Q}},$$
$$S_z(\mathbf{q}) = N\,S\,\delta_{\mathbf{q},\mathbf{Q}} - \frac{1}{2}\sum_{\mathbf{k}} \left(x_{\mathbf{k}}^{\dagger} x_{\mathbf{k}+\mathbf{q}-\mathbf{Q}} + p_{\mathbf{k}}^{\dagger} p_{\mathbf{k}+\mathbf{q}-\mathbf{Q}} - \delta_{\mathbf{q},\mathbf{Q}} \right). \tag{1.154}$$

We note that the expansion also looks like an expansion for large S. This is not surprising. Indeed, if we normalise the spin operators, $S_{i\mathbf{R}} \to s_{i\mathbf{R}}/S$, then the commutation relations for the operators $s_{i\mathbf{R}}$, which are $O(1)$ in S, read

$$\left[s_{i\mathbf{R}}\,,\, s_{j\mathbf{R}} \right] = \frac{1}{S}\, i\,\epsilon_{ijk}\, s_{k\mathbf{R}} \;\xrightarrow[S\to\infty]{}\; 0\,, \tag{1.155}$$

showing that the spin operators become classical vectors in the large S limit.[2]

1.8.1   Hamiltonian of Quantum Fluctuations

Now we substitute the expressions in (1.154) in the Hamiltonian

$$H = \frac{1}{N}\sum_{\mathbf{q}} J(\mathbf{q})\, S(\mathbf{q})^{\dagger} \cdot S(\mathbf{q})\,, \tag{1.158}$$

---

[2] There are actually more rigorous ways to implement the mapping from spin operators to bosonic annihilation and creation ones. However, we preferred the less rigorous approach presented, since it is more flexible and can be adopted in more general cases than just spin models. However, for completeness, we here mention one of those rigorous mappings, known as the Holstein-Primakoff transformation.

Suppose that one of the equivalent classical configurations corresponds to the spin $S_{\mathbf{R}}$ at site $\mathbf{R}$ polarised along the z-direction. One can readily demonstrate that the following way of writing spin operators in terms of bosonic one does preserve the spin commutation relations:

$$S_{\mathbf{R}}^z = S - d_{\mathbf{R}}^{\dagger} d_{\mathbf{R}},\quad S_{\mathbf{R}}^+ = \sqrt{2S - d_{\mathbf{R}}^{\dagger} d_{\mathbf{R}}}\; d_{\mathbf{R}},\quad S_{\mathbf{R}}^- = d_{\mathbf{R}}^{\dagger}\sqrt{2S - d_{\mathbf{R}}^{\dagger} d_{\mathbf{R}}}\,, \tag{1.156}$$

where the classical configuration is the vacuum of the bosonic operators $d_{\mathbf{R}}$ and $d_{\mathbf{R}}^{\dagger}$. The square roots in (1.156) assure that it is not possible to create more than $2S$ bosons at any given site, so that, correctly, $\langle S_{\mathbf{R}}^z \rangle = -S,\dots,S$. The spin-wave theory is recovered by assuming that the number of excited bosons is negligible with respect to S, namely that the actual quantum ground state is close to the classical one. In that case $\sqrt{2S - d_{\mathbf{R}}^{\dagger} d_{\mathbf{R}}} \simeq \sqrt{2S}$, and thus

$$S_{\mathbf{R}}^z = S - d_{\mathbf{R}}^{\dagger} d_{\mathbf{R}},\quad S_{\mathbf{R}}^+ \simeq \sqrt{2S}\; d_{\mathbf{R}},\quad S_{\mathbf{R}}^- = \sqrt{2S}\; d_{\mathbf{R}}^{\dagger}\,. \tag{1.157}$$

Substituting those expressions in the Hamiltonian reproduces the spin-wave theory we discuss.

consistently within an expansion for large N, or, more correctly, large S. We note that $S_x(\mathbf{q})$ and $S_y(\mathbf{q})$ are of order $O(\sqrt{S})$, $S_z(\mathbf{Q})$ of order $O(S)$ plus a correction $O(1)$, while $S_z(\mathbf{q})$ for $\mathbf{q} \neq \mathbf{Q}$ of order $O(1)$. Therefore, the Hamiltonian up to order $O(S)$ is simply

$$
\begin{aligned}
H &= \frac{1}{N} \sum_{\mathbf{q}} J(\mathbf{q}) \left(S_x(\mathbf{q})^\dagger S_x(\mathbf{q}) + S_y(\mathbf{q})^\dagger S_y(\mathbf{q}) \right) + \frac{J(\mathbf{Q})}{N} S_z(\mathbf{Q})^\dagger S_z(\mathbf{Q}) + O(S^0) \\
&\simeq S \sum_{\mathbf{q}} \left(J(\mathbf{q}) x_{\mathbf{q}}^\dagger x_{\mathbf{q}} + J(\mathbf{q}+\mathbf{Q}) p_{\mathbf{q}}^\dagger p_{\mathbf{q}} \right) \\
&\quad + J(\mathbf{Q}) N S^2 - J(\mathbf{Q}) S \sum_{\mathbf{q}} \left(x_{\mathbf{q}}^\dagger x_{\mathbf{q}} + p_{\mathbf{q}}^\dagger p_{\mathbf{q}} - 1 \right) \\
&= E^{\text{class}} + N S J(\mathbf{Q}) + S \sum_{\mathbf{q}} \left[\left(J(\mathbf{q}) - J(\mathbf{Q}) \right) x_{\mathbf{q}}^\dagger x_{\mathbf{q}} + \left(J(\mathbf{q}+\mathbf{Q}) - J(\mathbf{Q}) \right) p_{\mathbf{q}}^\dagger p_{\mathbf{q}} \right] \\
&= J(\mathbf{Q}) N S(S+1) + S \sum_{\mathbf{q}} \left[\left(J(\mathbf{q}) - J(\mathbf{Q}) \right) x_{\mathbf{q}}^\dagger x_{\mathbf{q}} + \left(J(\mathbf{q}+\mathbf{Q}) - J(\mathbf{Q}) \right) p_{\mathbf{q}}^\dagger p_{\mathbf{q}} \right].
\end{aligned}
$$
(1.159)

Note the appearance of the quantum spin magnitude $S(S+1)$ that corrects the large S classical limit, i.e., S^2.

We need to diagonalise the Hamiltonian, which looks like the Hamiltonian of coupled harmonic oscillators. For that, we recall that the Hamiltonian of a set of harmonic oscillators in the canonical basis is equal to

$$
H = \frac{1}{2} \sum_{\mathbf{q}} \omega_{\mathbf{q}} \left(x_{\mathbf{q}}^\dagger x_{\mathbf{q}} + p_{\mathbf{q}}^\dagger p_{\mathbf{q}} \right),
$$
(1.160)

with $\omega_{\mathbf{q}} = \omega_{-\mathbf{q}}$, so that, upon introducing the bosonic creation and annihilation operators through

$$
x_{\mathbf{q}} = \frac{1}{\sqrt{2}} \left(a_{\mathbf{q}} + a_{-\mathbf{q}}^\dagger \right), \quad p_{\mathbf{q}} = -\frac{i}{\sqrt{2}} \left(a_{\mathbf{q}} - a_{-\mathbf{q}}^\dagger \right),
$$
(1.161)

then

$$
H = \sum_{\mathbf{q}} \omega_{\mathbf{q}} \left(a_{\mathbf{q}}^\dagger a_{\mathbf{q}} + \frac{1}{2} \right),
$$
(1.162)

which is the known diagonal form. In order to bring H in (1.159) to the canonical form (1.160), we apply the transformation

$$
x_{\mathbf{q}} = \sqrt{K_{\mathbf{q}}} \, x_{\mathbf{q}}, \quad p_{\mathbf{q}} = \frac{1}{\sqrt{K_{\mathbf{q}}}} \, p_{\mathbf{q}},
$$
(1.163)

which preserves the commutation relations, and find

$$
H = J(\mathbf{Q}) N S(S+1) + S \sum_{\mathbf{q}} \left[\left(J(\mathbf{q}) - J(\mathbf{Q}) \right) K_{\mathbf{q}} \, x_{\mathbf{q}}^\dagger x_{\mathbf{q}} + \frac{J(\mathbf{q}+\mathbf{Q}) - J(\mathbf{Q})}{K_{\mathbf{q}}} \, p_{\mathbf{q}}^\dagger p_{\mathbf{q}} \right].
$$

We fix $K_{\mathbf{q}}$ so that

$$\left(J(\mathbf{q}) - J(\mathbf{Q})\right) K_{\mathbf{q}} = \frac{J(\mathbf{q}+\mathbf{Q}) - J(\mathbf{Q})}{K_{\mathbf{q}}} \quad \Rightarrow \quad K_{\mathbf{q}}^2 = \frac{J(\mathbf{q}+\mathbf{Q}) - J(\mathbf{Q})}{J(\mathbf{q}) - J(\mathbf{Q})} .$$

$$(1.164)$$

With that choice H does acquire a canonical expression

$$H = J(\mathbf{Q}) \, N \, S(S+1) + \frac{1}{2} \sum_{\mathbf{q}} \omega_{\mathbf{q}} \left(x_{\mathbf{q}}^{\dagger} x_{\mathbf{q}} + p_{\mathbf{q}}^{\dagger} p_{\mathbf{q}} \right)$$

$$= J(\mathbf{Q}) \, N \, S(S+1) + \sum_{\mathbf{q}} \omega_{\mathbf{q}} \left(a_{\mathbf{q}}^{\dagger} a_{\mathbf{q}} + \frac{1}{2} \right) ,$$

$$(1.165)$$

with frequency

$$\omega_{\mathbf{q}} = 2S \sqrt{\left(J(\mathbf{q}) - J(\mathbf{Q})\right) \left(J(\mathbf{q}+\mathbf{Q}) - J(\mathbf{Q})\right)} ,$$

$$(1.166)$$

which is the so-called spin-wave dispersion.

1.8.2 Spin-Wave Dispersion and Goldstone Theorem

Let us first assume that $\mathbf{Q} \neq \mathbf{\Gamma} = \mathbf{0}$. In this case, upon expanding in Taylor series for small q, assuming for simplicity an isotropic system,

$$J(\mathbf{q}+\mathbf{Q}) - J(\mathbf{Q}) \simeq \gamma \, q^2 ,$$

$$(1.167)$$

where $\gamma > 0$ since $J(\mathbf{Q})$ is the minimum of $J(\mathbf{q})$. It follows that

$$\omega_{\mathbf{q}} \xrightarrow[q \to 0]{} 2q \sqrt{\gamma \left(J(0) - J(\mathbf{Q})\right)} \equiv v q ,$$

$$(1.168)$$

namely the excitations show an acoustic dispersion, with velocity v becoming massless at $q = 0$. If $\mathbf{Q} = \mathbf{\Gamma} = \mathbf{0}$, which is the case of a ferromagnet, then

$$\omega_{\mathbf{q}} \xrightarrow[q \to 0]{} 2\gamma \, q^2 ,$$

$$(1.169)$$

the energy still vanishing at $q = 0$, but now quadratically. The difference between the two cases is that in the ferromagnetic one, $\mathbf{Q} = \mathbf{0}$, the classical 'order parameter', i.e., $S(0)$, is actually a conserved operator of the Hamiltonian, the total spin. On the contrary, for $\mathbf{Q} \neq \mathbf{0}$, $S(\mathbf{Q})$ is not a conserved quantity.

The vanishing of the energy at $\mathbf{q} \to \mathbf{0}$ is just a consequence of the Goldstone theorem, which states that whenever the ground state is not invariant under a continuous symmetry, in this case the spin $SU(2)$ symmetry, quantum counterpart of the classical $O(3)$, and the Hamiltonian is short-ranged, then the excitation spectrum has to become massless at $q = 0$. The reason is that $S(0)$ is just the generator of the $SU(2)$ rotations, which transform a state with $S(Q)$ directed, e.g., along the z-direction, into a state where it is directed along another direction. Since all these states have the same energy, by symmetry, then there is no energy cost in such global rotation. We mention that despite one can continuously rotate the direction of $S(Q)$, states with different directions are orthogonal in the thermodynamic limit $N \to \infty$, however small is the rotation, which is the essence of a symmetry breaking.

We further note that ω_q in (1.166) for $Q \neq 0$ also vanishes when $q \to Q$. The reason is that the magnetic configuration breaks the translational symmetry, so that Q folds to the Γ point. Even more, in case the star of Q contains other momenta distinct from Q, the spin-wave dispersion vanishes at all $g(Q) \neq Q$, with g in the coset S/S'.

1.8.3 Validity of the Approximation and the Mermin-Wagner Theorem

We need now to check the validity of the initial assumption that the ground state with quantum fluctuation is not much different from the classical one, i.e., the vacuum $| 0 \rangle$. The new ground state is actually the vacuum of the bosonic operators a_q in (1.165), which are different from the original ones, b_q in (1.148). One way to assess this assumption is to evaluate the expectation value of the order parameter and compare it with its classical counterpart. We note that

$$
\begin{aligned}
\langle S_z(Q) \rangle &= N S - \frac{1}{2} \sum_q \langle x_q^\dagger x_q + p_q^\dagger p_q - 1 \rangle \\
&= N S - \frac{1}{2} \sum_q \langle K_q \, x_q^\dagger x_q + K_q^{-1} \, p_q^\dagger p_q - 1 \rangle \\
&= N S - \frac{1}{2} \sum_q \left(\frac{K_q}{2} + \frac{1}{2 K_q} - 1 \right).
\end{aligned}
\tag{1.170}
$$

Therefore, the quantum fluctuations reduce the value of the order parameter.[3] We expect that the correction does not kill completely the classical order parameter

---

[3] If $Q = 0$, then $K_q = 1$ and the quantum fluctuations have no effect on the value of the order parameter. This is so since $S(0)$ is conserved.

unless the sum is divergent. Therefore, let us analyse

$$\Sigma = \frac{1}{2N} \sum_{\mathbf{q}} \left(\frac{K_{\mathbf{q}}}{2} + \frac{1}{2K_{\mathbf{q}}} - 1 \right)$$

$$= \frac{1}{4N} \sum_{\mathbf{q}} \left[\sqrt{\frac{J(\mathbf{q}+\mathbf{Q}) - J(\mathbf{Q})}{J(\mathbf{q}) - J(\mathbf{Q})}} + \sqrt{\frac{J(\mathbf{q}) - J(\mathbf{Q})}{J(\mathbf{q}+\mathbf{Q}) - J(\mathbf{Q})}} - 2 \right].$$

$$(1.171)$$

Convergence problems may arise only if $\mathbf{Q} \neq \mathbf{0}$ and in the points where the denominators vanish, which are actually when the spin-wave energy $\omega_{\mathbf{q}}$ vanishes, as, e.g., $\mathbf{q} = \mathbf{0}$. Therefore, the convergence of the sum is equivalent to the following d-dimensional integral close to the origin:

$$I_d = \int d^d q \, \frac{1}{\omega_{\mathbf{q}}} \sim \int d^d q \, \frac{1}{q} \, . \tag{1.172}$$

This integral is singular in $d = 1$, while converges for any $d > 1$. It follows that, however large S is, namely however close we are to the classical limit, the ground state of the quantum Hamiltonian in one dimension cannot have a finite expectation value of $S(\mathbf{Q})$ if $\mathbf{Q} \neq \mathbf{0}$. This result is consistent with the Mermin-Wagner theorem stating that a continuous symmetry cannot be spontaneously broken at zero temperature, $T = 0$, i.e., in the ground state, in one dimension.

Let us consider now the effect of a finite temperature, $T > 0$, i.e., of thermal fluctuations. In this case,

$$\langle x_{\mathbf{q}}^{\dagger} x_{\mathbf{q}} \rangle = \frac{1}{2} \langle (a_{\mathbf{q}}^{\dagger} + a_{-\mathbf{q}})(a_{\mathbf{q}} + a_{-\mathbf{q}}^{\dagger}) \rangle = 2n(\omega_{\mathbf{q}}) + 1$$

$$= \frac{2}{e^{\beta\omega_{\mathbf{q}}} - 1} + 1 = \coth \frac{\omega_{\mathbf{q}}}{2T} = \langle p_{\mathbf{q}}^{\dagger} p_{\mathbf{q}} \rangle .$$

$$(1.173)$$

It follows that

$$\Sigma(T) = \frac{1}{4N} \sum_{\mathbf{q}} \left[\coth \frac{\omega_{\mathbf{q}}}{2T} \sqrt{\frac{J(\mathbf{q}+\mathbf{Q}) - J(\mathbf{Q})}{J(\mathbf{q}) - J(\mathbf{Q})}} \right.$$

$$\left. + \coth \frac{\omega_{\mathbf{q}}}{2T} \sqrt{\frac{J(\mathbf{q}) - J(\mathbf{Q})}{J(\mathbf{q}+\mathbf{Q}) - J(\mathbf{Q})}} - 2 \right].$$

We observe that, since $\coth \omega_{\mathbf{q}}/2T$ grows like T for large T, also $\Sigma(T)$ grows and at some temperature unavoidably cancels the classical value, even if $\mathbf{Q} = \mathbf{0}$. This simply means that the symmetry will be restored above a critical temperature T_c that can be estimated through

$$\Sigma(T_c) \simeq S \, . \tag{1.174}$$

We also note that the integral I_d is now replaced by

$$I_d(T) = \int d^d q \, \frac{1}{\omega_{\mathbf{q}}} \, \coth \frac{\omega_{\mathbf{q}}}{2T} \simeq T \int d^d q \, \frac{1}{\omega_{\mathbf{q}}^2} \simeq T \int d^d q \, \frac{1}{q^2} \,, \qquad (1.175)$$

which is singular in $d = 1, 2$. It follows that in $d \leq 2$ the equilibrium state at any $T \neq 0$ cannot break the $SU(2)$ symmetry. Once again this is consistent with the Mermin-Wagner theorem stating that a continuous symmetry cannot be broken spontaneously at any finite temperature in $d \leq 2$.

1.8.4 Order from Disorder

If the ordering wave vector is not invariant under the space group S, it means that there is a set $\mathbf{Q}_i, i = 1, \ldots, n_*$, such that $J(\mathbf{Q}_i)$ is the same for all $\mathbf{Q}_i$. In this case, the classical configuration is actually not unique, since we can build any configuration that is a combination of all $S(\mathbf{Q}_i)$ and get the same classical energy.

Let us consider the simplest case of $n_* = 2$, thus a star that contains just two wave vectors, $\mathbf{Q}_1$ and $\mathbf{Q}_2$, which implies that there is an element of the space group, g, such that

$$g(\mathbf{Q}_1) = \mathbf{Q}_2 \,, \qquad (1.176)$$

and vice versa. For simplicity, we still assume $2\mathbf{Q}_i = \mathbf{0}, i = 1, 2$, apart from a reciprocal lattice vector. We take as classical configuration

$$S_{\mathbf{R}} = \cos \phi \, S \, \boldsymbol{u}_1 \cos \mathbf{Q}_1 \cdot \mathbf{R} + \sin \phi \, S \, \boldsymbol{u}_2 \cos \mathbf{Q}_2 \cdot \mathbf{R} \,, \qquad (1.177)$$

with $\phi \in [0, \pi/2]$, and $\boldsymbol{u}_i \cdot \boldsymbol{u}_j = \delta_{ij}$, which guarantees that $S_{\mathbf{R}} \cdot S_{\mathbf{R}} = 1$. It follows that

$$S(\mathbf{q}) = N S \cos \phi \, \boldsymbol{u}_1 \, \delta_{\mathbf{q}, \mathbf{Q}_1} + N S \sin \phi \, \boldsymbol{u}_2 \, \delta_{\mathbf{q}, \mathbf{Q}_2} \,, \qquad (1.178)$$

so that the classical energy

$$E = \frac{1}{N} \sum_{\mathbf{q}} J(\mathbf{q}) \, S(\mathbf{q})^\dagger \cdot S(\mathbf{q}) = N S^2 \left(J(\mathbf{Q}_1) \cos^2 \phi + J(\mathbf{Q}_2) \sin^2 \phi \right) = N S^2 \, J(\mathbf{Q}_1)$$

$$(1.179)$$

is independent of ϕ, $\boldsymbol{u}_1$ and $\boldsymbol{u}_2$. To better understand the freedom we actually have at the classical level, we note that we can always write

$$\cos \phi \, \boldsymbol{u}_1 = \frac{\boldsymbol{u} + \boldsymbol{v}}{2} \qquad \sin \phi \, \boldsymbol{u}_2 = \frac{\boldsymbol{u} - \boldsymbol{v}}{2} \,, \qquad (1.180)$$

with $\boldsymbol{u}$ and $\boldsymbol{v}$ two arbitrary unit vectors such that $\boldsymbol{u} \cdot \boldsymbol{v} = \cos 2\phi$. It follows that the arbitrariness consists of two $O(3)$, one related to the global spin $O(3)$, and another that represents an internal symmetry. We further note that under the symmetry transformation g

$$g\left(S(\mathbf{q})\right) = N S \cos\phi\, \mathbf{u}_1\, \delta_{\mathbf{q},\mathbf{Q}_2} + N S \sin\phi\, \mathbf{u}_2\, \delta_{\mathbf{q},\mathbf{Q}_1}, \tag{1.181}$$

so that g exchanges the two components of the classical configuration.

In order to represent the quantum spin operators as an expansion from the classical case, we have to slightly modify the approach of the previous section. First of all, we define another unit vector $\mathbf{u}_3 \equiv \mathbf{u}_1 \wedge \mathbf{u}_2$, so that the three form a basis set. Then, we define the spin components in such basis, i.e., $S_{i\mathbf{R}} \equiv \mathbf{S}_{\mathbf{R}} \cdot \mathbf{u}_i$, $i = 1, 2, 3$, which evidently satisfy the standard commutation relations

$$\left[S_{i\mathbf{R}},\, S_{j\mathbf{R}'}\right] = i\, \delta_{\mathbf{R},\mathbf{R}'}\, \epsilon_{ijk}\, S_{k\mathbf{R}}, \tag{1.182}$$

or, in momentum space,

$$\left[S_i(\mathbf{q}),\, S_j(\mathbf{q}')\right] = i\, \epsilon_{ijk}\, S_k(\mathbf{q} + \mathbf{q}'). \tag{1.183}$$

Denoting the classical spin configuration as

$$S^{\text{class}}(\mathbf{q}) = N S \cos\phi\, \mathbf{u}_1\, \delta_{\mathbf{q},\mathbf{Q}_1} + N S \sin\phi\, \mathbf{u}_2\, \delta_{\mathbf{q},\mathbf{Q}_2}, \tag{1.184}$$

namely

$$S_i^{\text{class}}(\mathbf{q}) = N S \cos\phi\, \delta_{i,1}\, \delta_{\mathbf{q},\mathbf{Q}_1} + N S \sin\phi\, \delta_{i,2}\, \delta_{\mathbf{q},\mathbf{Q}_2}, \tag{1.185}$$

which is simply a c-number, we write, as before, the quantum operators

$$S_i(\mathbf{q}) = S_i^{\text{class}}(\mathbf{q}) + \delta S_i(\mathbf{q}), \quad i = 1, 2, 3, \tag{1.186}$$

where $\delta S_i(\mathbf{q})$ describes the quantum fluctuations and has vanishing expectation value on the 'vacuum', which is identified with the classical state. It follows that

$$\begin{aligned}
\left[S_3(\mathbf{q}),\, S_1(\mathbf{q}')\right] &= \left[\delta S_3(\mathbf{q}),\, \delta S_1(\mathbf{q}')\right] = i\, S_2(\mathbf{q} + \mathbf{q}') \\
&= i\, N S \sin\phi\, \delta_{\mathbf{q}+\mathbf{q}',\mathbf{Q}_2} + i\, \delta S_2(\mathbf{q} + \mathbf{q}'), \\
\left[S_2(\mathbf{q}),\, S_3(\mathbf{q}')\right] &= \left[\delta S_2(\mathbf{q}),\, \delta S_3(\mathbf{q}')\right] = i\, S_1(\mathbf{q} + \mathbf{q}') \\
&= i\, N S \cos\phi\, \delta_{\mathbf{q}+\mathbf{q}',\mathbf{Q}_1} + i\, \delta S_1(\mathbf{q} + \mathbf{q}'), \\
\left[S_1(\mathbf{q}),\, S_2(\mathbf{q}')\right] &= \left[\delta S_1(\mathbf{q}),\, \delta S_2(\mathbf{q}')\right] = i\, S_3(\mathbf{q} + \mathbf{q}') = i\, \delta S_3(\mathbf{q} + \mathbf{q}').
\end{aligned} \tag{1.187}$$

One can readily check that the above commutation relations are satisfied if we assume the following equivalences:

$$S_3(\mathbf{q}) \simeq \sqrt{N S}\, x(\mathbf{q}),$$
$$S_1(\mathbf{q}) \simeq \cos\phi\, N S\, \delta_{\mathbf{q},\mathbf{Q}_1} - \cos\phi\, \Pi(\mathbf{q} - \mathbf{Q}_1) + \sin\phi\, \sqrt{N S}\, p^\dagger(\mathbf{Q}_2 - \mathbf{q}),$$
$$S_2(\mathbf{q}) \simeq \sin\phi\, N S\, \delta_{\mathbf{q},\mathbf{Q}_2} - \sin\phi\, \Pi(\mathbf{q} - \mathbf{Q}_2) - \cos\phi\, \sqrt{N S}\, p^\dagger(\mathbf{Q}_1 - \mathbf{q}),$$
$$(1.188)$$

where

$$\Pi(\mathbf{q}) = \frac{1}{2} \sum_{\mathbf{k}} \left(p^\dagger(\mathbf{k})\, p(\mathbf{k} + \mathbf{q}) + x^\dagger(\mathbf{k})\, x(\mathbf{k} + \mathbf{q}) - \delta_{\mathbf{q},0} \right), \qquad (1.189)$$

while $x(\mathbf{q}) = x^\dagger(-\mathbf{q})$ and $p(\mathbf{q}) = p^\dagger(-\mathbf{q})$ are conjugate variables, i.e., they satisfy $\left[x(\mathbf{q}), p^\dagger(\mathbf{q}') \right] = i\, \delta_{\mathbf{q},\mathbf{q}'}$, and are defined for $\mathbf{q} \neq \mathbf{0}$, $\mathbf{Q}_1 - \mathbf{Q}_2$. The reason is that $S(\mathbf{0})$ is the generator of the global $SU(2)$ rotations, therefore connects equivalent states with the order parameter pointing in a different direction. Similarly, the unitary operator

$$U(\gamma) = e^{-i\gamma\, S_3(\mathbf{Q}_1 - \mathbf{Q}_2)} = e^{-i\gamma\, S_3(\mathbf{Q}_2 - \mathbf{Q}_1)}, \qquad (1.190)$$

since $\mathbf{Q}_1 - \mathbf{Q}_2 \equiv \mathbf{Q}_2 - \mathbf{Q}_1$, simply shifts ϕ by γ; hence, it is a global transformation that connects states with different values of ϕ.

At leading order, the equations (1.188) become

$$S_3(\mathbf{q}) \simeq \sqrt{N S}\, x(\mathbf{q}),$$
$$S_1(\mathbf{Q}_2 - \mathbf{q}) \simeq \sin\phi\, \sqrt{N S}\, p^\dagger(\mathbf{q}), \qquad S_2(\mathbf{Q}_1 - \mathbf{q}) \simeq -\cos\phi\, \sqrt{N S}\, p^\dagger(\mathbf{q}),$$
$$S_1(\mathbf{Q}_1) \simeq \cos\phi\, N S - \cos\phi\, \Pi(\mathbf{0}), \qquad S_2(\mathbf{Q}_2) \simeq \sin\phi\, N S - \sin\phi\, \Pi(\mathbf{0}),$$
$$(1.191)$$

and substituted in the Hamiltonian lead, up to second order in the quantum fluctuations,

$$
\begin{aligned}
H &= \frac{1}{N} \sum_{\mathbf{q}} J(\mathbf{q})\, S(\mathbf{q})^\dagger \cdot S(\mathbf{q}) \simeq N S^2 J(\mathbf{Q}_1) + S \sum_{\mathbf{q}} J(\mathbf{q})\, x^\dagger(\mathbf{q})\, x(\mathbf{q}) \\
&\quad + S \sum_{\mathbf{q}} \left(\cos^2\phi\, J(\mathbf{q} + \mathbf{Q}_1) + \sin^2\phi\, J(\mathbf{q} + \mathbf{Q}_2) \right) p^\dagger(\mathbf{q})\, p(\mathbf{q}) \\
&\quad - S J(\mathbf{Q}_1) \sum_{\mathbf{q}} \left(x^\dagger(\mathbf{q})\, x(\mathbf{q}) + p^\dagger(\mathbf{q})\, p(\mathbf{q}) - 1 \right) \\
&= N S^2 J(\mathbf{Q}_1) + N S J(\mathbf{Q}_1) \\
&\quad + S \sum_{\mathbf{q}} \Bigg\{ \left(\cos^2\phi\, J(\mathbf{q} + \mathbf{Q}_1) + \sin^2\phi\, J(\mathbf{q} + \mathbf{Q}_2) - J(\mathbf{Q}_1) \right) p^\dagger(\mathbf{q})\, p(\mathbf{q}) \\
&\qquad\qquad + \left(J(\mathbf{q}) - J(\mathbf{Q}_1) \right) x^\dagger(\mathbf{q})\, x(\mathbf{q}) \Bigg\}.
\end{aligned}
$$
$$(1.192)$$

The eigenvalues of the normal modes can be obtained as we did earlier and are simply

$$\omega(\mathbf{q}, \phi) = 2S \sqrt{\left(\cos^2 \phi \, J(\mathbf{q} + \mathbf{Q}_1) + \sin^2 \phi \, J(\mathbf{q} + \mathbf{Q}_2) - J(\mathbf{Q}_1) \right) \left(J(\mathbf{q}) - J(\mathbf{Q}_1) \right)} \,,$$

(1.193)

so that, if $E^{\text{class}} = N S^2 J(\mathbf{Q}_1) < 0$ is the classical energy, the quantum one, including the zero-point contribution, is

$$E(\phi) = E^{\text{class}} + N S J(\mathbf{Q}_1) + \frac{1}{2} \sum_{\mathbf{q}} \omega(\mathbf{q}, \phi) \equiv E^{\text{class}} + \delta E^{\text{quant}}(\phi) \,.$$

(1.194)

We note that the zero-point contribution to the total energy depends on ϕ, and thus, the minimum energy corresponds to specific values of that angle. This phenomenon is known under the name *order from disorder* [4] and occurs when the classical lowest energy state has an accidental degeneracy that is unavoidably lifted by short-wavelength fluctuations, both quantum, as we are discussing here, and thermal. We further observe that, if $g(\mathbf{Q}_1) = \mathbf{Q}_2$, then $J\big(g(\mathbf{q})\big) = J(\mathbf{q})$ by symmetry and $J\big(g(\mathbf{q}) - \mathbf{Q}_1\big) = J(\mathbf{q} - \mathbf{Q}_2)$, so that

$$\omega\big(g(\mathbf{q}), \phi\big) = \omega(\mathbf{q}, \pi/2 - \phi) \,,$$

(1.195)

and thus $E(\phi) = E(\pi/2 - \phi)$ is symmetric around $\pi/4$, which is therefore either a minimum or a maximum. Let us therefore expand the energy around $\pi/4$, when $\cos 2\phi \simeq 0$ can be taken as expansion parameter. We can write

$$\omega(\mathbf{q}, \phi) = \sqrt{\omega(\mathbf{q}, \pi/4)^2 + \cos 2\phi \, \Gamma(\mathbf{q})}$$

$$\simeq \omega(\mathbf{q}, \pi/4) + \frac{\Gamma(\mathbf{q})}{2\omega(\mathbf{q}, \pi/4)} \cos 2\phi - \frac{\Gamma(\mathbf{q})^2}{8 \, \omega(\mathbf{q}, \pi/4)^3} \cos^2 2\phi \,,$$

(1.196)

where

$$\Gamma(\mathbf{q}) = 2S^2 \left(J(\mathbf{q} + \mathbf{Q}_1) - J(\mathbf{q} + \mathbf{Q}_2) \right) \left(J(\mathbf{q}) - J(\mathbf{Q}_1) \right)$$

(1.197)

is odd under g, i.e., $\Gamma\big(g(\mathbf{q})\big) = -\Gamma(\mathbf{q})$. Therefore,

$$\frac{1}{2} \sum_{\mathbf{q}} \omega(\mathbf{q}, \phi) = \frac{1}{4} \sum_{\mathbf{q}} \left(\omega(\mathbf{q}, \phi) + \omega\big(g(\mathbf{q}), \phi\big) \right) = \frac{1}{2} \sum_{\mathbf{q}} \omega(\mathbf{q}, \pi/4)$$

$$- \frac{1}{16} \cos^2 2\phi \sum_{\mathbf{q}} \frac{\Gamma(\mathbf{q})^2}{\omega(\mathbf{q}, \pi/4)^3} \,,$$

(1.198)

which shows that the zero-point energy is actually maximum at $\pi/4$ and minimum at $\phi = 0, \pi/2$, which are therefore the optimal values selected by quantum fluctuations.

The final result is that the huge degeneracy of the classical ground state, related to the internal $O(3)$ symmetry, is resolved by quantum fluctuations, thus an order from disorder effect, which favour either the ordering at the wave vector $\mathbf{Q}_1$, or that at $\mathbf{Q}_2 = g(\mathbf{Q}_1)$, but disfavour any linear combination of them. This result is not surprising, since the physical symmetries involved are the spin $SU(2)$ and the transformation g of the space group, which, since $g^2 = 1$, is effectively a Z_2 symmetry corresponding to the exchange $\mathbf{Q}_1 \leftrightarrow \mathbf{Q}_2$, i.e., an Ising one. In other words, the underlying symmetry, which is broken in the ground state, is actually $SU(2) \times Z_2$.

In order to let the Z_2 degrees of freedom emerge more clearly, let us go back to the state at fixed ϕ that derives from the classical configuration

$$\mathbf{S_R} = \frac{S}{2}\left(\mathbf{u} + \mathbf{v}\right)\cos\mathbf{Q}_1 \cdot \mathbf{R} + \frac{S}{2}\left(\mathbf{u} - \mathbf{v}\right)\cos\mathbf{Q}_2 \cdot \mathbf{R}, \tag{1.199}$$

where $\mathbf{u} \cdot \mathbf{u} = \mathbf{v} \cdot \mathbf{v} = 1$, and

$$\mathbf{u} \cdot \mathbf{v} = \cos 2\phi \equiv \sigma \in [-1, 1]. \tag{1.200}$$

The parameter $\sigma \in [-1, 1]$ plays the role of an Ising order parameter. Through (1.196) we can write $\omega(\mathbf{q}, \phi) \equiv \omega(\mathbf{q}, \sigma)$, so that $\omega(g(\mathbf{q}), \sigma) = \omega(\mathbf{q}, -\sigma)$, and the total energy becomes

$$E(\sigma) = E^{\text{class}} + N\, S\, J(\mathbf{Q}_1) + \frac{1}{4}\sum_{\mathbf{q}}\left(\omega(\mathbf{q}, \sigma) + \omega(\mathbf{q}, -\sigma)\right), \tag{1.201}$$

which explicitly looks like a double well potential, as one does expect in an Ising model as a function of the average magnetisation. That suggests, e.g., an Ising transition at finite temperature in two dimensions, corresponding to the symmetry lowering $C_4 \rightarrow C_2$, despite the spins are always disordered at any $T \neq 0$ [5].

Finally, to understand the order-from-disorder phenomenon in simple terms, let us imagine a particle of mass m moving in a two-dimensional potential well $V(x, y)$, which has a flat minimum along a curve $(x(s), y(s))$ parametrised by s, i.e.,

$$V\left(x(s), y(s)\right) = V_{\min} = \min_{x,y} V(x, y). \tag{1.202}$$

The classical ground state corresponds to the particle localised at any point $(x(s), y(s))$ along the curve; thus, it is degenerate and parametrised by the variable s. Inclusion of thermal and/or quantum fluctuations requires, at first approximation, to study the potential within the harmonic approximation, namely to calculate the Hessian at any of the points $(x(s), y(s))$. Evidently, along the curve $(x(s), y(s))$ the potential is flat, and thus has no curvature. Therefore, the Hessian at a given s has one vanishing eigenvalue in the tangential direction of the curve $(x(s), y(s))$, and the other eigenvalue $K(s) > 0$ in the normal direction, which corresponds to a frequency of oscillations

$$\omega(s) = \sqrt{\frac{K(s)}{m}}. \tag{1.203}$$

It follows that both zero-point quantum fluctuations and thermal ones prefer the values of s with minimum frequency, i.e., minimum curvature $K(s)$.

Problems

1.1 Various commutation relations—Calculate the commutators

$$C_{a,nm} = \left[c_a, c_n^\dagger c_m\right], \quad C_{ab,nm} = \left[c_a^\dagger c_b, c_n^\dagger c_m\right], \quad C_{a,nmlp} = \left[c_a, c_n^\dagger c_m^\dagger c_l c_p\right],$$

$$D_{a,nm} = \left[c_a^\dagger, c_n^\dagger c_m\right], \quad D_{a,nmlp} = \left[c_a^\dagger, c_n^\dagger c_m^\dagger c_l c_p\right],$$

where the operators c_a are either fermionic or bosonic ones.

1.2 Energy of the Fermi sea and f-sum rule—Consider the Hamiltonian

$$H = \sum_{k\sigma} \frac{\hbar^2 k^2}{2m} c_{k\sigma}^\dagger c_{k\sigma} + \frac{1}{2V} \sum_{kk'q\sigma\sigma'} U(\mathbf{q}) c_{k\sigma}^\dagger c_{k'+q\sigma'}^\dagger c_{k'\sigma'} c_{k+q\sigma},$$

in a plane-wave basis, and calculate the formal expression of its expectation value over the Fermi sea wavefunction (1.83). Then, through the Heisenberg equation of motion,

$$i\hbar \frac{\partial \rho(\mathbf{r})}{\partial t} = \left[\rho(\mathbf{r}), H\right],$$

and the continuity equation relating the density to the current density $\mathbf{J}(\mathbf{r})$, i.e.,

$$\frac{\partial \rho(\mathbf{r})}{\partial t} + \nabla \cdot \mathbf{J}(\mathbf{r}) = 0,$$

calculate the expression of the Fourier transform of the current $\mathbf{J}(\mathbf{q})$, and the following commutator

$$\left[\rho(\mathbf{q}, t), \mathbf{J}(-\mathbf{q}, t)\right], \tag{1.204}$$

which is commonly known as the f-sum rule.

1.3 A simple Hamiltonian to diagonalise—Consider the Hamiltonian

$$H = -\epsilon\left(c_1^\dagger c_1 - c_2^\dagger c_2\right) - t\left(c_1^\dagger c_2 + c_2^\dagger c_1\right) - \Delta\left(c_1^\dagger c_2^\dagger + c_2 c_1\right).$$

Find the Hermitian operator

$$\varphi = \mathbf{c}^\dagger \hat{\phi} \mathbf{c} + \left(\mathbf{c}^T \hat{\psi} \mathbf{c} + H.c.\right),$$

with $c^\dagger = \left(c_1^\dagger, c_2^\dagger\right)$, such that, if $U = e^{i\varphi}$, then

$$U^\dagger H U = \sum_{i=1}^{2} \epsilon_i c_i^\dagger c_i \,,$$

and calculate the eigenvalues ϵ_i. Determine under which conditions, in the case of bosons, $\epsilon_i \geq 0$, $i = 1, 2$.

1.4 Heisenberg antiferromagnet in a magnetic field—Consider the antiferromagnetic Heisenberg model

$$H = J \sum_{<RR'>} S_R \cdot S_{R'} \,, \tag{1.205}$$

where $J > 0$, and the sum runs over nearest neighbour bonds on a hypercubic lattice in $d \geq 3$ dimensions. Calculate the spin-wave spectrum and the spin-wave contribution to the specific heat.

Add to the Hamiltonian (1.205) with symmetry breaking along the z-direction a magnetic field along the x-direction with Fourier components B_q, i.e., a term

$$\delta H = - \sum_q B_{-q} S_q^x \simeq -\sqrt{SN} \sum_q B_{-q} x_q.$$

Diagonalise the Hamiltonian in the presence of this term and calculate the ground state energy.

1.5 Heisenberg ferromagnet—Calculate the spin-wave spectrum of the Heisenberg ferromagnetic Hamiltonian

$$H = -J \sum_{<RR'>} S_R \cdot S_{R'},$$

with $J > 0$, and on a hypercubic lattice in d dimensions.

References

1. P.W. Anderson, Phys. Rev. **124**, 41 (1961). https://doi.org/10.1103/PhysRev.124.41
2. J. Hubbard, Proc. R. Soc. Lond. A **276**, 238–257 (1963)
3. N.F. Mott, Proc. Phys. Soc. Sect. A **62**(7), 416 (1949). https://doi.org/10.1088/0370-1298/62/7/303
4. J. Villain, R. Bidaux, J.P. Carton, R. Conte, J. Phys. France **41**(11), 1263 (1980). https://doi.org/10.1051/jphys:0198000410110126300
5. P. Chandra, P. Coleman, A.I. Larkin, Phys. Rev. Lett. **64**, 88 (1990). https://doi.org/10.1103/PhysRevLett.64.88

Linear Response Theory

In order to access the physical properties of a system, one has to act on it with some external probe. This amounts to add to the unperturbed Hamiltonian $\hat{H}_0$ a time-dependent perturbation of the general form

$$\hat{V}(t) = \int d\mathbf{r} \, \hat{A}(\mathbf{r}) \, v(\mathbf{r}, t) \,, \tag{2.1}$$

where $v(\mathbf{r}, t)$ represents the external probe that couples to the Hermitian operator $\hat{A}(\mathbf{r})$. Our scope is to study the effects of $\hat{V}(t)$ on some measurable quantity described, e.g., by an operator $\hat{B}(\mathbf{r})$, namely to calculate

$$B(\mathbf{r}, t) \equiv \text{Tr}\left(\hat{\rho}(t) \, \hat{B}(\mathbf{r})\right),$$

being $\hat{\rho}(t)$ the time-dependent density matrix in presence of the perturbation.

2.1 Linear Response Functions

We assume that the perturbation is switched on at time $t \to -\infty$. Initially, the system is in thermal equilibrium, so that the density matrix

$$\lim_{t \to -\infty} \hat{\rho}(t) = \hat{\rho}_0 \,, \tag{2.2}$$

with

$$\hat{\rho}_0 = \frac{1}{Z_0} e^{-\beta \hat{H}_0} = \frac{1}{Z_0} \sum_n e^{-\beta E_n} \, |\phi_n\rangle \langle \phi_n| \,. \tag{2.3}$$

© The Author(s), under exclusive license to Springer Nature Switzerland AG 2022
M. Fabrizio, *A Course in Quantum Many-Body Theory*, Graduate Texts in Physics,
https://doi.org/10.1007/978-3-031-16305-0_2

Here $\hat{H}_0 |\phi_n\rangle = E_n |\phi_n\rangle$ and $Z_0 = \sum_n \exp(-\beta E_n)$. Therefore, the time evolution of the density matrix in the presence of $\hat{V}(t)$ is given by

$$\hat{\rho}(t) = \frac{1}{Z_0} \sum_n e^{-\beta E_n} |\phi_n(t)\rangle \langle \phi_n(t)|, \tag{2.4}$$

where

$$i\hbar \frac{\partial}{\partial t} |\phi_n(t)\rangle = \left[\hat{H}_0 + \hat{V}(t)\right] |\phi_n(t)\rangle \tag{2.5}$$

is the Shrœdinger equation which determines the evolution of the eigenstates of the unperturbed Hamiltonian in presence of the perturbation. The meaning of (2.4) is that initially the system is described by a statistical ensemble of sub-systems, each in a given eigenstate of $\hat{H}_0$ and weighed by its Boltzmann factor. After we switch on the perturbation, $|\phi_n\rangle$ ceases to be an eigenstate of the perturbed Hamiltonian, so it acquires a non-trivial time evolution.

Through (2.5) one readily finds the equation of motion for the density matrix

$$i\hbar \frac{\partial}{\partial t} \hat{\rho}(t) = \left[\hat{H}_0 + \hat{V}(t), \hat{\rho}(t)\right]. \tag{2.6}$$

We introduce the Dirac, also called interaction, representation of the density matrix as

$$\hat{\rho}_D(t) = e^{i\hat{H}_0 t/\hbar} \, \hat{\rho}(t) \, e^{-i\hat{H}_0 t/\hbar},$$

which satisfies

$$\begin{aligned} i\hbar \frac{\partial}{\partial t} \hat{\rho}_D(t) &= -\left[\hat{H}_0, \hat{\rho}_D(t)\right] + e^{i\hat{H}_0 t/\hbar} \left[\hat{H}_0 + \hat{V}(t), \hat{\rho}(t)\right] e^{-i\hat{H}_0 t/\hbar} \\ &= \left[\hat{V}_D(t), \hat{\rho}_D(t)\right], \end{aligned} \tag{2.7}$$

where

$$\hat{V}_D(t) = \int d\mathbf{r} \, e^{i\hat{H}_0 t/\hbar} \, \hat{A}(\mathbf{r}) \, e^{-i\hat{H}_0 t/\hbar} \, v(\mathbf{r}, t) = \int d\mathbf{r} \, \hat{A}(\mathbf{r}, t) \, v(\mathbf{r}, t),$$

being $\hat{A}(\mathbf{r}, t)$ the Heisenberg evolution of $\hat{A}(\mathbf{r})$ with the unperturbed Hamiltonian. We solve (2.7) perturbatively in $v(\mathbf{x}, t)$, i.e., $\rho_D(t) = \rho_D^{(0)}(t) + \rho_D^{(1)}(t) + \cdots$, where $\rho_D^{(n)}(t)$ contains n-powers of the perturbation. Obviously,

$$\lim_{t \to -\infty} \hat{\rho}_D(t) = \hat{\rho}_0 = \hat{\rho}_D^{(0)} \quad \Rightarrow \quad \lim_{t \to -\infty} \hat{\rho}_D^{(n)}(t) = \delta_{n0} \, \hat{\rho}_0. \tag{2.8}$$

We will limit our analysis to the linear response; hence, we just need the first-order term that satisfies

$$i\hbar \frac{\partial}{\partial t} \hat{\rho}_D^{(1)}(t) = \left[\hat{V}_D(t), \hat{\rho}_D^{(0)}(t)\right],$$

with solution

$$\hat{\rho}_D^{(1)}(t) = -\frac{i}{\hbar} \int_{-\infty}^{t} dt' \left[\hat{V}_D(t'), \hat{\rho}_0 \right]. \tag{2.9}$$

Therefore, at linear order,

$$
\begin{aligned}
B(\mathbf{r}, t) &= \text{Tr} \left[\hat{\rho}(t) \, \hat{B}(\mathbf{r}) \right] = \text{Tr} \left[\hat{\rho}_D(t) \, e^{i\hat{H}_0 t/\hbar} \, \hat{B}(\mathbf{r}) \, e^{i\hat{H}_0 t/\hbar} \right] \\
&= \text{Tr} \left[\hat{\rho}_D(t) \, \hat{B}(\mathbf{r}, t) \right] = \text{Tr} \left[\hat{\rho}_0 \, \hat{B}(\mathbf{r}, t) \right] + \text{Tr} \left[\hat{\rho}_D^{(1)}(t) \, \hat{B}(\mathbf{r}, t) \right] \\
&= B_0(\mathbf{r}) + \text{Tr} \left[\hat{\rho}_D^{(1)}(t) \, \hat{B}(\mathbf{r}, t) \right],
\end{aligned}
$$

where we used the fact that

$$\text{Tr} \left[\hat{\rho}_0 \, \hat{B}(\mathbf{r}, t) \right] = \text{Tr} \left[\hat{\rho}_0 \, \hat{B}(\mathbf{r}) \right] = B_0(\mathbf{r})$$

is the unperturbed expectation value. We thus find that the variation of the latter is simply given by

$$
\begin{aligned}
B(\mathbf{r}, t) - B_0(\mathbf{r}) &= -\frac{i}{\hbar} \int_{-\infty}^{t} dt' \, \text{Tr} \left(\left[\hat{V}_D(t'), \hat{\rho}_0 \right] \hat{B}(\mathbf{r}, t) \right) \\
&= -\frac{i}{\hbar} \int_{-\infty}^{t} dt' \int d\mathbf{r}' \, \text{Tr} \left(\hat{\rho}_0 \left[\hat{B}(\mathbf{r}, t), \hat{A}(\mathbf{r}', t') \right] \right) v(\mathbf{r}', t') \\
&= -\frac{i}{\hbar} \int_{-\infty}^{\infty} dt' \int d\mathbf{r}' \, \theta(t - t') \, \text{Tr} \left(\hat{\rho}_0 \left[\hat{B}(\mathbf{r}, t), \hat{A}(\mathbf{r}', t') \right] \right) v(\mathbf{r}', t') \\
&\equiv \int_{-\infty}^{\infty} dt' \int d\mathbf{r}' \, \chi_{BA}(\mathbf{r}, \mathbf{r}', t - t') \, v(\mathbf{r}', t'),
\end{aligned} \tag{2.10}
$$

with the linear response function defined through

$$\boxed{\chi_{BA}(\mathbf{r}, \mathbf{r}', t - t') = -\frac{i}{\hbar} \theta(t - t') \left\langle \left[\hat{B}(\mathbf{r}, t), \hat{A}(\mathbf{r}', t') \right] \right\rangle}, \tag{2.11}$$

where $\langle \ldots \rangle$ means thermal average with the unperturbed density matrix of operators evolved in time with $\hat{H}_0$, and $\chi_{BA}(\mathbf{r}, \mathbf{r}', t - t')$ only depends on the time difference since the Schrœdinger equation is time-translationally invariant. Indeed,

$$
\begin{aligned}
\langle \hat{B}(t) \, \hat{A}(t') \rangle &= \frac{1}{Z} \, \text{Tr} \left(e^{-\beta H} \, e^{i H t/\hbar} \, \hat{B} \, e^{-i H t/\hbar} \, e^{i H t'/\hbar} \, \hat{A} \, e^{-i H t'/\hbar} \right) \\
&= \frac{1}{Z} \, \text{Tr} \left(e^{-\beta H} \, e^{i H (t-t')/\hbar} \, \hat{B} \, e^{-i H (t-t')/\hbar} \, \hat{A} \right) = \langle \hat{B}(t - t') \, \hat{A}(0) \rangle \\
&= \frac{1}{Z} \, \text{Tr} \left(e^{-\beta H} \, \hat{B} \, e^{i H (t'-t)/\hbar} \, \hat{A} \, e^{-i H (t'-t)/\hbar} \right) = \langle \hat{B}(0) \, \hat{A}(t' - t) \rangle.
\end{aligned}
$$

Equation (2.10) shows that, at linear order, the variation of any measurable quantity is obtained through the linear response function (2.11) which is only related to expectation values on the unperturbed system.

2.2   Kramers-Kronig Relations

Let us now study the analytical properties of the response function in the frequency domain. In the following, we drop the space-coordinate dependence of the response function, which is not relevant for what we are going to demonstrate. The response of an operator $\hat{A}$ in the presence of an external probe that couples to $\hat{B}$ is therefore

$$\chi_{AB}(t) = -\frac{i}{\hbar}\,\theta(t)\,\langle[\hat{A}(t),\hat{B}]\rangle\,, \tag{2.12}$$

where

$$\hat{A}(t) = e^{i\frac{\hat{H}_0 t}{\hbar}}\,\hat{A}\,e^{-i\frac{\hat{H}_0 t}{\hbar}}\,.$$

The response function (2.12) vanishes for $t < 0$, which is a consequence of *causality*. We introduce the Fourier transform through

$$\chi_{AB}(t) = \int_{-\infty}^{\infty} \frac{d\omega}{2\pi}\,e^{-i\omega t}\,\chi_{AB}(\omega)\,, \tag{2.13}$$

as well as its analytical continuation in the complex frequency plane $\chi_{AB}(z)$. If we assume, as it is always the case, that $\chi_{AB}(z)$ does not diverge exponentially for $|z| \to \infty$, we can regard (2.13) as the result of a contour integral

$$\chi_{AB}(t) = \oint \frac{dz}{2\pi}\,e^{-izt}\,\chi_{AB}(z)\,,$$

where the contour is in the upper half-plane for $t < 0$ and in the lower one for $t > 0$. The integral thus catches all poles and branch cuts lying inside the contour. Since $\chi_{AB}(t) = 0$ for $t < 0$, it follows that

> **Analytic property of the response function**

As a consequence of causality $\chi_{AB}(z)$ is analytic in the upper half-plane.

Let us now consider the contour drawn in Fig. 2.1. Since there are no poles enclosed by the contour, the integral

$$\oint_C dz\,\frac{\chi_{AB}(z)}{\omega - z} = 0\,. \tag{2.14}$$

On the other hand, the above integral is also equal to the line integral along the lower edge, hence

$$0 = \left\{ \int_{-\infty}^{\omega-\epsilon} + \int_{\omega+\epsilon}^{\infty} \right\} d\omega'\,\frac{\chi_{AB}(\omega')}{\omega-\omega'} + \oint_{z=\omega+\epsilon\,\exp(i\theta):\,\theta\in[\pi,0]} dz\,\frac{\chi_{AB}(z)}{\omega-z}$$

$$= \fint d\omega'\,\frac{\chi_{AB}(\omega')}{\omega-\omega'} - i\int_{\pi}^{0} d\theta\,\chi_{AB}\left(\omega + \epsilon\,e^{i\theta}\right)\,,$$

Fig. 2.1 Integration contour
in (2.14)

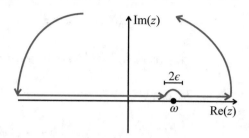

the symbol $\fint \ldots$ denoting the Cauchy principal value of the integral. In the limit $\epsilon \to 0$, the above expression simplifies into

$$\fint d\omega' \, \frac{\chi_{AB}(\omega')}{\omega - \omega'} + i\pi \, \chi_{AB}(\omega) = 0, \qquad (2.15)$$

which implies that

$$\mathrm{Im}\,\chi_{AB}(\omega) = \fint \frac{d\omega'}{\pi} \, \frac{\mathrm{Re}\,\chi_{AB}(\omega')}{\omega - \omega'}, \qquad (2.16)$$

$$\mathrm{Re}\,\chi_{AB}(\omega) = -\fint \frac{d\omega'}{\pi} \, \frac{\mathrm{Im}\,\chi_{AB}(\omega')}{\omega - \omega'}, \qquad (2.17)$$

known as the Kramers-Kronig relations. Therefore, because of causality, the real
and imaginary parts of the response function are not independent of each other. It is
possible to rewrite both expressions as

$$\chi_{AB}(\omega) = \int_{-\infty}^{\infty} \frac{d\omega'}{\pi} \, \frac{\mathrm{Im}\,\chi_{AB}(\omega')}{\omega' - \omega - i\eta}, \qquad (2.18)$$

with η an infinitesimal positive number.

2.2.1   Symmetries

Let us introduce back the space dependence, so that

$$\chi_{AB}(\mathbf{r}, \mathbf{r}', t - t') = -\frac{i}{\hbar} \, \theta(t - t') \, \langle [\hat{A}(\mathbf{r}, t), \hat{B}(\mathbf{r}', t')] \rangle. \qquad (2.19)$$

Since both operators are Hermitian, it follows that

$$\chi_{AB}(\mathbf{r}, \mathbf{r}', t - t')^* = \frac{i}{\hbar} \, \theta(t - t') \, \langle [\hat{B}(\mathbf{r}', t'), \hat{A}(\mathbf{r}, t)] \rangle = \chi_{AB}(\mathbf{r}, \mathbf{r}', t - t'). \qquad (2.20)$$

By definition

$$\chi_{AB}(\mathbf{r}, \mathbf{r}', \omega) = \int dt \, e^{i\omega t} \, \chi_{AB}(\mathbf{r}, \mathbf{r}', t),$$

therefore, through (2.20), we find that

$$\chi_{AB}(\mathbf{r}, \mathbf{r}', \omega)^* = \int dt\, e^{-i\omega t}\, \chi_{AB}(\mathbf{r}, \mathbf{r}', t)^* = \chi_{AB}(\mathbf{r}, \mathbf{r}', -\omega)\,.$$

This implies that

$$\mathrm{Re}\,\chi_{AB}(\mathbf{r}, \mathbf{r}', \omega) = \frac{1}{2}\left[\chi_{AB}(\mathbf{r}, \mathbf{r}', \omega) + \chi_{AB}(\mathbf{r}, \mathbf{r}', -\omega)\right]$$

is even in frequency, while

$$\mathrm{Im}\,\chi_{AB}(\mathbf{r}, \mathbf{r}', \omega) = \frac{1}{2i}\left[\chi_{AB}(\mathbf{r}, \mathbf{r}', \omega) - \chi_{AB}(\mathbf{r}, \mathbf{r}', -\omega)\right]$$

is odd.

2.3   Fluctuation-Dissipation Theorem

Let us introduce other types of functions. The first is the so-called structure factor, defined through

$$S_{AB}(\mathbf{r}, \mathbf{r}', t) = \frac{1}{\hbar}\langle \hat{A}(\mathbf{r}, t)\, \hat{B}(\mathbf{r}')\rangle\,. \tag{2.21}$$

In addition, we introduce the dissipation function

$$\chi''_{AB}(\mathbf{r}, \mathbf{r}', t) = \frac{1}{2\hbar}\langle\left[\hat{A}(\mathbf{r}, t), \hat{B}(\mathbf{r}')\right]\rangle = \frac{1}{2}\left[S_{AB}(\mathbf{r}, \mathbf{r}', t) - S_{BA}(\mathbf{r}', \mathbf{r}, -t)\right], \tag{2.22}$$

whose meaning will be explained in the following section, as well as the fluctuation one

$$F_{AB}(\mathbf{r}, \mathbf{r}', t) = \frac{1}{2}\langle\left\{\hat{A}(\mathbf{r}, t), \hat{B}(\mathbf{r}')\right\}\rangle = \frac{\hbar}{2}\left[S_{AB}(\mathbf{r}, \mathbf{r}', t) + S_{BA}(\mathbf{r}', \mathbf{r}, -t)\right]. \tag{2.23}$$

One readily verifies that the former is related to the response function through

$$\chi''_{AB}(\mathbf{r}, \mathbf{r}', t) = \frac{i}{2}\left[\chi_{AB}(\mathbf{r}, \mathbf{r}', t) - \chi_{BA}(\mathbf{r}', \mathbf{r}, -t)\right],$$

which in the frequency domain reads

$$\chi''_{AB}(\mathbf{r}, \mathbf{r}', \omega) = \frac{i}{2}\left[\chi_{AB}(\mathbf{r}, \mathbf{r}', \omega) - \chi_{BA}(\mathbf{r}', \mathbf{r}, -\omega)\right]. \tag{2.24}$$

In particular,

$$\chi''_{AA}(\mathbf{r}, \mathbf{r}, \omega) = -\mathrm{Im}\,\chi_{AA}(\mathbf{r}, \mathbf{r}, \omega)\,. \tag{2.25}$$

Through the definition (2.21), we find that

$$
S_{BA}(\mathbf{r}', \mathbf{r}, -t) = \frac{1}{Z_0} \mathrm{Tr}\left(e^{-\beta \hat{H}_0}\, e^{-i\hat{H}_0 \frac{t}{\hbar}}\, \hat{B}(\mathbf{r}')\, e^{i\hat{H}_0 \frac{t}{\hbar}}\, \hat{A}(\mathbf{r})\right)
$$

$$
= \frac{1}{Z_0} \mathrm{Tr}\left(e^{-\beta \hat{H}_0}\, e^{i\hat{H}_0 \frac{(t-i\beta\hbar)}{\hbar}}\, \hat{A}(\mathbf{r})\, e^{-i\hat{H}_0 \frac{(t-i\beta\hbar)}{\hbar}}\, \hat{B}(\mathbf{r}')\right)
$$

$$
= S_{AB}(\mathbf{r}, \mathbf{r}', t - i\hbar\beta)\,.
$$

Therefore,

$$
S_{BA}(\mathbf{r}', \mathbf{r}, -\omega) = \int dt\, e^{-i\omega t}\, S_{BA}(\mathbf{r}', \mathbf{r}, t) = \int dt\, e^{i\omega t}\, S_{BA}(\mathbf{r}', \mathbf{r}, -t)
$$

$$
= \int dt\, e^{i\omega t}\, S_{AB}(\mathbf{r}, \mathbf{r}', t - i\hbar\beta) = e^{-\beta\hbar\omega}\, S_{AB}(\mathbf{r}, \mathbf{r}', \omega)\,,
$$

namely,

$$
\chi''_{AB}(\mathbf{r}, \mathbf{r}', \omega) = \frac{1}{2} S_{AB}(\mathbf{r}, \mathbf{r}', \omega)\left(1 - e^{-\beta\hbar\omega}\right)\,,
$$

$$
F_{AB}(\mathbf{r}, \mathbf{r}', \omega) = \frac{\hbar}{2} S_{AB}(\mathbf{r}, \mathbf{r}', \omega)\left(1 + e^{-\beta\hbar\omega}\right)\,.
$$

In other words, the following relation holds:

$$
F_{AB}(\mathbf{r}, \mathbf{r}', \omega) = \hbar \coth\left(\frac{\beta\hbar\omega}{2}\right)\, \chi''_{AB}(\mathbf{r}, \mathbf{r}', \omega)\,, \qquad (2.26)
$$

known as 'fluctuation-dissipation theorem'. Indeed, if $\hat{A} = \hat{B}$ and $\mathbf{r} = \mathbf{r}'$, $F_{AA}(\mathbf{r}, \mathbf{r}, t = 0)$ is an estimate of the fluctuations of $\hat{A}$. On the other hand,

$$
F_{AA}(\mathbf{r}, \mathbf{r}, t = 0) = \int \frac{d\omega}{2\pi}\, F_{AA}(\mathbf{r}, \mathbf{r}, \omega) = \hbar \int \frac{d\omega}{2\pi}\, \coth\left(\frac{\beta\hbar\omega}{2}\right)\, \chi''_{AA}(\mathbf{r}, \mathbf{r}, \omega)\,,
$$
$$
(2.27)
$$

which relates fluctuations to dissipation.

2.4   Spectral Representation

The spectral representation of the response functions gives instructive information about their physical meaning. Let us start from the structure factor (2.21), which can be written as (in the following we do not explicitly indicate the space dependence)

$$
S_{AB}(t) = \frac{1}{\hbar Z} \sum_n e^{-\beta E_n}\, \langle n|e^{i\hat{H}t/\hbar}\, \hat{A}\, e^{-i\hat{H}t/\hbar}\, \hat{B}|n\rangle
$$

$$
= \frac{1}{\hbar Z} \sum_{nm} e^{-\beta E_n}\, e^{i(E_n - E_m)t/\hbar}\, \langle n|\hat{A}|m\rangle \langle m|\hat{B}|n\rangle\,.
$$

The Fourier transform in frequency thus reads

$$S_{AB}(\omega) = \frac{2\pi}{Z} \sum_{if} e^{-\beta E_i} \langle i \mid \hat{A} \mid f \rangle \langle f \mid \hat{B} \mid i \rangle \, \delta(\hbar\omega + E_i - E_f). \qquad (2.28)$$

The meaning is now self-evident. The matrix element $\langle f \mid \hat{B} \mid i \rangle$ is the transition amplitude of the excitation from the initial state $\mid i \rangle$ into the final one $\mid f \rangle$ induced by the operator $\hat{B}$, while $\langle i \mid \hat{A} \mid f \rangle$ describes the reverse process but now induced by $\hat{A}$. The excitation followed by the relaxation process is weighed by the Boltzmann factor of the initial state and contributes to $S_{AB}(\omega)$ only if the energy difference $E_f - E_i$ is equal to $\hbar\omega$. Thus $S_{AB}(\omega)$ is a spectral function that measures the transition amplitude for excitations induced by $\hat{B}$ and de-excitation induced by $\hat{A}$ with a given energy $\hbar\omega$.

Through (2.26) we also find that

$$
\begin{aligned}
\chi''_{AB}(\omega) &= \frac{\pi}{Z} \sum_{if} e^{-\beta E_i} \left(1 - e^{-\beta\hbar\omega}\right) \langle i \mid \hat{A} \mid f \rangle \langle f \mid \hat{B} \mid i \rangle \, \delta\left(\hbar\omega + E_i - E_f\right) \\
&= \frac{\pi}{Z} \sum_{if} \left(e^{-\beta E_i} - e^{-\beta E_f}\right) \langle i \mid \hat{A} \mid f \rangle \langle f \mid \hat{B} \mid i \rangle \, \delta\left(\hbar\omega + E_i - E_f\right),
\end{aligned}
$$

$$(2.29)$$

which means that $\chi''_{AB}(\omega)$ is the transition amplitude for $\mid i \rangle \to \mid f \rangle$ induced by $\hat{B}$ and $\mid f \rangle \to \mid i \rangle$ induced by $\hat{A}$ weighted by the occupation probability

$$p_i = \frac{e^{-\beta E_i}}{Z}$$

of the initial state $\mid i \rangle$ minus the one

$$p_f = \frac{e^{-\beta E_f}}{Z}$$

of the final state $\mid f \rangle$. We note that

$$p_i - p_f = p_i \left(1 - p_f\right) - p_f \left(1 - p_i\right),$$

namely it is the probability of i being occupied and f empty, minus the opposite. In other words, $\chi''_{AB}(\omega)$ measures the absorption minus the emission probability of energy $\hbar\omega$, namely the total absorption probability.

Finally, one can analogously derive the spectral representation of the response function χ_{AB},

$$\chi_{AB}(t) = -\frac{i}{\hbar}\theta(t) \frac{1}{Z} \sum_{nm} \langle n \mid \hat{A} \mid m \rangle \langle m \mid \hat{B} \mid n \rangle \, e^{i(E_n - E_m)t/\hbar} \left(e^{-\beta E_n} - e^{-\beta E_m}\right).$$

To obtain the Fourier transform, one has to evaluate the integral

$$-i \int_0^\infty dt \, e^{i\omega t} \, e^{i(E_n - E_m)t/\hbar} \, .$$

Since the perturbation has been assumed to be switched on at very early times, namely at time difference $t \to \infty$, a meaningful regularisation of the above integral is

$$-i \int_0^\infty dt\, e^{i\omega t}\, e^{i(E_n - E_m)t/\hbar}\, e^{-\eta t/\hbar} = \frac{\hbar}{\hbar\omega - (E_m - E_n) + i\eta}\,,$$

where $\eta/\hbar > 0$ is the switching rate of the perturbation, and is taken to be an infinitesimal positive number. As a result, we find that

$$\chi_{AB}(\omega) = \frac{1}{Z} \sum_{nm} \langle n \mid \hat{A} \mid m \rangle \langle m \mid \hat{B} \mid n \rangle\, \frac{e^{-\beta E_n} - e^{-\beta E_m}}{\hbar\omega - (E_m - E_n) + i\eta}\,. \qquad (2.30)$$

We note that η correctly makes the function analytic in the upper half-plane.

2.5 Power Dissipation

Till now we have formally introduced several response functions. In this section and in the following ones, we are going to show how those functions emerge in real experiments.

Let us first analyse the power dissipated in presence of the perturbation. Given our starting assumption about the time evolution of the density matrix (2.4), it is clear that the entropy defined through the phase space occupied by the statistical ensemble remains constant and equal to the thermal equilibrium one, S_0. Therefore, the system free energy is

$$F(t) = U(t) - T S_0 = (U(t) - U_0) + F_0\,,$$

so that

$$\frac{\partial F(t)}{\partial t} = \frac{\partial U(t)}{\partial t}\,.$$

On the other hand, since

$$U(t) = \mathrm{Tr}\left(\hat{\rho}(t)\, \hat{H}_0\right),$$

it follows that

$$\frac{\partial U(t)}{\partial t} = -\frac{i}{\hbar} \mathrm{Tr}\left(\left[\hat{H}_0 + \hat{V}(t), \hat{\rho}(t)\right] \hat{H}_0\right) = -\frac{i}{\hbar} \mathrm{Tr}\left(\hat{\rho}(t)\left[\hat{H}_0, \hat{H}_0 + \hat{V}(t)\right]\right)$$

$$= -\frac{i}{\hbar} \mathrm{Tr}\left(\hat{\rho}(t)\left[\hat{H}_0, \hat{V}(t)\right]\right).$$

$$(2.31)$$

We assume that the perturbation has the general form

$$\hat{V}(t) = \sum_J \hat{A}_J\, v_J(t)\,,$$

so that

$$\frac{\partial U(t)}{\partial t} = -\frac{i}{\hbar} \sum_J v_J(t) \, \mathrm{Tr}\left(\hat{\rho}(t)\left[\hat{H}_0, \hat{A}_J\right]\right). \tag{2.32}$$

We further note that

$$A_J(t) = \mathrm{Tr}\left(\hat{\rho}(t)\,\hat{A}_J\right) = \sum_{J'} \int dt' \, \chi_{JJ'}(t-t')\,v_{J'}(t'),$$

where

$$\chi_{JJ'}(t-t') = -\frac{i}{\hbar}\,\theta(t-t')\left\langle\left[\hat{A}_J(t), \hat{A}_{J'}(t')\right]\right\rangle.$$

Therefore, by recalling that the operators evolve with the unperturbed Hamiltonian, we find that

$$i\hbar \sum_{J'} \int dt' \, \frac{\partial}{\partial t}\chi_{JJ'}(t-t')\,v_{J'}(t') = \sum_{J'} \int dt' \, \delta(t-t')\left\langle\left[\hat{A}_J(t), \hat{A}_{J'}(t')\right]\right\rangle v_{J'}(t')$$

$$-\sum_{J'} \int dt' \, \frac{i}{\hbar}\theta(t-t')\left\langle\left[\left[\hat{A}_J(t), \hat{H}_0\right], \hat{A}_{J'}(t')\right]\right\rangle v_{J'}(t')$$

$$= \sum_{J'} \left\langle\left[\hat{A}_J(t), \hat{A}_{J'}(t)\right]\right\rangle v_{J'}(t) + \left\langle\left[\hat{A}_J(t), \hat{H}_0\right]\right\rangle.$$

After inserting into (2.32), we finally get

$$\frac{\partial U(t)}{\partial t} = -\sum_{JJ'} \int dt' \, \frac{\partial}{\partial t}\chi_{JJ'}(t-t')\,v_J(t)\,v_{J'}(t')$$

$$-\frac{i}{\hbar}\sum_{JJ'}\left\langle\left[\hat{A}_J(t), \hat{A}_{J'}(t)\right]\right\rangle v_J(t)\,v_{J'}(t).$$

The last term vanishes since the commutator is odd by interchanging J with J', while $v_J(t)\,v_{J'}(t)$ is even. Hence,

$$\frac{\partial U(t)}{\partial t} = -\sum_{JJ'} \int dt' \, \frac{\partial}{\partial t}\chi_{JJ'}(t-t')\ v_J(t)\,v_{J'}(t'). \tag{2.33}$$

Let us write

$$v_J(t) = \frac{1}{2}\left(v_J\,e^{-i\omega t} + v_J^*\,e^{i\omega t}\right), \tag{2.34}$$

and define the power dissipated within a cycle, W, through

$$W = \frac{\omega}{2\pi} \int_0^{2\pi/\omega} dt \, \frac{\partial U(t)}{\partial t}. \tag{2.35}$$

By performing the integral and by means of (2.33) and (2.34), we obtain for W the expression

$$W = i \frac{\omega}{4} \sum_{JJ'} v_J^* v_{J'} \left[\chi_{JJ'}(\omega) - \chi_{J'J}(-\omega) \right] = \frac{\omega}{2} \sum_{JJ'} v_J^* \chi_{JJ'}''(\omega) v_{J'} , \quad (2.36)$$

where we use (2.24). The power dissipated during a cycle is proportional to what we denoted as the dissipation response function, thus explaining its name. Indeed, as we showed in the previous section, $\chi_{JJ'}''(\omega)$ measures the probability of energy absorption during the process, hence its appearance in (2.36) it is not unexpected. We further note that, since $W > 0$, it follows that $\omega \chi_{JJ'}''(\omega)$ is a positive-definite quadratic form. In particular,

$$\omega \chi_{JJ}''(\omega) = -\omega \operatorname{Im} \chi_{JJ}(\omega) > 0 ,$$

namely the imaginary part of $\chi_{JJ}(\omega)$ is positive for $\omega < 0$ and negative otherwise.

2.5.1 Absorption/Emission Processes

The power dissipation is related to the absorption minus emission probability. However, there are other measurements where only absorption or emission is revealed. For instance, one can shot on a sample with a beam of particles, either photons, neutrons, electrons, etc..., and measure the absorption probability of an energy $\hbar\omega$. If the coupling between the beam and the sample is represented by an operator $\hat{A}$, the Fermi golden rule tells us that the absorption rate per unit time of an ensemble at thermal equilibrium is

$$P_A(\omega) = \frac{2\pi}{Z} \sum_{if} e^{-\beta E_i} \left| \langle f | \hat{A} | i \rangle \right|^2 \delta \left(\hbar\omega + E_i - E_f \right) = S_{AA}(\omega) , \quad (2.37)$$

which enlightens the meaning of the structure factors.

2.5.2 Thermodynamic Susceptibilities

Let us consider again a perturbation of the form

$$\hat{V}(t) = \sum_J \int d\mathbf{r} \, \hat{A}_J(\mathbf{r}) \, v_J(\mathbf{r}, t) .$$

Let us further assume that the only time-dependence of the external probes $v_J(\mathbf{r}, t)$ comes from a very slow switching rate that just sets the proper regularisation of time-integrals as in (2.30). Therefore, for times far away from the time at which the perturbation is switched on, the external probes become constant in time, $v_J(\mathbf{r}, t) \to v_J(\mathbf{r})$, and the thermodynamic averages lose any time-dependence. In this limit,

$$
A_J(\mathbf{r}) - A_{0J}(\mathbf{r}) = \sum_{J'} \int_{-\infty}^{\infty} dt' \int d\mathbf{r}' \, \chi_{J\,J'}(\mathbf{r}, \mathbf{r}', t - t') \, v_{J'}(\mathbf{r}')
$$
$$
\to \sum_{J'} \int d\mathbf{r}' \, \chi_{J\,J'}(\mathbf{r}, \mathbf{r}', \omega = 0) \, v_{J'}(\mathbf{r}') \, .
$$

Let us now consider a generically perturbed Hamiltonian of the form

$$
\hat{H} = \hat{H}_0 + \sum_J \int d\mathbf{r}' \, \hat{A}_J(\mathbf{r}') \, v_J(\mathbf{r}') \, ,
$$

which therefore admits stationary eigenstates. The perturbed free energy turns out to be a functional of the static external fields $v_J(\mathbf{r}')$, i.e., $F = F\left[v_J(\mathbf{r}')\right]$. Standard thermodynamics tells us that

$$
\langle \hat{A}_J(\mathbf{r}) \rangle = \frac{\delta F}{\delta v_J(\mathbf{r})} \, ,
$$

which is in general different from its expectation value, $A_{0J}(\mathbf{r}) \equiv \langle \hat{A}_J(\mathbf{r}) \rangle_0$, in the absence of external fields. When the latter are very small, one finds at linear order that

$$
\langle \hat{A}_J(\mathbf{r}) \rangle - \langle \hat{A}_J(\mathbf{r}) \rangle_0 = \sum_{J'} \int d\mathbf{r}' \left(\frac{\delta^2 F}{\delta v_J(\mathbf{r}) \, \delta v_{J'}(\mathbf{r}')} \right)_{v=0} v_{J'}(\mathbf{r}')
$$
$$
\equiv -\sum_{J'} \int d\mathbf{r}' \, \kappa_{JJ'}(\mathbf{r}, \mathbf{r}') \, v_{J'}(\mathbf{r}') \, ,
$$
(2.38)

where $\kappa_{JJ'}(\mathbf{r}, \mathbf{r}')$ are the thermodynamic susceptibilities. Comparing (2.38) with (2.5.2) one obtains that

$$
\kappa_{JJ'}(\mathbf{r}, \mathbf{r}') = -\chi_{J\,J'}(\mathbf{r}, \mathbf{r}', \omega = 0) \, ,
$$
(2.39)

which relates thermodynamic susceptibilities to the response functions at zero frequency. Notice that thermodynamic stability implies that $\kappa_{JJ'}(\mathbf{r}, \mathbf{r}')$ is positive definite.

2.6 Application: Linear Response to an Electromagnetic Field

Let us consider a system made of interacting electrons and of immobile positive ions with charge density $e \, \rho_{ions}(\mathbf{r})$ such that

$$\int d\mathbf{r} \, \langle \rho(\mathbf{r}) \rangle = \sum_{\sigma} \int d\mathbf{r} \, \langle \Psi_{\sigma}^{\dagger}(\mathbf{r}) \, \Psi_{\sigma}(\mathbf{r}) \rangle = \int d\mathbf{r} \, \rho_{ions}(\mathbf{r}) \,,$$

where $\Psi_{\sigma}(\mathbf{r})$ is the Fermi field for spin-σ electrons, which guarantees overall charge neutrality. We assume that at equilibrium the electron density is homogeneous so that $\langle \rho(\mathbf{r}) \rangle_0 = n$, with n the average density. The system is coupled to the electromagnetic field (EMF), whose source is provided by the same system's charges as well as by an external time- and space-dependent source that we assume classical. Hereafter, we decompose any space-dependent vector $\mathbf{v}(\mathbf{r})$ as

$$\mathbf{v}(\mathbf{r}) = \mathbf{v}_{||}(\mathbf{r}) + \mathbf{v}_{\perp}(\mathbf{r}) \,,$$

where the longitudinal component $\mathbf{v}_{||}(\mathbf{r})$ is curl free, i.e., $\nabla \wedge \mathbf{v}_{||}(\mathbf{r}) = 0$, and the transverse one $\mathbf{v}_{\perp}(\mathbf{r})$ divergence-less, i.e., $\nabla \cdot \mathbf{v}_{\perp}(\mathbf{r}) = 0$. With this convention, the Hamiltonian that describes the system plus the EMF in the Coulomb gauge and in CGS units, taking $\hbar = 1$, reads

$$
\begin{aligned}
H = & \frac{1}{8\pi} \int d\mathbf{r} \left(\mathbf{E}_{\perp}(\mathbf{r}) \cdot \mathbf{E}_{\perp}(\mathbf{r}) + \mathbf{B}(\mathbf{r}) \cdot \mathbf{B}(\mathbf{r}) \right) - \frac{1}{c} \int d\mathbf{r} \, \mathbf{J}_{\perp ext}(t, \mathbf{r}) \cdot \mathbf{A}_{\perp}(\mathbf{r}) \\
& + \frac{1}{2m} \sum_{\sigma} \int d\mathbf{r} \, \Psi_{\sigma}^{\dagger}(\mathbf{r}) \left(-i \nabla + \frac{e}{c} \mathbf{A}_{\perp}(\mathbf{r}) + \frac{e}{c} \mathbf{A}_{|| ext}(t, \mathbf{r}) \right)^2 \Psi_{\sigma}(\mathbf{r}) \\
& + \frac{e^2}{2} \iint d\mathbf{r} \, d\mathbf{r}' \left(\rho(\mathbf{r}) - \rho_{ion}(\mathbf{r}) \right) \frac{1}{|\mathbf{r} - \mathbf{r}'|} \left(\rho(\mathbf{r}') - \rho_{ion}(\mathbf{r}') \right) \\
& - e \int d\mathbf{r} \, \phi_{ext}(t, \mathbf{r}) \left(\rho(\mathbf{r}) - \rho_{ion}(\mathbf{r}) \right) - \int d\mathbf{r} \, \mathbf{M}(\mathbf{r}) \cdot \mathbf{B}(\mathbf{r}) \,.
\end{aligned}
$$

$$(2.40)$$

In the Hamiltonian (2.40), $\mathbf{J}_{\perp ext}(t, \mathbf{r})$ is an external transverse current, source of an external transverse vector potential defined through the conventional wave-equation for the EMF, i.e.,

$$\left(\frac{\partial^2}{\partial t^2} - c^2 \nabla^2 \right) \mathbf{A}_{\perp ext}(t, \mathbf{r}) = 4\pi c \, \mathbf{J}_{\perp ext}(t, \mathbf{r}) \,, \qquad (2.41)$$

while $\mathbf{A}_{|| ext}(t, \mathbf{r})$ and $\phi_{ext}(t, \mathbf{r})$ are, respectively, the longitudinal component of the external vector potential and the external scalar potential. We recall that only their combination

$$\mathbf{E}_{|| ext}(t, \mathbf{r}) \equiv -\frac{1}{c} \frac{\partial \mathbf{A}_{|| ext}(t, \mathbf{r})}{\partial t} - \nabla \phi_{ext}(t, \mathbf{r}) \qquad (2.42)$$

has physical meaning, $E_{\|\,ext}(t, \mathbf{r})$ being just the longitudinal component of the external electric field.[1] The system operator $M(\mathbf{r})$ is instead the magnetisation density contributed by the electron spins, defined by

$$M(\mathbf{r}) = g_e\,\mu_B \sum_{\alpha\beta} \Psi_\alpha^\dagger(\mathbf{r})\,S_{\alpha\beta}\,\Psi_\beta(\mathbf{r})\,, \tag{2.43}$$

where $\mu_B = e/2mc$ and $g_e \simeq 2$ are the electron Bohr magneton and spin gyromagnetic ratio, and S is the spin vector matrix.

---

[1] That statement is just a consequence of gauge invariance. Specifically, let us consider the Schrœdinger equation

$$i\,|\,\dot\psi\rangle = H\,|\,\psi\rangle\,,$$

and apply on both sides a unitary transformation

$$U(t) \equiv \exp\left[-ie\int d\mathbf{r}\,\varphi(t, \mathbf{r})\left(\rho(\mathbf{r}) - \rho_{ion}(\mathbf{r})\right)\right],$$

so that

$$i\,U(t)\,|\,\dot\psi\rangle = i\,\frac{\partial}{\partial t}\Big(U(t)\,|\,\psi\rangle\Big) - i\,\dot U(t)\,U^\dagger(t)\Big(U(t)\,|\,\psi\rangle\Big) = U(t)\,H\,U^\dagger(t)\Big(U(t)\,|\,\psi\rangle\Big).$$

Therefore, if we define $|\,\Psi\rangle \equiv U(t)\,|\,\psi\rangle$, that wavefunction satisfies $i\,|\,\dot\Psi\rangle = H_*\,|\,\Psi\rangle$, where

$$H_* = U(t)\,H\,U^\dagger(t) + i\,\dot U(t)\,U^\dagger(t) = U(t)\,H\,U^\dagger(t) + e\int d\mathbf{r}\,\dot\varphi(t, \mathbf{r})\left(\rho(\mathbf{r}) - \rho_{ion}(\mathbf{r})\right).$$

Since

$$U(t)\,\Psi_\sigma(\mathbf{r})\,U^\dagger(t) = e^{ie\,\varphi(t,\mathbf{r})}\,\Psi_\sigma(\mathbf{r})\,,$$

as one can readily demonstrate, H_* has the same expression as H apart from

$$\phi_{ext}(t, \mathbf{r}) \to \phi_{ext}(t, \mathbf{r}) - \dot\varphi(t, \mathbf{r})\,, \qquad A_{\|ext}(t, \mathbf{r}) \to A_{\|ext}(t, \mathbf{r}) + c\,\nabla\varphi(t, \mathbf{r})\,,$$

which implies that the combination

$$E_{\|ext}(t, \mathbf{r}) = -\frac{1}{c}\,\dot A_{\|ext}(t, \mathbf{r}) - \nabla\phi_{ext}(t, \mathbf{r})$$

remains invariant, as well as any gauge invariant operator, like the density $\rho(\mathbf{r})$ or its corresponding current; see (2.53). In other words, given a gauge invariant operator A such that $U(t)\,A\,U^\dagger(t) = A$,

$$\langle\psi\,|\,A\,|\,\psi\rangle = \langle\Psi\,|\,U(t)\,A\,U^\dagger(t)\,|\,\Psi\rangle = \langle\Psi\,|\,A\,|\,\Psi\rangle\,,$$

so that we can work with either wavefunctions $|\,\psi\rangle$ or $|\,\Psi\rangle$, provided the transformation leaves $E_{\|ext}(t, \mathbf{r})$ invariant.

It is worth remarking that, since we have explicitly included the mutual Coulomb interaction among the system charges, the longitudinal component of the electric field that enters the Hamiltonian refers exclusively to the external source and we assume is a purely classical field. With that choice, the total longitudinal electric field, which we shall denote as 'internal' field, is not an independent dynamical quantity, but is defined through

$$E_{\parallel}(\mathbf{r}, t) = E_{\parallel ext}(\mathbf{r}, t) + E_{\parallel sys}(\mathbf{r}), \tag{2.44}$$

where $E_{\parallel sys}(\mathbf{r}, t)$ is the longitudinal field generated by the system charges through the Gauss law

$$\nabla \cdot E_{\parallel sys}(\mathbf{r}) = -4\pi \, e \left(\langle \rho(\mathbf{r}) \rangle - \rho_{ion}(\mathbf{r}) \right). \tag{2.45}$$

On the contrary, the transverse fields $A_{\perp}(\mathbf{r})$, $E_{\perp}(\mathbf{r})$ and $B(\mathbf{r}) \equiv \nabla \wedge A_{\perp}(\mathbf{r})$ are, so far, genuine quantum fields, not to be confused with the external ones, whose dynamics is defined through the commutation relations between $A_{\perp}(\mathbf{r})$ and $E_{\perp}(\mathbf{r})$, which we now introduce.

2.6.1 Quantisation of the Electromagnetic Field

The transverse $E_{\perp}(\mathbf{r})$ is actually conjugate to $A_{\perp}(\mathbf{r})$, specifically,

$$\left[A_{i\perp}(\mathbf{r}), \, E_{j\perp}(\mathbf{r}') \right] = -i \, 4\pi c \left(\delta_{ij} - \frac{\partial_i \, \partial_j}{\nabla^2} \right) \delta(\mathbf{r} - \mathbf{r}'), \tag{2.46}$$

or, in Fourier components,

$$\left[A_{i\perp}(\mathbf{q}), \, E_{j\perp}(-\mathbf{q}') \right] = -i \, 4\pi c \, \delta_{\mathbf{q},\mathbf{q}'} \left(\delta_{ij} - \frac{q_i \, q_j}{q^2} \right). \tag{2.47}$$

We introduce the polarisation vectors $\epsilon_s(\mathbf{q})$, $s = 1, 2$, which are orthogonal unit vectors satisfying

$$\epsilon_1(\mathbf{q}) \wedge \epsilon_2(\mathbf{q}) = \frac{\mathbf{q}}{q}, \quad \epsilon_s(-\mathbf{q}) = (-1)^s \, \epsilon_s(\mathbf{q}), \quad \sum_{s=1}^{2} \epsilon_{is}(\mathbf{q}) \, \epsilon_{js}(\mathbf{q}) = \delta_{ij} - \frac{q_i \, q_j}{q^2}.$$

The transverse vector potential and electric field can thus be written as

$$
A_\perp(\mathbf{r}) = \frac{1}{\sqrt{V}} \sum_{\mathbf{q}} e^{i\mathbf{q}\cdot\mathbf{r}} A_\perp(\mathbf{q}) = -\frac{1}{\sqrt{V}} \sum_{s=1}^{2} \sum_{\mathbf{q}} e^{i\mathbf{q}\cdot\mathbf{r}} (-i)^s \, \boldsymbol{\epsilon}_s(\mathbf{q}) \, x_s(\mathbf{q}),
$$

$$
E_\perp(\mathbf{r}) = \frac{1}{\sqrt{V}} \sum_{\mathbf{q}} e^{i\mathbf{q}\cdot\mathbf{r}} E_\perp(\mathbf{q}) = 4\pi c \frac{1}{\sqrt{V}} \sum_{s=1}^{2} \sum_{\mathbf{q}} e^{i\mathbf{q}\cdot\mathbf{r}} (-i)^s \, \boldsymbol{\epsilon}_s(\mathbf{q}) \, p_s(\mathbf{q}),
$$

$$(2.48)$$

where, since both $A_\perp(\mathbf{r})$ and $E_\perp(\mathbf{r})$ are Hermitian, $x_s(-\mathbf{q}) = x_s(\mathbf{q})^\dagger$ and $p_s(-\mathbf{q}) = p_s(\mathbf{q})^\dagger$, which, through (2.46), satisfy

$$
\left[x_s(\mathbf{q}), \, p_{s'}^\dagger(\mathbf{q}') \right] = i \, \delta_{ss'} \, \delta_{\mathbf{q},\mathbf{q}'},
$$

namely they are conventional conjugate variables. Using that representation, and defining, accordingly,

$$
J_{\perp ext}(\mathbf{r}) = -\frac{1}{\sqrt{V}} \sum_{s=1}^{2} \sum_{\mathbf{q}} e^{i\mathbf{q}\cdot\mathbf{r}} (-i)^s \, \boldsymbol{\epsilon}_s(\mathbf{q}) \, J_{s\,ext}(\mathbf{q}),
$$

we find that

$$
\begin{aligned}
H_{EMF} &\equiv \frac{1}{8\pi} \int d\mathbf{r} \left(E_\perp(\mathbf{r}) \cdot E_\perp(\mathbf{r}) + B(\mathbf{r}) \cdot B(\mathbf{r}) \right) \\
&\quad - \frac{1}{c} \int d\mathbf{r} \, J_{\perp ext}(t, \mathbf{r}) \cdot A_\perp(\mathbf{r}) \\
&= \frac{1}{2} \sum_{s=1}^{2} \sum_{\mathbf{q}} \left(4\pi c^2 \, p_s^\dagger(\mathbf{q}) \, p_s(\mathbf{q}) + \frac{q^2}{4\pi} \, x_s^\dagger(\mathbf{q}) \, x_s(\mathbf{q}) \right) \\
&\quad - \frac{1}{c} \sum_{s=1}^{2} \sum_{\mathbf{q}} J_{s\,ext}(t, \mathbf{q}) \, x_s^\dagger(\mathbf{q}) \\
&= \sum_{s=1}^{2} \sum_{\mathbf{q}} \omega_{\mathbf{q}} \left(b_{s\mathbf{q}}^\dagger b_{s\mathbf{q}} + \frac{1}{2} \right) - \frac{1}{c} \sum_{s=1}^{2} \sum_{\mathbf{q}} J_{s\,ext}(t, \mathbf{q}) \, x_s^\dagger(\mathbf{q}),
\end{aligned}
$$

$$(2.49)$$

where $\omega_{\mathbf{q}} = c\,q$, and we define

$$
x_s(\mathbf{q}) \equiv \sqrt{\frac{2\pi c}{q}} \left(b_{s\mathbf{q}} + b_{s-\mathbf{q}}^\dagger \right), \qquad p_s(\mathbf{q}) \equiv -i \sqrt{\frac{q}{8\pi c}} \left(b_{s\mathbf{q}} - b_{s-\mathbf{q}}^\dagger \right),
$$

in terms of the photon annihilation, $b_{s\mathbf{q}}$, and creation, $b_{s\mathbf{q}}^\dagger$, operators with polarisation s.

We observe that, through (2.49), the following Heisenberg equations of motion readily follow:

$$\dot{p}_s(\mathbf{q}) = -\frac{q^2}{4\pi} x_s(\mathbf{q}) + \frac{1}{c} J_{s\,ext}(t, \mathbf{q}), \quad \dot{x}_s(\mathbf{q}) = 4\pi c^2 p_s(\mathbf{q}),$$

so that, since $\mathbf{q} \wedge \mathbf{q} \wedge \boldsymbol{\epsilon}_s(\mathbf{q}) = -q^2 \boldsymbol{\epsilon}_s(\mathbf{q})$,

$$\dot{\mathbf{A}}_\perp(\mathbf{r}) = -\frac{1}{\sqrt{V}} \sum_{s=1}^{2} \sum_{\mathbf{q}} e^{i\mathbf{q}\cdot\mathbf{r}} (-i)^s \boldsymbol{\epsilon}_s(\mathbf{q}) 4\pi c^2 p_s(\mathbf{q}) = -c\,\mathbf{E}_\perp(\mathbf{r}),$$

$$\dot{\mathbf{E}}_\perp(\mathbf{r}) = 4\pi c \frac{1}{\sqrt{V}} \sum_{s=1}^{2} \sum_{\mathbf{q}} e^{i\mathbf{q}\cdot\mathbf{r}} (-i)^s \boldsymbol{\epsilon}_s(\mathbf{q}) \left(-\frac{q^2}{4\pi} x_s(\mathbf{q}) + \frac{1}{c} J_{s\,ext}(t, \mathbf{q}) \right)$$

$$= c\,\nabla \wedge \nabla \wedge \mathbf{A}_\perp(\mathbf{r}) - 4\pi\,\mathbf{J}_{\perp ext}(t, \mathbf{r}) = c\,\nabla \wedge \mathbf{B}(\mathbf{r}) - 4\pi\,\mathbf{J}_{\perp ext}(t, \mathbf{r}),$$

$$(2.50)$$

thus the well-known Maxwell equations

$$\mathbf{E}_\perp(\mathbf{r}) = -\frac{1}{c} \frac{\partial \mathbf{A}_\perp(\mathbf{r})}{\partial t}, \quad \nabla \wedge \mathbf{B}(\mathbf{r}) = \frac{1}{c} \frac{\partial \mathbf{E}_\perp(\mathbf{r})}{\partial t} + \frac{4\pi}{c} \mathbf{J}_{\perp ext}(t, \mathbf{r}). \quad (2.51)$$

2.6.2 System's Sources for the Electromagnetic Field

Through the continuity equation,[2] the electron current is defined by

$$-e\,\dot{\rho}(\mathbf{r}) = -i\left[-e\,\rho(\mathbf{r}), H \right] \equiv -\nabla \cdot \mathbf{J}(\mathbf{r}),$$

---

[2] Let us consider a conserved quantity Q, and the corresponding density $\rho_Q(\mathbf{r})$, so that

$$Q = \int d\mathbf{r}\,\rho_Q(\mathbf{r}) \equiv \rho_Q(\mathbf{q} = 0),$$

where $\rho_Q(\mathbf{q})$ is the Fourier transform of $\rho(\mathbf{r})$. Since Q is conserved

$$\dot{Q} = -i\left[Q, H \right] = 0 = \int d\mathbf{r}\,\dot{\rho}_Q(\mathbf{r}) = -i \int d\mathbf{r}\left[\rho_Q(\mathbf{r}), H \right].$$

Since the commutator of $\rho_Q(\mathbf{r})$ with the Hamiltonian does not vanish, the only possibility for the integral over the whole volume to vanish is that

$$\dot{\rho}_Q(\mathbf{r}) = -i\left[\rho_Q(\mathbf{r}), H \right] \equiv -\nabla \cdot \mathbf{J}_Q(\mathbf{r}),$$

which defines the current $\mathbf{J}_Q(\mathbf{r})$ associated to $\rho_Q(\mathbf{r})$, under the assumption that the flux of $\mathbf{J}_Q(\mathbf{r})$ through the surface of the system vanishes. The equation

$$\dot{\rho}_Q(\mathbf{r}) + \nabla \cdot \mathbf{J}_Q(\mathbf{r}) = 0 \quad (2.52)$$

is the *continuity equation* corresponding to the conserved quantity Q.

which, given the Hamiltonian (2.40), leads to

$$J(\mathbf{r}) = -e \left\{ \frac{1}{2m} \sum_{\sigma} \left[\Psi_{\sigma}^{\dagger}(\mathbf{r}) \left(-i \nabla \Psi_{\sigma}(\mathbf{r}) \right) + \left(i \nabla \Psi_{\sigma}^{\dagger}(\mathbf{r}) \right) \Psi_{\sigma}(\mathbf{r}) \right] + \frac{e}{mc} A(t, \mathbf{r}) \rho(\mathbf{r}) \right\}$$

$$\equiv -e \left(j(\mathbf{r}) + \frac{e}{mc} A(t, \mathbf{r}) \rho(\mathbf{r}) \right),$$

(2.53)

where $A(t, \mathbf{r}) = A_{\perp}(\mathbf{r}) + A_{\parallel ext}(t, \mathbf{r})$, and $j(\mathbf{r})$ is the current density in absence of the EMF.

The total charge current, defined through

$$J_{tot}(\mathbf{r}) \equiv -c \frac{\delta H}{\delta A(\mathbf{r})},$$

(2.54)

is contributed also from the spin magnetisation. Indeed,

$$-\int d\mathbf{r} M(\mathbf{r}) \cdot B(\mathbf{r}) = -\int d\mathbf{r} M(\mathbf{r}) \cdot \left(\nabla \wedge A(\mathbf{r}) \right) = -\int d\mathbf{r} A(\mathbf{r}) \cdot \left(\nabla \wedge M(\mathbf{r}) \right),$$

thus

$$\delta J_{tot}(\mathbf{r}) = c \nabla \wedge M(\mathbf{r}).$$

Finally, also the external transverse current contributes, so that,

$$J_{tot}(\mathbf{r}) = -e \left(j(\mathbf{r}) + \frac{e}{mc} A(t, \mathbf{r}) \rho(\mathbf{r}) \right) + c \nabla \wedge M(\mathbf{r}) + J_{\perp ext}(t, \mathbf{r}),$$

(2.55)

or, decomposed in longitudinal and transverse components,

$$J_{\perp tot}(\mathbf{r}) = -e \left(j_{\perp}(\mathbf{r}) + \frac{e}{mc} A_{\perp}(\mathbf{r}) \rho(\mathbf{r}) \right) + c \nabla \wedge M(\mathbf{r}) + J_{\perp ext}(t, \mathbf{r}),$$

$$J_{\parallel tot}(\mathbf{r}) = -e \left(j_{\parallel}(\mathbf{r}) + \frac{e}{mc} A_{\parallel ext}(t, \mathbf{r}) \rho(\mathbf{r}) \right).$$

(2.56)

With the above definitions, the Hamiltonian can be rewritten as

$$H = \frac{1}{8\pi} \int d\mathbf{r} \left(E_{\perp}(\mathbf{r}) \cdot E_{\perp}(\mathbf{r}) + B(\mathbf{r}) \cdot B(\mathbf{r}) \right)$$

$$+ \frac{1}{2m} \sum_{\sigma} \int d\mathbf{r} \Psi_{\sigma}^{\dagger}(\mathbf{r}) \left(-\nabla^2 \Psi_{\sigma}(\mathbf{r}) \right)$$

$$+ \frac{e^2}{2} \iint d\mathbf{r} d\mathbf{r}' \left(\rho(\mathbf{r}) - \rho_{ion}(\mathbf{r}) \right) \frac{1}{|\mathbf{r} - \mathbf{r}'|} \left(\rho(\mathbf{r}') - \rho_{ion}(\mathbf{r}') \right)$$

$$- \frac{1}{c} \int d\mathbf{r} A(\mathbf{r}) \cdot \left(-e j(\mathbf{r}) + c \nabla \wedge M(\mathbf{r}) + J_{\perp ext}(t, \mathbf{r}) - \frac{e^2}{2mc} A(\mathbf{r}) \rho(\mathbf{r}) \right)$$

$$- e \int d\mathbf{r} \phi_{ext}(t, \mathbf{r}) \left(\rho(\mathbf{r}) - \rho_{ion}(\mathbf{r}) \right)$$

$$\equiv H_0 - e \int d\mathbf{r}\, \phi_{ext}(t, \mathbf{r}) \left(\rho(\mathbf{r}) - \rho_{ion}(\mathbf{r}) \right)$$

$$- \frac{1}{c} \int d\mathbf{r}\, A(\mathbf{r}) \cdot \left(-e\, j(\mathbf{r}) + c\, \nabla \wedge M(\mathbf{r}) + J_{\perp ext}(t, \mathbf{r}) - \frac{e^2}{2mc} A(\mathbf{r}) \rho(\mathbf{r}) \right)$$

$$= H_0 + \delta H,$$

$$(2.57)$$

so that, through the Heisenberg equation of motion for $E_\perp(\mathbf{r})$, we obtain the Maxwell equation

$$\nabla \wedge B(\mathbf{r}) = \frac{1}{c} \frac{\partial E_\perp(\mathbf{r})}{\partial t} + \frac{4\pi}{c} J_{\perp tot}(t, \mathbf{r}). \qquad (2.58)$$

2.6.3 Optical Constants

Hereafter, all quantum fields and density operators are to be considered as their expectation values calculated within linear response theory.

Within the linear response, in frequency and momentum space and assuming translational invariance, we write, see (2.53),

$$J(\omega, \mathbf{q}) \equiv \hat{\sigma}(\omega, \mathbf{q})\, E(\omega, \mathbf{q}), \qquad (2.59)$$

where $E(\omega, \mathbf{q})$ is the internal electric field, and $\hat{\sigma}(\omega, \mathbf{q})$ is the optical conductivity tensor with components $\sigma_{ij}(\omega, \mathbf{q})$, $i, j = x, y, z$. If we further assume spatial isotropy, then

$$\sigma_{ij}(\omega, \mathbf{q}) = \frac{q_i q_j}{q^2} \sigma_\parallel(\omega, \mathbf{q}) + \left(\delta_{ij} - \frac{q_i q_j}{q^2} \right) \sigma_\perp(\omega, \mathbf{q}),$$

where $\sigma_\parallel(\omega, \mathbf{q})$ and $\sigma_\perp(\omega, \mathbf{q})$ are, respectively, the longitudinal and transverse components of the optical conductivity, which are the first two optical constants we shall be interested in. It follows that

$$J_\perp(\omega, \mathbf{q}) = \sigma_\perp(\omega, \mathbf{q})\, E_\perp(\omega, \mathbf{q}), \quad J_\parallel(\omega, \mathbf{q}) = \sigma_\parallel(\omega, \mathbf{q})\, E_\parallel(\omega, \mathbf{q}). \qquad (2.60)$$

We next define a 'displacement' field through

$$\frac{\partial D(\mathbf{r})}{\partial t} \equiv \frac{\partial E(\mathbf{r})}{\partial t} + 4\pi\, J_{tot}(t, \mathbf{r}). \qquad (2.61)$$

We note that, after taking the divergence of both sides, thus selecting the longitudinal components, and using the continuity equation, the Gauss law, as well as the fact that $\rho_{ion}(\mathbf{r})$ is time independent,

$$
\begin{aligned}
\frac{\partial \nabla \cdot \mathbf{D}_{\parallel}(\mathbf{r})}{\partial t} &= \frac{\partial \nabla \cdot \mathbf{E}_{\parallel}(\mathbf{r})}{\partial t} + 4\pi \, \nabla \cdot \mathbf{J}(t, \mathbf{r}) \\
&= \frac{\partial \nabla \cdot \mathbf{E}_{\parallel sys}(\mathbf{r})}{\partial t} + \frac{\partial \nabla \cdot \mathbf{E}_{\parallel ext}(\mathbf{r})}{\partial t} - 4\pi \, \frac{\partial \rho(\mathbf{r})}{\partial t} \\
&= \frac{\partial}{\partial t} \left(\nabla \cdot \mathbf{E}_{\parallel sys}(\mathbf{r}) + 4\pi e \left(\rho(\mathbf{r}) - \rho_{ion}(\mathbf{r}) \right) \right) + \frac{\partial \nabla \cdot \mathbf{E}_{\parallel ext}(\mathbf{r})}{\partial t} \\
&= \frac{\partial \nabla \cdot \mathbf{E}_{\parallel ext}(\mathbf{r})}{\partial t} \, ,
\end{aligned}
$$

which is satisfied if we assume that $\mathbf{D}_{\parallel}(\mathbf{r})$ is just the external $\mathbf{E}_{\parallel ext}(\mathbf{r})$. In frequency and momentum space (2.61) reads

$$
\mathbf{D}(\omega, \mathbf{q}) \equiv \mathbf{E}(\omega, \mathbf{q}) + i \, \frac{4\pi}{\omega} \, \mathbf{J}(\omega, \mathbf{q}) \,, \tag{2.62}
$$

or, through (2.59),

$$
\mathbf{D}(\omega, \mathbf{q}) = \left(1 + i \, \frac{4\pi}{\omega} \, \hat{\sigma}(\omega, \mathbf{q}) \right) \mathbf{E}(\omega, \mathbf{q}) \,. \tag{2.63}
$$

Now, we define the 'dielectric constant' tensor through

$$
\mathbf{D}(\omega, \mathbf{q}) \equiv \hat{\epsilon}(\omega, \mathbf{q}) \, \mathbf{E}(\omega, \mathbf{q}) \,, \tag{2.64}
$$

whose components, as before, can be written as

$$
\epsilon_{ij}(\omega, \mathbf{q}) = \frac{q_i q_j}{q^2} \, \epsilon_{\parallel}(\omega, \mathbf{q}) + \left(\delta_{ij} - \frac{q_i q_j}{q^2} \right) \epsilon_{\perp}(\omega, \mathbf{q}) \,,
$$

in terms of the longitudinal, $\epsilon_{\parallel}(\omega, \mathbf{q})$, and transverse, $\epsilon_{\perp}(\omega, \mathbf{q})$, components. Since $\mathbf{D}_{\parallel}(\omega, \mathbf{q}) \equiv \mathbf{E}_{\parallel ext}(\omega, \mathbf{q})$, it follows that

$$
\mathbf{E}_{\parallel ext}(\omega, \mathbf{q}) = \epsilon_{\parallel}(\omega, \mathbf{q}) \, \mathbf{E}_{\parallel}(\omega, \mathbf{q}) \,, \tag{2.65}
$$

which relates to each other external and internal longitudinal components of the electric field.

Finally, comparing (2.64) with (2.63), we find the following relation between transverse and longitudinal dielectric constants and optical conductivities:

$$
\epsilon_{\perp}(\omega, \mathbf{q}) = 1 + i \, \frac{4\pi}{\omega} \, \sigma_{\perp}(\omega, \mathbf{q}) \,, \quad \epsilon_{\parallel}(\omega, \mathbf{q}) = 1 + i \, \frac{4\pi}{\omega} \, \sigma_{\parallel}(\omega, \mathbf{q}) \,. \tag{2.66}
$$

2.6.4 Linear Response in the Longitudinal Case

Suppose that the external source generates only a longitudinal electric field, which is represented in the gauge such that

$$E_{\parallel ext}(t, \mathbf{r}) = -\nabla \phi_{ext}(t, \mathbf{r}) \implies E_{\parallel ext}(\omega, \mathbf{q}) = -i \mathbf{q} \phi_{ext}(\omega, \mathbf{q}). \tag{2.67}$$

Through the Hamiltonian (2.57), the perturbation is in this case

$$\delta H = -e \int d\mathbf{r} \, \phi_{ext}(t, \mathbf{r}) \left(\rho(\mathbf{r}) - \rho_{ion}(\mathbf{r}) \right). \tag{2.68}$$

We recall that at equilibrium $\rho_0(\mathbf{q}) = 0$ but at $\mathbf{q} = 0$, when its value is n times the volume V. Therefore, in presence of the external potential, a finite value of $\rho(\mathbf{q})$ for $\mathbf{q} \neq 0$ is already the deviation from equilibrium, and thus, in linear response theory,

$$\rho(\omega, \mathbf{q}) = -e \chi(\omega, \mathbf{q}) \phi_{ext}(\omega, \mathbf{q}),$$

where $\chi(\omega, \mathbf{q})$ is the Fourier transform of the density-density response function

$$\chi(t, \mathbf{q}) = -\frac{i}{V} \theta(t) \langle \left[\rho(t, \mathbf{q}), \rho(-\mathbf{q}, 0) \right] \rangle. \tag{2.69}$$

Through (2.65), we can also write

$$\begin{aligned} \rho(\omega, \mathbf{q}) &= -e \chi(\omega, \mathbf{q}) \phi_{ext}(\omega, \mathbf{q}) = -e \chi(\omega, \mathbf{q}) \epsilon_{\parallel}(\omega, \mathbf{q}) \phi(\omega, \mathbf{q}) \\ &\equiv -e \chi_*(\omega, \mathbf{q}) \phi(\omega, \mathbf{q}), \end{aligned} \tag{2.70}$$

where $\phi(\omega, \mathbf{q})$ is the internal scalar potential, and

$$\chi_*(\omega, \mathbf{q}) = \chi(\omega, \mathbf{q}) \epsilon_{\parallel}(\omega, \mathbf{q})$$

is the so-called *proper* response function as opposed to $\chi(\omega, \mathbf{q})$, which is denoted as the *improper* response function. Since $\phi(\omega, \mathbf{q}) = \phi_{ext}(\omega, \mathbf{q}) + \phi_{sys}(\omega, \mathbf{q})$ where the system field is obtained by the Gauss law

$$\begin{aligned} \phi_{sys}(\omega, \mathbf{q}) &= \phi(\omega, \mathbf{q}) - \phi_{ext}(\omega, \mathbf{q}) = -\frac{4\pi e}{q^2} \rho(\omega, \mathbf{q}) \\ &= \left(1 - \epsilon_{\parallel}(\omega, \mathbf{q}) \right) \phi(\omega, \mathbf{q}) = \frac{4\pi e^2}{q^2} \chi_*(\omega, \mathbf{q}) \phi(\omega, \mathbf{q}) \\ &= \left(\frac{1}{\epsilon_{\parallel}(\omega, \mathbf{q})} - 1 \right) \phi_{ext}(\omega, \mathbf{q}) = \frac{4\pi e^2}{q^2} \chi(\omega, \mathbf{q}) \phi_{ext}(\omega, \mathbf{q}), \end{aligned}$$

we arrive at the following equations:

$$\epsilon_{\|}(\omega, \mathbf{q}) = 1 - \frac{4\pi e^2}{q^2}\, \chi_*(\omega, \mathbf{q})\,, \qquad \frac{1}{\epsilon_{\|}(\omega, \mathbf{q})} = 1 + \frac{4\pi e^2}{q^2}\, \chi(\omega, \mathbf{q})\,,$$

$$\chi(\omega, \mathbf{q}) = \frac{\chi_*(\omega, \mathbf{q})}{1 - \dfrac{4\pi e^2}{q^2}\, \chi_*(\omega, \mathbf{q})}\,. \tag{2.71}$$

We note that, since $\rho(\mathbf{q} \to \mathbf{0})$ is the total number N of electrons, which is conserved, i.e., $N(t) = N$, then

$$V \lim_{\mathbf{q}\to 0} \chi(t, \mathbf{q}) = -i\,\theta(t) \lim_{\mathbf{q}\to 0} \left\langle \left[\rho(\mathbf{q}, t), \rho(-\mathbf{q})\right]\right\rangle = -i\,\theta(t)\left\langle \left[N(t), N\right]\right\rangle$$

$$= -i\,\theta(t)\left\langle \left[N, N\right]\right\rangle = 0\,.$$

Accordingly,

$$\lim_{\mathbf{q}\to 0} \chi(\omega, \mathbf{q}) = \lim_{\mathbf{q}\to 0} \chi_*(\omega, \mathbf{q}) = 0\,, \tag{2.72}$$

which derive from the conservation of N.

2.6.4.1 Consequences of Gauge Invariance

Equivalently, we can represent the external longitudinal field only through a vector potential, i.e.,

$$\mathbf{E}_{\|\,ext}(t, \mathbf{r}) = -\nabla\phi_{ext}(t, \mathbf{r}) \equiv -\frac{1}{c}\frac{\partial \mathbf{A}_{\|\,ext}(t, \mathbf{r})}{\partial t}\,,$$

implying

$$\mathbf{A}_{\|\,ext}(\omega, \mathbf{q}) = -\frac{\mathbf{q}\,c}{\omega}\,\phi_{ext}(\omega, \mathbf{q})\,,$$

in which case, see (2.57), the perturbation is

$$\delta H = -\frac{1}{c}\int d\mathbf{r}\,\mathbf{A}_{\|\,ext}(t, \mathbf{r})\cdot\left(-e\,\mathbf{j}_{\|}(\mathbf{r}) - \frac{e^2}{2mc}\mathbf{A}_{\|\,ext}(t, \mathbf{r})\,\rho(\mathbf{r})\right). \tag{2.73}$$

The Fourier transform $\chi_{ij}(\omega, \mathbf{q})$ of current-current linear response function

$$\chi_{ij}(t, \mathbf{q}) = -\frac{i}{V}\,\theta(t)\left\langle \left[J_i(\mathbf{q}, t)\,, J_j(-\mathbf{q}, 0)\right]\right\rangle\,, \quad i, j = x, y, z\,, \tag{2.74}$$

can also be written as

$$\chi_{ij}(\omega, \mathbf{q}) = \frac{q_i q_j}{q^2}\, \chi_{\|}(\omega, \mathbf{q}) + \left(\delta_{ij} - \frac{q_i q_j}{q^2}\right)\chi_{\perp}(\omega, \mathbf{q})\,. \tag{2.75}$$

It follows that, in linear response,

$$j_\parallel(\omega, \mathbf{q}) = \frac{e}{c} \chi_\parallel(\omega, \mathbf{q}) A_{\parallel ext}(\omega, \mathbf{q}). \tag{2.76}$$

Considering $J(\mathbf{r})$ in (2.53), its expectation value at first order in $A_{\parallel ext}$, which amounts to $\rho(\mathbf{r}) \to n$, is

$$\begin{aligned}
J_\parallel(\omega, \mathbf{q}) &= -\frac{e^2}{c} \left(\chi_\parallel(\omega, \mathbf{q}) + \frac{n}{m} \right) A_{\parallel ext}(\omega, \mathbf{q}) \\
&= \frac{e^2 \mathbf{q}}{\omega} \left(\chi_\parallel(\omega, \mathbf{q}) + \frac{n}{m} \right) \phi_{ext}(\omega, \mathbf{q}) \\
&= -\frac{e^2}{i\omega} \left(\chi_\parallel(\omega, \mathbf{q}) + \frac{n}{m} \right) E_{\parallel ext}(\omega, \mathbf{q}) \\
&= -\frac{e^2}{i\omega} \left(\chi_\parallel(\omega, \mathbf{q}) + \frac{n}{m} \right) \epsilon_\parallel(\omega, \mathbf{q}) E_\parallel(\omega, \mathbf{q}) \equiv \sigma_\parallel(\omega, \mathbf{q}) E_\parallel(\omega, \mathbf{q}).
\end{aligned} \tag{2.77}$$

The continuity equation, $e\dot{\rho} = \nabla \cdot J$ implies that

$$\begin{aligned}
e\,\omega\,\rho(\omega, \mathbf{q}) &= -e^2\,\omega\,\chi(\omega, \mathbf{q})\,\phi_{ext}(\omega, \mathbf{q}) \\
&= -\mathbf{q} \cdot J_\parallel(\omega, \mathbf{q}) = -\frac{e^2 q^2}{\omega} \left(\chi_\parallel(\omega, \mathbf{q}) + \frac{n}{m} \right) \phi_{ext}(\omega, \mathbf{q}),
\end{aligned} \tag{2.78}$$

leading to the equivalence

$$\omega^2 \chi(\omega, \mathbf{q}) = q^2 \left(\chi_\parallel(\omega, \mathbf{q}) + \frac{n}{m} \right), \tag{2.79}$$

which is merely a consequence of gauge invariance. In particular, if $\omega = 0$, since $\chi(0, \mathbf{q})$ is finite being a thermodynamic susceptibility, (2.79) implies that

$$\chi_\parallel(0, \mathbf{q}) = -\frac{n}{m}, \quad \forall \mathbf{q}. \tag{2.80}$$

Moreover, through (2.77), we find that the longitudinal component of the optical conductivity

$$\begin{aligned}
\sigma_\parallel(\omega, \mathbf{q}) &= -\frac{e^2}{i\omega} \left(\chi_\parallel(\omega, \mathbf{q}) + \frac{n}{m} \right) \epsilon_\parallel(\omega, \mathbf{q}) = -\frac{e^2}{i\omega} \frac{\omega^2}{q^2} \chi(\omega, \mathbf{q}) \epsilon_\parallel(\omega, \mathbf{q}) \\
&= -\frac{e^2}{i\omega} \frac{\omega^2}{q^2} \chi_*(\omega, \mathbf{q}).
\end{aligned} \tag{2.81}$$

2.6.4.2 Screening of a Static Long-Wavelength Electric Field

Let us assume to change the chemical potential and compensate for the change of electron number by a change of ionic charge so as to maintain charge neutrality. In this case, the perturbation is simply

$$\delta H = -\mu \int d\mathbf{r} \, \rho(\mathbf{r}) - e \int d\mathbf{r} \, \delta\phi_{ion}(\mathbf{r}) \, \rho(\mathbf{r}) \, ,$$

where $\delta\phi_{ion}(\mathbf{r})$ satisfies $\nabla^2 \delta\phi_{ion}(\mathbf{r}) = -4\pi e \, \delta\rho_{ion}(\mathbf{r})$, and is the additional potential generated by the change of the ionic charge. The induced field generated by the change in electron density satisfies instead $\nabla^2 \delta\phi_{els}(\mathbf{r}) = 4\pi e \, \delta\rho(\mathbf{r})$, so that the internal potential is

$$\phi(\mathbf{r}) = \frac{\mu}{e} + \delta\phi_{ion}(\mathbf{r}) + \delta\phi_{els}(\mathbf{r}) = \frac{\mu}{e} + \delta\phi_{sys}(\mathbf{r})$$

and is time independent. Therefore, only the $\omega = 0$ component of the response function is involved and we obtain, through the proper response,

$$\delta\rho(\mathbf{q}) = -e \, \chi_*(0, \mathbf{q}) \, \phi(0, \mathbf{q}) \, . \tag{2.82}$$

In the limit $\mathbf{q} \to \mathbf{0}$, since we assumed that charge neutrality is preserved then

$$\delta\phi_{sys}(\mathbf{q}) = \delta\phi_{ion}(\mathbf{q}) + \delta\phi_{els}(\mathbf{q}) \xrightarrow[\mathbf{q}\to\mathbf{0}]{} 0 \, ,$$

and, being $\delta\rho(\mathbf{q} \to \mathbf{0}) = \delta N$ just the change in electron number, we find

$$\delta N = -\chi_*(0, \mathbf{q} \to \mathbf{0}) \, \delta\mu \, ,$$

or, equivalently,

$$\kappa \equiv \frac{\delta N}{\delta\mu} = -\lim_{\mathbf{q}\to\mathbf{0}} \chi_*(0, \mathbf{q}) \, , \tag{2.83}$$

where $\kappa \geq 0$ is the uniform electron compressibility. Through (2.71) we find that

$$\epsilon_\parallel(0, \mathbf{q} \sim \mathbf{0}) = 1 + \frac{4\pi e^2}{q^2} \kappa \, , \tag{2.84}$$

so that the internal field felt by the electrons gets screened with respect to a long-wavelength static external one as

$$E_\parallel(0, \mathbf{q} \sim \mathbf{0}) = \frac{q^2}{q^2 + 4\pi e^2 \kappa} \, E_{\parallel \, ext}(0, \mathbf{q} \sim \mathbf{0}) \, .$$

Note that at $\omega = 0$ and in the limit $\mathbf{q} \to \mathbf{0}$ longitudinal and transverse components of the field coincide. Therefore, in the above equation, we can safely omit the label $\|$ and write

$$E(0, \mathbf{q} \sim \mathbf{0}) = \frac{q^2}{q^2 + 4\pi e^2 \kappa} \, E_{ext}(0, \mathbf{q} \sim \mathbf{0}) . \tag{2.85}$$

We now summarise through all the above results the behaviour of the longitudinal optical constants separately for insulators and metals.

2.6.4.3 Longitudinal Response of an Insulator

In an insulator, the chemical potential μ lies inside the gap, so that an infinitesimal variation $\delta\mu$ does not yield any variation δN in the electron number, thus $\kappa = 0 = -\chi_*(0, \mathbf{q} \to \mathbf{0})$. In particular, since $\chi_*(\omega, \mathbf{q} \to \mathbf{0}) \to 0$, see (2.72), then

$$\lim_{\omega \to 0} \lim_{\mathbf{q} \to \mathbf{0}} \chi_*(\omega, \mathbf{q}) = \lim_{\mathbf{q} \to \mathbf{0}} \lim_{\omega \to 0} \chi_*(\omega, \mathbf{q}) = 0 ,$$

namely $\chi_*(\omega, \mathbf{q})$ is analytic at $\omega = q = 0$. Since $-\chi_*(0, \mathbf{q} \sim \mathbf{0}) > 0$ is the thermodynamic susceptibility to a slowly varying chemical potential variation $\delta\mu(\mathbf{q} \sim \mathbf{0})$, we can conclude that, for small ω and q,

$$\chi_*(\omega, \mathbf{q}) \simeq -\kappa' q^2 , \quad \kappa' > 0 , \tag{2.86}$$

so that

$$\lim_{\mathbf{q} \to \mathbf{0}} \epsilon_\|(0, \mathbf{q}) = \lim_{\mathbf{q} \to \mathbf{0}} \epsilon_\perp(0, \mathbf{q}) \equiv \epsilon(0, \mathbf{0}) = 1 + 4\pi e^2 \kappa' > 1 , \tag{2.87}$$

namely the static long-wavelength limit of the dielectric constant is finite and greater than one, and, through (2.85),

$$E(0, \mathbf{0}) = \frac{1}{1 + 4\pi e^2 \kappa'} \, E_{ext}(0, \mathbf{0}) , \tag{2.88}$$

thus the internal field is reduced with respect to the external one but not totally suppressed.

Moreover, (2.81) implies that the DC conductivity, i.e., $\mathrm{Re}\,\sigma_\|(\omega \to 0, \mathbf{0})$, vanishes, as expected in an insulator.

2.6.4.4 Longitudinal Response of a Metal

Let us now consider a metal. We remark that, from the point of view of the longitudinal response, it makes no difference whether the metal is normal or superconducting.

In a metal, the chemical potential lies in the conduction band and so $\kappa \neq 0$. It follows that

$$\lim_{\mathbf{q} \to \mathbf{0}} \epsilon_\|(0, \mathbf{q}) = \lim_{\mathbf{q} \to \mathbf{0}} \left(1 + \frac{4\pi e^2}{q^2} \kappa \right) \to \infty , \tag{2.89}$$

Fig. 2.2 Hypothetical scenario in which the conduction electrons are rigidly shifted by x

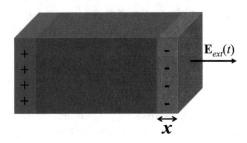

and thus, through (2.85), we conclude that a static long-wavelength external field is totally screened inside the metal. We further note, through (2.72), that

$$\lim_{\omega \to 0} \lim_{\mathbf{q} \to \mathbf{0}} \chi_*(\omega, \mathbf{q}) = 0, \quad \lim_{\mathbf{q} \to \mathbf{0}} \lim_{\omega \to 0} \chi_*(\omega, \mathbf{q}) = -\kappa < 0,$$

hence that $\chi_*(\omega, \mathbf{q})$ is not analytic at $\omega = q = 0$, which is a key property of metals. Since in a metal the energy scale associated with q is $v_F q$, where v_F is the Fermi velocity, the absence of analyticity implies that we cannot find the behaviour of $\chi_*(\omega, \mathbf{q})$ for $\omega \gg v_F q$ through that at $v_F q \gg \omega$.

In order to proceed, we note that at $\mathbf{q} = \mathbf{0}$ a change in the density of conduction electrons corresponds to a rigid shift of the latter with respect to the positive background of the ions plus all bound electrons, core and valence ones; see Fig. 2.2. In that limit, the electrons thus behave as a macroscopic object that can be regarded as classical plasma. If that shift is driven by an external time-dependent field $\boldsymbol{E}_{ext}(t)$, as shown in the figure, and assuming that the conduction electrons have an effective mass m_*, then the classical equation of motion reads

$$m_* \ddot{x}(t) = -4\pi e \, \sigma(x) - e \, E_{ext}(t),$$

where $\sigma(x) = n_* e \, x$ is the surface charge with n_* the density of conduction electrons. In Fourier space,

$$-m_* \omega^2 x(\omega) = -4\pi e^2 n_* x(\omega) - e E_{ext}(\omega) \implies x(\omega) = -\frac{e}{4\pi n_* e^2 - m_* \omega^2} E_{ext}(\omega),$$

which implies that the internal field

$$E(\omega) = 4\pi \, \sigma(x(\omega)) + E_{ext}(\omega) = -\frac{4\pi n_* e^2}{4\pi n_* e^2 - m_* \omega^2} E_{ext}(\omega) + E_{ext}(\omega)$$

$$= -\frac{m_* \omega^2}{4\pi n_* e^2 - m_* \omega^2} E_{ext}(\omega) = \frac{1}{1 - \dfrac{\omega_p^2}{\omega^2}} E_{ext}(\omega)$$

$$\equiv \frac{1}{\epsilon(\omega, \mathbf{0})} E_{ext}(\omega), \tag{2.90}$$

where

$$\omega_p^2 \equiv \frac{4\pi n_* e^2}{m_*} \tag{2.91}$$

is the so-called plasma frequency, namely that

$$\epsilon(\omega, \mathbf{0}) = 1 - \frac{\omega_p^2}{\omega^2} = 1 - \lim_{\mathbf{q} \to \mathbf{0}} \frac{4\pi e^2}{q^2} \chi_*(\omega, \mathbf{q}), \tag{2.92}$$

and thus

$$\chi_*(\omega, \mathbf{q} \sim \mathbf{0}) = \frac{n_*}{m_*} \frac{q^2}{\omega^2}. \tag{2.93}$$

Equations (2.90) and (2.92) imply that an infinitesimally small external field with frequency $\omega \simeq \omega_p$ produces a huge internal one. That corresponds to the frequency being in resonance with internal collective excitations, the plasmons. We remark that the above result is valid provided at $\omega \simeq \omega_p$ only the conduction electrons are free to move, while valence electrons are still bound, which is justified if ω_p is small enough with respect to the gap between valence and conduction band.

If that is the case, through (2.81), we find that

$$\sigma(\omega, \mathbf{q} \sim \mathbf{0}) = -\frac{e^2}{i\omega} \frac{\omega^2}{q^2} \chi_*(\omega, \mathbf{q} \sim \mathbf{0}) = -\frac{1}{4\pi} \frac{\omega_p^2}{i\omega}.$$

We recall that $\omega = \omega + i0^+$, so that

$$\sigma(\omega, \mathbf{0}) = i \frac{1}{4\pi} \omega_p^2 \frac{1}{\omega + i0^+} = \frac{\omega_p^2}{4} \delta(\omega) + i \frac{1}{4\pi} \frac{\omega_p^2}{\omega}, \tag{2.94}$$

It follows that $\mathrm{Re}\, \sigma(\omega, \mathbf{0})$ is a δ-peak at zero frequency, so-called Drude peak, with weight $\omega_p^2/4$, the Drude weight.

When there are channels that dissipate current, provided, for instance, by impurities or by electron-electron and electron-phonon umklapp scattering, one conventionally assumes a Drude-Lorentz expression

$$\sigma(\omega, \mathbf{0}) = i \frac{1}{4\pi} \omega_p^2 \frac{1}{\omega + i\gamma} = \frac{\omega_p^2}{4} \frac{1}{\pi} \frac{\gamma}{\omega^2 + \gamma^2} + i \frac{\omega_p^2}{4} \frac{1}{\pi} \frac{\omega}{\omega^2 + \gamma^2}, \tag{2.95}$$

so that the real part becomes a Lorentzian.

2.6.5   Linear Response in the Transverse Case

In the presence of a transverse electromagnetic field, the perturbation we must add
to the Hamiltonian is

$$
\delta H = \frac{e}{c} \int d\mathbf{r}\, A_\perp(\mathbf{r}) \cdot \mathbf{j}(\mathbf{r}) + \frac{e^2}{2mc^2} \int d\mathbf{r}\, \rho(\mathbf{r}) A_\perp(\mathbf{r}) \cdot A_\perp(\mathbf{r})
$$
$$
- \int d\mathbf{r}\, A_\perp(\mathbf{r}) \cdot \left(\nabla \wedge M(\mathbf{r}) \right) - \frac{1}{c} \int d\mathbf{r}\, A_\perp(\mathbf{r}) \cdot J_{\perp ext}(t, \mathbf{r}),
$$

where, we remark, $A_\perp(\mathbf{r})$ is the internal field, contributed both by the external source
$J_{\perp ext}(t, \mathbf{r})$ as well as by the electrons. Specifically, $A_\perp(\mathbf{r})$ satisfies the equation

$$
\left(\frac{\partial^2}{\partial t^2} - c^2 \nabla^2 \right) A_\perp(t, \mathbf{r}) = 4\pi c\, J_\perp(t, \mathbf{r}) + 4\pi c^2\, \nabla \wedge M(\mathbf{r}) + 4\pi c\, J_{\perp ext}(t, \mathbf{r}),
$$
(2.96)

where, see (2.53),

$$
J_\perp(t, \mathbf{r}) = -e \left(\mathbf{j}_\perp(t, \mathbf{r}) + \frac{e}{mc} \rho(t, \mathbf{r}) A_\perp(t, \mathbf{r}) \right).
$$

In linear response, and after Fourier transform in frequency and momentum,

$$
\mathbf{j}_\perp(\omega, \mathbf{q}) = \frac{e}{c} \chi_\perp(\omega, \mathbf{q}) A_\perp(\omega, \mathbf{q}),
$$
(2.97)

where $\chi_\perp(\omega, \mathbf{q})$ is the transverse component of the current-current response function,
see (2.74) and (2.75), and so

$$
J_\perp(\omega, \mathbf{q}) = -\frac{e^2}{c} \left(\chi_\perp(\omega, \mathbf{q}) + \frac{n}{m} \right) A_\perp(\omega, \mathbf{q}) = -\frac{e^2}{i\omega} \left(\chi_\perp(\omega, \mathbf{q}) + \frac{n}{m} \right) E_\perp(\omega, \mathbf{q})
$$
$$
\equiv \sigma_\perp(\omega, \mathbf{q}) E_\perp(\omega, \mathbf{q}) = \frac{i\omega}{c} \sigma_\perp(\omega, \mathbf{q}) A_\perp(\omega, \mathbf{q}).
$$
(2.98)

Similarly,

$$
M(\omega, \mathbf{q}) = -i\, \chi_\sigma(\omega, \mathbf{q})\, \mathbf{q} \wedge A_\perp(\omega, \mathbf{q}),
$$
(2.99)

where $\chi_\sigma(\omega, \mathbf{q})$ is the spin density-density linear response function multiplied by
$g^2 \mu_B^2$, so that the Fourier transform of $\nabla \wedge M(\mathbf{r})$ becomes in linear response
$-q^2 \chi_\sigma(\omega, \mathbf{q}) A_\perp(\omega, \mathbf{q})$, so that (2.96) reads in frequency and momentum

$$
\left(c^2 q^2 - \omega^2 \right) A_\perp(\omega, \mathbf{q}) = 4\pi\, i\, \omega\, \sigma_\perp(\omega, \mathbf{q}) A_\perp(\omega, \mathbf{q}) - 4\pi\, c^2 q^2\, \chi_\sigma(\omega, \mathbf{q}) A_\perp(\omega, \mathbf{q})
$$
$$
+ 4\pi c\, J_{\perp ext}(\omega, \mathbf{q}).
$$

Since

$$
\left(c^2 q^2 - \omega^2 \right) A_{\perp ext}(\omega, \mathbf{q}) = 4\pi c\, J_{\perp ext}(\omega, \mathbf{q})
$$

defines the external transverse field, we can write

$$
\begin{aligned}
A_\perp(\omega, \mathbf{q}) &= \frac{c^2 q^2 - \omega^2}{c^2 q^2 - \omega^2 - 4\pi\, i\, \omega\, \sigma_\perp(\omega, \mathbf{q}) + 4\pi\, c^2 q^2\, \chi_\sigma(\omega, \mathbf{q})}\, A_{\perp ext}(\omega, \mathbf{q}) \\
&= \frac{c^2 q^2 - \omega^2}{c^2 q^2 - \omega^2\, \epsilon_\perp(\omega, \mathbf{q}) + 4\pi\, c^2 q^2\, \chi_\sigma(\omega, \mathbf{q})}\, A_{\perp ext}(\omega, \mathbf{q})\,,
\end{aligned}
$$

(2.100)

which relates the internal vector potential to the external one, and where we use (2.66). The root with respect to ω of the denominator in (2.100) determines the dispersion of the light inside the system, which in the vacuum is simply $cq \gg v_F\, q$. For that reason, we can safely assume that $\omega \gg v_F\, q$, in which case, since the magnetisation is conserved, $\chi_\sigma(\omega, \mathbf{q}) \simeq \chi_\sigma(\omega, \mathbf{0}) = 0$, and $\epsilon_\perp(\omega, \mathbf{q}) \simeq \epsilon_\perp(\omega, \mathbf{0}) = \epsilon_\parallel(\omega, \mathbf{0}) = \epsilon(\omega, \mathbf{0})$, so that

$$
\omega \simeq \frac{c}{\sqrt{\epsilon(\omega, \mathbf{0})}}\, q\,.
$$

(2.101)

It follows that

- In an insulator, and for ω small compared to the gap, $\epsilon(0, \mathbf{0}) > 1$, see (2.87), so that light propagates, though with generally a smaller velocity than c. The insulator is therefore transparent in the visible.
- In a metal, (2.92) implies that $\epsilon(\omega, \mathbf{0})$ is negative for $\omega < \omega_p$, where ω_p is generally in the ultraviolet. Therefore, light cannot propagate as long as $\omega < \omega_p$, so that the metal is opaque in the visible range. Only when ω exceeds ω_p, light starts to propagate. The reflectivity is therefore close to one at small ω and bends down around the plasma frequency.

Concerning the response to a uniform $q = 0$ field, i.e., to a time-dependent electric field, the transverse optical constants become equal to the longitudinal ones, which we earlier discussed.

2.6.5.1 Response to a Static Non-uniform Field

If $\omega \to 0$ but $q \neq 0$, though small, i.e., in presence of a static non-uniform magnetic field, (2.100) becomes

$$
A_\perp(\omega \to 0, \mathbf{q}) = \frac{c^2 q^2}{c^2 q^2 - 4\pi\, i\, \omega\, \sigma_\perp(\omega, \mathbf{q}) + 4\pi\, c^2 q^2\, \chi_\sigma(0, \mathbf{q})}\, A_{\perp ext}(0, \mathbf{q})\,.
$$

In this case, $\chi_\sigma(0, \mathbf{q}) \simeq -\kappa_\sigma$, where $\kappa_\sigma > 0$ is the uniform magnetic susceptibility, while

$$
\lim_{\omega \to 0} i\, \omega\, \sigma_\perp(\omega, \mathbf{q}) = -e^2 \left(\chi_\perp(0, \mathbf{q}) + \frac{n}{m} \right).
$$

We recall that, as a consequence of gauge invariance, $\chi_{\parallel}(0, \mathbf{q}) = -n/m$. It would be tempting to assume that

$$\chi_{\perp}(0, \mathbf{q} \simeq \mathbf{0}) \simeq \chi_{\parallel}(0, \mathbf{0}) + \alpha\, q^2 = -\frac{n}{m} + \alpha\, c^2\, q^2 , \qquad (2.102)$$

with $\alpha > 0$ since $-\chi_{\perp}(0, \mathbf{q} \simeq \mathbf{0}) > 0$ is a thermodynamic susceptibility that is supposedly maximum at $\mathbf{q} = \mathbf{0}$, so that

$$
\begin{aligned}
A_{\perp}(0, \mathbf{q} \sim \mathbf{0}) &= \frac{c^2 q^2}{c^2 q^2 + 4\pi\, e^2\, \alpha\, c^2\, q^2 - 4\pi\, c^2 q^2\, \kappa_{\sigma}}\, A_{\perp ext}(0, \mathbf{q} \sim \mathbf{0}) \\
&= \frac{1}{1 + 4\pi\, e^2\, \alpha - 4\pi\, \kappa_{\sigma}}\, A_{\perp ext}(0, \mathbf{q} \sim \mathbf{0}) ,
\end{aligned}
\qquad (2.103)
$$

which implies that the magnetic field propagates freely inside the system, and is enhanced with respect to the external one if $\kappa_{\sigma} > e^2\, \alpha$ and reduced otherwise, corresponding, respectively, to paramagnetic and diamagnetic behaviour.

However, the assumption (2.102) is valid only in insulators and normal metals. In a superconducting metal,

$$\chi_{\perp}(0, \mathbf{q} \to \mathbf{0}) = -\frac{n - n_c}{m} > -\frac{n}{m} , \qquad (2.104)$$

where $n_c \leq n$ is the density of the superconducting fraction, the condensate. In this case,

$$A_{\perp}(0, \mathbf{q} \sim \mathbf{0}) = \frac{c^2 q^2}{c^2 q^2 + \omega_c^2 - 4\pi\, c^2 q^2\, \kappa_{\sigma}}\, A_{\perp ext}(0, \mathbf{q} \sim \mathbf{0}) ,$$

where $\omega_c = \sqrt{4\pi n_c e^2/m}$ is the plasma frequency of the condensate. It follows that a magnetic field cannot penetrate in a superconductor and decays from the surface over a distance $\lambda_L \sim c/\omega_c$, the so-called London penetration depth. A superconductor is therefore a perfect diamagnet, a phenomenon known as Meissner effect.

2.6.6   Power Dissipated by the Electromagnetic Field

In presence of an external longitudinal field, the power dissipated according to the general formula is

$$
\begin{aligned}
W_{\parallel} &= \left(\frac{e}{c}\right)^2 \frac{\omega}{2}\, |A_{\parallel ext}(\mathbf{q}, \omega)|^2\, \chi_{\parallel}''(\omega, \mathbf{q}) = -\left(\frac{e}{c}\right)^2 \frac{\omega}{2}\, |A_{\parallel ext}(\mathbf{q}, \omega)|^2\, \mathrm{Im}\, \chi_{\parallel}(\omega, \mathbf{q}) \\
&\quad - \left(\frac{e}{\omega}\right)^2 \frac{\omega}{2}\, |E_{\parallel ext}(\mathbf{q}, \omega)|^2\, \mathrm{Im}\, \chi_{\parallel}(\omega, \mathbf{q}) ,
\end{aligned}
$$

where $\chi_{\parallel}''(\omega, \mathbf{q})$ is the dissipative current-current function which is equal to minus the imaginary part of the linear response function. On the other hand,

$$\text{Im}\,\chi_{\parallel}(\omega, \mathbf{q}) = \text{Im}\left(-i\frac{\omega}{e^2}\frac{\sigma_{\parallel}(\omega, \mathbf{q})}{\epsilon_{\parallel}(\omega, \mathbf{q})}\right) = -\frac{\omega}{e^2}\,\text{Re}\,\frac{\sigma_{\parallel}(\omega, \mathbf{q})}{\epsilon_{\parallel}(\omega, \mathbf{q})}$$

$$= -\frac{\omega}{e^2}\,\text{Re}\left(\frac{\omega}{4\pi i}\left(\epsilon_{\parallel}(\omega, \mathbf{q}) - 1\right)\frac{1}{\epsilon_{\parallel}(\omega, \mathbf{q})}\right) = \frac{\omega^2}{4\pi e^2}\,\text{Im}\,\frac{1}{\epsilon_{\parallel}(\omega, \mathbf{q})}$$

$$= -\frac{\omega}{e^2}\,\text{Re}\left(\frac{\dfrac{\sigma_{\parallel}(\omega, \mathbf{q})}{4\pi i}}{1 + \dfrac{4\pi i}{\omega}\sigma_{\parallel}(\omega, \mathbf{q})}\right) = -\frac{\omega}{e^2}\frac{1}{|\epsilon_{\parallel}(\omega, \mathbf{q})|^2}\,\text{Re}\,\sigma_{\parallel}(\omega, \mathbf{q}),$$

which, since $\left|E_{\parallel ext}\right| = \left|\epsilon_{\parallel}\right|\left|E_{\parallel}\right|$, implies that

$$W_{\parallel} = -\frac{\omega}{8\pi}\,\text{Im}\,\frac{1}{\epsilon_{\parallel}(\omega, \mathbf{q})}\left|E_{\parallel ext}(\omega, \mathbf{q})\right|^2 = \frac{1}{2}\,\text{Re}\,\sigma_{\parallel}(\omega, \mathbf{q})\left|E_{\parallel}(\omega, \mathbf{q})\right|^2.$$

$$(2.105)$$

In the case of a transverse field

$$W_{\perp} = -\left(\frac{e}{c}\right)^2\frac{\omega}{2}\,|A_{\perp}(\omega, \mathbf{q})|^2\,\text{Im}\,\chi_{\perp}(\omega, \mathbf{q}),$$

where

$$\text{Im}\,\chi_{\perp}(\omega, \mathbf{q}) = -\frac{\omega}{e^2}\,\text{Re}\,\sigma_{\perp}(\omega, \mathbf{q}) = -\frac{\omega^2}{4\pi e^2}\,\text{Im}\,\epsilon_{\perp}(\omega, \mathbf{q}).$$

Therefore,

$$W_{\perp} = \frac{\omega^2}{2c^2}\,\text{Re}\,\sigma_{\perp}(\omega, \mathbf{q})\,|A_{\perp}(\omega, \mathbf{q})|^2 = \frac{1}{2}\,\text{Re}\,\sigma_{\perp}(\omega, \mathbf{q})\,|E_{\perp}(\omega, \mathbf{q})|^2$$

$$= \frac{\omega}{8\pi}\,\text{Im}\,\epsilon_{\perp}(\omega, \mathbf{q})\,|E_{\perp}(\omega, \mathbf{q})|^2.$$

$$(2.106)$$

If $q = 0$, longitudinal and transverse responses become equal, so that

$$W_{\parallel} = W_{\perp} = -\frac{\omega}{8\pi}\,\text{Im}\,\frac{1}{\epsilon_{\parallel}(\omega, 0)}\left|E_{ext}(\omega, 0)\right|^2 = \frac{1}{2}\,\text{Re}\,\sigma_{\parallel}(\omega, 0)\left|E(\omega, 0)\right|^2.$$

$$(2.107)$$

Problems

2.1 Linear response of free electrons—Consider a free electron Hamiltonian

$$\mathcal{H}_0 = \sum_{k\sigma} \epsilon_k\, c_{k\sigma}^{\dagger} c_{k\sigma},$$

in the presence of a Zeeman splitting due to a slowly varying magnetic field oriented along the z-direction $B(\mathbf{q}, t)$. The perturbation is

$$V(\mathbf{q}, t) = \frac{1}{V} \mu_B \, g \, B(\mathbf{q}, t) \, S_z(-\mathbf{q}),$$

where V is the volume, μ_B the Bohr magneton, $g \simeq 2$ the electron gyromagnetic ratio, and $S_z(\mathbf{q})$ the spin-density operator at momentum $\mathbf{q}$ defined through

$$S_z(\mathbf{q}) = \frac{1}{2} \sum_{\mathbf{k}} \left(c^\dagger_{\mathbf{k}\uparrow} c_{\mathbf{k}+\mathbf{q}\uparrow} - c^\dagger_{\mathbf{k}\downarrow} c_{\mathbf{k}+\mathbf{q}\downarrow} \right).$$

By linear response theory, the expectation value of $S_z(\mathbf{q})$ is

$$\langle S_z(\mathbf{q}) \rangle (t) = g \, \mu_B \int dt' \, \chi(\mathbf{q}, t - t') \, B(\mathbf{q}, t'),$$

where

$$\chi(\mathbf{q}, t - t') = -\frac{i}{\hbar} \frac{1}{V} \, \theta(t - t') \, \langle [S_z(\mathbf{q}, 0), S_z(-\mathbf{q})] \rangle$$

is the magnetic response function per unit volume.

- Find the formal expression of

$$\chi(\mathbf{q}, \omega) = \int dt \, e^{i\omega t} \, \chi(\mathbf{q}, t).$$

- Calculate the uniform magnetic susceptibility, namely

$$\chi = - \lim_{\mathbf{q} \to 0} \chi(\mathbf{q}, \omega = 0),$$

at zero temperature in terms of the density of states

$$N(\epsilon) \equiv \frac{1}{V} \sum_{\mathbf{k}} \delta(\epsilon - \epsilon_{\mathbf{k}}).$$

- Prove that

$$\chi(\mathbf{q} = \mathbf{0}, \omega) = 0,$$

and discuss why.

Fig. 2.3 Geometry of the tunnelling problem (2)

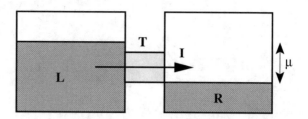

2.2 Tunnelling between two leads—Consider two disconnected metallic leads, a right one (R) and a left one (L), see Fig. 2.3, described by the unperturbed Hamiltonian

$$
\begin{aligned}
H_0 &= \sum_{\mathbf{k}\sigma} \epsilon_{\mathbf{k}} \left(c^\dagger_{R\,\mathbf{k}\sigma} c_{R\,\mathbf{k}\sigma} + c^\dagger_{L\,\mathbf{k}\sigma} c_{L\,\mathbf{k}\sigma} \right) + \frac{\mu}{2} \sum_{\mathbf{k}\sigma} \left(c^\dagger_{R\,\mathbf{k}\sigma} c_{R\,\mathbf{k}\sigma} - c^\dagger_{L\,\mathbf{k}\sigma} c_{L\,\mathbf{k}\sigma} \right) \\
&\equiv \sum_{\mathbf{k}\sigma} \epsilon_{\mathbf{k}} \left(c^\dagger_{R\,\mathbf{k}\sigma} c_{R\,\mathbf{k}\sigma} + c^\dagger_{L\,\mathbf{k}\sigma} c_{L\,\mathbf{k}\sigma} \right) + \frac{\mu}{2} (N_R - N_L)
\end{aligned}
$$

$$(2.108)$$

where $\mu = eV$ is the bias difference between the leads. At some given time, a tunnelling between the two leads is switched on, by the perturbation

$$
V = - \sum_{\mathbf{k}\sigma} T(\epsilon_{\mathbf{k}}, \epsilon_{\mathbf{p}}) \, c^\dagger_{R\,\mathbf{k}\sigma} c_{L\,\mathbf{p}\sigma} + H.c. \,,
$$

where we assume that the tunnelling amplitudes $T(\epsilon_{\mathbf{k}}, \epsilon_{\mathbf{p}})$ depend only on the energies. Therefore, the fully perturbed Hamiltonian is

$$
H = H_0 + V \,. \tag{2.109}
$$

- Using the Heisenberg equation of motion with the Hamiltonian (2.109), calculate the expression of the operator of the particle current flowing from the L to the R lead, which is defined through

$$
I = \frac{\partial}{\partial t} (N_R - N_L) = -\frac{i}{\hbar} \left[(N_R - N_L), H \right].
$$

- Calculate at the leading non-zero order in the tunnelling, actually second order, the expectation value $I(V)$ of the current operator, assuming that the density of states of both leads is $N(\epsilon)$.

Hartree-Fock Approximation

3

In this chapter, we describe the simplest approximation to tackle interacting fermions: the Hartree-Fock approximation. This technique is variational and can be applied both at zero and at finite temperature. Essentially, within the Hartree-Fock approximation, the effects of the particle-particle interaction are described by an effective single-particle potential that is self-consistently determined. This is also the reason why the Hartree-Fock approximation is a Mean Field theory.

3.1 Hartree-Fock Approximation for Fermions at Zero Temperature

> The Hartree-Fock (HF) approximation at zero temperature consists in searching for a Slater determinant that minimises the expectation value of the total energy.

Since in general the ground state wavefunction is not a single Slater determinant, the HF approach is variational; hence, the HF energy is an upper bound to the true ground state energy.

The trial HF wavefunction is by definition a Slater determinant, namely

$$\Phi_{HF}(x_1, \ldots, x_N) = \frac{1}{\sqrt{N!}} \begin{vmatrix} \phi_1(x_1) & \cdots & \phi_1(x_N) \\ \phi_2(x_1) & \cdots & \phi_2(x_N) \\ \vdots & \vdots\vdots\vdots & \vdots \\ \phi_N(x_1) & \cdots & \phi_N(x_N) \end{vmatrix}, \tag{3.1}$$

© The Author(s), under exclusive license to Springer Nature Switzerland AG 2022
M. Fabrizio, *A Course in Quantum Many-Body Theory*, Graduate Texts in Physics,
https://doi.org/10.1007/978-3-031-16305-0_3

where N is the electron number, and the single-particle wavefunctions $\phi_i(x)$ are to be determined variationally. However, they are assumed to belong to a complete set of orthonormal wavefunctions

$$\int dx\, \phi_i(x)^*\, \phi_j(x) = \delta_{ij}. \tag{3.2}$$

The Hamiltonian in first quantisation is

$$H = \sum_{i=1}^{N} T(x_i, p_i) + \frac{1}{2} \sum_{i \neq j} U(x_i, x_j),$$

where

$$T(x, p) = -\frac{\hbar^2}{2m}\nabla^2 + V(x)$$

is the non-interacting single-particle contribution that includes the kinetic energy as well as a potential.

The expectation value of the Hamiltonian over the wavefunction (3.1) is

$$E_{HF} = \frac{\langle \Phi_{HF}|\hat{H}|\Phi_{HF}\rangle}{\langle \Phi_{HF}|\Phi_{HF}\rangle}. \tag{3.3}$$

One has to impose that the variation of E_{HF} vanishes upon varying the trial wave-function, keeping it still of the form of a Slater determinant. This amounts to change one of the single particle wavefunctions, namely

$$\phi_i(x) \rightarrow \mathcal{N}_i\, (\phi_i(x) + \delta\phi_i(x)), \tag{3.4}$$

with $\mathcal{N}_i$ a normalisation constant. Clearly, if the variation $\delta\phi_i(x)$ has a finite overlap with any of the $\phi_k(x)$'s already present in (3.1), the Slater determinant does not change; hence, such variation is irrelevant. Therefore, the only meaningful possibility is that

$$\delta\phi_i(x) = \eta\, \phi_j(x), \tag{3.5}$$

where η is an infinitesimal quantity and ϕ_j belongs to the set of wavefunctions (3.2) but does not appear in (3.1), namely $j > N$. The normalisation $\mathcal{N}_i$ in (3.4) is given by

$$\mathcal{N}_i^{-2} = \int dx\, \left(\phi_i(x)^* + \eta^*\, \phi_j(x)^*\right)\left(\phi_i(x) + \eta\, \phi_j(x)\right) = 1 + |\eta|^2,$$

and thus $\mathcal{N}_i \simeq 1$ at linear order in η. This implies that also the normalisation of the Slater determinant $\Phi_{HF} + \eta\, \delta\Phi_{HF}$ remains one at linear order in η; hence,

$$\delta E_{HF} = \eta^* \langle\delta\Phi_{HF} \mid H \mid \Phi_{HF}\rangle + \eta \langle\Phi_{HF} \mid H \mid \delta\Phi_{HF}\rangle$$

$$= \operatorname{Re}\eta\, \operatorname{Re}\!\left(\langle\delta\Phi_{HF} \mid H \mid \Phi_{HF}\rangle\right) - \operatorname{Im}\eta\, \operatorname{Im}\!\left(\langle\delta\Phi_{HF} \mid H \mid \Phi_{HF}\rangle\right) = 0.$$

Since η is an arbitrary infinitesimal number, this implies that

$$\mathrm{Re}\left(\langle\delta\Phi_{HF}\mid H\mid\Phi_{HF}\rangle\right) = \mathrm{Im}\left(\langle\delta\Phi_{HF}\mid H\mid\Phi_{HF}\rangle\right) = 0\,,$$

namely

$$\langle\delta\Phi_{HF}\mid H\mid\Phi_{HF}\rangle = 0\,. \tag{3.6}$$

Let us rephrase (3.6) in second quantisation. We associate to any of the wavefunctions $\phi_i(x)$ an annihilation and a creation operator, c_i and $c_i^\dagger$, respectively. In second quantisation, the wavefunction (3.1) corresponds to the Fock state

$$|\Phi_{HF}\rangle = \prod_{i=1}^{N} c_i^\dagger\,|0\rangle\,,$$

while the variation

$$|\delta\Phi_{HF}\rangle = c_j^\dagger c_i\,|\Phi_{HF}\rangle\,,$$

with $j > N$ while $i \le N$. The Hamiltonian in second quantised form is

$$\hat{H} = \sum_{ij} t_{ij}\,c_i^\dagger c_j + \frac{1}{2}\sum_{ijkl} U_{ijkl}\,c_i^\dagger c_j^\dagger c_k c_l\,, \tag{3.7}$$

where the parameters are the matrix elements over the basis set (3.2). What we need to solve is therefore the equation

$$\langle\Phi_{HF}\mid c_i^\dagger c_j\,H\mid\Phi_{HF}\rangle = 0\,, \quad i \le N\,, \quad j > N\,.$$

For that we need the following two equalities, which can be easily derived:

$$\sum_{kl} t_{kl}\,\langle\Phi_{HF}\mid c_i^\dagger c_j\,c_k^\dagger c_l\mid\Phi_{HF}\rangle = t_{ji}\,,$$

$$\frac{1}{2}\sum_{klmn} U_{klmn}\,\langle\Phi_{HF}\mid c_i^\dagger c_j\,c_k^\dagger c_l^\dagger c_m c_n\mid\Phi_{HF}\rangle$$

$$= \frac{1}{2}\sum_{m=1}^{N} U_{jmmi} + U_{mjim} - U_{jmim} - U_{mjmi} = \sum_{m=1}^{N} U_{jmmi} - U_{jmim}\,,$$

where the last expression comes from the symmetry relation $U_{ijkl} = U_{jilk}$. Equation (3.6) thus implies that

$$t_{ji} + \sum_{m=1}^{N} U_{jmmi} - U_{jmim} = 0\,, \tag{3.8}$$

for any $j > N$ and $i \leq N$. In other words, the Slater determinant which minimises the total energy is constructed by single-particle wavefunctions $\phi_i(x)$'s that have matrix elements obeying (3.8). Let us suppose we find instead a set of wavefunctions ϕ_i's satisfying

$$t_{ji} + \sum_{m=1}^{N} U_{jmmi} - U_{jmim} = \epsilon_i \, \delta_{ij}. \tag{3.9}$$

This set automatically satisfies also (3.8), hence it does solve our variational problem. In first quantisation, (3.9) reads

$$T(x, p) \, \phi_i(x) + \sum_{m=1}^{N} \int dy \, U(x, y) \Big(\phi_m(y)^* \, \phi_m(y) \, \phi_i(x) - \phi_m(y)^* \, \phi_m(x) \, \phi_i(y) \Big) = \epsilon_i \, \phi_i(x),$$
$$\tag{3.10}$$

which is the standard Hartree-Fock set of equations.

Notice that there might be several Slater determinants built up using N of the wavefunctions solving (3.9) and which satisfy the HF variational principle. Among them, one has to choose the Slater determinant with the minimum total energy for N electrons, which can be easily found to be

$$E_{HF}(N) = \langle \Phi_{HF} \mid H \mid \Phi_{HF} \rangle = \sum_{i=1}^{N} t_{ii} + \frac{1}{2} \sum_{i,m=1}^{N} \Big(U_{immi} - U_{imim} \Big)$$
$$\tag{3.11}$$
$$= \sum_{i=1}^{N} \epsilon_i - \frac{1}{2} \sum_{i,m=1}^{N} \Big(U_{immi} - U_{imim} \Big).$$

If for $N - 1$ particles the Hartree-Fock single-particle wavefunctions stay approximately invariant, then

$$E_{HF}(N-1) \simeq \sum_{i=1}^{N-1} t_{ii} + \frac{1}{2} \sum_{i,m=1}^{N-1} \Big(U_{immi} - U_{imim} \Big)$$
$$= \sum_{i=1}^{N} t_{ii} \, (1 - \delta_{iN}) + \frac{1}{2} \sum_{i,m=1}^{N} \Big(U_{immi} - U_{imim} \Big) (1 - \delta_{iN})(1 - \delta_{mN})$$
$$= E_{HF}(N) - \epsilon_N,$$
$$\tag{3.12}$$

showing that the Hartree-Fock single-particle energies correspond approximately to the ionisation energies.

3.1.1   Alternative Approach

The Hartree-Fock equations (3.10) are complicated non-linear integral-differential equations. This is especially true if one does not impose any constraint on the form of the variational Slater determinant dictated for instance by the symmetry properties of the Hamiltonian, what is called *unrestricted Hartree-Fock approximation*. However, very often one expects that the true ground state has well-defined properties under symmetry transformations that leave the Hamiltonian invariant. For instance, if the Hamiltonian is translationally and spin-rotationally invariant, one may expect that the true ground state is an eigenstate of the total spin and of the total momentum. For this reason, one would like to search for a variational wavefunction within the subspace of Slater determinants which are eigenstates of the total spin and momentum. This amounts to impose symmetry constraints on the general form (3.1) of the variational wavefunction. Yet, this is not a simple task if one keeps working within first quantisation. The second quantisation approach to the Hartree-Fock approximation that we describe in the following has the big advantage to allow an easy implementation of such symmetry constraints.

Suppose that our Hamiltonian is written in a basis of single-particle wavefunctions $\{\phi_\alpha(x)\}$ which is more convenient to work with, for instance Block waves, and reads

$$H = \sum_{\alpha\beta} t_{\alpha\beta} \, c_\alpha^\dagger c_\beta + \frac{1}{2} \sum_{\alpha\beta\gamma\delta} U_{\alpha\beta\gamma\delta} \, c_\alpha^\dagger c_\beta^\dagger c_\gamma \, c_\delta \, . \tag{3.13}$$

Our scope is to find the basis $\{\phi_i(x)\}$ which solves the Hartree-Fock equations (3.9). The Hartree-Fock wavefunction (3.1), being a Slater determinant, should be the ground state of a single-particle Hamiltonian, which we define as $\hat{H}_{HF}$. We write such a Hamiltonian in the following general form:

$$\hat{H}_{HF} = \sum_{\alpha\beta} h_{\alpha\beta} \, c_\alpha^\dagger c_\beta \, , \tag{3.14}$$

where we have introduced a set of unknown variational parameters satisfying $h_{\alpha\beta} = h_{\beta\alpha}^*$ for the Hamiltonian to be Hermitian. The ignorance about the basis set $\{\phi_i(x)\}$ is reflected in the ignorance about the $h_{\alpha\beta}$'s. The Hartree-Fock wavefunction satisfies

$$\hat{H}_{HF} \mid \Phi_{HF}\rangle = E \mid \Phi_{HF}\rangle, \tag{3.15}$$

namely is the lowest energy eigenstate of $\hat{H}_{HF}$. We define

$$\Delta_{\alpha\beta} = \langle \Phi_{HF} \mid c_\alpha^\dagger c_\beta \mid \Phi_{HF}\rangle, \tag{3.16}$$

the expectation values of all bilinear operators on the wavefunction, which are therefore functional of the variational parameters $h_{\alpha\beta}$. Since

$$E = \langle \Phi_{HF} \mid \hat{H}_{HF} \mid \Phi_{HF}\rangle = \sum_{\alpha\beta} h_{\alpha\beta} \, \Delta_{\alpha\beta} \, , \tag{3.17}$$

it follows that

$$\frac{\partial E}{\partial h_{\mu\nu}} = \Delta_{\mu\nu} + \sum_{\alpha\beta} h_{\alpha\beta} \, \frac{\partial \Delta_{\alpha\beta}}{\partial h_{\mu\nu}} \, . \tag{3.18}$$

On the other hand, since $| \Phi_{HF} \rangle$ is an eigenstate of the Hamiltonian $\hat{H}_{HF}$, the Hellmann-Feynman theorem holds,[1] according to which

$$\frac{\partial E}{\partial h_{\mu\nu}} = \langle \Phi_{HF} \mid \frac{\partial \hat{H}_{HF}}{\partial h_{\mu\nu}} \mid \Phi_{HF} \rangle = \Delta_{\mu\nu} . \tag{3.19}$$

By comparing (3.18) with (3.19), we thus conclude that

$$\sum_{\alpha\beta} h_{\alpha\beta} \frac{\partial \Delta_{\alpha\beta}}{\partial h_{\mu\nu}} = 0 , \tag{3.20}$$

for any $h_{\mu\nu}$.

On the other hand, the expectation value of the original Hamiltonian (3.13) on the Hartree-Fock wavefunction is readily found to be

$$\begin{aligned} E_{HF} &= \langle \Phi_{HF} \mid \hat{H} \mid \Phi_{HF} \rangle \\ &= \sum_{\alpha\beta} t_{\alpha\beta} \, \Delta_{\alpha\beta} + \frac{1}{2} \sum_{\alpha\beta\gamma\delta} \Delta_{\alpha\beta} \Delta_{\gamma\delta} \Big(U_{\alpha\gamma\delta\beta} - U_{\alpha\gamma\beta\delta} \Big), \end{aligned} \tag{3.21}$$

which is also a functional of the $h_{\alpha\beta}$'s. Minimisation of E_{HF} requires as a necessary condition that

$$\frac{\partial E_{HF}}{\partial h_{\mu\nu}} = \sum_{\alpha\beta} \frac{\partial \Delta_{\alpha\beta}}{\partial h_{\mu\nu}} \left[t_{\alpha\beta} + \sum_{\gamma\delta} \Delta_{\gamma\delta} \Big(U_{\alpha\gamma\delta\beta} - U_{\alpha\gamma\beta\delta} \Big) \right] = 0. \tag{3.22}$$

The Hartree-Fock wavefunction, hence the variational parameters $h_{\mu\nu}$, must therefore satisfy both (3.20) and (3.22) for any $h_{\mu\nu}$, which thus implies that

$$h_{\alpha\beta} = t_{\alpha\beta} + \sum_{\gamma\delta} \Delta_{\gamma\delta}[h] \Big(U_{\alpha\gamma\delta\beta} - U_{\alpha\gamma\beta\delta} \Big) . \tag{3.23}$$

---

[1] The Hellmann-Feynman theorem states that, if $| \psi \rangle$ is the normalised eigenstate with eigenvalue E of a Hamiltonian $\hat{H}$ that depends on some parameter λ, then

$$\frac{\partial E}{\partial \lambda} = \langle \psi \mid \frac{\partial \hat{H}}{\partial \lambda} \mid \psi \rangle .$$

This theorem can be easily proved. Indeed

$$\begin{aligned} \frac{\partial E}{\partial \lambda} &= \langle \psi \mid \frac{\partial \hat{H}}{\partial \lambda} \mid \psi \rangle + \langle \frac{\partial \psi}{\partial \lambda} \mid \hat{H} \mid \psi \rangle + \langle \psi \mid \hat{H} \mid \frac{\partial \psi}{\partial \lambda} \rangle \\ &= \langle \psi \mid \frac{\partial \hat{H}}{\partial \lambda} \mid \psi \rangle + E \left(\langle \frac{\partial \psi}{\partial \lambda} \mid \psi \rangle + \langle \psi \mid \frac{\partial \psi}{\partial \lambda} \rangle \right) \\ &= \langle \psi \mid \frac{\partial \hat{H}}{\partial \lambda} \mid \psi \rangle + E \frac{\partial \langle \psi \mid \psi \rangle}{\partial \lambda} = \langle \psi \mid \frac{\partial \hat{H}}{\partial \lambda} \mid \psi \rangle , \end{aligned}$$

since $|\psi\rangle$ is normalised for any λ.

Since the average values $\Delta_{\gamma\delta}[h]$ are themselves functional of the h's, the above set of equations is actually a self-consistency equation. These equations are fully equivalent to solving the Hartree-Fock non-linear differential equations that are found in first quantisation with one major advantage. Indeed, the symmetry properties of $|\Phi_{HF}\rangle$ are completely determined by the parameters $h_{\alpha\beta}$. In turn, this implies that we can implement any desired symmetry operation $\hat{O}$ by imposing that

$$\left[\hat{O}, \hat{H}_{HF}\right] = 0,$$

which leads to simple conditions to be imposed on the $h_{\alpha\beta}$. For instance, if the quantum number $\alpha = (a, \sigma)$ includes an orbital index a and spin σ, and we would like to enforce full spin-rotational symmetry, then we must assume that

$$h_{(a\sigma)(b\sigma')} = \delta_{\sigma\sigma'}\, h_{ab}$$

does not depend on the spin indices. If we expect that spin $SU(2)$ symmetry is lowered down to $U(1)$, namely the rotation around the magnetisation axis, which we can always assume to be also the quantisation axis, then we have to impose that, while $h_{(a\sigma)(b\sigma)}$ are finite and depend on σ, $h_{(a\uparrow)(b\downarrow)}$ and $h_{(a\downarrow)(b\uparrow)}$ are identically zero.

We note that (3.23) could be also regarded as the definition of $h_{\alpha\beta}$ in terms of new variational parameters $\Delta_{\gamma\beta}$ that, only at the end, must be fixed in such a way that

$$\Delta_{\alpha\beta} = \langle \Phi_{HF} \mid c_{\alpha}^{\dagger} c_{\beta} \mid \Phi_{HF}\rangle, \tag{3.24}$$

the expectation value being done on a chosen eigenstate of

$$\hat{H}_{HF} = \sum_{\alpha\beta} c_{\alpha}^{\dagger} c_{\beta} \left[t_{\alpha\beta} + \sum_{\gamma\delta} \Delta_{\gamma\delta} \left(U_{\alpha\gamma\delta\beta} - U_{\alpha\gamma\beta\delta} \right) \right], \tag{3.25}$$

which depends parametrically on the Δ's. Moreover, it follows that the variational energy

$$\begin{aligned}
E_{HF}[\Delta] &= \langle \Phi_{HF} \mid \hat{H} \mid \Phi_{HF}\rangle \\
&= \langle \Phi_{HF} \mid \hat{H}_{HF} \mid \Phi_{HF}\rangle - \frac{1}{2} \sum_{\alpha\beta\gamma\delta} \Delta_{\alpha\beta}\, \Delta_{\gamma\delta} \left(U_{\alpha\gamma\delta\beta} - U_{\alpha\gamma\beta\delta} \right) \\
&= E\big[h[\Delta]\big] - \frac{1}{2} \sum_{\alpha\beta\gamma\delta} \Delta_{\alpha\beta}\, \Delta_{\gamma\delta} \left(U_{\alpha\gamma\delta\beta} - U_{\alpha\gamma\beta\delta} \right)
\end{aligned} \tag{3.26}$$

also becomes functional of the Δ's. Through (3.19) and (3.23), we find that

$$\frac{\partial E}{\partial \Delta_{\gamma\delta}} = \sum_{\alpha\beta} \frac{\partial E}{\partial h_{\alpha\beta}} \frac{\partial h_{\alpha\beta}}{\partial \Delta_{\gamma\delta}} = \sum_{\alpha\beta} \Delta_{\alpha\beta} \left(U_{\alpha\gamma\delta\beta} - U_{\alpha\gamma\beta\delta} \right),$$

which also implies

$$\frac{\partial E}{\partial \Delta_{\gamma\delta}} - \sum_{\alpha\beta} \Delta_{\alpha\beta} \left(U_{\alpha\gamma\delta\beta} - U_{\alpha\gamma\beta\delta} \right) = 0 \,,$$

or, equivalently,

$$\frac{\delta}{\delta\Delta_{\gamma\delta}} \left(E[\Delta] - \frac{1}{2} \sum_{\alpha\beta\alpha'\beta'} \Delta_{\alpha\beta} \Delta_{\alpha'\beta'} \left(U_{\alpha\alpha'\beta'\beta} - U_{\alpha\alpha'\beta\beta'} \right) \right) = 0 \,. \tag{3.27}$$

The equations (3.23), (3.25), (3.26), and (3.27) suggest two alternative ways of stating the Hartree-Fock variational problem, which are both easy to implement:

Given the interacting Hamiltonian

$$H = \sum_{\alpha\beta} t_{\alpha\beta} \, c_\alpha^\dagger c_\beta + \frac{1}{2} \sum_{\alpha\beta\gamma\delta} U_{\alpha\beta\gamma\delta} \, c_\alpha^\dagger c_\beta^\dagger c_\gamma c_\delta \,, \tag{3.28}$$

the Hartree-Fock variational wavefunction $| \Phi_{HF} \rangle$ is the ground state with energy E of the single-particle Hamiltonian

$$\hat{H}_{HF} = \sum_{\alpha\beta} t_{\alpha\beta} \, c_\alpha^\dagger c_\beta + \sum_{\alpha\beta\gamma\delta} c_\alpha^\dagger c_\beta \Delta_{\gamma\delta} \left(U_{\alpha\gamma\delta\beta} - U_{\alpha\gamma\beta\delta} \right) \,, \tag{3.29}$$

where the parameters $\Delta_{\alpha\beta}$ have to be determined self-consistently by imposing either of the following two conditions:

Condition 1.

$$\Delta_{\alpha\beta} = \langle \Phi_{HF} | c_\alpha^\dagger c_\beta | \Phi_{HF} \rangle \,. \tag{3.30}$$

Condition 2.

$$\frac{\delta}{\delta\Delta_{\mu\nu}} \left(E[\Delta] - \frac{1}{2} \sum_{\alpha\beta\gamma\delta} \Delta_{\alpha\beta} \Delta_{\gamma\delta} \left(U_{\alpha\gamma\delta\beta} - U_{\alpha\gamma\beta\delta} \right) \right) = 0 \,. \tag{3.31}$$

3.2   Hartree-Fock Approximation for Fermions at Finite Temperature

A variational Hartree-Fock approximation can be defined also at finite temperature T (we take $k_B = 1$).

Consider again the interacting Hamiltonian

$$H = \sum_{\alpha\beta} t_{\alpha\beta}\, c_\alpha^\dagger c_\beta + \frac{1}{2} \sum_{\alpha\beta\gamma\delta} U_{\alpha\beta\gamma\delta}\, c_\alpha^\dagger c_\beta^\dagger c_\gamma c_\delta\,, \tag{3.32}$$

represented in a given basis of single-particle wavefunctions, and define the free-energy functional

$$F[\rho, T] \equiv U[\rho] - T\, S[\rho] = \mathrm{Tr}\!\left(\rho\, H\right) + T\, \mathrm{Tr}\!\left(\rho\, \ln \rho\right), \tag{3.33}$$

where $U[\rho]$ is the internal energy, $S[\rho]$ the von Neumann entropy, and ρ a density matrix, i.e., a linear Hermitian and positive-definite operator in the many-body Hilbert space such that $\mathrm{Tr}(\rho) = 1$. The true free energy $F(T)$ of the system described by the Hamiltonian (3.32) satisfies

$$F(T) \equiv \min_\rho F[\rho, T]\,. \tag{3.34}$$

However, even though we do know what is the density matrix that minimises $F[\rho, T]$, namely, the Boltzmann distribution

$$\rho_{\min} \equiv \frac{e^{-\beta H}}{\mathrm{Tr}\!\left(e^{-\beta H}\right)}\,,$$

that knowledge is useless for any practical purpose, since $\rho_{\min}$ is a complicated multiparticle operator, which can be dealt with only at very high-temperature through an expansion in $\beta = 1/T \to 0$.

Alternatively, we could search for a minimum in a restricted subspace S of all density matrices. In that way, we can only find an upper limit to the true free energy, namely

$$\boxed{F(T) \leq \min_{\rho \in S} F[\rho, T]}\,. \tag{3.35}$$

Equation (3.35) is the desired variational principle at $T \neq 0$.

Since we can easily deal only with density matrices of non-interacting systems, we can take the subspace S as that of all Boltzmann distributions ρ_{HF} of non-interacting Hamiltonians, namely

$$\rho_{HF} \equiv \frac{e^{-\beta H_{HF}}}{\mathrm{Tr}\!\left(e^{-\beta H_{HF}}\right)}\,, \qquad H_{HF} = \sum_{\alpha\beta} h_{\alpha\beta}\, c_\alpha^\dagger c_\beta\,. \tag{3.36}$$

Therefore, the optimisation procedure amounts to search for the parameters $h_{\alpha\beta}$ that minimise $F[\rho, T]$ in (3.33). We note that

$$
\begin{aligned}
F[\rho_{HF}, T] &= \mathrm{Tr}\left(\rho_{HF}\, H\right) + T\,\mathrm{Tr}\left(\rho_{HF}\,\ln\rho_{HF}\right) \\
&= \mathrm{Tr}\left(\rho_{HF}\, H_{HF}\right) + T\,\mathrm{Tr}\left(\rho_{HF}\,\ln\rho_{HF}\right) + \mathrm{Tr}\left(\rho\left(H - H_{HF}\right)\right) \\
&= F_{HF}(T) + \mathrm{Tr}\left(\rho_{HF}\left(H - H_{HF}\right)\right) = F_{HF}(T) + \langle\left(H - H_{HF}\right)\rangle_{HF},
\end{aligned}
\tag{3.37}
$$

where

$$
F_{HF}(T) = -T\,\ln Z_{HF}, \qquad Z_{HF} = \mathrm{Tr}\left(e^{-\beta H_{HF}}\right)
\tag{3.38}
$$

is the true free energy of the system described by H_{HF}, and $\langle\ldots\rangle_{HF}$ is the thermal expectation value on the density matrix ρ_{HF}.

Since H_{HF} in (3.36) depends on the variational parameters $h_{\alpha\beta}$, so do ρ_{HF} and $F_{HF}(T)$. Therefore, the minimum must first of all be a saddle point, i.e., satisfy

$$
\frac{\partial F[\rho_{HF}, T]}{\partial h_{\alpha\beta}} = 0,
\tag{3.39}
$$

and, to be a minimum rather than a maximum, its Hessian, i.e., the tensor with components

$$
\mathcal{H}_{\alpha\beta,\gamma\delta} \equiv \frac{\partial^2 F[\rho_{HF}, T]}{\partial h_{\alpha\beta}\partial h_{\gamma\delta}},
\tag{3.40}
$$

has to be positive definite.

3.2.1 Saddle Point Solution

Let us start by solving the saddle point equation (3.39). We note that

$$
V \equiv H - H_{HF} = \frac{1}{2}\sum_{\alpha\beta\gamma\delta} U_{\alpha\beta\gamma\delta}\, c_\alpha^\dagger c_\beta^\dagger c_\gamma c_\delta + \sum_{\alpha\beta}\left(t_{\alpha\beta} - h_{\alpha\beta}\right) c_\alpha^\dagger c_\beta,
\tag{3.41}
$$

so that we must first learn how to calculate $\langle V \rangle_{HF}$. Since H_{HF} is quadratic in the fermionic operators, it can be diagonalised by some unitary transformation

$$
c_\alpha = \sum_i U_{\alpha i}\, c_i,
$$

after which

$$
H_{HF} = \sum_i \epsilon_i\, c_i^\dagger c_i.
$$

Any many-body eigenstate of H_{HF}, e.g., $|n\rangle$ with energy E_n, can be written as a single Slater determinant, i.e., a Fock state with $|n\rangle \equiv |\{n_i\}\rangle$. Through (3.36), the following property holds:

$$
\begin{aligned}
\langle c_\alpha^\dagger c_\beta^\dagger c_\gamma c_\delta \rangle_{HF} &= \frac{1}{Z_{HF}} \sum_{ijkl} U_{i\alpha}^\dagger U_{j\beta}^\dagger U_{\gamma k} U_{\delta l} \, \mathrm{Tr}\left(e^{-\beta H_{HF}} c_i^\dagger c_j^\dagger c_k c_l\right) \\
&= \frac{1}{Z_{HF}} \sum_{ijkl} U_{i\alpha}^\dagger U_{j\beta}^\dagger U_{\gamma k} U_{\delta l} \sum_n e^{-\beta E_n} \langle n | c_i^\dagger c_j^\dagger c_k c_l | n\rangle \\
&= \frac{1}{Z_{HF}} \sum_{ijkl} U_{i\alpha}^\dagger U_{j\beta}^\dagger U_{\gamma k} U_{\delta l} \\
&\quad \sum_n e^{-\beta E_n} \left(\delta_{il}\delta_{jk} - \delta_{ik}\delta_{jl}\right) \langle n | c_i^\dagger c_i | n\rangle \langle n | c_j^\dagger c_j | n\rangle \\
&= \sum_{ijkl} U_{i\alpha}^\dagger U_{j\beta}^\dagger U_{\gamma k} U_{\delta l} \left(\delta_{il}\delta_{jk} - \delta_{ik}\delta_{jl}\right) f_i(T) f_j(T) \\
&= \langle c_\alpha^\dagger c_\delta \rangle_{HF} \langle c_\beta^\dagger c_\gamma \rangle_{HF} - \langle c_\alpha^\dagger c_\gamma \rangle_{HF} \langle c_\beta^\dagger c_\delta \rangle_{HF} \\
&\equiv \Delta_{\alpha\delta}(T) \Delta_{\beta\gamma}(T) - \Delta_{\alpha\gamma}(T) \Delta_{\beta\delta}(T),
\end{aligned}
\tag{3.42}
$$

where[2]

$$
f_i(T) = \frac{1}{e^{\beta \epsilon_i} + 1}
$$

is the Fermi distribution function, and we have introduced the thermal expectation values over the Hartree-Fock density matrix

$$
\Delta_{\alpha\beta}(T) = \langle c_\alpha^\dagger c_\beta \rangle_{HF}.
$$

---

[2] In the previous equation, we make use of the following trivial result. If

$$
H = \sum_i \epsilon_i c_i^\dagger c_i = \sum_i \epsilon_i n_i,
$$

then

$$
\rho = \frac{e^{-\beta H}}{\mathrm{Tr}\left(e^{-\beta H}\right)} = \prod_i \rho_i, \quad \rho_i = \frac{e^{-\beta \epsilon_i n_i}}{\mathrm{Tr}\left(e^{-\beta \epsilon_i n_i}\right)} = \frac{e^{-\beta \epsilon_i n_i}}{1 + e^{-\beta \epsilon_i}},
$$

the last expression deriving from the Pauli principle, according to which n_i can be either zero or one. Therefore, e.g.,

$$
\langle n_i n_j \rangle = \mathrm{Tr}\left(\rho \, n_i n_j\right) = \mathrm{Tr}\left(\rho_i \, n_i\right) \mathrm{Tr}\left(\rho_j \, n_j\right) = f_i(T) f_j(T),
$$

since

$$
\mathrm{Tr}\left(\rho_i \, n_i\right) = \sum_{n_i=0,1} \frac{e^{-\beta \epsilon_i n_i}}{1 + e^{-\beta \epsilon_i}} = \frac{e^{-\beta \epsilon_i}}{1 + e^{-\beta \epsilon_i}} = \frac{1}{e^{\beta \epsilon_i} + 1} \equiv f_i(T).
$$

$$\langle c_1^\dagger c_2 c_3^\dagger c_4 \rangle = \langle c_1^\dagger c_2 c_3^\dagger c_4 \rangle + \langle c_1^\dagger c_2 c_3^\dagger c_4 \rangle$$
$$= \langle c_1^\dagger c_2 \rangle \langle c_3^\dagger c_4 \rangle + \langle c_1^\dagger c_4 \rangle \langle c_2^\dagger c_3 \rangle$$

$$\langle c_1^\dagger c_2^\dagger c_3 c_4 \rangle = \langle c_1^\dagger c_2^\dagger c_3 c_4 \rangle + \langle c_1^\dagger c_2^\dagger c_3 c_4 \rangle$$
$$= \langle c_1^\dagger c_4 \rangle \langle c_2^\dagger c_3 \rangle - \langle c_1^\dagger c_3 \rangle \langle c_2^\dagger c_4 \rangle$$

Fig. 3.1 Graphical way to determine the sign of the product of the contractions and the type of each of them shown for two cases. One has to join the two operators, one annihilation, e.g., c_i, and one creation, e.g., $c_j^\dagger$, which are contracted together by an arrow line directed from the annihilation to the creation. If the line goes from left to right the contraction means $\langle c_i c_j^\dagger \rangle$, while, if the line goes from the right to the left, the contraction means $\langle c_j^\dagger c_i \rangle$. The sign of each product of contractions is $(-1)^n$, where n is the number of crossings of the lines, which is zero in all shown cases but in the last one, when $n = 1$, the green dot in the figure

The previous result is a special case of the following general result:

The thermal average of a product of n creation and n annihilation operators with a bilinear non-interacting Hamiltonian H_0, namely

$$\langle c_{i_1}^\dagger c_{i_2}^\dagger \ldots c_{i_n}^\dagger c_{j_1} c_{j_2} \ldots c_{j_n} \rangle_0 = \frac{1}{Z_0} \mathrm{Tr} \left(e^{-\beta H_0} c_{i_1}^\dagger c_{i_2}^\dagger \ldots c_{i_n}^\dagger c_{j_1} c_{j_2} \ldots c_{j_n} \right)$$
$$= \sum_P (-1)^P \langle c_{i_1}^\dagger c_{j_{P_1}} \rangle_0 \langle c_{i_2}^\dagger c_{j_{P_2}} \rangle_0 \ldots \langle c_{i_n}^\dagger c_{j_{P_n}} \rangle_0 ,$$

$$(3.43)$$

where $(j_{P_1}, j_{P_2}, \ldots, j_{P_n}) = P(j_1, j_2, \ldots, j_n)$ is a permutation of the n j-indices, and P is the order of the permutation.

The most general case where the operators are not ordered in such a way that creation operators appear on the left can be straightforwardly derived from the above result. Essentially one has to consider all possible contractions between a creation, $c_i^\dagger$, and an annihilation, c_j, operator. If the former is on the left of the latter, the contraction means

$$\langle c_i^\dagger c_j \rangle_0;$$

otherwise, it means

$$\langle c_j c_i^\dagger \rangle_0 = \delta_{ij} - \langle c_i^\dagger c_j \rangle_0 .$$

The sign of a given product of contractions is plus if the number of fermionic hops one has to perform so to bring each pair of operators to be contracted close together is even, and minus otherwise. A simple graphical implementation of those rules is shown in Fig. 3.1.

We mention that the above rule is required to demonstrate the so-called Wick's theorem, which we shall discuss later.

Coming back to our original task, we need to solve the saddle point equation (3.39), namely

$$\frac{\partial F[\rho_{HF}, T]}{\partial h_{\alpha\beta}} = \frac{\partial F_{HF}(T)}{\partial h_{\alpha\beta}} + \frac{\partial \langle V \rangle_{HF}}{\partial h_{\alpha\beta}} = 0 . \tag{3.44}$$

Let us start from the first term on the right-hand side. We note that

$$\frac{\partial F_{HF}(T)}{\partial h_{\alpha\beta}} = -T \frac{\partial \ln Z_{HF}(T)}{\partial h_{\alpha\beta}} = -\frac{T}{Z_{HF}} \frac{\partial Z_{HF}}{\partial h_{\alpha\beta}} . \tag{3.45}$$

On the other hand, if E_n are the eigenvalues of the many-body eigenstates $\mid n \rangle$ of $\hat{H}_{HF}$, then

$$\frac{\partial Z_{HF}}{\partial h_{\alpha\beta}} = \sum_n \frac{\partial e^{-\beta E_n}}{\partial h_{\alpha\beta}}$$

$$= -\beta \sum_n e^{-\beta E_n} \frac{\partial E_n}{\partial h_{\alpha\beta}} = -\beta \sum_n e^{-\beta E_n} \langle n \mid \frac{\partial \hat{H}_{HF}}{\partial h_{\alpha\beta}} \mid n \rangle \tag{3.46}$$

$$= -\beta \sum_n e^{-\beta E_n} \langle n \mid c_\alpha^\dagger c_\beta \mid n \rangle = -\beta Z_{HF} \Delta_{\alpha\beta}(T),$$

where we made use of the aforementioned Hellmann-Feynman theorem. Therefore, through (3.45) and (3.46), we find that

$$\frac{\partial F_{HF}(T)}{\partial h_{\alpha\beta}} = \Delta_{\alpha\beta}(T), \tag{3.47}$$

a simple extension of the Hellmann-Feynman theorem at finite temperature.

Considering now the last term on the right-hand side of (3.44), we find, by means of (3.42), that before the derivative it reads

$$\langle \hat{V} \rangle_{HF} = \frac{1}{2} \sum_{\alpha\beta\gamma\delta} \Delta_{\alpha\beta}(T) \Delta_{\gamma\delta}(T) \left(U_{\alpha\gamma\delta\beta} - U_{\alpha\gamma\beta\delta} \right) + \sum_{\alpha\beta} \left(t_{\alpha\beta} - h_{\alpha\beta} \right) \Delta_{\alpha\beta}(T),$$

hence

$$\frac{\partial \langle V \rangle_{HF}}{\partial h_{\alpha\beta}} = -\Delta_{\alpha\beta}(T) + \sum_{\gamma\delta} \left(t_{\gamma\delta} - h_{\gamma\delta} \right) \frac{\partial \Delta_{\gamma\delta}(T)}{\partial h_{\alpha\beta}}$$

$$+ \sum_{\gamma\delta\mu\nu} \frac{\partial \Delta_{\gamma\delta}(T)}{\partial h_{\alpha\beta}} \Delta_{\mu\nu}(T) \left(U_{\gamma\mu\nu\delta} - U_{\gamma\mu\delta\nu} \right). \tag{3.48}$$

The sum of (3.47) and (3.48) yields

$$\sum_{\gamma\delta} \frac{\partial \Delta_{\gamma\delta}(T)}{\partial h_{\alpha\beta}} \left[t_{\gamma\delta} - h_{\gamma\delta} + \sum_{\mu\nu} \Delta_{\mu\nu}(T) \left(U_{\gamma\mu\nu\delta} - U_{\gamma\mu\delta\nu} \right) \right] = 0, \quad (3.49)$$

which must hold for every pair of indices $(\alpha\beta)$. The solution of this equation thus reads

$$h_{\gamma\delta} = t_{\gamma\delta} + \sum_{\mu\nu} \Delta_{\mu\nu}(T) \left(U_{\gamma\mu\nu\delta} - U_{\gamma\mu\delta\nu} \right)$$
$$= t_{\gamma\delta} + \frac{1}{Z_{HF}} \sum_{\mu\nu} \left(U_{\gamma\mu\nu\delta} - U_{\gamma\mu\delta\nu} \right) \mathrm{Tr}\left(e^{-\beta\hat{H}_{HF}} c_\mu^\dagger c_\nu \right). \quad (3.50)$$

Therefore, the single-particle Hamiltonian H_{HF} that minimises $F[\rho_{HF}, T]$, hence providing an upper bound to the exact free energy, is defined through variational parameters $h_{\alpha\beta}$ that have to be determined self-consistently through (3.50). We note that this equation is just the extension of the Hartree-Fock Hamiltonian (3.14) at finite temperature, where the variational parameters correspond now to thermal averages and not anymore to ground state averages.

Correspondingly, if we instead write

$$H_{HF} = \sum_{\alpha\beta} \left(t_{\alpha\beta} + \sum_{\gamma\delta} \Delta_{\gamma\delta} \left(U_{\alpha\gamma\delta\beta} - U_{\alpha\gamma\beta\delta} \right) \right) c_\alpha^\dagger c_\beta, \quad (3.51)$$

as functional of the parameters $\Delta_{\alpha\beta}$, so that

$$F[\rho_{HF}, T] = F_{HF}(T) + \langle \hat{V} \rangle_{HF}$$
$$= F_{HF}(T) - \frac{1}{2} \sum_{\alpha\beta\gamma\delta} \Delta_{\alpha\beta} \Delta_{\gamma\delta} \left(U_{\alpha\gamma\delta\beta} - U_{\alpha\gamma\beta\delta} \right) \quad (3.52)$$

also becomes functional of the $\Delta_{\alpha\beta}$'s, then the saddle point corresponds to imposing

$$\Delta_{\alpha\beta} = \langle c_\alpha^\dagger c_\beta \rangle_{HF} = \frac{\mathrm{Tr}\left(e^{-\beta H_{HF}} c_\alpha^\dagger c_\beta \right)}{\mathrm{Tr}\left(e^{-\beta H_{HF}} \right)}, \quad (3.53)$$

or, equivalently, to the solution of

$$\frac{\partial F[\rho_{HF}, T]}{\partial \Delta_{\alpha\beta}} = 0, \quad (3.54)$$

which are just the finite temperature generalisation of (3.30) and (3.31).

However, we still need to determine under which conditions the above saddle point is a minimum. Through (3.49), we readily find that the elements of the Hessian matrix are

$$\frac{\partial^2 F[\rho_{HF}, T]}{\partial h_{\alpha\beta} \partial h_{\eta\xi}} = \frac{\partial}{\partial h_{\eta\xi}} \left\{ \sum_{\gamma\delta} \frac{\partial \Delta_{\gamma\delta}(T)}{\partial h_{\alpha\beta}} \left[t_{\gamma\delta} - h_{\gamma\delta} + \sum_{\mu\nu} \Delta_{\mu\nu}(T) \left(U_{\gamma\mu\nu\delta} - U_{\gamma\mu\delta\nu} \right) \right] \right\}.$$

The derivative with respect to $h_{\eta\xi}$ of $\partial\Delta_{\gamma\delta}(T)/\partial h_{\alpha\beta}$ vanishes since the term in square brackets is zero at the saddle point. It follows that

$$\begin{aligned}
\frac{\partial^2 F[\rho_{HF}, T]}{\partial h_{\alpha\beta} \partial h_{\eta\xi}} &= \sum_{\gamma\delta} \frac{\partial \Delta_{\gamma\delta}(T)}{\partial h_{\alpha\beta}} \left[-\frac{\partial h_{\gamma\delta}}{\partial h_{\eta\xi}} + \sum_{\mu\nu} \frac{\partial \Delta_{\mu\nu}(T)}{\partial h_{\eta\xi}} \left(U_{\gamma\mu\nu\delta} - U_{\gamma\mu\delta\nu} \right) \right] \\
&= -\frac{\partial \Delta_{\eta\xi}(T)}{\partial h_{\alpha\beta}} + \sum_{\gamma\delta\mu\nu} \frac{\partial \Delta_{\gamma\delta}(T)}{\partial h_{\alpha\beta}} \frac{\partial \Delta_{\mu\nu}(T)}{\partial h_{\eta\xi}} \left(U_{\gamma\mu\nu\delta} - U_{\gamma\mu\delta\nu} \right) \\
&= -\frac{\partial^2 F_{HF}(T)}{\partial h_{\alpha\beta} \partial h_{\eta\xi}} + \sum_{\gamma\delta\mu\nu} \frac{\partial^2 F_{HF}(T)}{\partial h_{\alpha\beta} \partial h_{\gamma\delta}} \frac{\partial^2 F_{HF}(T)}{\partial h_{\mu\nu} \partial h_{\eta\xi}} \left(U_{\gamma\mu\nu\delta} - U_{\gamma\mu\delta\nu} \right).
\end{aligned}$$

We note that

$$-\frac{\partial^2 F_{HF}(T)}{\partial h_{\alpha\beta} \partial h_{\eta\xi}} \equiv \kappa_{\alpha\beta,\eta\xi}^{HF} \tag{3.55}$$

are the components of the thermodynamic susceptibility tensor $\hat{\kappa}^{HF}$ of the Hartree-Fock Hamiltonian, which is positive definite. Therefore, the Hessian components are defined through

$$\mathcal{H}_{\alpha\beta,\eta\xi} = \kappa_{\alpha\beta,\eta\xi}^{HF} + \sum_{\gamma\delta\mu\nu} \kappa_{\alpha\beta,\gamma\delta}^{HF} \left(U_{\gamma\mu\nu\delta} - U_{\gamma\mu\delta\nu} \right) \kappa_{\mu\nu,\eta\xi}^{HF},$$

thus, in matrix notations,

$$\hat{\mathcal{H}} = \hat{\kappa}^{HF} + \hat{\kappa}^{HF} \hat{U} \hat{\kappa}^{HF}, \tag{3.56}$$

which also defines the tensor $\hat{U}$. It follows that the saddle point is a minimum if $\hat{\mathcal{H}}$ in (3.56) is positive definite, which is surely the case if $\hat{U} \hat{\kappa}^{HF}$ is a weak correction to the identity matrix.

We end noticing that the Hartree-Fock Hamiltonian H_{HF} represents non-interacting electrons in the presence of a *fictitious* external field that, through (3.50), is indeed the self-consistent field generated by the same electrons. This is the reason why the Hartree-Fock approximation is essentially a mean-field approximation.

3.3   Time-Dependent Hartree-Fock Approximation

Suppose we have solved the time-independent Hartree-Fock problem, namely, we have found a set of single-particle wavefunctions $\phi_i(x)$ that diagonalise

$$t_{ij} + \sum_{k=1}^{N} \left(U_{ikkj} - U_{ikjk} \right) = \epsilon_i \, \delta_{ij} . \tag{3.57}$$

The Hartree-Fock wavefunction is just the Fock state

$$|\Phi_{HF}\rangle = \prod_{i=1}^{N} c_i^\dagger \, | \, 0 \rangle , \tag{3.58}$$

where $c_i^\dagger$ creates a fermion in state ϕ_i. The original Hamiltonian in the Hartree-Fock basis is given by

$$H = \sum_{ij} t_{ij} c_i^\dagger c_j + \frac{1}{2} \sum_{ijkl} U_{ij,kl} \, c_i^\dagger c_j^\dagger c_k c_l . \tag{3.59}$$

Let us suppose to perturb the Hamiltonian by a time-dependent perturbation

$$V(t) = \sum_{ij} V_{ij}(t) \, c_i^\dagger c_j \tag{3.60}$$

and study the response of the system to linear order. Therefore, the full Hamiltonian $H(t) = H + V(t)$ is time-dependent and the variational principle now involves the time-dependent Schrœdinger equation. Namely, we are now going to search within the subspace of time-dependent Slater determinants for a $|\Phi_{HF}(t)\rangle$ that satisfies

$$\delta \left[\frac{\langle \Phi_{HF}(t) \mid i\hbar \dfrac{\partial}{\partial t} - H(t) \mid \Phi_{HF}(t) \rangle}{\langle \Phi_{HF}(t) \mid \Phi_{HF}(t) \rangle} \right] = 0 . \tag{3.61}$$

We assume that the time-dependent Slater determinant is build by N single-particle normalised wavefunctions $\phi_\alpha(x, t)$. By analogy with the conventional derivation of the time-independent Hartree-Fock equations, one finds that

$$i\hbar \frac{\partial}{\partial t} \phi_\alpha(x, t) = [T(x, p) + V(x, p, t)] \, \phi_\alpha(x, t)$$

$$+ \sum_{\beta=1}^{N} \int dy \, U(x, y) \left[\, |\phi_\beta(y, t)|^2 \, \phi_\alpha(x, t) - \phi_\beta(y, t)^* \, \phi_\beta(x, t) \, \phi_\alpha(y, t) \right] , \tag{3.62}$$

where $T(x, p)$ is the single-particle term of the Hamiltonian, $V(x, p, t)$ the perturbation, and $U(x, y)$ the interaction. We assume that the $\phi_\alpha(x, t)$'s are related to the $\phi_i(x)$'s by the time-dependent unitary transformation

$$\phi_\alpha(x, t) = \sum_i U_{i\alpha}(t)\, \phi_i(x) , \quad U_{i\alpha}(t) = \int dx\, \phi_i(x)^* \, \phi_\alpha(x, t) ,$$

which, exploiting the invariance of the Fermi fields, imply the following transformations of the corresponding annihilation operators:

$$c_i = \sum_\alpha U_{i\alpha}(t)\, c_\alpha , \quad c_i^\dagger = \sum_\alpha U_{i\alpha}(t)^* \, c_\alpha^\dagger . \tag{3.63}$$

In terms of the matrix elements, the time-dependent equations (3.62) read

$$
\begin{aligned}
i\hbar \frac{\partial}{\partial t} U_{i\alpha}(t) &= \sum_j \left[t_{ij} + V_{ij}(t) \right] U_{j\alpha}(t) \\
&\quad + \sum_{\beta=1}^{N} \sum_{jkl} U_{k\beta}(t)^* \, U_{l\beta}(t) \, U_{j\alpha}(t) \left(U_{iklj} - U_{ikjl} \right), \\
i\hbar \frac{\partial}{\partial t} U_{i\alpha}(t)^* &= -\sum_j \left[t_{ji} + V_{ji}(t) \right] U_{j\alpha}(t)^* \\
&\quad - \sum_{\beta=1}^{N} \sum_{jkl} U_{k\beta}(t) \, U_{l\beta}(t)^* \, U_{j\alpha}(t)^* \left(U_{jlki} - U_{ljki} \right).
\end{aligned}
\tag{3.64}
$$

Through (3.63), the time-dependent Hartree-Fock wavefunction is uniquely identified by the expectation values

$$\Delta_{ij}(t) = \langle \Phi_{HF}(t) \mid c_i^\dagger c_j \mid \Phi_{HF}(t) \rangle = \sum_{\alpha=1}^{N} U_{i\alpha}(t)^* \, U_{j\alpha}(t) , \tag{3.65}$$

which, calculated on the time-independent wavefunction (3.58), are simply

$$\Delta_{ij}^{(0)} = \delta_{ij}\, n_i , \tag{3.66}$$

with $n_i = 1$ if $i \leq N$ and $n_i = 0$ otherwise. Through (3.64) and (3.65), we obtain the following equations for the Δ_{ij}'s:

$$
\begin{aligned}
i\hbar \frac{\partial}{\partial t} \Delta_{ij}(t) &= \sum_k \left[-\left(t_{ki} + V_{ki}(t) \right) \Delta_{kj}(t) + \left(t_{jk} + V_{jk}(t) \right) \Delta_{ik}(t) \right] \\
&\quad + \sum_{klm} \left[-\Delta_{kl}(t)\, \Delta_{mj}(t) \left(U_{mkli} - U_{kmli} \right) \right. \\
&\quad\quad\quad\quad \left. + \Delta_{il}(t)\, \Delta_{km}(t) \left(U_{jkml} - U_{jklm} \right) \right].
\end{aligned}
\tag{3.67}
$$

We solve the above equation up to the first order in the perturbation $V(t)$. One can readily check that the zeroth order time-independent term is indeed given by (3.66). The first-order terms instead satisfy the equations

$$
i\hbar \frac{\partial}{\partial t} \Delta_{ij}^{(1)}(t) = \sum_k - \left(t_{ki} + \sum_{m=1}^{N} \left(U_{kmmi} - U_{mkmi} \right) \right) \Delta_{kj}^{(1)}(t)
$$

$$
+ \sum_k \left(t_{jk} + \sum_{m=1}^{n} \left(U_{jmmk} - U_{jmkm} \right) \right) \Delta_{ik}^{(1)}(t)
$$

$$
+ \left(n_i - n_j \right) \sum_{lm} \Delta_{ml}^{(1)}(t) \left(U_{jmli} - U_{jmil} \right) + V_{ji}(t) \left(n_i - n_j \right).
$$

$$(3.68)$$

Through (3.57) we finally obtain that, for any pair of indices i and j, the following equation must be satisfied:

$$
\left(i\hbar \frac{\partial}{\partial t} - \epsilon_j + \epsilon_i \right) \Delta_{ij}^{(1)}(t) = \left(n_i - n_j \right) \left[V_{ji}(t) + \sum_{kl} \Delta_{kl}^{(1)}(t) \left(U_{jkli} - U_{jkil} \right) \right],
$$

$$(3.69)$$

which is the desired result. We note that (3.69) is just the equation of motion of non-interacting electrons described by the diagonalised Hartree-Fock Hamiltonian

$$
H_{HF} = \sum_i \epsilon_i c_i^\dagger c_i ,
$$

$$(3.70)$$

at linear order in a self-consistent time-dependent potential

$$
V_*(t) = \sum_{ij} V_{*ij}(t) c_i^\dagger c_i , \quad V_{*ij}(t) \equiv V_{ji}(t) + \sum_{kl} \Delta_{kl}^{(1)}(t) \left(U_{jkli} - U_{jkil} \right),
$$

$$(3.71)$$

once again showing the mean-field character of the approximation. Indeed, since

$$
i\hbar \frac{\partial c_j}{\partial t} = \left[c_j , H_{HF} + V_*(t) \right] = \epsilon_j c_j + \sum_k V_{*jk}(t) c_k ,
$$

$$
i\hbar \frac{\partial c_i^\dagger}{\partial t} = \left[c_i^\dagger , H_{HF} + V_*(t) \right] = -\epsilon_i c_i^\dagger + \sum_k V_{*ki}(t) c_k^\dagger ,
$$

then

$$
i\hbar \frac{\partial \langle c_i^\dagger c_j \rangle}{\partial t} \equiv i\hbar \frac{\partial \Delta_{ij}(t)}{\partial t} = \left(\epsilon_j - \epsilon_i \right) \Delta_{ij}(t) + \sum_k \left(V_{*jk}(t) \Delta_{ik}(t) - V_{*ki}(t) \Delta_{kj}(t) \right).
$$

At zeroth order in V, $\Delta_{ij}^{(0)} = \delta_{ij} n_i$, while consistently at first order,

$$i\hbar \frac{\partial \Delta_{ij}^{(1)}(t)}{\partial t} = \left(\epsilon_j - \epsilon_i\right) \Delta_{ij}^{(1)}(t) + \sum_k \left(V_{*jk}^{(1)}(t) \Delta_{ik}^{(0)}(t) - V_{*ki}^{(1)}(t) \Delta_{kj}^{(0)}(t)\right)$$

$$= \left(\epsilon_j - \epsilon_i\right) \Delta_{ij}^{(1)}(t) + \left(n_i - n_j\right) V_{*ji}^{(1)}(t) \,,$$

(3.72)

where

$$V_{*ji}^{(1)}(t) = V_{ji}(t) + \sum_{kl} \Delta_{kl}^{(1)}(t) \left(U_{jkli} - U_{jkil}\right) .$$

Indeed, (3.72) coincides with (3.69). This observation allows us to easily calculate linear response functions within the time-dependent Hartree-Fock approximation. Let us consider again the non-interacting Hamiltonian H_{HF} perturbed by $V_*(t)$. The linear response function of H_{HF} is by definition

$$\chi_{ij,kl}^{HF}(t) = -i\,\theta(t)\,\langle\, \left[c_i^\dagger(t)\, c_j(t)\,,\ c_k^\dagger c_l \right] \rangle$$

and is generally a tensor. Its Fourier transform in frequency can be readily obtained through (3.72), and it is

$$\chi_{ij,kl}^{HF}(\omega) = \delta_{il}\,\delta_{jk}\,\frac{n_i - n_j}{\omega - \epsilon_j + \epsilon_i} \equiv \delta_{il}\,\delta_{jk}\,\chi_{ij}^{HF}(\omega)\,,$$

(3.73)

where ω has an infinitesimal imaginary part $\eta > 0$ that enforces causality. It follows that, in linear response,

$$\Delta_{ij}^{(1)}(\omega) = \sum_{kl} \chi_{ij,kl}^{HF}(\omega)\, V_{*kl}^{(1)}(\omega) = \chi_{ij}^{HF}(\omega)\, V_{*ji}^{(1)}(\omega)$$

$$= \chi_{ij}^{HF}(\omega) \left[V_{ji}(\omega) + \sum_{kl} \Delta_{kl}^{(1)}(t) \left(U_{jkli} - U_{jkil}\right) \right],$$

namely

$$\sum_{kl} \left[\delta_{ij,kl} - \chi_{ij}^{HF}(\omega) \left(U_{jkli} - U_{jkil}\right) \right] \Delta_{kl}^{(1)}(\omega) = \chi_{ij}^{HF}(\omega)\, V_{ji}(\omega)\,,$$

which, upon defining the vector, $\mathbf{\Delta}(\omega)$ and $\mathbf{V}(\omega)$ with components $\Delta_{ij}^{(1)}(\omega)$ and $V_{ij}(\omega)$, respectively, and the tensors $\hat{\chi}^{HF}(\omega)$ and $\hat{Q}(\omega) \equiv \hat{I} - \hat{\chi}^{HF}(\omega)\,\hat{U}$ with components, respectively,

$$\chi_{ij,kl}^{HF}(\omega) = \delta_{il}\,\delta_{jk}\,\chi_{ij}^{HF}(\omega)\,, \quad Q_{ij,kl}(\omega) = \delta_{ij,kl} - \chi_{ij}^{HF}(\omega) \left(U_{jkli} - U_{jkil}\right),$$

can be shortly written as

$$\hat{Q}(\omega)\,\Delta(\omega) = \hat{\chi}^{HF}(\omega)\,V(\omega),$$

namely,

$$\Delta(\omega) = \hat{Q}(\omega)^{-1}\,\hat{\chi}^{HF}(\omega)\,V(\omega) = \left[\,\hat{I} - \hat{\chi}^{HF}(\omega)\,\hat{U}\,\right]^{-1}\hat{\chi}^{HF}(\omega)\,V(\omega)$$

$$\equiv \hat{\chi}^{\,tD-HF}(\omega)\,V(\omega),$$

(3.74)

which defines the linear response functions within the time-dependent Hartree-Fock approximation in terms of the simple Hartree-Fock one $\hat{\chi}^{HF}(\omega)$.

We have till now worked at zero temperature, $T = 0$. It is straightforward to extend all previous results at $T \neq 0$ and find that the only difference is the expression of $\chi_{ij}^{HF}(\omega)$ in (3.73), which at finite temperature becomes

$$\chi_{ij}^{HF}(\omega) = \frac{n_i(T) - n_j(T)}{\omega - \epsilon_j(T) + \epsilon_i(T)},$$

(3.75)

where $\epsilon_i(T)$ and $\epsilon_j(T)$ are eigenvalues of the Hartree-Fock Hamiltonian at $T \neq 0$, and

$$n_i(T) = \frac{1}{e^{-\beta\epsilon_i(T)} + 1}.$$

We finally observe that the time-dependent Hartree-Fock approximation, being based on the variational principle directly applied to the Schrœding equation, does not spoil the symmetries of the interacting Hamiltonian, evidently preserved during the time evolution, unlike what may happen within the simple Hartree-Fock approximation. For that reason, the time-dependent Hartree-Fock approximation is the simplest symmetry-conserving approximation scheme, and it is historically named as the random-phase approximation (RPA).

3.3.1   Bosonization of the Low-Energy Particle-Hole Excitations

We note from (3.69) that the only expectation values that are affected at linear order by the perturbation are those where either $n_i = 1$ and $n_j = 0$, or viceversa, namely where one of the index refers to an occupied state within (3.58) and the other to an unoccupied one. Let us denote by greek letters $\alpha, \beta, \cdots > N$ the unoccupied states, hereafter dubbed 'particles', and by roman letters $a, b, \cdots \leq N$ the occupied ones, named 'holes'. We denote by

$$b_{\alpha a}^{\dagger} = c_{\alpha}^{\dagger}\,c_a,$$

(3.76)

the particle-hole creation operator, which destroys one electron, i.e., creates one hole, inside the Slater determinant, and creates it outside; as well as its Hermitian conjugate

$$b_{\alpha a} = c_a^{\dagger}\,c_{\alpha}.$$

(3.77)

The commutators

$$\left[b_{\alpha a}, b_{\beta b}\right] = \left[b_{\alpha a}^{\dagger}, b_{\beta b}^{\dagger}\right] = 0$$

vanish, while

$$\left[b_{\alpha a}, b_{\beta b}^{\dagger}\right] = \delta_{\alpha\beta}\, c_a^{\dagger}\, c_b - \delta_{ab}\, c_{\beta}^{\dagger}\, c_{\alpha}$$

has expectation value on (3.58)

$$\langle\Phi_{HF} \mid \left[b_{\alpha a}, b_{\beta b}^{\dagger}\right] \mid \Phi_{HF}\rangle = \delta_{\alpha\beta}\,\delta_{ab}\,(n_a - n_\alpha) = \delta_{\alpha\beta}\delta_{ab}\,,$$

as if the particle-hole creation and annihilation operators were bosonic particles. If we assume that, at linear order in $V(t)$, the Hartree-Fock wavefunction is very slightly modified with respect to its time-independent value; in other words, we make a similar assumption as we did in deriving spin-wave theory, then we can approximate the commutator by its average value, thus obtaining

$$\left[b_{\alpha a}, b_{\beta b}^{\dagger}\right] = \delta_{\alpha\beta}\,\delta_{ab}\,, \quad \left[b_{\alpha a}, b_{\beta b}\right] = \left[b_{\alpha a}^{\dagger}, b_{\beta b}^{\dagger}\right] = 0\,, \qquad (3.78)$$

which are indeed bosonic commutation relations. We observe that in this approximation the bosonic vacuum is actually the time-independent Hartree-Fock wavefunction, whereas, in the spin-wave approximation, it is the classical spin configuration. Since

$$\Delta_{a\alpha}(t) = \langle\Phi_{HF}(t) \mid b_{\alpha a} \mid \Phi_{HF}(t)\rangle\,,$$

Equation (3.69) corresponds to the following equations of motion for the bosonic operators:

$$\left(i\hbar\frac{\partial}{\partial t} - \epsilon_\alpha + \epsilon_a\right) b_{\alpha a} = V_{\alpha a}(t)$$
$$+ \sum_{\beta b}\left[b_{\beta b}\left(U_{\alpha b\beta a} - U_{\alpha b a\beta}\right) + b_{\beta b}^{\dagger}\left(U_{\alpha\beta b a} - U_{\alpha\beta a b}\right)\right]. \qquad (3.79)$$

We may now ask the following question: What would be a bosonic Hamiltonian leading to the above equation of motion? The answer can be readily found and reads

$$\sum_{\alpha a}\left(\epsilon_\alpha - \epsilon_a\right) b_{\alpha a}^{\dagger}\,b_{\alpha a} + \sum_{\alpha\beta ab} b_{\alpha a}^{\dagger}\,b_{\beta b}\left(U_{\alpha b\beta a} - U_{\alpha b a\beta}\right)$$
$$+ \frac{1}{2}\sum_{\alpha\beta ab}\left[b_{\alpha a}^{\dagger}\,b_{\beta b}^{\dagger}\left(U_{\alpha\beta b a} - U_{\alpha\beta a b}\right) + b_{\beta b}b_{\alpha a}\left(U_{ab\beta\alpha} - U_{ab\alpha\beta}\right)\right] \qquad (3.80)$$
$$+ \sum_{\alpha a}\left[V_{\alpha a}(t)\,b_{\alpha a}^{\dagger} + V_{a\alpha}(t)\,b_{\alpha a}\right].$$

It is now clear that the external field simply probes particle-hole excitations, which have their own dynamics provided by the Hamiltonian

$$
H_{tD-HF} = \sum_{\alpha a} \left(\epsilon_\alpha - \epsilon_a \right) b^\dagger_{\alpha a} b_{\alpha a} + \sum_{\alpha\beta ab} b^\dagger_{\alpha a} b_{\beta b} \left(U_{\alpha b\beta a} - U_{\alpha b a\beta} \right)
$$
$$
+ \frac{1}{2} \sum_{\alpha\beta ab} \left[b^\dagger_{\alpha a} b^\dagger_{\beta b} \left(U_{\alpha\beta ba} - U_{\alpha\beta ab} \right) + b_{\beta b} b_{\alpha a} \left(U_{ab\beta a} - U_{ab\alpha\beta} \right) \right].
$$

$$(3.81)$$

Due to the presence of the last two terms, the true ground state of (3.81) is not the vacuum, namely the Hartree-Fock wavefunction, but another state

$$
| \Phi_{tD-HF} \rangle = e^{iA} | \Phi_{HF} \rangle ,
$$

with A an Hermitian operator quadratic in the bosonic operators. The above wavefunction is actually an improvement of the Hartree-Fock one that includes zero-point fluctuations of the particle-hole excitations.

The above results can be rederived, still in the spirit of the spin-wave theory, without even invoking the time-dependent variational principle. If we assume, to be checked a posteriori, that the Hartree-Fock wavefunction is not strongly modified by the inclusion of quantum fluctuations, we can assume that the commutators

$$
\left[c^\dagger_a c_\alpha , c^\dagger_\beta c_b \right] = \delta_{\alpha\beta} c^\dagger_a c_b - \delta_{ab} c^\dagger_\beta c_\alpha , \qquad \left[c^\dagger_\alpha c_a , c^\dagger_b c_\beta \right] = \delta_{ab} c^\dagger_\alpha c_\beta - \delta_{\alpha\beta} c^\dagger_b c_a ,
$$
$$
\left[c^\dagger_a c_b , c^\dagger_c c_d \right] = \delta_{bc} c^\dagger_a c_d - \delta_{ad} c^\dagger_c c_b , \qquad \left[c^\dagger_\alpha c_\beta , c^\dagger_\gamma c_\delta \right] = \delta_{\beta\gamma} c^\dagger_\alpha c_\delta - \delta_{\alpha\delta} c^\dagger_\gamma c_\beta
$$

can be effectively substituted by their expectation values on the Hartree-Fock wavefunction, i.e.,

$$
c^\dagger_a c_b \simeq \langle c^\dagger_a c_b \rangle_{HF} = \delta_{ab} , \qquad c^\dagger_\alpha c_\beta \simeq \langle c^\dagger_\alpha c_\beta \rangle_{HF} = 0 ,
$$

so that

$$
\left[c^\dagger_a c_\alpha , c^\dagger_\beta c_b \right] \simeq \delta_{\alpha\beta} \delta_{ab} , \qquad \left[c^\dagger_\alpha c_a , c^\dagger_b c_\beta \right] \simeq -\delta_{\alpha\beta} \delta_{ab} ,
$$
$$
\left[c^\dagger_a c_b , c^\dagger_c c_d \right] \simeq 0 , \qquad \left[c^\dagger_\alpha c_\beta , c^\dagger_\gamma c_\delta \right] \simeq 0 ,
$$

which justifies the identification with bosonic operators. The non-interacting Hamiltonian plus the component of interaction that enters in the Hartree-Fock calculation define the Hartree-Fock Hamiltonian, which can be represented as

$$
H_{HF} \simeq E_{HF} + \sum_{\alpha a} \left(\epsilon_\alpha - \epsilon_a \right) b^\dagger_{\alpha a} b_{\alpha a} ,
$$

$$(3.82)$$

where E_{HF} is the total energy of the Hartree-Fock Slater determinant, i.e., the vacuum of the bosons, and the other term is just the cost of a particle-hole excitation.

However, the interaction includes also matrix elements that couple holes with particles. The only matrix elements that may have non-vanishing effect when applied either on the right or on the left to the Hartree-Fock wavefunction are those that contain two particles and two holes, namely

$$\delta H_{int} \simeq \frac{1}{2} \sum_{ab\alpha\beta} \left\{ U_{ba\beta a}\, c_b^\dagger c_\alpha^\dagger c_\beta c_a + U_{\alpha ba\beta}\, c_\alpha^\dagger c_b^\dagger c_a c_\beta \right.$$
$$+ U_{ba\alpha\beta}\, c_b^\dagger c_\alpha^\dagger c_a c_\beta + U_{\alpha b\beta a}\, c_\alpha^\dagger c_b^\dagger c_\beta c_a$$
$$\left. + U_{\alpha\beta ba}\, c_\alpha^\dagger c_\beta^\dagger c_b c_a + U_{ab\beta a}\, c_a^\dagger c_b^\dagger c_\beta c_\alpha \right\}$$
$$= \frac{1}{2} \sum_{ab\alpha\beta} \left[2\left(U_{\alpha b\beta a} - U_{\alpha ba\beta} \right) c_\alpha^\dagger c_b^\dagger c_\beta c_a + U_{\alpha\beta ba}\, c_\alpha^\dagger c_\beta^\dagger c_b c_a \right.$$
$$\left. + U_{ab\beta\alpha}\, c_a^\dagger c_b^\dagger c_\beta c_\alpha \right].$$

We have to express δH_{int} in terms of bosons. For the first term that is simple, i.e., it remains the same

$$\delta H_{1\,int} = \sum_{ab\alpha\beta} \left(U_{\alpha b\beta a} - U_{\alpha ba\beta} \right) b_{\alpha a}^\dagger\, b_{\beta b}. \tag{3.83}$$

The second term is more delicate to express in terms of the bosonic particle-hole operators, since one can at will pair α with a or b, thus getting either $b_{\alpha a}^\dagger b_{\beta b}^\dagger$ or $-b_{\alpha b}^\dagger b_{\beta a}^\dagger$. Although the two terms correspond to the same fermionic operator, they are independent from the bosonic point of view; Hence, they both have to be taken into account, leading to

$$\delta H_{2\,int} = \frac{1}{2} \sum_{ab\alpha\beta} \left[\left(U_{\alpha\beta ba} - U_{\alpha\beta ab} \right) b_{\alpha a}^\dagger\, b_{\beta b}^\dagger + H.c. \right]. \tag{3.84}$$

Indeed, the sum of (3.82), (3.83), and (3.84) is just H_{tD-HF} of (3.81). Correspondingly, the linear response functions obtained within the time-dependent Hartree-Fock approximation are simply the linear response functions of the bosonic Hamiltonian (3.81) upon consistently representing the perturbation in terms of bosons. Specifically,

$$V(t) = \sum_{ij} V_{ij}(t)\, c_i^\dagger c_j \simeq \sum_{a\alpha} \left(V_{a\alpha}(t)\, c_a^\dagger c_\alpha + V_{\alpha a}(t)\, c_\alpha^\dagger c_a \right)$$
$$\rightarrow \sum_{a\alpha} \left(V_{a\alpha}(t)\, b_{\alpha a} + V_{\alpha a}(t)\, b_{\alpha a}^\dagger \right), \tag{3.85}$$

where, since $V(t)$ must be Hermitian, then $V_{\alpha a}(t) = V_{a\alpha}(t)^*$ so that the bosonic representation is Hermitian, too.

We remark that the bosonization of the particle-hole excitation has been constructed strictly at zero temperature, and therefore cannot be used at $T \neq 0$. Indeed, the basic assumption we made was

$$\left[b_{\alpha a}, b_{\alpha a}^{\dagger} \right] \simeq \langle c_a^{\dagger} c_a \rangle_{HF} - \langle c_{\alpha}^{\dagger} c_{\alpha} \rangle_{HF} = n_a(T) - n_{\alpha}(T),$$

which does reproduce the correct bosonic commutation relation only at $T = 0$, when $n_a(0) = 1$ and $n_{\alpha}(0) = 0$, but not at $T \neq 0$, when $n_a(T) - n_{\alpha}(T)$ is strictly less than one, and vanishes for $T \rightarrow \infty$.

3.4   Application: Antiferromagnetism in the Half-Filled Hubbard Model

Let us consider the simplest version of a Hubbard model on a hypercubic lattice in d-dimensions:

$$H = -t \sum_{<\mathbf{RR'}>} \sum_{\sigma} \left(c_{\mathbf{R}\sigma}^{\dagger} c_{\mathbf{R'}\sigma} + H.c. \right) + U \sum_{\mathbf{R}} \hat{n}_{\mathbf{R}\uparrow} \hat{n}_{\mathbf{R}\downarrow}, \qquad (3.86)$$

where

$$\hat{n}_{\mathbf{R}\sigma} = c_{\mathbf{R}\sigma}^{\dagger} c_{\mathbf{R}\sigma},$$

and $< \mathbf{RR'} >$ means that the sum runs over nearest neighbour bonds, i.e., that $\mathbf{R}$ and $\mathbf{R'}$ are nearest neighbour sites. We already showed that for very large U/t and one electron per site, i.e., at half-filling, that model describes a Mott insulator with a Neèl magnetic order in $d > 1$ at zero temperature and in $d > 2$ at finite temperature below a critical T_N. Let us now try to recover that result within the Hartree-Fock approximation. We use directly the $T \neq 0$ version, since the zero temperature Hartree-Fock approximation is just a special case.

The first step is to identify the variational parameters. The interaction, being on-site, leads to local variational parameters

$$\Delta_{\mathbf{R},\sigma\sigma'} = \langle c_{\mathbf{R}\sigma}^{\dagger} c_{\mathbf{R}\sigma'} \rangle_{HF}. \qquad (3.87)$$

We search for a Hartree-Fock Hamiltonian that describes a Neèl magnetic order with the antiferromagnetic order parameter along the z-direction, namely we assume that the spin $SU(2)$ symmetry is lowered down to a $U(1)$ symmetry that describes just spin-rotations around the z-axis. As a consequence, only the diagonal variational parameters $\Delta_{\mathbf{R},\sigma\sigma} \equiv n_{\mathbf{R}\sigma}$ in (3.87) are allowed by the remaining $U(1)$ symmetry. Since the average number of electrons per site is $\sum_{\sigma} n_{\mathbf{R}\sigma} = 1$, we can use the following parametrisation:

$$n_{\mathbf{R}\uparrow} = \frac{1}{2} + m\,(-1)^R, \quad n_{\mathbf{R}\downarrow} = \frac{1}{2} - m\,(-1)^R, \qquad (3.88)$$

where, if the vector $\mathbf{R} = a(n_1, n_2, \ldots, n_d)$, a being the lattice spacing, then $R = \sum_{i=1}^{d} n_i$. In other words, the expectation values of components of the magnetisation density are assumed to be

$$S_{z\,\mathbf{R}} = \frac{1}{2}\left(n_{\mathbf{R}\uparrow} - n_{\mathbf{R}\downarrow}\right) = m\,(-1)^R, \quad S_{x\,\mathbf{R}} = S_{y\,\mathbf{R}} = 0.$$

The Hartree-Fock Hamiltonian is therefore

$$H_{HF} = -t \sum_{<\mathbf{RR'}>\sigma} \left(c^{\dagger}_{\mathbf{R}\sigma} c_{\mathbf{R'}\sigma} + H.c.\right) + U \sum_{\mathbf{R}} \left(\hat{n}_{\mathbf{R}\uparrow} n_{\mathbf{R}\downarrow} + \hat{n}_{\mathbf{R}\downarrow} n_{\mathbf{R}\uparrow}\right)$$

$$= -t \sum_{<\mathbf{RR'}>\sigma} \left(c^{\dagger}_{\mathbf{R}\sigma} c_{\mathbf{R'}\sigma} + H.c.\right) - Um \sum_{\mathbf{R}} (-1)^R \left(\hat{n}_{\mathbf{R}\uparrow} - \hat{n}_{\mathbf{R}\downarrow}\right) + \frac{U}{2}\sum_{\mathbf{R}}\left(\hat{n}_{\mathbf{R}\uparrow} + \hat{n}_{\mathbf{R}\downarrow}\right).$$

The last term is proportional to the total number of electrons N, which is conserved. Therefore, that term, actually equal to $UN/2$, is just a constant that can be dropped leading to the Hartree-Fock Hamiltonian

$$H_{HF} = -t \sum_{<\mathbf{RR'}>}\sum_{\sigma} \left(c^{\dagger}_{\mathbf{R}\sigma} c_{\mathbf{R'}\sigma} + H.c.\right) - U\,m \sum_{\mathbf{R}}(-1)^R \left(\hat{n}_{\mathbf{R}\uparrow} - \hat{n}_{\mathbf{R}\downarrow}\right).$$

$$(3.89)$$

Let us rewrite this Hamiltonian in momentum space. Since

$$(-1)^R = e^{i\mathbf{Q}\cdot\mathbf{R}},$$

where $\mathbf{Q} = \pi(1, 1, \ldots, 1)/a$, one readily finds that

$$H_{HF} = \sum_{\mathbf{k}\sigma} \epsilon_{\mathbf{k}}\, c^{\dagger}_{\mathbf{k}\sigma}\, c_{\mathbf{k}\sigma} - U\,m \sum_{\mathbf{k}} \left(c^{\dagger}_{\mathbf{k}\uparrow} c_{\mathbf{k}+\mathbf{Q}\uparrow} - c^{\dagger}_{\mathbf{k}\downarrow} c_{\mathbf{k}+\mathbf{Q}\downarrow}\right), \qquad (3.90)$$

where, if $\mathbf{k} = (k_1, k_2, \ldots, k_d)$, the bare dispersion is simply

$$\epsilon_{\mathbf{k}} = -2t \sum_{i=1}^{d} \cos k_i a.$$

We observe that, since $\mathbf{Q} \equiv -\mathbf{Q}$ apart from a reciprocal lattice vector, then $(\mathbf{k} + \mathbf{Q}) + \mathbf{Q} \equiv \mathbf{k}$, which implies a one-to-one correspondence between $\mathbf{k}$ and $\mathbf{k} + \mathbf{Q}$. Moreover, it trivially holds that $\epsilon_{\mathbf{k}+\mathbf{Q}} = -\epsilon_{\mathbf{k}}$. Therefore, if we define a Magnetic Brillouin Zone (MBZ) as the volume that encloses all $\mathbf{k}$-points such that $\epsilon_{\mathbf{k}} \leq 0$, the rest of the Brillouin zone includes all partners $\mathbf{k} + \mathbf{Q}$, with $\epsilon_{\mathbf{k}+\mathbf{Q}} > 0$. It follows that the volume of the MBZ is half that of the Brillouin zone, which also implies that at half-filling the Fermi volume is just the MBZ, i.e., it includes all $\mathbf{k}$'s with $\epsilon_{\mathbf{k}} \leq 0$, and the Fermi surface is thus defined through $\mathbf{k}_F : \epsilon_{\mathbf{k}_F} = 0$. In Fig. 3.2, we show the case of a square lattice.

Fig. 3.2 Brillouin zone, the outer square, magnetic Brillouin zone, the inner rotated yellow square, and Fermi surface, in red, of a tight-binding Hamiltonian with nearest neighbour hopping at half-filling and on a square lattice

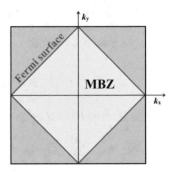

Since the Hartree-Fock Hamiltonian breaks translational symmetry—the new unit cell is twice as large—the actual Brillouin zone is just the MBZ. Because of the one-to-one correspondence between $\mathbf{k}$ and $\mathbf{k} + \mathbf{Q}$, we are allowed to define new fermionic operators with $\mathbf{k} \in MBZ$ by

$$a_{\mathbf{k}\sigma} = c_{\mathbf{k}\sigma}, \qquad b_{\mathbf{k}\sigma} = c_{\mathbf{k}+\mathbf{Q}\sigma},$$

and spinor operators

$$\psi_{\mathbf{k}\sigma} = \begin{pmatrix} a_{\mathbf{k}\sigma} \\ b_{\mathbf{k}\sigma} \end{pmatrix}. \tag{3.91}$$

The Hamiltonian (3.90) in terms of those spinors can be written as

$$H_{HF} = \sum_{\mathbf{k}\in MBZ\,\sigma} \epsilon_{\mathbf{k}} \, \psi_{\mathbf{k}\sigma}^{\dagger} \, \tau_3 \, \psi_{\mathbf{k}\sigma} - U m \sum_{\mathbf{k}\in MBZ} \left(\psi_{\mathbf{k}\uparrow}^{\dagger} \, \tau_1 \, \psi_{\mathbf{k}\uparrow} - \psi_{\mathbf{k}\downarrow}^{\dagger} \, \tau_1 \, \psi_{\mathbf{k}\downarrow} \right),$$

$$\tag{3.92}$$

where τ_i's, $i = 1, 2, 3$, are the Pauli matrices acting in the spinor basis. We observe that[3]

$$\epsilon_{\mathbf{k}} \, \tau_3 \mp U m \, \tau_1 = -E_{\mathbf{k}} \, \tau_3 \left(-\frac{\epsilon_{\mathbf{k}}}{E_{\mathbf{k}}} \pm i \, \frac{U m}{E_{\mathbf{k}}} \, \tau_2 \right)$$

$$= -E_{\mathbf{k}} \, \tau_3 \, e^{\pm 2 i \, \theta_{\mathbf{k}} \, \tau_2} = -E_{\mathbf{k}} \, e^{\mp i \, \theta_{\mathbf{k}} \, \tau_2} \, \tau_3 \, e^{\pm i \, \theta_{\mathbf{k}} \, \tau_2},$$

---

[3] We recall that the Pauli matrices σ_i, $i = 1, 2, 3$ satisfy

$$\{\sigma_i, \sigma_i\} = \delta_{ij}, \quad \sigma_i \sigma_j = \delta_{ij} + i \, \epsilon_{ijk} \sigma_k, \quad \sigma_i \sigma_j \sigma_i = \delta_{ij} \sigma_i - (1 - \delta_{ij}) \sigma_j,$$

where ϵ_{ijk} is the antisymmetric tensor. It follows that $\sigma_i^{2n} = 1$ while $\sigma_i^{2n+1} = \sigma_i$, and thus that

$$e^{i\theta\sigma_i} = \sum_{n\geq 0} \frac{(i\theta)^n}{n!} \sigma_i^n = \sum_{n\geq 0} \frac{(i\theta)^{2n}}{(2n)!} + \sigma_i \sum_{n\geq 0} \frac{(i\theta)^{2n+1}}{(2n+1)!} = \cos\theta + i \, \sigma_i \, \sin\theta.$$

Moreover, since $\sigma_i = \sigma_i^{\dagger} = \sigma_i^{-1}$ is Hermitian and unitary, for $i \neq j$ the following result holds:

$$e^{-i\theta\sigma_i} \, \sigma_j = \sigma_j \sigma_i e^{-i\theta\sigma_i} \, \sigma_j = \sigma_j \, e^{-i\theta \, \sigma_j \, \sigma_i \, \sigma_j} = \sigma_j \, e^{i\theta\sigma_i}.$$

where

$$E_{\mathbf{k}} = \sqrt{\epsilon_{\mathbf{k}}^2 + U^2 m^2} \,, \tag{3.93}$$

and

$$\cos 2\theta_{\mathbf{k}} = -\frac{\epsilon_{\mathbf{k}}}{E_{\mathbf{k}}} \,, \quad \sin 2\theta_{\mathbf{k}} = \frac{Um}{E_{\mathbf{k}}} \,.$$

Therefore, if we define two new spinors through

$$\phi_{\mathbf{k}\uparrow} = \begin{pmatrix} \alpha_{\mathbf{k}\uparrow} \\ \beta_{\mathbf{k}\uparrow} \end{pmatrix} = e^{i\theta_{\mathbf{k}} \tau_2} \, \psi_{\mathbf{k}\uparrow} \,, \quad \phi_{\mathbf{k}\downarrow} = \begin{pmatrix} \alpha_{\mathbf{k}\downarrow} \\ \beta_{\mathbf{k}\downarrow} \end{pmatrix} = e^{-i\theta_{\mathbf{k}} \tau_2} \, \psi_{\mathbf{k}\downarrow} \,, \tag{3.94}$$

the Hamiltonian becomes

$$H_{HF} = -\sum_{\mathbf{k}\in MBZ\,\sigma} E_{\mathbf{k}} \, \phi_{\mathbf{k}\sigma}^\dagger \, \tau_3 \, \phi_{\mathbf{k}\sigma} = -\sum_{\mathbf{k}\in MBZ\,\sigma} E_{\mathbf{k}} \left(\alpha_{\mathbf{k}\sigma}^\dagger \alpha_{\mathbf{k}\sigma} - \beta_{\mathbf{k}\sigma}^\dagger \beta_{\mathbf{k}\sigma} \right) \,, \tag{3.95}$$

namely it acquires a diagonal form. Since $E_{\mathbf{k}} > 0$ the ground state with a number of electrons equal to the number of sites N is simply obtained by filling completely the α-band (notice that the MBZ contains $N/2$ $\mathbf{k}$-points, hence each band can accommodate $2 N/2 = N$ electrons). The Hartree-Fock Hamiltonian thus describes a band insulator with valence and conduction bands separated by a gap of minimal value $2Um$. It follows that

$$\langle \phi_{\mathbf{k}\sigma}^\dagger \, \tau_i \, \phi_{\mathbf{k}\sigma} \rangle_{HF} = \delta_{i3} \, \langle \alpha_{\mathbf{k}\sigma}^\dagger \alpha_{\mathbf{k}\sigma} - \beta_{\mathbf{k}\sigma}^\dagger \beta_{\mathbf{k}\sigma} \rangle_{HF}$$
$$= \delta_{i3} \left(f(-E_{\mathbf{k}}) - f(E_{\mathbf{k}}) \right) = \delta_{i3} \, \tanh \frac{E_{\mathbf{k}}}{2T} \,, \tag{3.96}$$

with $f(x)$ the Fermi distribution function.

We still have to impose the self-consistency condition

$$m = \frac{1}{2V} \sum_{\mathbf{R}} (-1)^R \langle \hat{n}_{\mathbf{R}\uparrow} - \hat{n}_{\mathbf{R}\downarrow} \rangle_{HF} = \frac{1}{2V} \sum_{\mathbf{k}\in MBZ} \langle \psi_{\mathbf{k}\uparrow}^\dagger \, \tau_1 \, \psi_{\mathbf{k}\uparrow} - \psi_{\mathbf{k}\downarrow}^\dagger \, \tau_1 \, \psi_{\mathbf{k}\downarrow} \rangle_{HF}$$
$$= \frac{1}{2V} \sum_{\mathbf{k}\in MBZ} \langle \psi_{\mathbf{k}\uparrow}^\dagger \, e^{i\theta_{\mathbf{k}} \tau_2} \, \tau_1 \, e^{-i\theta_{\mathbf{k}} \tau_2} \, \psi_{\mathbf{k}\uparrow} - \psi_{\mathbf{k}\downarrow}^\dagger \, e^{-i\theta_{\mathbf{k}} \tau_2} \, \tau_1 \, e^{i\theta_{\mathbf{k}} \tau_2} \, \psi_{\mathbf{k}\downarrow} \rangle_{HF}$$
$$= \frac{1}{2V} \sum_{\mathbf{k}\in MBZ} \langle \phi_{\mathbf{k}\uparrow}^\dagger \, (\cos 2\theta_{\mathbf{k}} \, \tau_1 + \sin 2\theta_{\mathbf{k}} \, \tau_3) \, \phi_{\mathbf{k}\uparrow} - \phi_{\mathbf{k}\downarrow}^\dagger \, (\cos 2\theta_{\mathbf{k}} \, \tau_1 - \sin 2\theta_{\mathbf{k}} \, \tau_3) \, \phi_{\mathbf{k}\downarrow} \rangle_{HF}$$
$$= \frac{1}{V} \sum_{\mathbf{k}\in MBZ} \sin 2\theta_{\mathbf{k}} \, \tanh \frac{E_{\mathbf{k}}}{2T} \,,$$

which corresponds to the self-consistency equation

$$m \left(1 - \frac{U}{2V} \sum_{\mathbf{k}} \frac{1}{E_{\mathbf{k}}} \tanh \frac{E_{\mathbf{k}}}{2T} \right) \equiv m \, K(U, m, T) = 0 \,, \tag{3.97}$$

which has a trivial solution $m = 0$, corresponding to a non-magnetic state, and a non-trivial one, namely the real root of the term in parenthesis, which we denote as

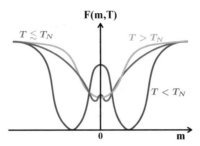

Fig. 3.3 Sketch of the Hartree-Fock free energy $F(m, T)$ as function of m at different temperatures. For $T < T_N$, blue curve, $F(m, T)$ has two opposite minima and a maximum at $m = 0$. Increasing T, the two minima approach each other till they merge into a single minimum with vanishing curvature at $m = 0$ when $T = T_N$. Above T_N, the minimum, with finite curvature, remains at $m = 0$. This is the conventional behaviour of a second-order phase transition

$K(U, m, T)$. We note that $K(U, m, T)$ actually depends on m^2, so that when a root exists with $m > 0$, there is an equivalent one at $-m < 0$, and they both describe the desired Néel state. Specifically, $m > 0$ implies that at sites with R even/odd the spin is up/down, and the reverse for $m < 0$. Moreover, one can easily realise that, if that root exists, it also corresponds to the global minimum of the variational free energy, i.e.,

$$F[\rho_{HF}, T] = -2T \sum_{\mathbf{k}} \ln\left(1 + e^{\beta E_{\mathbf{k}}}\right) + V U m^2,$$

whose saddle point equation with respect to m is exactly (3.97), as expected, while the saddle point at $m = 0$ is a maximum. We also note that, if a non-trivial solution with $m \neq 0$ exists at $T = 0$, since $\tanh E_{\mathbf{k}}/2T$ decreases with increasing T, and vanishes for $T \to \infty$, above some temperature one cannot find anymore a root, so that only the trivial solution $m = 0$ remains. The situation is depicted in Fig. 3.3 and resembles what happens, e.g., in an Ising model. Indeed, once within Hartree-Fock one chooses from the start a symmetry-breaking axis, what remains of the $SU(2)$ symmetry is the $U(1)$ rotations around that axis, but also the Z_2 symmetry $\uparrow \leftrightarrow \downarrow$ that is broken below T_N. However, since the choice of the symmetry-breaking axis is arbitrary, the actually symmetry that is broken is the original $SU(2)$.

We can gain a rough idea of the non-trivial solution of (3.97) by assuming a constant density of states in the interval $[-D, D]$ and zero outside, namely,

$$\frac{1}{V} \sum_{\mathbf{k}} \delta(\epsilon - \epsilon_{\mathbf{k}}) \simeq \rho_0 \, \theta(D - \epsilon) \, \theta(\epsilon + D),$$

where $\rho_0 = 1/2D$, with $D \sim 2dt$. With that assumption and at zero temperature, the root of $K(U, m, T)$ is the solution of

$$1 = \frac{U \rho_0}{2} \int_{-D}^{D} d\epsilon \, \frac{1}{\sqrt{\epsilon^2 + U^2 m^2}},$$

which exists for any positive U and reads

$$m = \pm \frac{D}{U} \sinh^{-1} \frac{1}{U \rho_0} .$$

In particular, for $U \ll D$,

$$m \simeq \pm \frac{D}{U} \exp \left(-\frac{1}{U \rho_0} \right) . \tag{3.98}$$

This result implies that the Hubbard model at half-filling on a hypercubic lattice and within the Hartree-Fock approximation is always an antiferromagnetic insulator whatever is the value of the Hubbard $U > 0$.[4] Note that in the opposite limit of $U \gg D$, $m \sim \pm D \rho_0 = \pm 1/2$, which is the expected result since, for large U, electrons localise, one per site, hence behave like local spin-1/2 moments.

We can also determine the Neèl temperature, which is the temperature T_N at which $K(U, 0, T_N) = 0$, i.e., at which a non-trivial solution of (3.97) exists at $m = 0$, namely

$$1 = \frac{U}{2V} \sum_{\mathbf{k}} \frac{1}{\epsilon_{\mathbf{k}}} \tanh \frac{\epsilon_{\mathbf{k}}}{2T_N} \simeq \frac{U \rho_0}{2} \int_{-D}^{D} \frac{d\epsilon}{\epsilon} \tanh \frac{\epsilon}{2T_N} . \tag{3.99}$$

Upon expanding (3.97) for T smaller but very close to T_N, and $m \ll 1$, one finds that $m(T) \sim \pm \sqrt{T_N - T}$, the standard mean-field behaviour. The Néel temperature defined by the solution of (3.99) is proportional to $U m(T = 0)$, namely

$$T_N \sim D \exp \left(-\frac{1}{U \rho_0} \right) ,$$

for small U, but $T_N \sim U$ at large U. The latter results contradict what we previously found in the large U limit by mapping the Hubbard model onto a Heisenberg model.

---

[4] This conclusion actually reflects the nesting property of the non-interacting Fermi surface, i.e., the set of points $\mathbf{k}_F : \epsilon_{\mathbf{k}_F} = \mu$, where μ is the chemical potential and is zero in our case. A Fermi surface has nesting when there exists a single momentum $\mathbf{Q}$, defined apart from a reciprocal lattice vector, such that $\epsilon_{\mathbf{k}+\mathbf{Q}} - \mu = -(\epsilon_{\mathbf{k}} - \mu)$ for a set of momenta W with non-zero measure around the Fermi surface. In our case, nesting holds throughout the whole Brillouin zone. When the Fermi surface has nesting, the non-interacting value of the thermodynamic susceptibility $\kappa(\mathbf{Q})$ to a static field with momentum $\mathbf{Q}$,

$$\chi(\mathbf{Q}) \sim -\frac{1}{V} \sum_{\mathbf{k}} \frac{f(\epsilon_{\mathbf{k}+\mathbf{Q}} - \mu) - f(\epsilon_{\mathbf{k}} - \mu)}{(\epsilon_{\mathbf{k}+\mathbf{Q}} - \mu) - (\epsilon_{\mathbf{k}} - \mu)} \sim -\frac{1}{V} \sum_{\mathbf{k} \in W} \frac{1 - 2f(\epsilon_{\mathbf{k}} - \mu)}{-2(\epsilon_{\mathbf{k}} - \mu)}$$

diverges logarithmically at $T = 0$ when the density of states at the Fermi energy is finite. It follows that the non-interacting model may become unstable to a modulation at momentum $\mathbf{Q}$ in presence of interaction. For instance, within the Hartree-Fock approximation, that occurs when the Hessian (3.56) becomes negative definite, which is always the case when $\chi(\mathbf{Q})$ diverges and the matrix $\hat{U}$ in the channel transferring momentum $\mathbf{Q}$ has a negative eigenvalue.

There, within spin-wave theory, we showed that $T_N \sim J \sim t^2/U$, hence vanishes for large U. The origin of the disagreement comes from the fact that the Hartree-Fock approximation is valid only at weak interaction compared to the electron bandwidth. Essentially, the Mott phenomenon that dominates at large U/t escapes any description using a single Slater determinant, which is only able to mimic a band insulator.

3.4.1 Spin-Wave Spectrum by Time-Dependent Hartree-Fock

Rigorously speaking, the Hartree-Fock approximation only allows accessing static equilibrium properties. Therefore, it is not legitimate to interpret the spectrum of the Hartree-Fock Hamiltonian as an approximation of the actual spectrum. For instance, we showed that the Heisenberg antiferromagnet, which corresponds to the large U-limit of the Hubbard model at half-filling, has gapless spin-wave excitations, which are simply a consequence of Goldstone's theorem and thus reflect conservation laws. On the contrary, the Hartree-Fock Hamiltonian describes a band insulator and thus has a gap for all excitations.

We already mentioned that the time-dependent Hartree-Fock approximation is supposed not to spoil the Hamiltonian symmetries. Therefore, the half-filled Hubbard model is the ideal example to check that assumption.

We define

$$S(\mathbf{q}) \equiv \frac{1}{2} \sum_{\mathbf{k}} c_{\mathbf{k}\alpha}^{\dagger} \, \sigma_{\alpha\beta} \, c_{\mathbf{k}+\mathbf{q},\beta} \, ,$$

the Fourier transform at momentum $\mathbf{q}$ of the spin density $\mathbf{S_R} = (S_{x\mathbf{R}}, S_{y\mathbf{R}}, S_{z\mathbf{R}})$, with $\boldsymbol{\sigma} = (\sigma_1, \sigma_2, \sigma_3)$, where σ_i are the Pauli matrices in the spin-up and spin-down space. At $\mathbf{q} = \mathbf{0}$, $S_a(\mathbf{0})$ are the generators of the spin $SU(2)$ symmetry, thus commute with the Hamiltonian,

$$\left[S_a(\mathbf{0}) \, , \, H \right] = 0 \, , \quad a = x, y, z \, .$$

Therefore,

$$S_a(\mathbf{q}, t) \equiv e^{iHt} \, S_a(\mathbf{q}) \, e^{-iHt}$$

becomes time independent at $\mathbf{q} = \mathbf{0}$, i.e., $S_a(\mathbf{0}, t) = S_a(\mathbf{0})$. It follows that the spin-spin linear response functions, whose only non-zero components by translation and spin $SU(2)$ invariance are

$$\chi_a(t, \mathbf{q}) = -i\theta(t) \left\langle \left[S_a(\mathbf{q}, t) \, S_a(-\mathbf{q}) \right] \right\rangle ,$$

at $\mathbf{q} = \mathbf{0}$,

$$\chi_a(t, \mathbf{0}) = -i\theta(t) \left\langle \left[S_a(\mathbf{0}, t) \, S_a(\mathbf{0}) \right] \right\rangle = -i\theta(t) \left\langle \left[S_a(\mathbf{0}) \, S_a(\mathbf{0}) \right] \right\rangle = 0 \, , \tag{3.100}$$

must vanish.

Within the Hartree-Fock approximation, we decided to break spin $SU(2)$ symmetry along the z-direction, but we could have equally well chosen any other direction and get right the same results. The rotation of the symmetry-breaking axis from the chosen one, i.e., z, is accomplished by the generators $S_x(\mathbf{0})$ and $S_y(\mathbf{0})$, or, equivalently, their rising/lowering combinations

$$
S^+(\mathbf{q}) = \sum_{\mathbf{k}} c^\dagger_{\mathbf{k}\uparrow} c_{\mathbf{k}+\mathbf{q}\downarrow} \,, \quad S^-(\mathbf{q}) = \sum_{\mathbf{k}} c^\dagger_{\mathbf{k}\downarrow} c_{\mathbf{k}+\mathbf{q}\uparrow} = \left(S^+(-\mathbf{q}) \right)^\dagger \,,
$$

at $\mathbf{q} = \mathbf{0}$. It follows that, e.g., the expectation value of H over the wavefunction $S^+(\mathbf{q})\,|\,\phi_{HF}\rangle$, where $|\,\phi_{HF}\rangle$ is the Hartree-Fock Slater determinant with symmetry breaking along z, must vanish for $\mathbf{q} \to \mathbf{0}$. In terms of the Fourier transform in time $\chi_\pm(\omega, \mathbf{q})$ of the corresponding response function

$$
\chi_\pm(t, \mathbf{q}) = -i\theta(t)\,\Big\langle \Big[S^+(\mathbf{q}, t)\, S^-(-\mathbf{q}) \Big] \Big\rangle \,,
$$

that implies the existence of a pole in $\chi_\pm(\omega, \mathbf{q})$ for $\omega = \omega(\mathbf{q})$ with $\omega(\mathbf{q} \to \mathbf{0}) \to 0$, which is just the spin-wave dispersion in momentum space. We will therefore check whether this condition as well as (3.100) are indeed satisfied by the time-dependent Hartree-Fock approximation.

We define four components spinors

$$
\Psi_{\mathbf{k}} = \begin{pmatrix} \Psi_{\mathbf{k}\uparrow} \\ \Psi_{\mathbf{k}\downarrow} \end{pmatrix} \,, \quad \Phi_{\mathbf{k}} = \begin{pmatrix} \Phi_{\mathbf{k}\uparrow} \\ \Phi_{\mathbf{k}\downarrow} \end{pmatrix} \,,
$$

which are related to each other through (3.94), namely

$$
\Psi_{\mathbf{k}} = e^{-i\theta_{\mathbf{k}}\, \sigma_3\, \tau_1}\, \Phi_{\mathbf{k}} \,, \quad \Psi^\dagger_{\mathbf{k}} = \Phi^\dagger_{\mathbf{k}}\, e^{i\theta_{\mathbf{k}}\, \sigma_3\, \tau_1} \,.
$$

For $q = |\mathbf{q}| \ll |\mathbf{Q}|$, the following spin-density operators in terms of those spinors become

$$
S(\mathbf{q}) = \frac{1}{2} \sum_{\mathbf{k}\alpha\beta} c^\dagger_{\mathbf{k}\alpha}\, \sigma_{\alpha\beta}\, c^\dagger_{\mathbf{k}+\mathbf{q}\beta} \simeq \frac{1}{2} \sum_{\mathbf{k}\in MBZ} \Psi^\dagger_{\mathbf{k}}\, \sigma\, \Psi^\dagger_{\mathbf{k}+\mathbf{q}}
$$

$$
= \frac{1}{2} \sum_{\mathbf{k}\in MBZ} \Phi^\dagger_{\mathbf{k}}\, e^{i\theta_{\mathbf{k}}\tau_2\sigma_3}\, \sigma\, e^{-i\theta_{\mathbf{k}+\mathbf{q}}\tau_2\sigma_3}\, \Phi^\dagger_{\mathbf{k}+\mathbf{q}} \,,
$$

$$
S(\mathbf{q}+\mathbf{Q}) = \frac{1}{2} \sum_{\mathbf{k}\alpha\beta} c^\dagger_{\mathbf{k}\alpha}\, \sigma_{\alpha\beta}\, c^\dagger_{\mathbf{k}+\mathbf{Q}+\mathbf{q}\beta} \simeq \frac{1}{2} \sum_{\mathbf{k}\in MBZ} \Psi^\dagger_{\mathbf{k}}\, \sigma\, \tau_1\, \Psi^\dagger_{\mathbf{k}+\mathbf{q}}
$$

$$
= \frac{1}{2} \sum_{\mathbf{k}\in MBZ} \Phi^\dagger_{\mathbf{k}}\, e^{i\theta_{\mathbf{k}}\tau_2\sigma_3}\, \sigma\, \tau_1\, e^{-i\theta_{\mathbf{k}+\mathbf{q}}\tau_2\sigma_3}\, \Phi^\dagger_{\mathbf{k}+\mathbf{q}} \,.
$$

Hereafter, unless necessary, we omit to specify that the momenta are within the MBZ, and thus all summations over momenta are implicitly assumed to be restricted within the MBZ.

In the spirit of the time-dependent Hartree-Fock approximation, we need to keep only particle-hole excitations with respect to the Hartree-Fock state, i.e., terms like $\Phi^\dagger \tau_1 \Phi$ and $\Phi^\dagger \tau_2 \Phi$. Therefore, for small q,[5]

$$S_z(\mathbf{q}) \simeq -\frac{i}{2}\mathbf{q} \cdot \sum_\mathbf{k} \nabla_\mathbf{k}\theta_\mathbf{k} \; \Phi_\mathbf{k}^\dagger \tau_2 \Phi_{\mathbf{k}+\mathbf{q}}$$

and vanishes at $q = 0$, thus automatically guaranteeing the validity of (3.100) for $a = z$.

On the contrary, for $a = x, y$,

$$S_a(\mathbf{q}) = \frac{1}{2} \sum_\mathbf{k} \Phi_\mathbf{k}^\dagger e^{i\theta_\mathbf{k}\tau_2\sigma_3} \sigma_a e^{-i\theta_{\mathbf{k}+\mathbf{q}}\tau_2\sigma_3} \Phi_{\mathbf{k}+\mathbf{q}\beta}^\dagger$$

$$= \frac{1}{2} \sum_\mathbf{k} \Phi_\mathbf{k}^\dagger \sigma_a e^{-i(\theta_{\mathbf{k}+\mathbf{q}}+\theta_\mathbf{k})\tau_2\sigma_3} \Phi_{\mathbf{k}+\mathbf{q}\beta}^\dagger$$

$$\simeq -\frac{i}{2} \sum_\mathbf{k} \sin\left(\theta_{\mathbf{k}+\mathbf{q}} + \theta_\mathbf{k}\right) \Phi_\mathbf{k}^\dagger \sigma_a \sigma_3 \tau_2 \Phi_{\mathbf{k}+\mathbf{q}}$$

$$= \frac{1}{2} \epsilon_{a3b} \sum_\mathbf{k} \sin\left(\theta_{\mathbf{k}+\mathbf{q}} + \theta_\mathbf{k}\right) \Phi_\mathbf{k}^\dagger \sigma_b \tau_2 \Phi_{\mathbf{k}+\mathbf{q}} \; ,$$

$$S_a(\mathbf{q} + \mathbf{Q}) = \frac{1}{2} \sum_\mathbf{k} \Phi_\mathbf{k}^\dagger e^{i\theta_\mathbf{k}\tau_2\sigma_3} \sigma_a \tau_1 e^{-i\theta_{\mathbf{k}+\mathbf{q}}\tau_2\sigma_3} \Phi_{\mathbf{k}+\mathbf{q}\beta}^\dagger$$

$$= \frac{1}{2} \sum_\mathbf{k} \Phi_\mathbf{k}^\dagger \sigma_a \tau_1 e^{-i(\theta_{\mathbf{k}+\mathbf{q}}-\theta_\mathbf{k})\tau_2\sigma_3} \Phi_{\mathbf{k}+\mathbf{q}\beta}^\dagger$$

$$\simeq \frac{1}{2} \sum_\mathbf{k} \cos\left(\theta_{\mathbf{k}+\mathbf{q}} - \theta_\mathbf{k}\right) \Phi_\mathbf{k}^\dagger \sigma_a \tau_1 \Phi_{\mathbf{k}+\mathbf{q}} \; ,$$

---

[5] If we deal with two sets of Pauli matrices, σ_i and τ_i, $i = 0, 1, 2, 3$, assuming that σ_0 and τ_0 are the identity matrices, it is straightforward to demonstrate that, as before,

$$e^{i\theta\,\tau_i\,\sigma_j} = \cos\theta + i\,\sin\theta\,\tau_i\,\sigma_j \; ,$$

since it is still true that $\left(\tau_i\,\sigma_j\right)^{2n} = 1$ and $\left(\tau_i\,\sigma_j\right)^{2n+1} = \tau_i\,\sigma_j$. Moreover, also in this case, $\sigma_k\,\tau_n$ is a unitary Hermitian operator and therefore

$$e^{-i\theta\,\sigma_i\,\tau_j}\,\sigma_k\,\tau_n = \sigma_k\,\tau_n\,\sigma_k\,\tau_n\,e^{-i\theta\,\sigma_i\,\tau_j}\,\sigma_k\,\tau_n = \sigma_k\,\tau_n\,\exp\left(i\theta\,\sigma_k\,\tau_n\,\sigma_i\,\tau_j\,\sigma_k\,\tau_n\right),$$

which can be easily calculated using the properties of the Pauli matrices.

so that

$$S^+(\mathbf{q}) = i \sum_{\mathbf{k}} \sin\left(\theta_{\mathbf{k+q}} + \theta_{\mathbf{k}}\right) \Phi^\dagger_{\mathbf{k}\uparrow} \tau_2 \Phi_{\mathbf{k+q}\downarrow}$$

$$\equiv i\sqrt{2} \sum_{\mathbf{k}} \sin\left(\theta_{\mathbf{k+q}} + \theta_{\mathbf{k}}\right) P_{\mathbf{k};\mathbf{q}} \,,$$

$$S^+(\mathbf{q}+\mathbf{Q}) = \sum_{\mathbf{k}} \cos\left(\theta_{\mathbf{k+q}} - \theta_{\mathbf{k}}\right) \Phi^\dagger_{\mathbf{k}\uparrow} \tau_1 \Phi_{\mathbf{k+q}\downarrow} \tag{3.101}$$

$$\equiv \sqrt{2} \sum_{\mathbf{k}} \cos\left(\theta_{\mathbf{k+q}} - \theta_{\mathbf{k}}\right) X_{\mathbf{k};\mathbf{q}} \,,$$

where we have introduced the following operators:

$$X_{\mathbf{k};\mathbf{q}} = \sqrt{\frac{1}{2}} \, \phi^\dagger_{\mathbf{k}\uparrow} \tau_1 \phi_{\mathbf{k+q}\downarrow} \,, \quad X^\dagger_{\mathbf{k};\mathbf{q}} = \sqrt{\frac{1}{2}} \, \phi^\dagger_{\mathbf{k+q}\downarrow} \tau_1 \phi_{\mathbf{k}\uparrow} \,,$$

$$P_{\mathbf{k};\mathbf{q}} = \sqrt{\frac{1}{2}} \, \phi^\dagger_{\mathbf{k}\uparrow} \tau_2 \phi_{\mathbf{k+q}\downarrow} \,, \quad P^\dagger_{\mathbf{k};\mathbf{q}} = \sqrt{\frac{1}{2}} \, \phi^\dagger_{\mathbf{k+q}\downarrow} \tau_2 \phi_{\mathbf{k}\uparrow} \,.$$

According to the time-dependent Hartree-Fock approximation, we can assume that

$$\left[X_{\mathbf{k};\mathbf{q}}, P^\dagger_{\mathbf{k'};\mathbf{q'}}\right] = \frac{i}{2}\left(\delta_{\mathbf{k+q},\mathbf{k'+q'}} \, \phi^\dagger_{\mathbf{k}\uparrow} \tau_3 \phi_{\mathbf{k'}\uparrow} + \delta_{\mathbf{k},\mathbf{k'}} \, \phi^\dagger_{\mathbf{k'+q'}\downarrow} \tau_3 \phi_{\mathbf{k+q}\downarrow}\right)$$

$$\simeq \left\langle \left[X_{\mathbf{k};\mathbf{q}}, P^\dagger_{\mathbf{k'};\mathbf{q'}}\right] \right\rangle_{HF} = i \, \delta_{\mathbf{k},\mathbf{k'}} \, \delta_{\mathbf{q},\mathbf{q'}} \,,$$

which shows that X and $P^\dagger$ are like conjugate variables. The energy cost of the spin-flip excitations are described by the free Hamiltonian

$$H_0 = \sum_{\mathbf{k}\mathbf{q}} \omega_{\mathbf{k};\mathbf{q}} \left(X^\dagger_{\mathbf{k};\mathbf{q}} X_{\mathbf{k};\mathbf{q}} + P^\dagger_{\mathbf{k};\mathbf{q}} P_{\mathbf{k};\mathbf{q}}\right) \,, \tag{3.102}$$

where

$$\omega_{\mathbf{k};\mathbf{q}} = E_{\mathbf{k+q}} + E_{\mathbf{k}}$$

is simply the energy required to destroy an α particle at $\mathbf{k} + \mathbf{q}$ and create a β one at $\mathbf{k}$.

Following the prescriptions of time-dependent Hartree-Fock, we still need to express the Hubbard interaction in terms of these spin-flip bosonic excitations. Since

$$\hat{n}_{\mathbf{R}\uparrow}\hat{n}_{\mathbf{R}\downarrow} = c^\dagger_{\mathbf{R}\uparrow} c_{\mathbf{R}\uparrow} c^\dagger_{\mathbf{R}\downarrow} c_{\mathbf{R}\downarrow} = -c^\dagger_{\mathbf{R}\uparrow} c_{\mathbf{R}\downarrow} c^\dagger_{\mathbf{R}\downarrow} c_{\mathbf{R}\uparrow} = -S^+_{\mathbf{R}} S^-_{\mathbf{R}} \,,$$

then

$$H_{int} = U \sum_{\mathbf{R}} \hat{n}_{\mathbf{R}\uparrow}\hat{n}_{\mathbf{R}\downarrow} = -U \sum_{\mathbf{R}} S^+_{\mathbf{R}} S^-_{\mathbf{R}} = -\frac{U}{V} \sum_{\mathbf{q}} S^+(\mathbf{q}) S^-(-\mathbf{q}) \,,$$

or, considering all momenta restricted within the MBZ,

$$H_{int} \simeq -\frac{U}{V} \sum_{\mathbf{q}} \left(S^+(\mathbf{q}) \, S^-(-\mathbf{q}) + S^+(\mathbf{q} + \mathbf{Q}) \, S^-(-\mathbf{q} - \mathbf{Q}) \right).$$

The dynamical behaviour of $S^+(\mathbf{q})$ and $S^-(-\mathbf{q}) = S^+(\mathbf{q})^\dagger$ is determined just by replacing the spin-density operators with the bosonic representation (3.101), which leads to

$$H_{int} \simeq -\frac{2U}{V} \sum_{\mathbf{q}} \sum_{\mathbf{kp}} \left[\, \sin\left(\theta_{\mathbf{k}+\mathbf{q}} + \theta_{\mathbf{k}}\right) \sin\left(\theta_{\mathbf{k}+\mathbf{q}} + \theta_{\mathbf{k}}\right) P^\dagger_{\mathbf{k};\mathbf{q}} P_{\mathbf{p};\mathbf{q}} \right.$$

$$\left. + \cos\left(\theta_{\mathbf{k}+\mathbf{q}} - \theta_{\mathbf{k}}\right) \cos\left(\theta_{\mathbf{k}+\mathbf{q}} - \theta_{\mathbf{k}}\right) X^\dagger_{\mathbf{k};\mathbf{q}} X_{\mathbf{p};\mathbf{q}} \right]. \tag{3.103}$$

The full Hamiltonian for the spin-flip excitations is therefore

$$\mathcal{H} = \sum_{\mathbf{k}\,\mathbf{q}} \omega_{\mathbf{k};\mathbf{q}} \left(X^\dagger_{\mathbf{k};\mathbf{q}} X_{\mathbf{k};\mathbf{q}} + P^\dagger_{\mathbf{k};\mathbf{q}} P_{\mathbf{k};\mathbf{q}} \right)$$

$$- \frac{2U}{V} \sum_{\mathbf{q}} \sum_{\mathbf{kp}} \left[\, \sin\left(\theta_{\mathbf{k}+\mathbf{q}} + \theta_{\mathbf{k}}\right) \sin\left(\theta_{\mathbf{p}+\mathbf{q}} + \theta_{\mathbf{k}}\right) P^\dagger_{\mathbf{k};\mathbf{q}} P_{\mathbf{p};\mathbf{q}} \right. \tag{3.104}$$

$$\left. + \cos\left(\theta_{\mathbf{k}+\mathbf{q}} - \theta_{\mathbf{k}}\right) \cos\left(\theta_{\mathbf{p}+\mathbf{q}} - \theta_{\mathbf{k}}\right) X^\dagger_{\mathbf{k};\mathbf{q}} X_{\mathbf{p};\mathbf{q}} \right].$$

One may diagonalise this Hamiltonian to obtain the full spin-flip excitation spectrum. Here, however, we just aim to show that (3.104) does contain the spin waves. For that, we add a perturbation

$$V(t) = \sum_{\mathbf{q}} V(\mathbf{q}, t) \, S^-(-\mathbf{q}) \simeq -i \sqrt{2} \sum_{\mathbf{kq}} \sin\left(\theta_{\mathbf{k}+\mathbf{q}} + \theta_{\mathbf{k}}\right) V(t, \mathbf{q}) \, P^\dagger_{\mathbf{k};\mathbf{q}},$$

in presence of which the equations of motion read

$$\frac{\partial X_{\mathbf{k};\mathbf{q}}}{\partial t} = \omega_{\mathbf{k};\mathbf{q}} P_{\mathbf{k};\mathbf{q}} - \sin\left(\theta_{\mathbf{k}+\mathbf{q}} + \theta_{\mathbf{k}}\right) \frac{2U}{V} \sum_{\mathbf{p}} \sin\left(\theta_{\mathbf{p}+\mathbf{q}} + \theta_{\mathbf{k}}\right) P_{\mathbf{p};\mathbf{q}}$$

$$- i \sqrt{2} \, \sin\left(\theta_{\mathbf{k}+\mathbf{q}} + \theta_{\mathbf{k}}\right) V(t, \mathbf{q}), \tag{3.105}$$

$$\frac{\partial P_{\mathbf{k};\mathbf{q}}}{\partial t} = -\omega_{\mathbf{k};\mathbf{q}} X_{\mathbf{k};\mathbf{q}} + \cos\left(\theta_{\mathbf{k}+\mathbf{q}} - \theta_{\mathbf{k}}\right) \frac{2U}{V} \sum_{\mathbf{p}} \cos\left(\theta_{\mathbf{p}+\mathbf{q}} - \theta_{\mathbf{k}}\right) X_{\mathbf{p};\mathbf{q}}.$$

In absence of $V(t)$, the expectation values of X and P vanish, while they become finite when the perturbation is present. Therefore, we take the expectation value of both right- and left-hand sides of the two equations and write, for simplicity,

$\langle X_{\mathbf{k};\mathbf{q}} \rangle \equiv X_{\mathbf{k};\mathbf{q}}$, and similarly for P. In that way, we can directly obtain the linear response function $\chi_\pm(\omega, \mathbf{q})$ we are interested in.

Taking the Fourier transform in frequency, we obtain

$$-i\omega\, X_{\mathbf{k};\mathbf{q}}(\omega) = \omega_{\mathbf{k};\mathbf{q}}\, P_{\mathbf{k};\mathbf{q}}(\omega) - \sin\left(\theta_{\mathbf{k}+\mathbf{q}} + \theta_{\mathbf{k}}\right) \frac{2U}{V} \sum_{\mathbf{p}} \sin\left(\theta_{\mathbf{p}+\mathbf{q}} + \theta_{\mathbf{k}}\right) P_{\mathbf{p};\mathbf{q}}(\omega)$$

$$- i\sqrt{2}\, \sin\left(\theta_{\mathbf{k}+\mathbf{q}} + \theta_{\mathbf{k}}\right) V(\omega, \mathbf{q})\,,$$

$$-i\omega\, P_{\mathbf{k};\mathbf{q}}(\omega) = -\omega_{\mathbf{k};\mathbf{q}}\, X_{\mathbf{k};\mathbf{q}}(\omega) + \cos\left(\theta_{\mathbf{k}+\mathbf{q}} - \theta_{\mathbf{k}}\right) \frac{2U}{V} \sum_{\mathbf{p}} \cos\left(\theta_{\mathbf{p}+\mathbf{q}} - \theta_{\mathbf{k}}\right) X_{\mathbf{p};\mathbf{q}}(\omega)\,.$$

We define

$$X(\omega, \mathbf{q}) \equiv \frac{2U}{V} \sum_{\mathbf{p}} \cos\left(\theta_{\mathbf{p}+\mathbf{q}} - \theta_{\mathbf{k}}\right) X_{\mathbf{p};\mathbf{q}}(\omega) \simeq U\sqrt{2}\, S^+(\omega, \mathbf{q} + \mathbf{Q})\,,$$

$$P(\omega, \mathbf{q}) \equiv \frac{2U}{V} \sum_{\mathbf{p}} \sin\left(\theta_{\mathbf{p}+\mathbf{q}} + \theta_{\mathbf{k}}\right) P_{\mathbf{p};\mathbf{q}}(\omega) \simeq -i\, U\sqrt{2}\, S^+(\omega, \mathbf{q})\,,$$

and the effective field

$$V_*(\omega, \mathbf{q}) \equiv V(\omega, \mathbf{q}) - \frac{i}{\sqrt{2}}\, P(\omega, \mathbf{q}) = V(\omega, \mathbf{q}) - U\, S^+(\omega, \mathbf{q})\,,$$

so that

$$\begin{pmatrix} -i\omega & -\omega_{\mathbf{k};\mathbf{q}} \\ \omega_{\mathbf{k};\mathbf{q}} & -i\omega \end{pmatrix} \begin{pmatrix} X_{\mathbf{k};\mathbf{q}}(\omega) \\ P_{\mathbf{k};\mathbf{q}}(\omega) \end{pmatrix} = \begin{pmatrix} -i\sqrt{2}\, \sin\left(\theta_{\mathbf{k}+\mathbf{q}} + \theta_{\mathbf{k}}\right) V_*(\omega, \mathbf{q}) \\ \cos\left(\theta_{\mathbf{k}+\mathbf{q}} - \theta_{\mathbf{k}}\right) X(\omega, \mathbf{q}) \end{pmatrix}\,,$$

namely

$$\begin{pmatrix} X_{\mathbf{k};\mathbf{q}}(\omega) \\ P_{\mathbf{k};\mathbf{q}}(\omega) \end{pmatrix} = \frac{1}{\omega^2 - \omega_{\mathbf{k};\mathbf{q}}^2} \begin{pmatrix} i\omega & -\omega_{\mathbf{k};\mathbf{q}} \\ \omega_{\mathbf{k};\mathbf{q}} & i\omega \end{pmatrix} \begin{pmatrix} -i\sqrt{2}\, \sin\left(\theta_{\mathbf{k}+\mathbf{q}} + \theta_{\mathbf{k}}\right) V_*(\omega, \mathbf{q}) \\ \cos\left(\theta_{\mathbf{k}+\mathbf{q}} - \theta_{\mathbf{k}}\right) X(\omega, \mathbf{q}) \end{pmatrix}\,.$$

If we use the above equation to calculate $X(\omega, \mathbf{q})$ and $P(\omega, \mathbf{q})$, we finally find, recalling the definition of $V_*(\omega, \mathbf{q})$,

$$\begin{pmatrix} 1 + \Sigma_c(\omega, \mathbf{q}) & \Sigma(\omega, \mathbf{q}) \\ -\Sigma(\omega, \mathbf{q}) & 1 + \Sigma_s(\omega, \mathbf{q}) \end{pmatrix} \begin{pmatrix} X(\omega, \mathbf{q}) \\ P(\omega, \mathbf{q}) \end{pmatrix} = -i\sqrt{2}\, V(\omega, \mathbf{q}) \begin{pmatrix} \Sigma(\omega, \mathbf{q}) \\ \Sigma_s(\omega, \mathbf{q}) \end{pmatrix}\,, \tag{3.106}$$

where we define

$$\Sigma(\omega, \mathbf{q}) \equiv i\omega\, \frac{2U}{V} \sum_{\mathbf{k}} \frac{\sin 2\theta_{\mathbf{k}}}{\omega^2 - \omega_{\mathbf{k};\mathbf{q}}^2}\,,$$

$$\Sigma_c(\omega, \mathbf{q}) \equiv \frac{2U}{V} \sum_{\mathbf{k}} \cos^2\left(\theta_{\mathbf{k}+\mathbf{q}} - \theta_{\mathbf{k}}\right) \frac{\omega_{\mathbf{k};\mathbf{q}}}{\omega^2 - \omega_{\mathbf{k};\mathbf{q}}^2}\,,$$

$$\Sigma_s(\omega, \mathbf{q}) \equiv \frac{2U}{V} \sum_{\mathbf{k}} \sin^2\left(\theta_{\mathbf{k}+\mathbf{q}} + \theta_{\mathbf{k}}\right) \frac{\omega_{\mathbf{k};\mathbf{q}}}{\omega^2 - \omega_{\mathbf{k};\mathbf{q}}^2}\,.$$

We shall analyse the behaviour of the solution for small q. We note, using (3.97) at zero temperature, that

$$\Sigma_c(\omega, \mathbf{0}) = \frac{2U}{V} \sum_{\mathbf{k}} \frac{2E_{\mathbf{k}}}{\omega^2 - 4E_{\mathbf{k}}^2} = \frac{U}{V} \sum_{\mathbf{k}} \frac{1}{E_{\mathbf{k}}} \left(-1 + \frac{\omega^2}{\omega^2 - 4E_{\mathbf{k}}^2} \right)$$

$$= -1 - \omega^2 \frac{U}{V} \sum_{\mathbf{k}} \frac{1}{E_{\mathbf{k}} \left(4E_{\mathbf{k}}^2 - \omega^2 \right)} \equiv -1 - \omega^2 \, A(\omega) \, ,$$

where $A(\omega)$ is positive at small ω. Similarly,

$$\Sigma(\omega, \mathbf{0}) \equiv i\omega \, \frac{2U}{V} \sum_{\mathbf{k}} \frac{\sin 2\theta_{\mathbf{k}}}{\omega^2 - 4E_{\mathbf{k}}^2} = -i\omega \, 2Um \, A(\omega) \, ,$$

$$\Sigma_s(\omega, \mathbf{0}) \equiv \frac{2U}{V} \sum_{\mathbf{k}} \sin^2 2\theta_{\mathbf{k}} \, \frac{2E_{\mathbf{k}}}{\omega^2 - 4E_{\mathbf{k}}^2} = -4U^2 m^2 \, A(\omega) \, .$$

At leading order in the small q, we can write

$$1 + \Sigma_c(\omega, \mathbf{q}) = 1 + \Sigma_c(\omega, \mathbf{0}) + \left(\Sigma_c(\omega, \mathbf{q}) - \Sigma_c(\omega, \mathbf{0}) \right) \simeq -\omega^2 \, A(\omega) + q^2 \, B(\omega) \, ,$$

where $B(\omega)$ is positive at small ω, and finite at $\omega = 0$, while we can safely take $\Sigma(\omega, \mathbf{q})$ and $\Sigma_s(\omega, \mathbf{q})$ at $\mathbf{q} = \mathbf{0}$. Therefore, (3.106) becomes

$$\begin{pmatrix} -\omega^2 \, A(\omega) + q^2 \, B(\omega) & -i\omega \, 2Um \, A(\omega) \\ i\omega \, 2Um \, A(\omega) & 1 - 4U^2 m^2 \, A(\omega) \end{pmatrix} \begin{pmatrix} X(\omega, \mathbf{q}) \\ P(\omega, \mathbf{q}) \end{pmatrix}$$

$$= -i \, \sqrt{2} \, V(\omega, \mathbf{q}) \begin{pmatrix} -i\omega \, 2Um \, A(\omega) \\ -4U^2 m^2 \, A(\omega) \end{pmatrix} \, ,$$

with solution

$$X(\omega, \mathbf{q}) = \sqrt{2} \, U \, S^+(\mathbf{q} + \mathbf{Q})$$

$$= -\sqrt{2} \, U \, \frac{2m \, \omega \, A(\omega)}{\left[1 - 4U^2 m^2 \, A(\omega) \right] B(\omega) \, q^2 - \omega^2 \, A(\omega)} \, V(\omega, \mathbf{q}) \, ,$$

$$P(\omega, \mathbf{q}) = -i \, \sqrt{2} \, U \, S^+(\omega, \mathbf{q})$$

$$= i \, \sqrt{2} \, U \, \frac{4Um^2 \, A(\omega) \, B(\omega) \, q^2}{\left[1 - 4U^2 m^2 \, A(\omega) \right] B(\omega) \, q^2 - \omega^2 \, A(\omega)} \, V(\omega, \mathbf{q}) \, ,$$

which implies that

$$\chi_{\pm}(\omega, \mathbf{q}) = -\frac{4Um^2 \, A(\omega) \, B(\omega) \, q^2}{\left[1 - 4U^2 m^2 \, A(\omega) \right] B(\omega) \, q^2 - \omega^2 \, A(\omega)} \, . \tag{3.107}$$

Indeed $\chi_\pm(\omega, \mathbf{q})$ vanishes at $q = 0$, consistently with (3.100) for $a = x, y$, and, for small q, has a pole at $\omega^2 = \omega^2(\mathbf{q}) \equiv \left(1 - 4U^2 m^2 A(0)\right) B(0) q^2$, an acoustic mode in agreement with Goldstone's theorem.

For $U \gg t$, $E_\mathbf{k} \simeq Um + \epsilon_\mathbf{k}^2/2Um$, and the root of $K(U, m, 0)$ in (3.97) yields

$$|m| \simeq \frac{1}{2} - 2d\frac{t^2}{U}\ .$$

In that same limit,

$$1 - 4U^2 m^2 A(0) \to 8d\frac{t^2}{U^2}\ , \quad B(0) \to 2\frac{t^2}{U^2}\ , \tag{3.108}$$

so that

$$\chi_\pm(\omega, \mathbf{q}) \to -\frac{1}{2}\frac{J q^2}{d\,J^2 q^2 - \omega^2}\ ,$$

where $J = 4t^2/U$ is the antiferromagnetic exchange of the Heisenberg model corresponding to the large-U limit of the half-filled Hubbard model, and therefore $\omega^2(\mathbf{q}) \to d\,J^2 q^2$. We recall that within the spin-wave approximation one finds

$$\omega_{SW}^2(\mathbf{q}) = 4S^2 \left(J(\mathbf{q}) - J(\mathbf{Q})\right)\left(J(\mathbf{q}+\mathbf{Q}) - J(\mathbf{Q})\right),$$

where, in a d-dimensional hypercubic lattice with nearest neighbour exchange, $J(\mathbf{q}) = J \sum_{i=1}^d \cos q_i$. For the present case of $S = 1/2$, one can easily realise that $\omega_{SW}^2(\mathbf{q} \to \mathbf{0}) \to d\,J^2 q^2$, remarkably the same result of the time-dependent Hartree-Fock approximation.

We further note that, if we set $\omega = 0$ and then send $q \to 0$,

$$\chi_\pm(0, \mathbf{q} \to \mathbf{0}) = -\frac{4U m^2 A(0)}{1 - 4U^2 m^2 A(0)} \equiv -\kappa_\pm\ , \tag{3.109}$$

where $\kappa_\pm$ is the thermodynamic susceptibility to a magnetic field perpendicular to the symmetry-breaking axis. It follows that $\kappa_\pm$ is positive and finite, contrary to κ_z that vanishes for $q \to 0$ irrespective whether we first send $\omega \to 0$ and then $q \to 0$ or vice versa. That is a well-known behaviour of antiferromagnets.

In conclusion, we have explicitly demonstrated in the case of a half-filled Hubbard model that the time-dependent Hartree-Fock approximation yields the linear response functions consistent with conservation laws.

Problems

3.1 BCS mean-field theory of superconductivity—Let us consider again the half-filled repulsive Hubbard model (3.86), thus $U > 0$, now in presence of a Zeeman field that splits spin-up electrons from spin-down electrons. The Hamiltonian is written as

$$H = -t \sum_{<\mathbf{RR'}>} \sum_{\sigma} \left(c^{\dagger}_{\mathbf{R}\sigma} c_{\mathbf{R'}\sigma} + H.c. \right) + U \sum_{\mathbf{R}} \left(\hat{n}_{\mathbf{R}\uparrow} - \frac{1}{2} \right) \left(\hat{n}_{\mathbf{R}\downarrow} - \frac{1}{2} \right)$$
$$- \mu \sum_{\mathbf{R}} \left(\hat{n}_{\mathbf{R}\uparrow} - \hat{n}_{\mathbf{R}\downarrow} \right).$$

(3.110)

We previously showed that the Hartree-Fock approximation at $\mu = 0$ yields an antiferromagnetic state that breaks spin-$SU(2)$ symmetry and is characterised by a staggered magnetisation that can be oriented in any direction as well as by a finite magnetic susceptibility to a Zeeman field perpendicular to the staggered magnetisation. In the case of the Hamiltonian (3.110) where the Zeeman field is along the z-axis, the staggered magnetisation must lie within the $x - y$ plane, which corresponds to an order parameter

$$\Delta \equiv (-1)^R \left(\cos \phi \, S_{x\mathbf{R}} + \sin \phi \, S_{y\mathbf{R}} \right) = \frac{(-1)^R}{2} \left(e^{-i\phi} \langle c^{\dagger}_{\mathbf{R}\uparrow} c_{\mathbf{R}\downarrow} \rangle + e^{i\phi} \langle c^{\dagger}_{\mathbf{R}\downarrow} c_{\mathbf{R}\uparrow} \rangle \right),$$

(3.111)

with arbitrary phase ϕ reflecting the residual spin-$U(1)$ symmetry at $\mu \neq 0$. In addition to the staggered in-plane magnetisation, the Hartree-Fock state has a finite uniform magnetisation along z.

Let us consider now the unitary transformation

$$c_{\mathbf{R}\uparrow} \to c_{\mathbf{R}\uparrow}, \quad c_{\mathbf{R}\downarrow} \to (-1)^R c^{\dagger}_{\mathbf{R}\downarrow},$$

under which

$$\hat{n}_{\mathbf{R}\uparrow} - \frac{1}{2} \to \hat{n}_{\mathbf{R}\uparrow} - \frac{1}{2}, \quad \hat{n}_{\mathbf{R}\downarrow} - \frac{1}{2} \to -\hat{n}_{\mathbf{R}\downarrow} + \frac{1}{2},$$

so that the half-filling condition

$$0 = N - V = \sum_{\mathbf{R}} \left(\hat{n}_{\mathbf{R}\uparrow} + \hat{n}_{\mathbf{R}\downarrow} - 1 \right) \to \sum_{\mathbf{R}} \left(\hat{n}_{\mathbf{R}\uparrow} - \hat{n}_{\mathbf{R}\downarrow} \right) \equiv S_z$$

transforms into the condition of vanishing magnetisation along z, while the Zeeman term

$$-\mu \sum_{\mathbf{R}} \left(\hat{n}_{\mathbf{R}\uparrow} - \hat{n}_{\mathbf{R}\downarrow} \right) \to -\mu \sum_{\mathbf{R}} \left(\hat{n}_{\mathbf{R}\uparrow} + \hat{n}_{\mathbf{R}\downarrow} - 1 \right) \equiv \mu \left(N - V \right),$$

becomes a chemical potential term, and the Hamiltonian (3.110) turns into

$$H = -t \sum_{<\mathbf{RR'}>} \sum_{\sigma} \left(c^{\dagger}_{\mathbf{R}\sigma} c_{\mathbf{R'}\sigma} + H.c. \right) - U \sum_{\mathbf{R}} \left(\hat{n}_{\mathbf{R}\uparrow} - \frac{1}{2} \right) \left(\hat{n}_{\mathbf{R}\downarrow} - \frac{1}{2} \right)$$
$$- \mu \sum_{\mathbf{R}} \left(\hat{n}_{\mathbf{R}\uparrow} + \hat{n}_{\mathbf{R}\downarrow} - 1 \right),$$

(3.112)

namely into the attractive Hubbard model in the subspace of zero magnetisation and with finite chemical potential.[6] Since the spectrum is invariant under a unitary transformation, the Hartree-Fock state transforms into a state characterised by a finite expectation value

$$\Delta = \frac{(-1)^R}{2} \left(e^{-i\phi} \langle c^{\dagger}_{\mathbf{R}\uparrow} c_{\mathbf{R}\downarrow} \rangle + e^{i\phi} \langle c^{\dagger}_{\mathbf{R}\downarrow} c_{\mathbf{R}\uparrow} \rangle \right) \rightarrow e^{-i\phi} \langle c^{\dagger}_{\mathbf{R}\uparrow} c^{\dagger}_{\mathbf{R}\downarrow} \rangle + e^{i\phi} \langle c_{\mathbf{R}\downarrow} c_{\mathbf{R}\uparrow} \rangle,$$

(3.114)

which corresponds to an order parameter breaking charge $U(1)$ symmetry. This state describes therefore a superconductor below a critical temperature T_c which corresponds to the Neél temperature T_N of the repulsive case.

- Show that this conclusion remains true even if further neighbour hopping terms are added to the nearest neighbour one in (3.112), yielding to a generic dispersion $\epsilon(\mathbf{k})$ in momentum space, provided the density of states at the chemical potential

$$\rho_0 \equiv \frac{1}{V} \sum_{\mathbf{k}} \delta\big(\epsilon(\mathbf{k}) - \mu\big)$$

is finite.

One way to convince oneself that the Hartree-Fock state indeed describes a super-conductor is to prove that it has Meissner effect, see Sect. 2.6.5. For that,

---

[6] We note that at $\mu = 0$ the Hamiltonian (3.112) is invariant under the particle-hole transformation

$$c_{\mathbf{R}\uparrow} \rightarrow (-1)^R c^{\dagger}_{\mathbf{R}\downarrow}, \quad c_{\mathbf{R}\downarrow} \rightarrow -(-1)^R c^{\dagger}_{\mathbf{R}\uparrow},$$

(3.113)

which leaves the spin operators invariant but transforms $N - V \rightarrow -(N - V)$. If the equilibrium state is also invariant under that transformation, then $N - V \equiv -(N - V)$, thus $N = V$, i.e., the model is at half-filling for $\mu = 0$. On the contrary, if $\mu \neq 0$, the Hamiltonian is not invariant and thus $N \neq V$. In reality, the Hamiltonian (3.112) at $\mu = 0$ is invariant, besides spin $SU(2)$, also under a charge $SU(2)$ symmetry with generators

$$I_z \equiv \frac{1}{2} \sum_{\mathbf{R}\sigma} \left(n_{\mathbf{R}\sigma} - \frac{1}{2} \right), \quad I^+ \equiv \sum_{\mathbf{R}} (-1)^R c^{\dagger}_{\mathbf{R}\uparrow} c_{\mathbf{R}\downarrow}, \quad I^- \equiv I^{+\dagger},$$

which implement the transformation (3.113). A finite chemical potential term $-\mu I^z$ lowers that charge $SU(2)$ symmetry into the common charge $U(1)$.

- Apply the time-dependent Hartree-Fock approximation and show that the transverse current-current response function vanishes for $\omega = 0$ and $\mathbf{q} \to 0$, corresponding to $n = n_c$ in (2.104), thus to a perfect diamagnetic behaviour.

In a lattice model with dispersion $\epsilon(\mathbf{k})$, the consequence of gauge invariance (2.79) changes.

- Find that equation calculating

$$i \frac{\partial}{\partial t} \, i \frac{\partial}{\partial t'} \, \chi(t - t', \mathbf{q}) = i \frac{\partial}{\partial t} \, i \frac{\partial}{\partial t'} \left(-i \, \theta(t - t') \left\langle \left[\rho(t, \mathbf{q}), \, \rho(-\mathbf{q}, t') \right] \right\rangle \right),$$

where

$$\rho(\mathbf{q}) = \sum_{\mathbf{k}\sigma} c_{\mathbf{k}\sigma}^{\dagger} \, c_{\mathbf{k}+\mathbf{q}\sigma} \,,$$

recalling that

$$i \frac{\partial \rho(\mathbf{q})}{\partial t} = \left[\rho(\mathbf{q}), \, H \right],$$

and taking the Fourier transform in frequency.
- Show that within the time-dependent Hartree-Fock approximation gauge invariance is satisfied, i.e., that equation holds, despite the Hartree-Fock solution breaks global charge $U(1)$ symmetry. To simplify the calculation, consider the Hamiltonian (3.112) at $\mu = 0$ so to use the results of Sect. 3.4.

Feynman Diagram Technique

4

Time-dependent Hartree-Fock is a relatively simple way to access excitation properties thus to go beyond Hartree-Fock theory. A more systematic scheme, which includes time-dependent Hartree-Fock as an approximation, is to perform perturbation theory. However, the perturbation expansion in a many-body problem is not as straightforward as in a single-body case but can be simplified substantially by means of the so-called Feynman diagram technique. In this chapter we present this technique at finite temperature in the so-called Matsubara frequency formalism. The relevant references of all this chapter are the books by Abrikosov, Gorkov, and Dzyaloshinskii [1] and by Noziéres [2], as well as the seminal work by Luttinger and Word [3].

4.1 Preliminaries

We first need some preliminary definitions and results, which we are going to present in this section.

4.1.1 Imaginary-Time Ordered Products

We introduce the imaginary time evolution of an operator $A(\mathbf{x})$ through

$$A(\mathbf{x}, \tau) = e^{H\tau} A(\mathbf{x}) e^{-H\tau} , \tag{4.1}$$

where H is the fully interacting Hamiltonian and $\tau \in [0, \beta]$, where $\beta = 1/T$ is the inverse temperature (we use $\hbar = 1$ and $K_B = 1$).

© The Author(s), under exclusive license to Springer Nature Switzerland AG 2022
M. Fabrizio, *A Course in Quantum Many-Body Theory*, Graduate Texts in Physics,
https://doi.org/10.1007/978-3-031-16305-0_4

Given two operators $A(\mathbf{x})$ and $B(\mathbf{y})$, we define their time-ordered product through

$$T_\tau\Big(A(\mathbf{x}, \tau)\, B(\mathbf{y}, \tau')\Big) \equiv -\theta(\tau - \tau')\, A(\mathbf{x}, \tau)\, B(\mathbf{y}, \tau') \mp \theta(\tau' - \tau)\, B(\mathbf{y}, \tau')\, A(\mathbf{x}, \tau),$$
(4.2)

where T_τ denotes the time-ordered product, and the minus sign applies when one or both the operators are bosonic-like, namely contain bosons or an even product of fermionic operators, while the plus sign when both operators are fermionic-like, i.e., contain an odd product of fermionic operators.

Analogously, given n operators, $A_i(\mathbf{x}_i, \tau_i), i = 1, \ldots, n$, we can define their time-ordered product

$$T_\tau\Big(A_1(\mathbf{x}_1, \tau_1) \ldots A_n(\mathbf{x}_n, \tau_n)\Big),$$
(4.3)

as the product in which operators with earlier times appear on the right of those at later times, and the sign is plus if the number of permutations needed to bring fermionic-like operators in the correct time-ordered sequence is even, and minus otherwise.

The imaginary time Green's function is defined by the expectation value of time-ordered products of operators. Specifically, for two operators $A(\mathbf{x})$ and $B(\mathbf{y})$, the Green's function is

$$\begin{aligned}
G_{AB}(\tau - \tau'; \mathbf{x}, \mathbf{y}) &= -\langle T_\tau\Big(A(\mathbf{x}, \tau)\, B(\mathbf{y}, \tau')\Big)\rangle \\
&= -\theta(\tau - \tau')\, \langle A(\mathbf{x}, \tau)\, B(\mathbf{y}, \tau')\rangle \mp \theta(\tau' - \tau)\, \langle B(\mathbf{y}, \tau')\, A(\mathbf{x}, \tau)\rangle.
\end{aligned}$$
(4.4)

We note that, because of time-translation invariance, the Green's function only depends on the time difference $\tau - \tau' \in [-\beta, \beta]$.

Similarly, the multi-operator Green's function is defined as

$$G_{A_1, \ldots, A_n}(\tau_1, \ldots, \tau_n; \mathbf{x}_1, \ldots, \mathbf{x}_n) = -\langle T_\tau\Big(A_1(\mathbf{x}_1, \tau_1) \ldots A_n(\mathbf{x}_n, \tau_n)\Big)\rangle.$$
(4.5)

4.1.2 Matsubara Frequencies

Let us consider the two-operator Green's function $G_{AB}(\mathbf{x}, \mathbf{y}; \tau)$, with $\tau \in [-\beta, \beta]$ being the time difference. Since the time domain is bounded, we can introduce a discrete Fourier transform, with frequencies $\omega_n = (2\pi/2\beta)\, n = \pi n T$, with n an integer. We define, dropping for simplicity the spatial dependence,

$$G_{AB}(i\omega_n) = \frac{1}{2} \int_{-\beta}^{\beta} d\tau\, e^{i\omega_n \tau}\, G_{AB}(\tau),$$

and, consequently,

$$G_{AB}(\tau) = T \sum_n e^{-i\omega_n \tau}\, G_{AB}(i\omega_n).$$

Since the trace is invariant under cyclic permutations, it follows that

$$\langle B(0) A(\tau) \rangle = \frac{1}{Z} \, \mathrm{Tr} \left[e^{-\beta H} B \, e^{H\tau} A \, e^{H-\tau} \right]$$
$$= \frac{1}{Z} \, \mathrm{Tr} \left[e^{-\beta H} e^{\beta H} e^{H\tau} A \, e^{-H\tau} e^{-\beta H} B \right] = \langle A(\beta + \tau) B(0) \rangle,$$

as well as

$$G_{AB}(i\omega_n) = \frac{1}{2} \int_{-\beta}^{\beta} d\tau \, e^{i\omega_n \tau} \, G_{AB}(\tau)$$

$$= -\frac{1}{2} \int_{0}^{\beta} d\tau \, e^{i\omega_n \tau} \, \langle A(\tau) B(0) \rangle \mp \frac{1}{2} \int_{-\beta}^{0} d\tau \, e^{i\omega_n \tau} \, \langle B(0) A(\tau) \rangle$$

$$= -\frac{1}{2} \int_{0}^{\beta} d\tau \, e^{i\omega_n \tau} \, \langle A(\tau) B(0) \rangle \mp \frac{1}{2} \int_{-\beta}^{0} d\tau \, e^{i\omega_n \tau} \, \langle A(\beta + \tau) B(0) \rangle$$

$$= -\frac{1}{2} \int_{0}^{\beta} d\tau \, e^{i\omega_n \tau} \, \langle A(\tau) B(0) \rangle \mp \frac{1}{2} e^{-i\omega_n \beta} \int_{0}^{\beta} d\tau \, e^{i\omega_n \tau} \, \langle A(\tau) B(0) \rangle$$

$$= \frac{1}{2} \left(1 \pm e^{-i\omega_n \beta} \right) \int_{0}^{\beta} d\tau \, e^{i\omega_n \tau} \, G_{AB}(\tau). \tag{4.6}$$

Since $\exp(-i\omega_n \beta) = (-1)^n$ then a non-zero Fourier transform requires for bosonic-like operators even integers $n = 2m$, and for fermionic-like odd $n = 2m + 1$. We thus define the bosonic Matsubara frequencies

$$\Omega_m = 2m \pi T , \tag{4.7}$$

and the fermionic Matsubara frequencies

$$\epsilon_m = (2m + 1) \pi T , \tag{4.8}$$

so that the bosonic- and fermionic-like Fourier transform of $G_{AB}(\tau)$ become, respectively,

$$G_{AB}(\Omega_m) = \int_{0}^{\beta} d\tau \, e^{i\Omega_m \tau} \, G_{AB}(\tau) , \quad G_{AB}(\epsilon_m) = \int_{0}^{\beta} d\tau \, e^{i\epsilon_m \tau} \, G_{AB}(\tau) , \tag{4.9}$$

hence just require the knowledge of the Green's functions for positive τ.

4.1.2.1 Connection Between Bosonic Green's Functions and Linear Response Functions

Let us assume that both A and B are observable quantities, which implies they are hermitian, hence bosonic-like operators. Then, for positive τ, we have

$$G_{AB}(\tau) = -\langle A(\tau)\,B(0)\rangle = -\frac{1}{Z}\sum_n e^{-\beta E_n}\,\langle n\,|\,e^{H\tau}\,A\,e^{-H\tau}\,B\,|\,n\rangle$$

$$= -\frac{1}{Z}\sum_{nm} e^{-\beta E_n}\,e^{-(E_m-E_n)\tau}\,\langle n\,|\,A\,|\,m\rangle\,\langle m\,|\,B\,|\,n\rangle\,.$$

Since

$$\int_0^\beta d\tau\;e^{i\Omega_l\tau}\,e^{-(E_m-E_n)\tau} = \frac{1}{i\Omega_l-(E_m-E_n)}\left(e^{-(E_m-E_n)\beta}-1\right),$$

we finally get

$$G_{AB}(i\Omega_l) = \frac{1}{Z}\sum_{nm}\left(e^{-\beta E_n}-e^{-\beta E_m}\right)\frac{1}{i\Omega_l-(E_m-E_n)}\,\langle n\,|\,A\,|\,m\rangle\,\langle m\,|\,B\,|\,n\rangle\,.$$

$$(4.10)$$

Comparing this expression with the Fourier transform in frequency of the linear response function $\chi_{AB}(\omega)$, see (2.30), we easily realize that the two coincide if we analytically continue $G_{AB}(i\Omega_l)$ on the real axis, $i\Omega_l \to \omega + i\eta$. In other words the knowledge of $G_{AB}(i\Omega_l)$ means that we know the value of the function $G_{AB}(z)$ of the complex variable z on an infinite set of points on the imaginary axis $z_l = i\Omega_l$. This is sufficient to determine $G_{AB}(z)$ in the complex plane given the supplementary condition that, since $G_{AB}(z) = \chi_{AB}(z)$, it has to be analytic in the upper half-plane.

4.1.2.2 Useful Formulas

There is a very useful trick to perform summations over Matsubara frequencies. Let us start from the fermionic case. Suppose we have to calculate

$$T\sum_n F(i\epsilon_n)\,,$$

where we assume that $F(z)$ vanishes faster than $1/|z|$ for $|z| \to \infty$ and has poles but on the imaginary axis. We note that the Fermi distribution function of a complex variable z,

$$f(z) = \frac{1}{e^{\beta z}+1}\,,$$

has poles at

$$z_n = i\,(2n+1)\,\frac{\pi}{\beta} = i\epsilon_n\,,$$

Fig. 4.1 Integration contour
that run anticlockwise
avoiding the imaginary axis

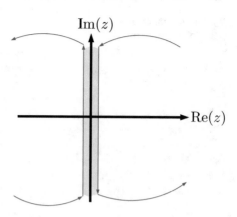

namely right at the Matsubara frequencies, with residue $-T$. Let us consider the
integral along the contour shown in Fig. 4.1,

$$I = \oint \frac{dz}{2\pi i} \, f(z) \, F(z) \, .$$

This integral can be calculated through all poles of the integrand in the region enclosed
by the contour that includes the imaginary axis, shaded area in Fig. 4.1. Since this
area is enclosed clockwise, the integral yields

$$I = -\sum_{z_n} \mathrm{Res}\Big(f(z_n) \Big) F(z_n) = T \sum_n F(i\epsilon_n) \, ,$$

which is just the sum we want to calculate. On the other hand, I can be equally
calculated by catching all poles of the integrand in the area which does not include
the imaginary axis, the non-shaded region, which, being enclosed anti-clockwise
gives

$$I = \sum_{z_* = \text{ poles of } F(z)} f(z_*) \, \mathrm{Res}\Big(F(z_*) \Big) \, .$$

Therefore

$$T \sum_n F(i\epsilon_n) = \sum_{z_* = \text{ poles of } F(z)} f(z_*) \, \mathrm{Res}\Big(F(z_*) \Big) \, . \tag{4.11}$$

If $F(z)$ has branch cuts the calculation is slightly more complicated since we have
to deform the contour so as to avoid them. In this case, instead of catching poles, we
have to integrate along branch cuts. For instance, suppose that $F(z)$ has a branch cut
along the horizontal axis $z = x + i\omega$, with $x \in [-\infty, \infty]$, as shown in Fig. 4.2. Let
us consider the non-shaded area enclosed inside the contour depicted in Fig. 4.2. In
this area the integrand is analytic, and thus the contour integral vanishes. On the other
hand, this contour integral is also equal to the contribution of the poles inside the

Fig. 4.2 Integration contour
in the presence of a branch
cut

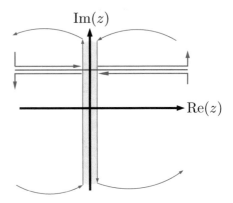

shaded area, which includes the imaginary axis, plus the integral along the contour
that encloses the branch cut. Therefore

$$T \sum_n F(i\epsilon_n) = - \int \frac{dx}{2\pi i} \, f(x + i\omega) \left[F(x + i\omega + i0^+) - F(x + i\omega - i0^+) \right].$$

(4.12)

Note that the above equation accounts also for the circumstance in which the branch
cut does not extend over the whole axis $z = x + i\omega$, since the term in square brackets
does vanishes only along the branch cut.

When $F(z)$ has both poles, at $z = z_n$, and branch cuts, at $z = x + i\omega_m$, with
$x \in [-\infty, \infty]$, the result of the summation over the Matsubara frequencies can be
readily found by the previous two examples, and is

$$T \sum_n F(i\epsilon_n) = \sum_n f(z_n) \, \mathrm{Res}\left(F(z_n) \right)$$
$$- \sum_m \int \frac{dx}{2\pi i} \, f(x + i\omega_m) \left[F(x + i\omega_m + i0^+) - F(x + i\omega_m - i0^+) \right].$$

(4.13)

In the case in which the summation is performed over bosonic frequencies, we can
proceed similarly once we recognise that the poles of the Bose distribution function

$$b(z) = \frac{1}{e^{\beta z} - 1} \, ,$$

coincide with the bosonic Matsubara frequencies,

$$z_n = i \, 2n \frac{\pi}{\beta} \equiv i \, \Omega_n \, ,$$

and have residue T.

4.1.3 Single-Particle Green's Functions

Among the expectation values of time-ordered products, an important role in the diagrammatic technique is played by the so-called single-particle Green's functions.

4.1.3.1 Fermionic Case

If $\Psi_\sigma(\mathbf{x}, \tau)$ is the imaginary-time evolution of the Fermi field, then the single-particle Green's functions is defined through

$$
\begin{aligned}
G_{\sigma\sigma'}(\tau; \mathbf{x}, \mathbf{y}) &= -\langle T_\tau \left(\Psi_\sigma(\mathbf{x}, \tau) \Psi_{\sigma'}^\dagger(\mathbf{y}) \right) \rangle \\
&= -\theta(\tau) \langle \Psi_\sigma(\mathbf{x}, \tau) \Psi_{\sigma'}^\dagger(\mathbf{y}) \rangle + \theta(-\tau) \langle \Psi_{\sigma'}^\dagger(\mathbf{y}) \Psi_\sigma(\mathbf{x}, \tau) \rangle .
\end{aligned}
\tag{4.14}
$$

In principle, we could also define *anomalous* Green's functions, as opposed to the above *normal* ones, as the expectation values of the time-ordered products $\Psi_\sigma(\mathbf{x}, \tau) \Psi_{\sigma'}(\mathbf{y})$ or $\Psi_\sigma^\dagger(\mathbf{x}, \tau) \Psi_{\sigma'}^\dagger(\mathbf{y})$. However, those expectation values can be finite only if the $U(1)$ symmetry that derives from the conservation of the number of particles,

$$
\Psi_\sigma(\mathbf{x}) \to e^{i\phi} \Psi_\sigma(\mathbf{x}), \qquad \Psi_\sigma^\dagger(\mathbf{x}) \to e^{-i\phi} \Psi_\sigma^\dagger(\mathbf{x}), \tag{4.15}
$$

is spontaneously broken, which occurs in superconductors, if the fermions are charged, or in superfluids, if they are neutral. Hereafter, we shall discard such possibility, and thus assume that only normal single-particle Green's functions can be different from zero.

If we perform a spectral representation of $G_{\sigma\sigma'}(\tau; \mathbf{x}, \mathbf{y})$ in (4.14) and calculate the Fourier transform in the fermionic Matsubara frequency $i\epsilon$, we readily find that

$$
G_{\sigma\sigma'}(i\epsilon; \mathbf{x}, \mathbf{y}) = \frac{1}{Z} \sum_{nm} \left(e^{-\beta E_n} + e^{-\beta E_m} \right) \frac{\langle n | \Psi_\sigma(\mathbf{x}) | m \rangle \langle m | \Psi_{\sigma'}^\dagger(\mathbf{y}) | n \rangle}{i\epsilon - \left(E_m - E_n \right)} .
$$

Let us take $\sigma = \sigma'$ and $\mathbf{x} = \mathbf{y}$, which correspond to the so-called local Green's function, and introduce the real and positive spectral function

$$
A_\sigma(\epsilon, \mathbf{x}) = \frac{1}{Z} \sum_{nm} \left(e^{-\beta E_n} + e^{-\beta E_m} \right) \left| \langle m | \Psi_\sigma^\dagger(\mathbf{x}) | n \rangle \right|^2 \delta\left(\epsilon - E_m + E_n \right),
$$

through which, after continuation in the complex plane $i\epsilon \to z$

$$
G_{\sigma\sigma}(z; \mathbf{x}, \mathbf{x}) = \int d\epsilon \, \frac{1}{z - \epsilon} \, A_\sigma(\epsilon, \mathbf{x}) .
$$

As function of the complex frequency, $G(z)$ has generally branch cut singularities along the real axis. Indeed

$$
G_{\sigma\sigma}\left(z = \epsilon + i0^+; \mathbf{x}, \mathbf{x}\right) - G_{\sigma\sigma}\left(z = \epsilon - i0^+; \mathbf{x}, \mathbf{x}\right) = -2\pi i \, A_\sigma(\epsilon, \mathbf{x}).
$$

What is the physical meaning of the spectral function? Let us rewrite $A_\sigma(\epsilon, \mathbf{x})$ in the following equivalent way:

$$
\begin{aligned}
A_\sigma(\epsilon, \mathbf{x}) = & \sum_{nm} \frac{e^{-\beta E_n}}{Z} \left| \langle m \mid \Psi_\sigma^\dagger(\mathbf{x}) \mid n \rangle \right|^2 \delta(\epsilon - E_m + E_n) \\
& + \sum_{nm} \frac{e^{-\beta E_n}}{Z} \left| \langle m \mid \Psi_\sigma(\mathbf{x}) \mid n \rangle \right|^2 \delta(\epsilon + E_m - E_n),
\end{aligned}
\tag{4.16}
$$

which shows that $A_\sigma(\epsilon, \mathbf{x})$ is just the probability of adding, first term in the right hand side, or removing, second term, a particle at position $\mathbf{x}$ with spin σ. For that reason, $A_\sigma(\epsilon, \mathbf{x})$ is commonly known as the single-particle local density of states. Finally we note that

$$
\begin{aligned}
\int d\epsilon \, A_\sigma(\epsilon, \mathbf{x}) &= \sum_{nm} \frac{e^{-\beta E_n}}{Z} \left[\left| \langle m|\Psi_\sigma^\dagger(\mathbf{x})|n \rangle \right|^2 + \left| \langle m|\Psi_\sigma(\mathbf{x})|n \rangle \right|^2 \right] \\
&= \sum_{n} \frac{e^{-\beta E_n}}{Z} \left[\langle n \mid \Psi_\sigma(\mathbf{x}) \Psi_\sigma^\dagger(\mathbf{x}) \mid n \rangle + \langle n \mid \Psi_\sigma^\dagger(\mathbf{x}) \Psi_\sigma(\mathbf{x}) \mid n \rangle \right] \\
&= \sum_{n} \frac{e^{-\beta E_n}}{Z} \langle n \mid \left\{ \Psi_\sigma(\mathbf{x}), \Psi_\sigma^\dagger(\mathbf{x}) \right\} \mid n \rangle = 1,
\end{aligned}
$$

namely that the integral of the spectral function is normalised to one.

Hereafter, we assume to work in the grand canonical ensemble, so that the Hamiltonian H contains a chemical potential term $-\mu N$. In the limit of zero temperature, $T = 0$, the sum over $| n \rangle$ in (4.16) reduces to just the ground state $| 0 \rangle$ (not to be confused with the vacuum), which contains N_0 electrons and has energy E_0. It follows that (4.16) becomes

$$
\begin{aligned}
A_\sigma(\epsilon, \mathbf{x}) = \sum_{m} \Big\{ & \left| \langle m \mid \Psi_\sigma^\dagger(\mathbf{x}) \mid 0 \rangle \right|^2 \delta(\epsilon - E_m + E_0) \\
& + \left| \langle m \mid \Psi_\sigma(\mathbf{x}) \mid 0 \rangle \right|^2 \delta(\epsilon + E_m - E_0) \Big\},
\end{aligned}
$$

and, since $E_m - E_0$ is positive by definition, $A_\sigma(\epsilon, \mathbf{x})$ at $\epsilon > 0$ is the density of states for adding one electrons, while at $\epsilon < 0$ for removing one electron. The local single-particle density of states is, e.g., measurable in tunnelling microscopy, see Fig. 4.3.

In a periodic system, and considering a metal with a partially filled conduction band and unbroken spin $SU(2)$, we can define the single-particle Green's function through the creation and annihilation operators corresponding to the conduction band Bloch waves, namely,

$$
G_{\sigma\sigma'}(\mathbf{k}, \mathbf{p}; \tau) = -\langle T_\tau \left(c_{\mathbf{k}\sigma}(\tau) c_{\mathbf{p}\sigma'}^\dagger \right) \rangle = \delta_{\mathbf{kp}} \, \delta_{\sigma\sigma'} \, G(\tau, \mathbf{k}),
$$

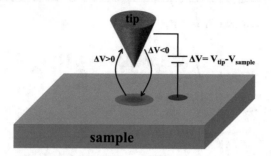

Fig. 4.3 Sketch of a tunnelling microscopy experiment. A tip at a certain distance from the sample surface is kept at a different electrochemical potential with respect to the sample below, the potential drop being ΔV. If $\Delta V > 0$, electrons from the surface of the sample flow to the tip, and the reverse if $\Delta V > 0$. The flowing current is proportional to the probability of removing an electron in the former case, thus to the local single-particle density of states at negative energy $\epsilon = -e\,\Delta V$, while to the probability of adding an electron in the latter case, i.e., the local single-particle density of states at positive energy $\epsilon = -e\,\Delta V > 0$

the δ-functions deriving from translational and spin-$SU(2)$ symmetries, where

$$G(\tau, \mathbf{k}) = -\langle T_\tau \left(c_{\mathbf{k}\sigma}(\tau)\, c_{\mathbf{k}\sigma}^\dagger \right) \rangle, \qquad \forall \sigma = \uparrow, \downarrow. \qquad (4.17)$$

The Fourier transform in fermionic Matsubara frequencies $i\epsilon$ has the spectral decomposition

$$G(i\epsilon, \mathbf{k}) = \frac{1}{Z} \sum_{nm} \left(e^{-\beta E_n} + e^{-\beta E_m} \right) \left| \langle m \mid c_{\mathbf{k}\sigma}^\dagger \mid n \rangle \right|^2 \frac{1}{i\epsilon - E_m + E_n}, \qquad (4.18)$$

which, upon defining the real and positive density of states for adding and removing a particle at momentum $\mathbf{k}$, i.e.,

$$A(\epsilon, \mathbf{k}) = \frac{1}{Z} \sum_{nm} \left(e^{-\beta E_n} + e^{-\beta E_m} \right) \left| \langle m \mid c_{\mathbf{k}\sigma}^\dagger \mid n \rangle \right|^2 \delta\left(\epsilon - E_m + E_n \right) \geq 0, \qquad (4.19)$$

satisfying, as before,

$$\int d\epsilon\, A(\epsilon, \mathbf{k}) = 1, \qquad (4.20)$$

can be simply written as the Hilbert transform

$$G(i\epsilon, \mathbf{k}) = \int d\omega\, \frac{A(\omega, \mathbf{k})}{i\epsilon - \omega}. \qquad (4.21)$$

Therefore

$$\operatorname{Re} G(i\epsilon, \mathbf{k}) = \operatorname{Re} G(-i\epsilon, \mathbf{k}) = -\int d\omega\, \frac{\omega}{\epsilon^2 + \omega^2}\, A(\omega, \mathbf{k}),$$

$$\operatorname{Im} G(i\epsilon, \mathbf{k}) = -\operatorname{Im} G(-i\epsilon, \mathbf{k}) = -\epsilon \int d\omega\, \frac{A(\omega, \mathbf{k})}{\epsilon^2 + \omega^2}, \qquad (4.22)$$

which also imply through (4.20) that

$$\operatorname{Re} G(i\epsilon, \mathbf{k}) \xrightarrow[|\epsilon|\to\infty]{} -\frac{1}{\epsilon^2} \int d\omega\, \omega\, A(\omega, \mathbf{k}) = -\frac{1}{\epsilon^2} \left\langle \left\{ \left[c_{\mathbf{k}\sigma}, H \right], c_{\mathbf{k}\sigma}^\dagger \right\} \right\rangle,$$

$$\operatorname{Im} G(i\epsilon, \mathbf{k}) \xrightarrow[|\epsilon|\to\infty]{} -\frac{1}{\epsilon} \int d\omega\, A(\omega, \mathbf{k}) = -\frac{1}{\epsilon}.$$

(4.23)

Moreover, we note that for $\epsilon > 0$,

$$-\frac{1}{\pi} \operatorname{Im} G(i\epsilon, \mathbf{k}) = \int d\omega\, \frac{1}{\pi}\, \frac{\epsilon}{\epsilon^2 + \omega^2}\, A(\omega, \mathbf{k}) \tag{4.24}$$

is the overlap integral between the density of states $A(\omega, \mathbf{k})$ and a Lorentzian centred at $\omega = 0$ and with width ϵ. In an insulator with a hard gap, $A(\omega, \mathbf{k}) = 0$ for $\omega \in [-\omega_1, \omega_2]$, where ω_1 and ω_2 are both positive and their sum is the gap. In this case, the overlap integral (4.24) is vanishingly small for $\epsilon \ll \min(\omega_1, \omega_2)$ and strictly zero at $\epsilon = 0$. In a metal or, more generally, in a system with gapless single-particle excitations, $A(\omega, \mathbf{k})$ is finite and smooth around $\omega = 0$, and $A(0, \mathbf{k})$ is different from zero in a bona fide metal, and zero in a pseudo-gapped anomalous one. Here, therefore, the overlap integral (4.24) is finite and smooth at small ϵ, eventually vanishing as a power law at $\epsilon = 0$ in the case of a pseudo gap.

We can straightforwardly continue $G(\mathbf{k}, i\epsilon)$ in the complex frequency plane, $i\epsilon \to z \in \mathbb{C}$,

$$G(z, \mathbf{k}) = \int d\epsilon\, \frac{1}{z - \epsilon}\, A(\epsilon, \mathbf{k}), \tag{4.25}$$

or impose particular analytic properties. For instance, if we define

$$G_\pm(z, \mathbf{k}) = \int d\epsilon\, \frac{1}{z - \epsilon \pm i\eta}\, A(\epsilon, \mathbf{k}), \tag{4.26}$$

with $\eta > 0$ infinitesimal, then $G_+(z, \mathbf{k})$ is analytic in the upper half plane, and thus is a causal function commonly named *retarded* Green's function, while $G_-(z, \mathbf{k})$ is analytic in the lower half plane, an anti-causal function known as *advanced* Green's function. It follows that $G(z, \mathbf{k})$ has generally branch cuts on the real axis, which are shifted below in $G_+(z, \mathbf{k})$ and above in $G_-(z, \mathbf{k})$. Specifically, if $\epsilon \in \mathbb{R}$,

$$G(\epsilon + i0^+, \mathbf{k}) - G(\epsilon - i0^+, \mathbf{k}) = G_+(\epsilon, \mathbf{k}) - G_-(\epsilon, \mathbf{k}) = -2\pi i\, A(\epsilon, \mathbf{k}).$$

(4.27)

In other words, since $G_-(\epsilon, \mathbf{k}) = G_+(\epsilon, \mathbf{k})^*$,

$$\operatorname{Re} G_+(\epsilon, \mathbf{k}) = \operatorname{Re} G_-(\epsilon, \mathbf{k}), \quad \operatorname{Im} G_+(\epsilon, \mathbf{k}) = -\operatorname{Im} G_-(\epsilon, \mathbf{k}) = -\frac{1}{\pi}\, A(\epsilon, \mathbf{k}).$$

(4.28)

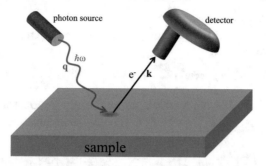

Fig. 4.4 Sketch of an ARPES experiment. A photon with momentum **q**, and thus frequency $\omega = cq$, hits the sample and provoke the emission of an electron. The detector allows extracting energy and momentum of the photoemitted electron. From energy and momentum of the photon and the emitted electron, one can trace back with controlled assumptions to the original energy ϵ and momentum **p** of that electron in the system. The intensity of the signal is therefore proportional to the probability of removing that electron, and thus to the density of states $A(\epsilon, \mathbf{k})$ times the Fermi distribution function $f(\epsilon)$

Throughout these lecture notes, whenever we write $G(\epsilon, \mathbf{k})$ with real ϵ we refer to the retarded component $G_+(\epsilon, \mathbf{k})$, unless otherwise specified. The advantage of dealing with $G_+(\epsilon, \mathbf{k})$ is that, being causal, its real and imaginary parts are related to each other by the Kramers-Krœnig equations.

We note that, because of translational symmetry,

$$A(\epsilon, \mathbf{x}) = A(\epsilon) = \frac{1}{V} \sum_{\mathbf{k}} A(\epsilon, \mathbf{k}) \,. \tag{4.29}$$

The single-particle density of states $A(\epsilon, \mathbf{k})$ for removing an electron can be measured, e.g., by Angle-Resolved Photoemission Spectroscopy (ARPES), as sketched in Fig. 4.4.

As an example, which will turn useful in what follows, let us consider non-interacting electrons described by the Hamiltonian

$$H_0 = \sum_{\mathbf{k}\sigma} \epsilon_{\mathbf{k}} \, c_{\mathbf{k}\sigma}^{\dagger} c_{\mathbf{k}\sigma} \,,$$

where, as discussed previously, $\epsilon_{\mathbf{k}}$ is measured with respect to the chemical potential μ, which is the Fermi energy at $T = 0$. In this case, it is easy to show that

$$G_0(i\epsilon, \mathbf{k}) = \frac{1}{i\epsilon - \epsilon_{\mathbf{k}}} \,, \tag{4.30}$$

while the spectral function is simply

$$A_0(\epsilon, \mathbf{k}) = \delta(\epsilon - \epsilon_{\mathbf{k}}) \,. \tag{4.31}$$

4.1.3.2 Bosonic Case

If $\Phi(\mathbf{x})$ is the Bose field for spinless bosons, the corresponding . single-particle Green's function reads

$$G(\mathbf{x}, \mathbf{y}; \tau) = -\langle T_\tau \left(\Phi(\mathbf{x}, \tau)\ \Phi(\mathbf{y})^\dagger \right) \rangle = -\theta(\tau)\langle\Phi(\mathbf{x}, \tau)\ \Phi(\mathbf{y})^\dagger\rangle - \theta(-\tau)\langle\Phi(\mathbf{y})^\dagger\ \Phi(\mathbf{x}, \tau)\ \rangle .$$

Let us again assume that the model is translationally invariant and introduce the bosonic operators $a_\mathbf{q}$ and $a_\mathbf{q}^\dagger$, as well as the conjugate variables

$$x_\mathbf{q} = \sqrt{\frac{1}{2}}\left(a_\mathbf{q} + a_{-\mathbf{q}}^\dagger\right) , \qquad\qquad p_\mathbf{q} = -i\sqrt{\frac{1}{2}}\left(a_\mathbf{q} - a_{-\mathbf{q}}^\dagger\right) .$$

In most situations, the object that appears in the perturbation theory is not the single-particle Green's function but the $x - x$ time-ordered product, namely

$$D(\tau, \mathbf{q}) = -\langle T_\tau \left(x_\mathbf{q}(\tau)\, x_{-\mathbf{q}}(0) \right) \rangle . \tag{4.32}$$

The reason why the two operators have opposite momentum is again translational symmetry, which implies that only momentum-zero operators can have finite expectation value.

As before, let us consider a bosonic non-interacting Hamiltonian

$$H_0 = \sum_\mathbf{q} \omega_\mathbf{q}\, a_\mathbf{q}^\dagger a_\mathbf{q} ,$$

with $\omega_\mathbf{q} = \omega_{-\mathbf{q}} \geq 0$. Then, for positive τ,

$$\begin{aligned}
D_0(\mathbf{q}, \tau) &= -\frac{1}{2}\langle a_\mathbf{q}(\tau)\, a_\mathbf{q}^\dagger(0)\rangle - \frac{1}{2}\langle a_{-\mathbf{q}}^\dagger(\tau)\, a_{-\mathbf{q}}(0)\rangle \\
&= -\frac{1}{2}\, e^{-\omega_\mathbf{q}\tau}\left(1 + b(\omega_\mathbf{q})\right) - \frac{1}{2}\, e^{\omega_\mathbf{q}\tau}\, b(\omega_\mathbf{q}) ,
\end{aligned}$$

where

$$b(\omega_\mathbf{q}) = \langle a_\mathbf{q}^\dagger a_\mathbf{q}\rangle = \frac{1}{e^{\beta\omega_\mathbf{q}} - 1} ,$$

is the Bose distribution function, and

$$\langle a_\mathbf{q}\, a_\mathbf{q}^\dagger\rangle = 1 + b(\omega_\mathbf{q}) = -\frac{1}{e^{-\beta\omega_\mathbf{q}} - 1} .$$

Since

$$\int_0^\beta d\tau\, e^{i\Omega_m\tau}\, e^{\mp\omega_\mathbf{q}\tau} = \frac{1}{i\Omega_m \mp \omega_\mathbf{q}}\left(e^{\mp\beta\omega_\mathbf{q}} - 1\right) ,$$

one readily finds that

$$D_0(i\Omega_m, \mathbf{q}) = \frac{1}{2}\left(\frac{1}{i\Omega_m - \omega_\mathbf{q}} - \frac{1}{i\Omega_m + \omega_\mathbf{q}}\right) = -\frac{\omega_\mathbf{q}}{\Omega_m^2 + \omega_\mathbf{q}^2} . \tag{4.33}$$

4.2   Perturbation Expansion in Imaginary Time

Let us suppose that the full Hamiltonian

$$H = H_0 + V,$$

where H_0 is a single particle Hamiltonian that can be exactly diagonalised, while V is a perturbation that makes H not solvable anymore, e.g., the electron-electron interaction.

We consider the Heisenberg evolution operator $e^{-H(\tau-\tau')}$ from τ' to τ, and write it as

$$e^{-H(\tau-\tau')} = e^{-H_0\tau}\, S(\tau,\tau')\, e^{H_0\tau'} \;\Rightarrow\; S(\tau,\tau') = e^{H_0\tau}\, e^{-H(\tau-\tau')}\, e^{-H_0\tau'}.$$
(4.34)

Note that, trivially,

$$S(\tau,\tau) = 1, \quad S(\tau,\tau_1)\, S(\tau_1,\tau') = S(\tau,\tau'), \quad S(\tau,\tau')\, S(\tau',\tau) = S(\tau,\tau) = 1,$$

namely, $S(\tau',\tau) = S^{-1}(\tau,\tau')$.

We observe that

$$\frac{\partial S(\tau,\tau')}{\partial \tau} = e^{H_0\tau}\, H_0\, e^{-H(\tau-\tau')}\, e^{-H_0\tau'} - e^{H_0\tau}\, H\, e^{-H(\tau-\tau')}\, e^{-H_0\tau'}$$

$$= -e^{H_0\tau}\, V\, e^{-H(\tau-\tau')}\, e^{-H_0\tau'} = -e^{H_0\tau}\, V\, e^{-H_0\tau}\, e^{H_0\tau}\, e^{-H(\tau-\tau')}\, e^{-H_0\tau'}$$

$$\equiv -V(\tau)\, S(\tau,\tau'),$$
(4.35)

where $V(\tau)$ is the perturbation evolved in imaginary time with the unperturbed Hamiltonian. The reason why we introduced $S(\tau,\tau')$ is because, through (4.35), it admits a very simple perturbative expansion. Suppose therefore that

$$S(\tau,\tau') = \sum_{n=0} S^{(n)}(\tau,\tau'),$$
(4.36)

where $S^{(n)}(\tau,\tau')$ contains n powers of V. By definition, the zeroth order term corresponds to $H = H_0$, and thus, through (4.34), $S^{(0)}(\tau,\tau') = 1$, $\forall\, \tau,\tau'$. It follows that the boundary condition $S(\tau,\tau) = 1$ is already satisfied by the zeroth order term, so that $S^{(n)}(\tau,\tau) = 0$ for any $n > 0$. The (4.35) for the nth term is, consistently,

$$\frac{\partial S^{(n)}(\tau,\tau')}{\partial \tau} = -V(\tau)\, S^{(n-1)}(\tau,\tau'), \qquad\qquad S^{(n)}(\tau,\tau) = 0,$$

with solution

$$S^{(n)}(\tau,\tau') = -\int_{\tau'}^{\tau} d\tau_1\, V(\tau_1)\, S^{(n-1)}(\tau_1,\tau').$$
(4.37)

Since we know $S^{(0)}(\tau, \tau') = 1$, we can iteratively find all $S^{(n)}(\tau, \tau')$. At first order, $n = 1$,

$$S^{(1)}(\tau, \tau') = -\int_{\tau'}^{\tau} d\tau_1 \, V(\tau_1) \, S^{(0)}(\tau_1, \tau') = -\int_{\tau'}^{\tau} d\tau_1 \, V(\tau_1). \qquad (4.38)$$

At second order, $n = 2$,

$$
\begin{aligned}
S^{(2)}(\tau, \tau') &= -\int_{\tau'}^{\tau} d\tau_1 \, V(\tau_1) \, S^{(1)}(\tau, \tau') = \int_{\tau'}^{\tau} d\tau_1 \int_{\tau'}^{\tau_1} d\tau_2 \, V(\tau_1) \, V(\tau_2) \\
&= \frac{1}{2} \int_{\tau'}^{\tau} d\tau_1 \int_{\tau'}^{\tau_1} d\tau_2 \, V(\tau_1) \, V(\tau_2) + \frac{1}{2} \int_{\tau'}^{\tau} d\tau_2 \int_{\tau'}^{\tau_2} d\tau_1 \, V(\tau_2) \, V(\tau_1) \\
&= \frac{1}{2} \int_{\tau'}^{\tau} d\tau_1 \, d\tau_2 \, T_\tau \Big(V(\tau_1) \, V(\tau_2) \Big).
\end{aligned}
$$

Iterating the procedure, one finds that

$$S^{(n)}(\tau, \tau') = \frac{(-1)^n}{n!} \int_{\tau'}^{\tau} d\tau_1 \, d\tau_2 \ldots d\tau_n \, T_\tau \Big(V(\tau_1) \, V(\tau_2) \ldots V(\tau_n) \Big).$$

It follows that

$$S(\tau, \tau') = T_\tau \left(e^{-\int_{\tau'}^{\tau} d\tau_1 \, V(\tau_1)} \right), \qquad (4.39)$$

whose meaning is the following. One has first to expand the exponential, and after move the time-ordering operator inside the integration.

Suppose we have to calculate the multi-operator Green's function (4.5), and further assume that, e.g., $\tau_1 \geq \tau_2 \geq \cdots \geq \tau_{n-1} \geq \tau_n$, then

$$
\begin{aligned}
G_{A_1, \ldots, A_n}(\tau_1, \ldots, \tau_n) &= -\langle A_1(\tau_1) \ldots A_n(\tau_n) \rangle \\
&= \frac{1}{Z} \text{Tr}\left[e^{-\beta H} e^{\tau_1 H} A_1 e^{-\tau_1 H} e^{\tau_2 H} A_2 e^{-\tau_2 H} \ldots e^{-\tau_{n-1} H} e^{\tau_n H} A_n e^{-\tau_n H} \right] \\
&= \frac{1}{Z} \text{Tr}\left[e^{-\beta H_0} S(\beta, \tau_1) e^{\tau_1 H_0} A_1 e^{-\tau_1 H_0} S(\tau_1, \tau_2) e^{\tau_2 H_0} A_2 e^{-\tau_2 H_0} \ldots \right. \\
&\qquad\qquad \left. \ldots S(\tau_{n-1}, \tau_n) e^{\tau_n H_0} A_n e^{-\tau_n H_0} S(\tau_n, 0) \right] \\
&= \frac{1}{Z} \text{Tr}\left[e^{-\beta H_0} S(\beta, \tau_1) A_1(\tau_1) S(\tau_1, \tau_2) A_2(\tau_2) \ldots S(\tau_{n-1}, \tau_n) A_n(\tau_n) S(\tau_n, 0) \right],
\end{aligned}
$$

where, unlike in the first line, in the last one the operators are evolved with the unperturbed Hamiltonian H_0, and the expectation value is over the same Hamiltonian H_0. We readily see that all times remain ordered. Indeed, e.g., $S(\tau_1, \tau_2)$ contains all times $\in [\tau_2, \tau_1]$ and therefore must be on the right of $A_1(\tau_1)$ but on the left of $A_2(\tau_2)$. In other words, since

$$S(\beta, 0) = S(\beta, \tau_1) \, S(\tau_1, \tau_2) \ldots S(\tau_n, 0),$$

then, for $\beta > \tau_1 > \tau_2 \cdots > \tau_n > 0$,

$$S(\beta, \tau_1) A_1(\tau_1) S(\tau_1, \tau_2) A_2(\tau_2) \ldots S(\tau_{n-1}, \tau_n) A_n(\tau_n) S(\tau_n, 0)$$
$$= T_\tau \Big(S(\beta) A_1(\tau_1) \ldots A_n(\tau_n) \Big),$$

so that, for a generic time order,

$$G_{A_1,\ldots,A_n}(\tau_1,\ldots,\tau_n) = -\langle T_\tau (A_1(\tau_1) \ldots A_n(\tau_n)) \rangle$$
$$= \frac{1}{Z} \text{Tr}\Big[e^{-\beta H_0} \, T_\tau \Big(S(\beta, 0) A_1(\tau_1) \ldots A_n(\tau_n) \Big) \Big],$$

where, we emphasise again, the time evolution of the A_i's operators in the last equation is controlled by the non-interacting Hamiltonian. Since, by definition,

$$e^{-\beta H} = e^{-\beta H_0} S(\beta, 0),$$

then

$$Z = \text{Tr}\Big(e^{-\beta H}\Big) = \text{Tr}\Big(e^{-\beta H_0} S(\beta, 0)\Big) = \text{Tr}\Big(e^{-\beta H_0}\Big) \frac{\text{Tr}\Big(e^{-\beta H_0} S(\beta, 0)\Big)}{\text{Tr}\Big(e^{-\beta H_0}\Big)} = Z_0 \, \langle S(\beta, 0) \rangle .$$

We therefore conclude that

$$G_{A_1,\ldots,A_n}(\tau_1,\ldots,\tau_n) = -\frac{\langle T_\tau (S(\beta, 0) A_1(\tau_1) \ldots A_n(\tau_n)) \rangle}{\langle S(\beta, 0) \rangle}, \tag{4.40}$$

where the thermal averages as well as the imaginary-time evolution are done with the non-interacting Hamiltonian H_0. This expression is now suitable for an expansion in V, which is the reason why we introduced the operator $S(\beta, 0)$.

4.2.1   Wick's Theorem

Upon expanding $S(\beta, 0)$ in powers of the perturbation V, the calculation of any Green's function reduces to evaluate the average value of a time-ordered product of Fermi or Bose fields with a non interacting Hamiltonian. It is therefore essential to know how to perform this calculation.

Suppose we have to evaluate the expectation value

$$-\langle T_\tau \left(\Psi(\mathbf{x}_1, \tau_1) \Psi(\mathbf{x}_2, \tau_2) \ldots \Psi(\mathbf{x}_n, \tau_n) \Psi^\dagger(\mathbf{x}'_n, \tau'_n) \ldots \Psi^\dagger(\mathbf{x}'_1, \tau'_1) \right) \rangle, \tag{4.41}$$

over the non-interacting Hamiltonian H_0. For any time-ordering this amounts to average a product of creation and annihilation operators. We already know that this is the sum of the products of all possible contractions of an annihilation with a

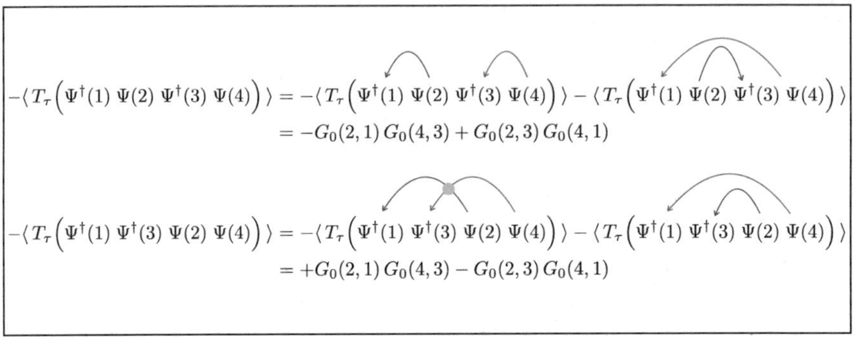

$$-\langle T_\tau\left(\Psi^\dagger(1)\,\Psi(2)\,\Psi^\dagger(3)\,\Psi(4)\right)\rangle = -\langle T_\tau\left(\Psi^\dagger(1)\,\Psi(2)\,\Psi^\dagger(3)\,\Psi(4)\right)\rangle - \langle T_\tau\left(\Psi^\dagger(1)\,\Psi(2)\,\Psi^\dagger(3)\,\Psi(4)\right)\rangle$$

$$= -G_0(2,1)\,G_0(4,3) + G_0(2,3)\,G_0(4,1)$$

$$-\langle T_\tau\left(\Psi^\dagger(1)\,\Psi^\dagger(3)\,\Psi(2)\,\Psi(4)\right)\rangle = -\langle T_\tau\left(\Psi^\dagger(1)\,\Psi^\dagger(3)\,\Psi(2)\,\Psi(4)\right)\rangle - \langle T_\tau\left(\Psi^\dagger(1)\,\Psi^\dagger(3)\,\Psi(2)\,\Psi(4)\right)\rangle$$

$$= +G_0(2,1)\,G_0(4,3) - G_0(2,3)\,G_0(4,1)$$

Fig. 4.5 Wick's theorem applied in two exemplary cases. We represent a contraction as an oriented line from the annihilation operator, e.g., $\Psi(2)$, where 2 stems for space and time coordinates, to the creation operator, e.g., $\Psi^\dagger(1)$ to which $\Psi(2)$ is contracted. Each contraction now represent a non-interacting Green's function, in the case of $\Psi(2)$ contracted with $\Psi^\dagger(1)$ is $G_0(2,1)$. The sign of each term is -1 from the definition of the expectation value, times $(-1)^{n_R}$, where n_R is the number of lines that go from left to right, and, finally, times $(-1)^{n_c}$, where n_c is the number of crossing, the green dot in the bottom panel

creation operator. Suppose that the operator $\Psi(\mathbf{x}_i, \tau_i)$ is contracted with $\Psi^\dagger(\mathbf{x}'_j, \tau'_j)$. If all other times but τ_i and τ'_j remain the same, there are two cases: if $\tau_i \geq \tau'_j$ we have to evaluate the contraction

$$\langle\,\Psi(\mathbf{x}_i, \tau_i)\,\Psi^\dagger(\mathbf{x}'_j, \tau'_j)\,\rangle\,,$$

while, if $\tau_i \leq \tau'_j$ we need to interchange the two operators in the time-ordering, which leads to a minus sign and thus to

$$-\langle\,\Psi^\dagger(\mathbf{x}'_j, \tau'_j)\,\Psi(\mathbf{x}_i, \tau_i)\,\rangle\,.$$

Both cases can be represented by a single quantity, namely,

$$-G_0\left(\tau_i - \tau'_j; \mathbf{x}_i, \mathbf{x}'_j\right).$$

This argument can be extended to all other contractions. As we did when discussing the Hartree-Fock approximation, it is more convenient to express the general rules graphically, as in Fig. 4.5 for two simple cases. The rule is very simple.

> Wick's theorem

The expectation value (4.41) is the sum of all possible ways to contract pairs of annihilation and creation operators. In this case, the contraction is just a non-interacting Green's function. The sign of each product of Green's functions can be found by drawing oriented lines from each annihilation operator to the creation operator which it is contracted to. In that case, the sign is simply $-(-1)^{n_R}\,(-1)^{n_c}$, where n_R is the number of lines that go from left to right, and n_c the number of crossings.

This is the just the so-called Wick's theorem.

In the case of bosons Wick's theorem does not hold rigorously. Let us consider, e.g., a periodic system, a non-interacting bosonic Hamiltonian

$$H_0 = \sum_{\mathbf{q}} \omega_{\mathbf{q}} b_{\mathbf{q}}^\dagger b_{\mathbf{q}},$$

and the corresponding Bose fields

$$\Phi(\mathbf{r}) = \frac{1}{\sqrt{V}} \sum_{\mathbf{q}} e^{i\mathbf{q}\cdot\mathbf{r}} b_{\mathbf{q}}, \qquad \Phi^\dagger(\mathbf{r}) = \frac{1}{\sqrt{V}} \sum_{\mathbf{q}} e^{-i\mathbf{q}\cdot\mathbf{r}} b_{\mathbf{q}}^\dagger.$$

We note that the expectation value over the non-interacting system

$$\begin{aligned}
\langle b_{\mathbf{q}_1}^\dagger b_{\mathbf{q}_2}^\dagger b_{\mathbf{q}_3} b_{\mathbf{q}_4} \rangle &= \left(1 - \delta_{\mathbf{q}_1\mathbf{q}_2}\right)\left(\delta_{\mathbf{q}_1\mathbf{q}_4}\delta_{\mathbf{q}_2\mathbf{q}_3} + \delta_{\mathbf{q}_1\mathbf{q}_3}\delta_{\mathbf{q}_2\mathbf{q}_4}\right) n_{\mathbf{q}_1} n_{\mathbf{q}_2} \\
&\quad + \delta_{\mathbf{q}_1\mathbf{q}_2}\delta_{\mathbf{q}_1\mathbf{q}_4}\delta_{\mathbf{q}_2\mathbf{q}_3} n_{\mathbf{q}_1}\left(n_{\mathbf{q}_1} - 1\right) \\
&= \langle b_{\mathbf{q}_1}^\dagger b_{\mathbf{q}_4} \rangle \langle b_{\mathbf{q}_2}^\dagger b_{\mathbf{q}_3} \rangle + \langle b_{\mathbf{q}_1}^\dagger b_{\mathbf{q}_3} \rangle \langle b_{\mathbf{q}_2}^\dagger b_{\mathbf{q}_4} \rangle \\
&\quad - \delta_{\mathbf{q}_1\mathbf{q}_2}\delta_{\mathbf{q}_1\mathbf{q}_4}\delta_{\mathbf{q}_2\mathbf{q}_3} n_{\mathbf{q}_1}\left(n_{\mathbf{q}_1} + 1\right),
\end{aligned}$$

where $n_{\mathbf{q}} = b(\omega_{\mathbf{q}})$ is the Bose distribution function, differs from simple pairwise contractions of creation and annihilation operators because of the last term where all four operators have the same momentum. That circumstance does not occur for fermions because of Pauli principle, but it does for bosons, thus invalidating Wick's theorem.

Indeed, assuming $\tau_1 > \tau_2 > \tau_3 > \tau_4$,

$$\begin{aligned}
&-\langle T_\tau\left(\Phi^\dagger(\mathbf{r}_1, \tau_1)\, \Phi^\dagger(\mathbf{r}_2, \tau_2)\, \Phi\,(\mathbf{r}_3, \tau_3)\, \Phi\,(\mathbf{r}_4, \tau_4)\right)\rangle \\
&= -\frac{1}{V^2} \sum_{\mathbf{q}_1\mathbf{q}_2\mathbf{q}_3\mathbf{q}_4} e^{-i(\mathbf{q}_1\cdot\mathbf{r}_1 + \mathbf{q}_2\cdot\mathbf{r}_2 - \mathbf{q}_3\cdot\mathbf{r}_3 - \mathbf{q}_4\cdot\mathbf{r}_4)}\, e^{\omega_{\mathbf{q}_1}\tau_1 + \omega_{\mathbf{q}_2}\tau_2 - \omega_{\mathbf{q}_3}\tau_3 - \omega_{\mathbf{q}_4}\tau_4}\, \langle b_{\mathbf{q}_1}^\dagger b_{\mathbf{q}_2}^\dagger b_{\mathbf{q}_3} b_{\mathbf{q}_4} \rangle \\
&= -G_0(\tau_4 - \tau_1, \mathbf{r}_4 - \mathbf{r}_1)\, G_0(\tau_3 - \tau_2, \mathbf{r}_3 - \mathbf{r}_2) \\
&\quad - G_0(\tau_4 - \tau_2, \mathbf{r}_4 - \mathbf{r}_2)\, G_0(\tau_3 - \tau_1, \mathbf{r}_3 - \mathbf{r}_1) \\
&\quad - \frac{1}{V^2} \sum_{\mathbf{q}} e^{-i\mathbf{q}\cdot(\mathbf{r}_1 + \mathbf{r}_2 - \mathbf{r}_3 - \mathbf{r}_4)}\, e^{\omega_{\mathbf{q}}(\tau_1 + \tau_2 - \tau_3 - \tau_4)}\, n_{\mathbf{q}}\left(n_{\mathbf{q}} + 1\right), \quad (4.42)
\end{aligned}$$

where

$$\begin{aligned}
G_0(\tau - \tau', \mathbf{r} - \mathbf{r}') &= -\langle T_\tau\left(\Phi(\mathbf{r}, \tau)\, \Phi^\dagger(\mathbf{r}', \tau')\right)\rangle \\
&= -\frac{1}{V} \sum_{\mathbf{q}} e^{i\mathbf{q}\cdot(\mathbf{r} - \mathbf{r}')}\, \langle T_\tau\left(b_{\mathbf{q}}(\tau)\, b_{\mathbf{q}}^\dagger(\tau')\right)\rangle.
\end{aligned}$$

The two products of Green's functions in (4.42) are those expected from Wick's theorem, while the last term in that equation is the deviation from that theorem. However, in the thermodynamic limit, $V \to \infty$, that deviation vanishes since it is of order $1/V$. It follows that Wick's theorem does hold also for bosons in periodic systems and in the thermodynamic limit. In that case, the rule is the same as for fermions, with the only difference that the sign of each product of n Green's functions is simply $(-1)^{n+1}$.

There is however a caveat. In some cases, Bose condensation occurs at a given $\mathbf{q}$, usually $\mathbf{q} = \mathbf{0}$, as we assume. In the condensed phase, $n_0 = \langle b_0^\dagger b_0 \rangle = \rho_c V$, where ρ_c is the finite density of the condensate, thus yielding a deviation from Wick's theorem that remains finite also in the thermodynamic limit. Nonetheless, a way out still exists. Since there is a macroscopic occupation at $\mathbf{q} = \mathbf{0}$, one can safely replace the operator b_0 by a c-number, $b_0 \sim e^{i\phi_c} \sqrt{V \rho_c}$, and similarly its hermitian conjugate, so that, e.g.,

$$\Phi(\mathbf{r}) \simeq e^{i\phi_c} \sqrt{\rho_c} + \frac{1}{\sqrt{V}} \sum_{\mathbf{q} \neq 0} e^{i\mathbf{q} \cdot \mathbf{r}} b_{\mathbf{q}} \equiv e^{i\phi_c} \sqrt{\rho_c} + \Phi'(\mathbf{r}) \,.$$

Under that assumption, Wick's theorem does work for $\Phi'(\mathbf{r})$ and its hermitian conjugate in the thermodynamic limit.

4.3 Perturbation Theory for the Single-Particle Green's Function and Feynman Diagrams

Let us consider the Hamiltonian for free electrons

$$H_0 = \sum_{\mathbf{k}\sigma} \epsilon_{\mathbf{k}} c_{\mathbf{k}\sigma}^\dagger c_{\mathbf{k}\sigma} \,, \tag{4.43}$$

with spin-independent non-interacting Green's function

$$G_0(\tau - \tau', \mathbf{r} - \mathbf{r}') = -\langle T_\tau \left(\Psi_\sigma(\mathbf{r}, \tau) \, \Psi_\sigma^\dagger(\mathbf{r}', \tau') \right) \rangle \,, \tag{4.44}$$

in presence of the interaction

$$H_{int} = \frac{1}{2} \sum_{\sigma\sigma'} \int d\mathbf{r} \, d\mathbf{r}' \, \Psi_\sigma^\dagger(\mathbf{r}) \, \Psi_{\sigma'}^\dagger(\mathbf{r}') \, U(\mathbf{r} - \mathbf{r}') \, \Psi_{\sigma'}(\mathbf{r}') \, \Psi_\sigma(\mathbf{r}) \,,$$

which plays the role of the perturbation V. The operator $S(\beta, 0)$ is built through

$$\int d\tau \, H_{int}(\tau) = \frac{1}{2} \sum_{\sigma\sigma'} \int d\mathbf{r} \, d\mathbf{r}' \int d\tau \, \Psi_\sigma^\dagger(\mathbf{r}, \tau) \, \Psi_{\sigma'}^\dagger(\mathbf{r}', \tau) \, U(\mathbf{r} - \mathbf{r}') \, \Psi_{\sigma'}(\mathbf{r}', \tau) \, \Psi_\sigma(\mathbf{r}, \tau)$$

$$= \frac{1}{2} \sum_{\sigma\sigma'} \int d\mathbf{r} \, d\mathbf{r}' \int d\tau \, d\tau' \, \Psi_\sigma^\dagger(\mathbf{r}, \tau) \, \Psi_{\sigma'}^\dagger(\mathbf{r}', \tau') \, U(\mathbf{r} - \mathbf{r}') \delta(\tau - \tau') \, \Psi_{\sigma'}(\mathbf{r}', \tau') \, \Psi_\sigma(\mathbf{r}, \tau)$$

$$\equiv \frac{1}{2} \sum_{\sigma\sigma'} \int dx\, dy\, \Psi_\sigma^\dagger(x)\, \Psi_{\sigma'}^\dagger(y)\, U(x-y)\, \Psi_{\sigma'}(y)\, \Psi_\sigma(x) \,,$$

where we have introduced the space-time coordinates $x = (\mathbf{r}, \tau)$, $y = (\mathbf{r}', \tau')$ and, by definition,

$$U(x-y) = U(\mathbf{r}-\mathbf{r}')\, \delta(\tau-\tau') \,,$$

so that

$$S(\beta, 0) = T_\tau \left[\exp\left(-\frac{1}{2} \sum_{\sigma\sigma'} \int dx_1\, dx_2\, \Psi_\sigma^\dagger(x_1)\, \Psi_{\sigma'}^\dagger(x_2)\, U(x_1-x_2)\, \Psi_{\sigma'}(x_2)\, \Psi_\sigma(x_1) \right) \right] .$$

The single-particle Green's function can be calculated perturbatively through

$$G(x, y) = -\frac{1}{\langle S(\beta, 0) \rangle}\, \left\langle T_\tau\left(S(\beta, 0)\, \Psi_\sigma(x)\, \Psi_\sigma^\dagger(y) \right) \right\rangle , \qquad (4.45)$$

where the operators evolve with H_0, the expectation value is over the non-interacting states, and we assume unbroken spin $SU(2)$ symmetry, implying that the Green's function is diagonal in spin and independent of it.

Let us perform the calculation up to first order in perturbation theory, namely approximating

$$S(\beta, 0) \simeq S^{(0)}(\beta, 0) + S^{(1)}(\beta, 0)$$
$$= 1 - \frac{1}{2} \sum_{\sigma\sigma'} \int dx_1\, dx_2\, \Psi_\sigma^\dagger(x_1)\, \Psi_{\sigma'}^\dagger(x_2)\, U(x_1-x_2)\, \Psi_{\sigma'}(x_2)\, \Psi_\sigma(x_1) \,.$$

We need to calculate up to first order both numerator and denominator in (4.45). About the numerator, by Wick's theorem we find

$$-\left\langle T_\tau\left(S(\beta)\, \Psi_\sigma(x)\, \Psi_\sigma^\dagger(y) \right) \right\rangle = G_0(x, y)$$
$$+ \frac{1}{2} \sum_{\alpha\beta} \int dx_1\, dx_2\, U(x_1-x_2)\, \left\langle T_\tau\left[\Psi_\sigma(x)\, \Psi_\sigma^\dagger(y)\, \Psi_\alpha^\dagger(x_1)\, \Psi_\beta^\dagger(x_2)\, \Psi_\beta(x_2)\, \Psi_\alpha(x_1) \right] \right\rangle$$
$$= G_0(x, y)\left(1 + \langle S^{(1)}(\beta, 0) \rangle \right)$$
$$+ \frac{1}{2} \sum_{\alpha\beta} \int dx_1\, dx_2\, U(x_1-x_2)\, \Big[\delta_{\sigma\alpha}\, G_0(x, x_1)\, G_0(x_1, y)\, G_0(x_2, x_2)$$
$$- \delta_{\sigma\alpha}\, \delta_{\alpha\beta}\, G_0(x, x_1)\, G_0(x_1, x_2)\, G_0(x_2, y) \Big]$$
$$+ \frac{1}{2} \sum_{\alpha\beta} \int dx_1\, dx_2\, U(x_1-x_2)\, \Big[\delta_{\sigma\beta}\, G_0(x, x_2)\, G_0(x_2, y)\, G_0(x_1, x_1)$$
$$- \delta_{\sigma\beta}\, \delta_{\beta\alpha}\, G_0(x, x_2)\, G_0(x_2, x_1)\, G_0(x_1, y) \Big] .$$

The term $G_0(x, y) \langle S^{(1)}(\beta, 0) \rangle$ is dubbed disconnected, which implies that the contractions factorise in separate terms that do not have common variables. We note that the last two integrals are actually equal since we can exchange $x_1 \leftrightarrow x_2$, so that

$$- \langle T_\tau \left(S(\beta, 0)\, \Psi_\sigma(x)\, \Psi_\sigma^\dagger(y) \right) \rangle = G_0(x, y) \left(1 + \langle S^{(1)}(\beta, 0) \rangle \right)$$

$$+ \int dx_1\, dx_2\, U(x_1 - x_2) \left[2\, G_0(x, x_1)\, G_0(x_1, y)\, G_0(x_2, x_2) - G_0(x, x_1)\, G_0(x_1, x_2)\, G_0(x_2, y) \right],$$

where the factor two comes from the summation over the free internal spin. The denominator is simply $1 + \langle S^{(1)}(\beta, 0) \rangle$ so that, up to first order, it cancels the disconnected term from the numerator and the final result reads

$$G(x, y) = G_0(x, y) + \int dx_1\, dx_2\, U(x_1 - x_2) \left[2\, G_0(x, x_1)\, G_0(x_1, y)\, G_0(x_2, x_2) \right.$$

$$\left. - G_0(x, x_1)\, G_0(x_1, x_2)\, G_0(x_2, y) \right].$$

$$(4.46)$$

Let us give a graphical representation of this result. We represent the fully interacting Green's function as a bold line directed from the coordinate of Ψ to that of $\Psi^\dagger$, the non-interacting one as a regular line directed as before, and the interaction as a wavy line with four legs, see Fig. 4.6. An incoming vertex represent a creation operator, while an outgoing one an annihilation operator. With these notations the Green's function up to first order in the interaction can be represented as in Fig. 4.7. The conventions are that any internal coordinate is integrated, the spin is conserved along a Green's function as well as at any interaction-vertex, and a wavy line is the interaction U. There are two first order diagrams. The tadpole one has a plus sign and a factor two, while the other diagram a minus sign. We note that both diagrams are fully connected; the disconnected terms cancel out with the denominator $\langle S(\beta, 0) \rangle$. One can go on and calculate the second order corrections to infer the rules for constructing diagrams. Here we just quote the final answer.

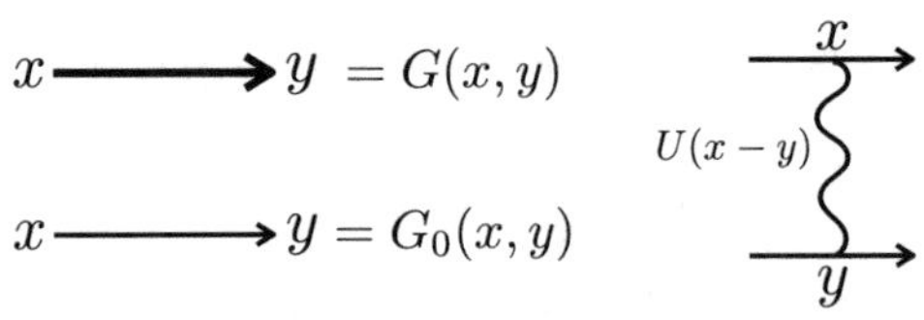

Fig. 4.6 Graphical representation of the Green's functions and the interaction

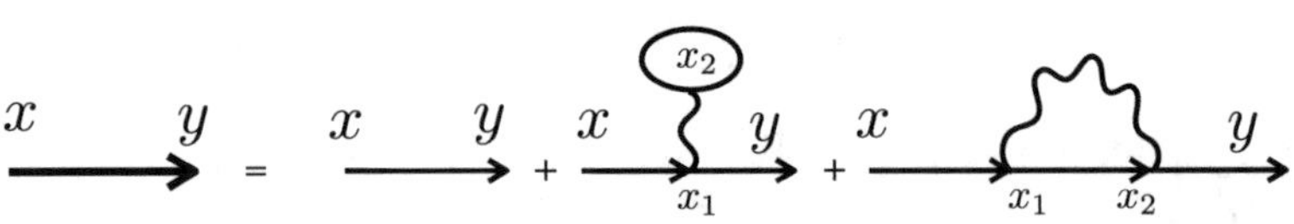

Fig. 4.7 Graphical representation of the Green's function up to first order. Note that sign and the loop multiplicity 2^L are implicit in the diagram definition, and thus are not indicated explicitly

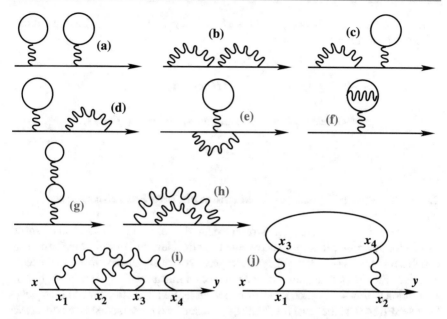

Fig. 4.8 Graphical representation of all second order corrections to the Green's function

The nth order diagrams for the single particle Green's functions are all fully connected and topologically inequivalent diagrams which can be constructed with n interaction lines and $2n + 1$ non-interacting Green's functions, with external points at x and y. All internal coordinates are integrated. The sign of each diagram is $(-1)^n (-1)^L$, where L is the number of internal loops. Spin is conserved at each vertex, hence each loop implies a spin sum, thus a prefactor 2^L. Finally, if a Green's function is connected by the same interaction line it has to be interpreted as the limit of $\tau \to 0^-$, since, in the interaction, creation operators are on the left of the annihilation ones.

One easily realises that the following rules reproduce the two first order corrections we just derived.

In Fig. 4.8 we draw all topologically inequivalent diagrams at second order in perturbation theory. Let us follow the above rules to find the expression of the last two, (i) and (l). The former has no loops, hence its sign is simply $(-1)^n = (-1)^2 = 1$. The spin is the same for all lines hence its expression is

$$(h) = \int \prod_{i=1}^{4} dx_i \, U(x_1 - x_3) \, U(x_2 - x_4)$$

$$G_0(x, x_1) \, G_0(x_1, x_2) \, G_0(x_2, x_3) \, G_0(x_3, x_4) \, G_0(x_4, y) \,.$$

Diagram (l) has a loop, hence $(-1)^n \, (-1)^L = (-1)^2 \, (-1)^1 = -1$. The internal loop implies a spin summation, hence

$$(i) = -2 \int \prod_{i=1}^{4} dx_i \, U(x_1 - x_3) \, U(x_2 - x_4)$$
$$G_0(x, x_1) \, G_0(x_1, x_2) \, G_0(x_2, y) \, G_0(x_3, x_4) \, G_0(x_4, x_3) \,.$$

4.3.1 Diagram Technique in Momentum and Frequency Space

Since there is time-translation invariance, and the Hamiltonian is also space-translation invariant, it is more convenient to consider the Fourier transform of the Green's function $G(i\epsilon_n, \mathbf{k})$, and derive its perturbative expansion. We further need to introduce the Fourier transform of the interaction line as well as we need to indicate a direction to this line, which is the direction along which momentum and frequency flow, see Fig. 4.9. Notice that the frequency carried by the interaction is the difference between two fermionic Matsubara frequencies, hence it is a bosonic one, namely an even multiple of $\pi \, T$. Moreover, since $U(x - y) = U(\mathbf{r} - \mathbf{r}') \delta(\tau - \tau')$, the Fourier transform is

$$U(i\omega, \mathbf{q}) = \int d\tau \int d\mathbf{r} \, e^{-i\mathbf{q}\cdot\mathbf{r}} \, e^{i\omega\tau} \, \delta(\tau) \, U(\mathbf{r}) = U(\mathbf{q}) \,,$$

independent of frequency. Conversely

$$U(x) \equiv U(\mathbf{r}) \delta(\tau) = T \sum_{\omega} \frac{1}{V} \sum_{\mathbf{q}} e^{i\mathbf{q}\cdot\mathbf{r}} \, e^{-i\omega\tau} \, U(i\omega, \mathbf{q}) \,, \qquad (4.47)$$

and, seemingly,

$$G_0(x - y) = T \sum_{\epsilon} \frac{1}{V} \sum_{\mathbf{k}} e^{i\mathbf{k}\cdot(\mathbf{r}-\mathbf{r}')} \, e^{-i\epsilon(\tau-\tau')} \, G_0(i\epsilon, \mathbf{k}) \,. \qquad (4.48)$$

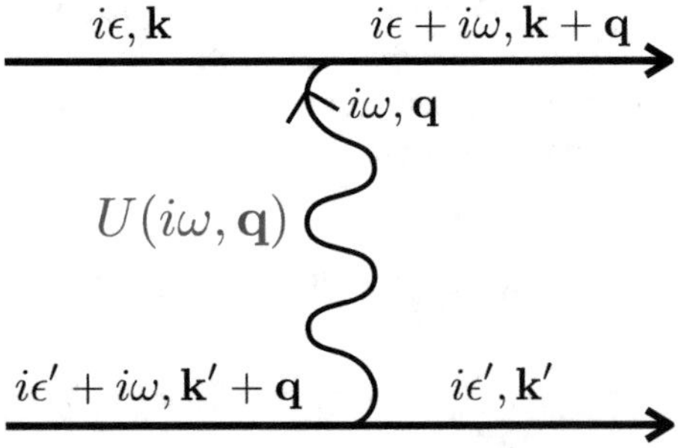

Fig. 4.9 Graphical representation of the interaction in Fourier space, where ϵ and ϵ' are fermionic Matsubara frequencies, while ω a bosonic one

Moving to the perturbation expansion, since the spatial and time coordinates of each internal vertex are integrated out, momentum and frequency are conserved at each vertex, namely the sum of the incoming values is equal to that of the outgoing ones.[1]

Therefore the rules are:

The nth order diagrams for the single particle Green's functions with momentum $\mathbf{k}$ and frequency $i\epsilon$ are all fully connected and topologically not equivalent diagrams which can be constructed with n interaction lines and $2n + 1$ non-interacting Green's functions, with two external lines $G_0(i\epsilon, \mathbf{k})$ and n internal frequencies and momenta. At each vertex, the sum of momenta and that of frequencies of the incoming lines must be equal to the those of outgoing lines. All internal momenta $\mathbf{k}_i$ and frequencies ϵ_i, $i = 1, \ldots, n$, are summed, i.e.,

$$\prod_{i=1}^{n} \frac{1}{V} \sum_{\mathbf{k}_i} T \sum_{\epsilon_i} \cdots .$$

The sign of each diagram is $(-1)^n (-1)^L$, where L is the number of internal loops. Spin is conserved at each vertex, hence each loop implies a factor 2. Finally, if a Green's function is connected by the same interaction line it has to be interpreted as the inverse Fourier transform at $\tau = 0^-$, i.e.

$$T \sum_{\epsilon} G_0(i\epsilon, \mathbf{k}) \, e^{-i\epsilon \, 0^-} = T \sum_{\epsilon} G_0(i\epsilon, \mathbf{k}) \, e^{i\epsilon \, 0^+} .$$

---

[1] Before integration over the internal coordinates, an nth diagram $D_n(x, y, x_1, y_1, \ldots, x_n, y_n)$ has $2n$ internal coordinates, $x_i = (\tau_i, \mathbf{r}_i)$ and $y_i = (\tau'_i, \mathbf{r}'_i)$, $i = 1, \ldots, n$, for each interaction line, and two external ones, $x = (\tau, \mathbf{r})$ and $y = (0, \mathbf{0})$, where we exploit space and time translation invariance. Therefore, it is composed by $2n + 1$ Green's functions G_0, and n interactions $U(x_i - y_i)$. The Fourier transform of $D_n(x, y, x_1, y_1, \ldots, x_n, y_n)$ is

$$D(i\epsilon, \mathbf{k}) = \int d(x - y) \, e^{i\epsilon\tau} \, e^{-i\mathbf{q}\cdot\mathbf{r}} \int \prod_{i=1}^{n} dx_i \, dy_i \, D_n(x, y, x_1, y_1, \ldots, x_n, y_n) . \quad (4.49)$$

If each of the $n + 1$ Green's functions is written as in (4.48), and each of the interaction as in (4.47), we have $2n + 1$ space-time integrations and $(2n + 1) + n$ frequency-momentum summations. It follows that just n internal frequency-momentum summations remains, the space-time integration yielding frequency-momentum conservation at each vertex.

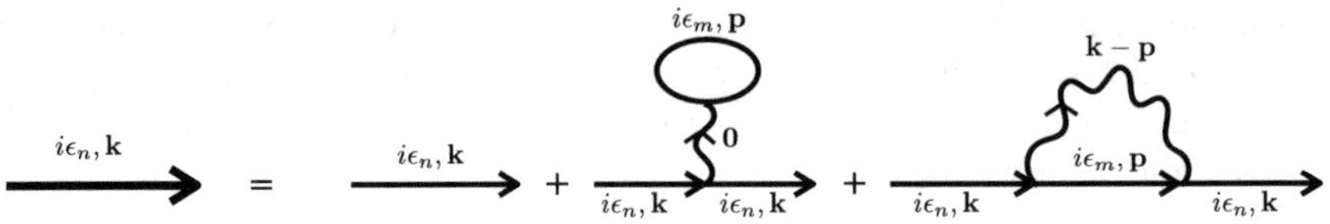

Fig. 4.10 Diagrams for the single-particle Green's function up to first order in Fourier space

For instance, the diagrams up to first order are those in Fig. 4.10, and their expression reads

$$G(i\epsilon_n, \mathbf{k}) = G_0(i\epsilon_n, \mathbf{k}) + 2\, G_0(i\epsilon_n, \mathbf{k})^2\, U(\mathbf{0})\, T \sum_m \frac{1}{V} \sum_{\mathbf{p}} G_0(i\epsilon_m, \mathbf{p})\, e^{i\epsilon_m\, 0^+}$$

$$- G_0(i\epsilon_n, \mathbf{k})^2\, T \sum_m \frac{1}{V} \sum_{\mathbf{p}} U(\mathbf{k} - \mathbf{p})\, G_0(i\epsilon_m, \mathbf{p})\, e^{i\epsilon_m\, 0^+}\,.$$

Using the contour in Fig. 4.1, we can write

$$T \sum_m G_0(i\epsilon_m, \mathbf{p})\, e^{i\epsilon_m\, 0^+} = T \sum_m e^{i\epsilon_m\, 0^+} \frac{1}{i\epsilon_m - \epsilon_{\mathbf{p}}}$$

$$= \oint \frac{dz}{2\pi i}\, f(z)\, e^{z\, 0^+} \frac{1}{i\epsilon_m - \epsilon_{\mathbf{p}}} = f(\epsilon_{\mathbf{p}})\,,$$

where the exponential $e^{z\, 0^+}$ guarantees a decay faster than $1/z$ for $|z| \to \infty$, and thus allows using Cauchy's theorem. Therefore, the diagrams in Fig. 4.10 reads explicitly

$$G(i\epsilon_n, \mathbf{k}) = G_0(i\epsilon_n, \mathbf{k}) + 2\, G_0^2(i\epsilon_n, \mathbf{k}) \frac{1}{V} \sum_{\mathbf{p}} \left(2\, U(\mathbf{0}) - U(\mathbf{k} - \mathbf{p}) \right) f(\epsilon_{\mathbf{p}})\,.$$

4.3.2 The Dyson Equation

The graphical representation of the single-particle Green's function perturbative expansion already represents a substantial simplification. However, the number of diagrams grow exponentially with the order n, see Table 4.1. Therefore, if the only way to proceed were to calculate order by order all diagrams, the Feynman diagram technique would be of little practical use. In reality, the technique is extremely useful since it gives hints how to sum up the perturbative series. Let us discuss how that works for the single-particle Green's function.

Among all possible diagrams for the single particle Green's function, we can distinguish two classes. The first includes all diagrams that, by cutting an internal Green's function line, transform in two lower-order diagrams for the Green's function. These kind of diagrams are called *single-particle reducible*. For instance diagrams (a), (b), (c) and (d) in Fig. 4.8 are single-particle reducible. The other class

Table 4.1 Number of topologically inequivalent diagrams up to order $n = 7$

n	# of diagrams
1	2
2	10
3	74
4	706
5	8162
6	110410
7	1708391

contains all diagrams which are not single-particle reducible, also called *single-particle irreducible*. For instance both first order diagrams as well as the second order ones (e) to (j) are irreducible. Let us define as the single-particle self-energy $\Sigma_\sigma(i\epsilon, \mathbf{k})$ the sum of all irreducible diagrams without the external legs. For instance, the self-energy diagrams up to second order are drawn in Fig. 4.11, where the self-energy is represented by a rounded box. It is not difficult to realise that, in terms of the self-energy, the perturbation expansion can be rewritten as in Fig. 4.12, which has the formal solution

$$G(i\epsilon_n, \mathbf{k}) = \frac{G_0(i\epsilon_n, \mathbf{k})}{1 - G_0(i\epsilon_n, \mathbf{k})\,\Sigma(i\epsilon_n, \mathbf{k})} = \frac{1}{i\epsilon_n - \epsilon_{\mathbf{k}} - \Sigma(i\epsilon, \mathbf{k})}. \tag{4.50}$$

This is the so-called Dyson equation for the single-particle Green's function. We note that, because of (4.22)

$$\mathrm{Re}\,\Sigma(i\epsilon, \mathbf{k}) = \mathrm{Re}\,\Sigma(-i\epsilon, \mathbf{k}), \qquad \mathrm{Im}\,\Sigma(i\epsilon, \mathbf{k}) = -\mathrm{Im}\,\Sigma(-i\epsilon, \mathbf{k}). \tag{4.51}$$

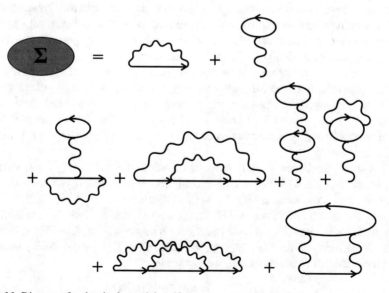

Fig. 4.11 Diagrams for the single-particle self-energy up to second order

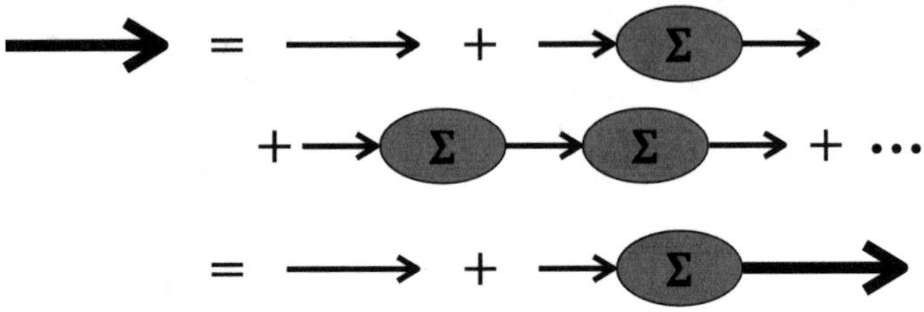

Fig. 4.12 Dyson equation for the single-particle Green's function

Moreover, through (4.23), and, since

$$\left\langle \left\{ \left[c_{\mathbf{k}\sigma}, H \right], c_{\mathbf{k}\sigma}^{\dagger} \right\} \right\rangle = \epsilon_{\mathbf{k}} + \frac{1}{V} \sum_{\mathbf{p}\sigma'} \left(U(\mathbf{0}) - \delta_{\sigma\sigma'} U(\mathbf{k} - \mathbf{p}) \right) \langle c_{\mathbf{p}\sigma'}^{\dagger} c_{\mathbf{p}\sigma'} \rangle ,$$

then

$$\operatorname{Im} \Sigma(i\epsilon, \mathbf{k}) \xrightarrow[|\epsilon| \to \infty]{} 0 ,$$

$$\operatorname{Re} \Sigma(i\epsilon, \mathbf{k}) \xrightarrow[|\epsilon| \to \infty]{} \frac{1}{V} \sum_{\mathbf{p}\sigma'} \left(U(\mathbf{0}) - \delta_{\sigma\sigma'} U(\mathbf{k} - \mathbf{p}) \right) \langle c_{\mathbf{p}\sigma'}^{\dagger} c_{\mathbf{p}\sigma'} \rangle \equiv \Sigma_{\mathrm{HF}}(\mathbf{k}) ,$$

$$(4.52)$$

which defines $\Sigma_{\mathrm{HF}}(\mathbf{k})$.

There are therefore two possible ways of doing perturbation theory. The simplest is just to calculate directly the Green's function, e.g., up to order n. The second is to calculate up to order n the self-energy, and insert its expression in the Dyson equation (4.50). This provides an approximate Green's function that contains all orders in perturbation theory. Therefore, (4.50) represents a way to sum up perturbation theory, and it is in reality physically more appropriate. Indeed, we showed that the Green's function analytically continued in the complex plane has a branch cut on the real axis that is just the single-particle density-of-states, and which is strongly contributed by the interaction, since the non-interacting Green's function has a simple pole on the real axis. It is easy to convince oneself that calculating the Green's function up to any order does not give access to that branch cut, unlike calculating the self-energy at that same order.

Similarly to the Green's function, we can continue the self-energy $\Sigma(i\epsilon, \mathbf{k})$ to the complex plane, $\Sigma(z, \mathbf{k})$ with $z \in \mathbb{C}$. The straight continuation has generally a branch cut on the real axis because $G(z, \mathbf{k})$ has it. Otherwise, as in (4.26) we can define retarded, $\Sigma_{+}(z, \mathbf{k})$, and advanced, $\Sigma_{-}(z, \mathbf{k})$, components analytic in the upper and lower half-plane, respectively. As for the Green's function, $\Sigma(\epsilon, \mathbf{k})$ with $\epsilon \in \mathbb{R}$ refers to the retarded component, i.e., $\Sigma(\epsilon, \mathbf{k}) \equiv \Sigma_{+}(\epsilon, \mathbf{k}) = \Sigma(z \to \epsilon + i\eta, \mathbf{k})$ with $\eta > 0$ infinitesimal. Since that is causal, we can write

$$\Sigma(\epsilon, \mathbf{k}) = -\int \frac{d\omega}{\pi} \frac{\operatorname{Im} \Sigma(\omega, \mathbf{k})}{\epsilon - \omega + i\eta} , \qquad (4.53)$$

Fig. 4.13 Skeleton diagrams for the single-particle self-energy up to second order. Differently from Fig. 4.11, now all internal Green's functions are the fully interacting ones

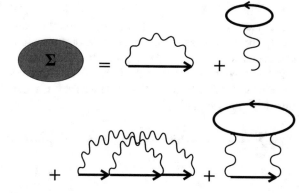

so that

$$\text{Re } \Sigma(\epsilon, \mathbf{k}) = -\int \frac{d\omega}{\pi} \frac{\text{Im } \Sigma(\omega, \mathbf{k})}{\epsilon - \omega} . \tag{4.54}$$

4.3.3   Skeleton Diagrams

The self-energy perturbative series can be recast in a simpler form. Indeed, some diagrams in the series are actually self-energy corrections of the internal Green's functions, like the first four second order diagrams in Fig. 4.11. Therefore, we can discard all such diagrams simply regarding the internal Green's functions as fully-interacting ones. The rules for constructing diagrams remain the same: each interaction yields a factor -1, each loop (-2), and equal-time Green's functions are to be evaluated at $\tau = 0^-$. However, in the skeleton expansion the order of a diagram only refers to the number of interaction lines explicitly drawn, ignoring that the internal Green's functions includes all orders in interactions. The skeleton diagram of the self-energy up to second order are shown in Fig. 4.13.

The skeleton expansion implies that the self-energy can be regarded as a functional of the interacting Green's function and of the interaction, that we shortly denote as U, thus $\Sigma = \Sigma[G, U]$. If follows that the Dyson equation (4.50) can be also written as the self-consistency equation

$$G = \frac{1}{G_0^{-1} - \Sigma[G, U]} . \tag{4.55}$$

Therefore, should we know $\Sigma[G, U]$, we would be able to calculate G solving (4.55). Moreover, as we will discuss in the following sections, the knowledge of $\Sigma[G, U]$ also allows calculating linear response function, thus providing the complete physical characterisation of the system, essentially its 'exact' solution.

However, the explicit expression of the self-energy functional is unknown. Nonetheless, one can assume an approximate expression, $\Sigma_{approx}[G, U]$, and with that calculate the Green's function solving the self-consistency equation (4.55) and all linear response functions. If that approximation captures the main physical ingredients, the outcomes provide sensible description of the system. For instance, if one assumes a self-energy functional that includes just the two first order diagrams in Fig. 4.13, that is equivalent to the Hartree-Fock approximation, as we shall explicitly show later. Explicitly, those first order diagrams, which involve equal-time Green's functions, read explicitly

$$\frac{1}{V}\sum_{\mathbf{p}\sigma'}\Big(U(0)-\delta_{\sigma\sigma'}\,U(\mathbf{k}-\mathbf{p})\Big)G(\tau=0^-,\mathbf{k})=\frac{1}{V}\sum_{\mathbf{p}\sigma'}\Big(U(0)-\delta_{\sigma\sigma'}\,U(\mathbf{k}-\mathbf{p})\Big)\langle c^\dagger_{\mathbf{p}\sigma'}c_{\mathbf{p}\sigma'}\rangle$$

$$=\Sigma_{\mathrm{HF}}(\mathbf{k})\,,$$

where $\Sigma_{\mathrm{HF}}(\mathbf{k})$ was earlier defined in (4.52), and explain the reason of the subscript HF. In particular, the tadpole diagram is known as Hartree term, while the other as Fock one.

4.3.4   Physical Meaning of the Self-energy

According to (4.27) and (4.28), and with our convention $G(\epsilon,\mathbf{k})\equiv G_+(\epsilon,\mathbf{k})$ and $\Sigma(\epsilon,\mathbf{k})\equiv\Sigma_+(\epsilon,\mathbf{k})$, the single-particle density of states (DOS) is

$$A(\epsilon,\mathbf{k})=-\frac{1}{\pi}\,\mathrm{Im}\,G(\epsilon,\mathbf{k})=\frac{1}{\pi}\,\frac{\eta-\mathrm{Im}\,\Sigma(\epsilon,\mathbf{k})}{\Big(\epsilon-\epsilon_\mathbf{k}-\mathrm{Re}\,\Sigma(\epsilon,\mathbf{k})\Big)^2+\Big(\eta-\mathrm{Im}\,\Sigma(\epsilon,\mathbf{k})\Big)^2}\,.$$

Since $A(\epsilon,\mathbf{k})>0$, and η is infinitesimal, it follows that $\mathrm{Im}\,\Sigma(\epsilon,\mathbf{k})<0$. For non-interacting electrons $\Sigma(\epsilon,\mathbf{k})=0$ and the spectral function becomes $\delta(\epsilon-\epsilon_\mathbf{k})$ for $\eta\to 0^+$. When the interaction is finite, and thus we can safely set $\eta=0$,

$$A(\epsilon,\mathbf{k})=\frac{1}{\pi}\,\frac{-\mathrm{Im}\,\Sigma(\epsilon,\mathbf{k})}{\Big(\epsilon-\epsilon_\mathbf{k}-\mathrm{Re}\,\Sigma(\epsilon,\mathbf{k})\Big)^2+\Big(\mathrm{Im}\,\Sigma(\epsilon,\mathbf{k})\Big)^2}\,. \qquad (4.56)$$

Equation (4.56) resembles a set of Lorentzian functions, each centred at $\epsilon=\epsilon_*(\mathbf{k})$, where $\epsilon_*(\mathbf{k})$ are the solutions of the equation

$$\epsilon_*(\mathbf{k})=\epsilon_\mathbf{k}+\mathrm{Re}\,\Sigma\big(\epsilon_*(\mathbf{k}),\mathbf{k}\big)\,,$$

and with width $-\mathrm{Im}\,\Sigma(\epsilon,\mathbf{k})$ calculated at $\epsilon=\epsilon_*(\mathbf{k})$. Indeed, let us assume ϵ close to a root $\epsilon_*(\mathbf{k})$, so that

$$\epsilon-\epsilon_\mathbf{k}-\mathrm{Re}\,\Sigma(\epsilon,\mathbf{k})\simeq Z\big(\epsilon_*(\mathbf{k}),\mathbf{k}\big)^{-1}\big(\epsilon-\epsilon_*(\mathbf{k})\big)\,,$$

where

$$Z(\epsilon,\mathbf{k})^{-1}\equiv 1-\frac{\partial\,\mathrm{Re}\,\Sigma(\epsilon,\mathbf{k})}{\partial\epsilon}\,, \qquad (4.57)$$

Fig. 4.14 A second-order
diagram contributing to the
imaginary part of the
self-energy

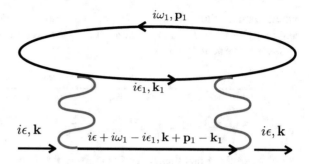

will be hereafter denoted as *quasiparticle residue* for reasons that will be clarified
later. It follows that

$$A(\epsilon \sim \epsilon_*(\mathbf{k}), \mathbf{k}) = \frac{Z(\epsilon_*(\mathbf{k}), \mathbf{k})}{\pi} \frac{\Gamma_*(\mathbf{k})}{(\epsilon - \epsilon_*(\mathbf{k}))^2 + \Gamma_*^2(\mathbf{k})},$$

indeed a Lorentzian, with weight $Z(\epsilon_*(\mathbf{k}), \mathbf{k})$ and width $\Gamma_*(\mathbf{k}) = -Z(\epsilon_*(\mathbf{k}), \mathbf{k})$
$\mathrm{Im}\, \Sigma(\epsilon_*(\mathbf{k}), \mathbf{k})$.

Therefore, a weak interaction at first shifts the position of the δ-peak of the non-
interacting DOS, $\epsilon_\mathbf{k} \to \epsilon_*(\mathbf{k})$, and yields a finite broadening. The latter reflects the
fact that a particle (hole), i.e., an electron outside (inside) the Fermi surface and
thus with $\epsilon_\mathbf{k} > 0$ ($\epsilon_\mathbf{k} < 0$), can decay because of interaction into a particle (hole)
plus several particle-hole pairs provided momentum is conserved. That process is
accounted for by $\mathrm{Im}\, \Sigma(\epsilon, \mathbf{k})$ that can be regarded as the decay rate of a particle
with momentum $\mathbf{k}$ into a composite multi-particle object with same momentum and
energy ϵ; an interpretation which can be justified by analysing the structure of the
self-energy diagrams that contribute to the imaginary part.

Let us consider, for instance, the second-order diagram drawn in Fig. 4.14, Assuming a short range interaction, $U(\mathbf{q}) \simeq U, \forall \mathbf{q}$, its expression is

$$\Sigma(i\epsilon, \mathbf{k}) = -2\, U^2\, \frac{1}{V^2} \sum_{\mathbf{k}_0\, \mathbf{k}_1\, \mathbf{p}_1} T^2 \sum_{\epsilon_1\, \omega_1} \delta_{\mathbf{k}+\mathbf{p}_1, \mathbf{k}_0+\mathbf{k}_1}$$

$$\frac{1}{i\epsilon_1 - \epsilon_{\mathbf{k}_1}}\; \frac{1}{i\omega_1 - \epsilon_{\mathbf{p}_1}}\; \frac{1}{i\epsilon + i\omega_1 - i\epsilon_1 - \epsilon_{\mathbf{k}_0}}$$

where the δ-function imposes the momentum conservation and the factor 2 comes
from the loop spin summation. Let us first sum over $i\epsilon_1$, which yields

$$T\sum_{\epsilon_1} \frac{1}{i\epsilon_1 - \epsilon_{\mathbf{k}_1}}\; \frac{1}{i\epsilon + i\omega_1 - i\epsilon_1 - \epsilon_{\mathbf{k}_0}} = \frac{f(\epsilon_{\mathbf{k}_1})}{i\epsilon + i\omega_1 - \epsilon_{\mathbf{k}_1} - \epsilon_{\mathbf{k}_0}} - \frac{f(i\epsilon + i\omega_1 - \epsilon_{\mathbf{k}_0})}{i\epsilon + i\omega_1 - \epsilon_{\mathbf{k}_0} - \epsilon_{\mathbf{k}_1}}$$

$$= \left(f(\epsilon_{\mathbf{k}_1}) - f(-\epsilon_{\mathbf{k}_0})\right) \frac{1}{i\epsilon + i\omega_1 - \epsilon_{\mathbf{k}_0} - \epsilon_{\mathbf{k}_1}}$$

$$= -\left(1 - f(\epsilon_{\mathbf{k}_1}) - f(\epsilon_{\mathbf{k}_0})\right) \frac{1}{i\epsilon + i\omega_1 - \epsilon_{\mathbf{k}_0} - \epsilon_{\mathbf{k}_1}},$$

where we used the fact the $i\epsilon + i\omega_1$ is a bosonic frequency, and thus $e^{i\beta(\epsilon+\omega_1)} = 1$, and that $f(-x) = 1 - f(x)$, where $f(x)$ is the Fermi distribution function.

Now let us sum over $i\omega_1$,

$$T \sum_{\omega_1} \frac{1}{i\omega_1 - \epsilon_{\mathbf{p}_1}} \frac{1}{i\epsilon + i\omega_1 - \epsilon_{\mathbf{k}_0} - \epsilon_{\mathbf{k}_1}} = \frac{f(\epsilon_{\mathbf{p}_1})}{i\epsilon + \epsilon_{\mathbf{p}_1} - \epsilon_{\mathbf{k}_0} - \epsilon_{\mathbf{k}_1}} + \frac{f(\epsilon_{\mathbf{k}_0} + \epsilon_{\mathbf{k}_1} - i\epsilon)}{\epsilon_{\mathbf{k}_0} + \epsilon_{\mathbf{k}_1} - i\epsilon - \epsilon_{\mathbf{p}_1}}$$

$$= \left(f(\epsilon_{\mathbf{p}_1}) + b(\epsilon_{\mathbf{k}_0} + \epsilon_{\mathbf{k}_1}) \right) \frac{1}{i\epsilon + \epsilon_{\mathbf{p}_1} - \epsilon_{\mathbf{k}_0} - \epsilon_{\mathbf{k}_1}} .$$

Here we used the fact that, since $i\epsilon$ is fermionic, then

$$f(\epsilon_{\mathbf{k}_0} + \epsilon_{\mathbf{k}_1} - i\epsilon) = \left(e^{\beta(\epsilon_{\mathbf{k}_0} + \epsilon_{\mathbf{k}_1} - i\epsilon)} + 1 \right)^{-1} = \left(- e^{\beta(\epsilon_{\mathbf{k}_0} + \epsilon_{\mathbf{k}_1})} + 1 \right)^{-1} = -b(\epsilon_{\mathbf{k}_0} + \epsilon_{\mathbf{k}_1}) .$$

Therefore,

$$\Sigma(i\epsilon, \mathbf{k}) = \frac{2U^2}{V^2} \sum_{\mathbf{k}_0 \mathbf{k}_1 \mathbf{p}_1} \delta_{\mathbf{k}+\mathbf{p}_1, \mathbf{k}_0 + \mathbf{k}_1} \frac{1}{i\epsilon + \epsilon_{\mathbf{p}_1} - \epsilon_{\mathbf{k}_0} - \epsilon_{\mathbf{k}_1}}$$
$$\left(1 - f(\epsilon_{\mathbf{k}_1}) - f(\epsilon_{\mathbf{k}_0}) \right) \left(f(\epsilon_{\mathbf{p}_1}) + b(\epsilon_{\mathbf{k}_0} + \epsilon_{\mathbf{k}_1}) \right) .$$

We note that, since,

$$e^{\beta x} = \frac{1 - f(x)}{f(x)} ,$$

then

$$b(x + y) = \frac{1}{e^{\beta x} e^{\beta y} - 1} = \frac{f(x) f(y)}{(1 - f(x)) (1 - f(y)) - f(x) f(y)} = \frac{f(x) f(y)}{1 - f(x) - f(y)} ,$$

and thus

$$f(\epsilon_{\mathbf{p}_1}) + b(\epsilon_{\mathbf{k}_0} + \epsilon_{\mathbf{k}_1}) = f(\epsilon_{\mathbf{p}_1}) + \frac{f(\epsilon_{\mathbf{k}_0}) f(\epsilon_{\mathbf{k}_1})}{1 - f(\epsilon_{\mathbf{k}_0}) - f(\epsilon_{\mathbf{k}_1})}$$

$$= f(\epsilon_{\mathbf{p}_1}) + \frac{f(\epsilon_{\mathbf{k}_0}) f(\epsilon_{\mathbf{k}_1})}{1 - f(\epsilon_{\mathbf{k}_0}) - f(\epsilon_{\mathbf{k}_1})} \left(1 - f(\epsilon_{\mathbf{p}_1}) + f(\epsilon_{\mathbf{p}_1}) \right)$$

$$= \frac{1}{1 - f(\epsilon_{\mathbf{k}_0}) - f(\epsilon_{\mathbf{k}_1})} \left[f(\epsilon_{\mathbf{p}_1}) \left(1 - f(\epsilon_{\mathbf{k}_0}) \right) \left(1 - f(\epsilon_{\mathbf{k}_1}) \right) \right.$$

$$\left. + \left(1 - f(\epsilon_{\mathbf{p}_1}) \right) f(\epsilon_{\mathbf{k}_0}) f(\epsilon_{\mathbf{k}_1}) \right] ,$$

implying that

$$\Sigma(i\epsilon, \mathbf{k}) = \frac{2U^2}{V^2} \sum_{\mathbf{k}_0 \mathbf{k}_1 \mathbf{p}_1} \delta_{\mathbf{k}+\mathbf{p}_1, \mathbf{k}_0+\mathbf{k}_1} \frac{1}{i\epsilon + \epsilon_{\mathbf{p}_1} - \epsilon_{\mathbf{k}_0} - \epsilon_{\mathbf{k}_1}}$$

$$\left[f(\epsilon_{\mathbf{p}_1}) \left(1 - f(\epsilon_{\mathbf{k}_0})\right) \left(1 - f(\epsilon_{\mathbf{k}_1})\right) + \left(1 - f(\epsilon_{\mathbf{p}_1})\right) f(\epsilon_{\mathbf{k}_0}) f(\epsilon_{\mathbf{k}_1}) \right].$$
(4.58)

Now, we send $i\epsilon \to \epsilon + i0^+$ and just consider the imaginary part,

$$\text{Im} \left(\frac{1}{\epsilon + i0^+ + \epsilon_{\mathbf{p}_1} - \epsilon_{\mathbf{k}_0} - \epsilon_{\mathbf{k}_1}} \right) = -\pi \, \delta(\epsilon + \epsilon_{\mathbf{p}_1} - \epsilon_{\mathbf{k}_0} - \epsilon_{\mathbf{k}_1}),$$

so that

$$\text{Im} \, \Sigma(\epsilon, \mathbf{k}) = -2\pi \, \frac{U^2}{V^2} \sum_{\mathbf{k}_0 \mathbf{k}_1 \mathbf{p}_1} \delta_{\mathbf{k}+\mathbf{p}_1, \mathbf{k}_0+\mathbf{k}_1} \, \delta(\epsilon + \epsilon_{\mathbf{p}_1} - \epsilon_{\mathbf{k}_0} - \epsilon_{\mathbf{k}_1})$$

$$\left[f(\epsilon_{\mathbf{p}_1}) \left(1 - f(\epsilon_{\mathbf{k}_0})\right) \left(1 - f(\epsilon_{\mathbf{k}_1})\right) + \left(1 - f(\epsilon_{\mathbf{p}_1})\right) f(\epsilon_{\mathbf{k}_0}) f(\epsilon_{\mathbf{k}_1}) \right],$$

which is correctly negative, and just represents the Fermi golden rule describing the decay of an electron at momentum $\mathbf{k}$ and energy ϵ into either a hole at $\mathbf{p}_1$ and two particles at $\mathbf{k}_0$ and $\mathbf{k}_1$, the term $f(\epsilon_{\mathbf{p}_1}) \left(1 - f(\epsilon_{\mathbf{k}_0})\right) \left(1 - f(\epsilon_{\mathbf{k}_1})\right)$, or a particle at $\mathbf{p}_1$ and two holes at $\mathbf{k}_0$ and $\mathbf{k}_1$, the term $\left(1 - f(\epsilon_{\mathbf{p}_1})\right) f(\epsilon_{\mathbf{k}_0}) f(\epsilon_{\mathbf{k}_1})$. This process requires momentum conservation, i.e., $\mathbf{k} = \mathbf{k}_0 + \mathbf{k}_1 - \mathbf{p}_1$, as well as energy conservation, $\epsilon = \epsilon_{\mathbf{k}_0} + \epsilon_{\mathbf{k}_1} - \epsilon_{\mathbf{p}_1}$. Upon defining the dimensionless function

$$S_{\mathbf{k}}(\epsilon_0, \epsilon_1, \omega_1) \equiv \frac{1}{\rho_0^3} \frac{1}{V^2} \sum_{\mathbf{k}_0 \mathbf{k}_1 \mathbf{p}_1} \delta(\epsilon_0 - \epsilon_{\mathbf{k}_0}) \delta(\epsilon_1 - \epsilon_{\mathbf{k}_1}) \delta(\omega_1 - \epsilon_{\mathbf{p}_1}) \delta_{\mathbf{k}+\mathbf{p}_1, \mathbf{k}_0+\mathbf{k}_1},$$
(4.59)

where

$$\rho_0 = \frac{1}{V} \sum_{\mathbf{k}} \delta(\epsilon_{\mathbf{k}}),$$

is the non-interacting single-particle density of states at the chemical potential, we can write

$$\text{Im} \, \Sigma(\epsilon, \mathbf{k}) = -2\pi \, U^2 \rho_0^3 \int d\epsilon_0 \, d\epsilon_1 \, d\omega_1 \, S_{\mathbf{k}}(\epsilon_0, \epsilon_1, \omega_1) \delta(\epsilon + \omega_1 - \epsilon_1 - \epsilon_0)$$

$$\left[f(\omega_1) \left(1 - f(\epsilon_0)\right) \left(1 - f(\epsilon_1)\right) + \left(1 - f(\omega_1)\right) f(\epsilon_0) f(\epsilon_1) \right].$$
(4.60)

We are interested in (4.60) evaluated at small ϵ and temperature T. Let us first take $T = 0$, thus $f(x) = \theta(-x)$, in which case $\epsilon > 0$ implies $\epsilon_0 > 0$, $\epsilon_1 > 0$ and $\omega_1 < 0$ (two particles and one hole), while $\epsilon < 0$ the opposite, i.e., $\epsilon_0 < 0$, $\epsilon_1 < 0$ and $\omega_1 > 0$ (two holes and one particle). For $\epsilon \gtrsim 0$, the energy conservation

$$\epsilon = \epsilon_0 + \epsilon_1 - \omega_1 = \epsilon_0 + \epsilon_1 + |\omega_1| \, ,$$

implies that all variables ϵ_0, ϵ_1 and $-\omega_1$ are not only positive but also smaller than ϵ, so that, after changing $\omega_1 \rightarrow -\omega_1$,

$$\mathrm{Im}\, \Sigma(\epsilon, \mathbf{k}) = -2\pi\, U^2\, \rho_0^3 \int_0^\epsilon d\epsilon_0\, d\epsilon_1\, d\omega_1\, S_\mathbf{k}(\epsilon_0, \epsilon_1, -\omega_1)\, \delta\big(\epsilon - \omega_1 - \epsilon_1 - \epsilon_0\big)\,.$$

Since the integration range already vanishes as $\epsilon \rightarrow 0$, at leading order we can make the approximation $S_\mathbf{k}(\epsilon_0, \epsilon_1, -\omega_1) \simeq S_\mathbf{k}(0, 0, 0)$, in which case

$$\mathrm{Im}\, \Sigma(\epsilon, \mathbf{k}) \simeq -2\pi\, U^2\, \rho_0^3\, S_\mathbf{k}(0, 0, 0) \int_0^\epsilon d\epsilon_0\, d\epsilon_1\, d\omega_1\, \delta\big(\epsilon - \omega_1 - \epsilon_1 - \epsilon_0\big)$$

$$= -2\pi\, U^2\, \rho_0^3\, S_\mathbf{k}(0, 0, 0)\, \frac{\epsilon^2}{2}\,,$$

$$(4.61)$$

vanishes quadratically as $\epsilon \rightarrow 0$. Equation (4.62) holds provided $S_\mathbf{k}(0, 0, 0)$ is not singular, which we assume for now and later discuss its validity.

Accordingly, if we are interested in the leading term in T, we can safely set $\epsilon = 0$ in (4.60) and still approximate $S_\mathbf{k}(\epsilon_0, \epsilon_1, \omega_1) \simeq S_\mathbf{k}(0, 0, 0)$, so that, at $T \gtrsim 0$, and recalling that $1 - f(x) = f(-x)$

$$\mathrm{Im}\, \Sigma(0, \mathbf{k}) \simeq -2\pi\, U^2\, \rho_0^3\, S_\mathbf{k}(0, 0, 0) \int_{-\infty}^{\infty} d\epsilon_0\, d\epsilon_1\, d\omega_1\, \delta\big(\omega_1 - \epsilon_1 - \epsilon_0\big)$$

$$\left[f(\omega_1)\, f(-\epsilon_0)\, f(-\epsilon_1) + f(-\omega_1)\, f(\epsilon_0)\, f(\epsilon_1) \right]$$

$$= -4\pi\, U^2\, \rho_0^3\, S_\mathbf{k}(0, 0, 0) \int_{-\infty}^{\infty} d\epsilon_0\, d\epsilon_1\, f(-\epsilon_0 - \epsilon_1)\, f(\epsilon_0)\, f(\epsilon_1)$$

$$= -2\pi\, U^2\, \rho_0^3\, S_\mathbf{k}(0, 0, 0)\, \frac{\pi^2}{2}\, T^2\,,$$

vanishes quadratically as $T \rightarrow 0$.[2] In conclusion, for small ϵ and T, and provided $S_\mathbf{k}(0, 0, 0)$ in (4.59) is not singular,

$$\mathrm{Im}\, \Sigma(0, \mathbf{k}) \simeq -\pi\, U^2\, \rho_0^3\, S_\mathbf{k}(0, 0, 0) \left(\epsilon^2 + \pi^2\, T^2\right), \qquad (4.62)$$

vanishes for both $\epsilon \rightarrow 0$ and $T \rightarrow 0$.

---

[2] If we define

$$F(x) = \frac{1}{e^x + 1}\,,$$

Fig. 4.15 Graphical representation of a generic self-energy diagram that contributes to the imaginary part

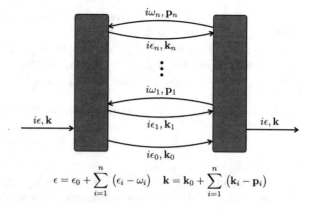

$$\epsilon = \epsilon_0 + \sum_{i=1}^{n} \left(\epsilon_i - \omega_i \right) \quad \mathbf{k} = \mathbf{k}_0 + \sum_{i=1}^{n} \left(\mathbf{k}_i - \mathbf{p}_i \right)$$

More generally, the diagrams which contribute to the imaginary part of the self-energy can be represented as in Fig. 4.15, where a particle with momentum $\mathbf{k}$ and Matsubara frequency $i\epsilon$ decays at the left box, which represents a generic matrix element, into a particle/hole, the line $(i\epsilon_0, \mathbf{k}_0)$, plus $n \geq 1$ particle-hole pairs, the pair of lines $(i\epsilon_i, \mathbf{k}_i) - (i\omega_i, \mathbf{p}_i)$, which recombine at the right box into the original particle/hole. Momentum and frequency conservation imply

$$\epsilon = \epsilon_0 + \sum_{i=1}^{n} \left(\epsilon_i - \omega_i \right), \quad \mathbf{k} = \mathbf{k}_0 + \sum_{i=1}^{n} \left(\mathbf{k}_i - \mathbf{p}_i \right).$$

As before, if we sum over all the independent internal frequencies by a contour integral, and assume that the leading terms are the poles of the Green's function, thus neglecting the frequency dependence of the matrix element, and finally send $i\epsilon \to \epsilon + i0^+$, the imaginary part at $T = 0$ is finite if, at $\epsilon > 0$, $\mathbf{k}_i$, $i = 0, \ldots, n$, are all particles, while $\mathbf{p}_j$, $j = 1, \ldots, n$, all holes, i.e., $\epsilon_{\mathbf{k}_i} > 0$ and $\epsilon_{\mathbf{p}_j} < 0$. On the contrary, if $\epsilon < 0$, $\mathbf{k}_i, i = 0, \ldots, n$, are all holes, while $\mathbf{p}_j, j = 1, \ldots, n$, all particles, i.e., $\epsilon_{\mathbf{k}_i} < 0$ and $\epsilon_{\mathbf{p}_j} > 0$. Energy conservation requires

$$\epsilon = \epsilon_{\mathbf{k}_0} + \sum_{i=1}^{n} \left(\epsilon_{\mathbf{k}_i} - \epsilon_{\mathbf{p}_i} \right),$$

where all terms on the right hand side have the same sign as ϵ. Therefore, if the latter is small in absolute value, each term on the right hand side is equally small in absolute value, i.e., $\left| \epsilon_{\mathbf{k}_i} \right| < |\epsilon|$, $i = 0, \ldots, n$, and $\left| \epsilon_{\mathbf{p}_j} \right| < |\epsilon|$, $j = 1, \ldots, n$. Proceeding as

then

$$\int_{-\infty}^{\infty} d\epsilon_0 \, d\epsilon_1 \, f\left(-\epsilon_0 - \epsilon_1 \right) f\left(\epsilon_0 \right) f\left(\epsilon_1 \right) = T^2 \int_{-\infty}^{\infty} dx \, dy \, F(-x - y) \, F(x) \, F(y) = T^2 \frac{3}{2} \zeta(2) = \frac{\pi^2}{2} \, T^2 \, ,$$

being $\zeta(n)$ the Riemann ζ-function.

before, we may realise that the phase space for the decay process vanishes at least as ϵ^{2n}, or, at $\epsilon = 0$ but $T \neq 0$, as T^{2n}.

Let us now discuss whether $S_{\mathbf{k}}(0, 0, 0)$ in (4.59) is indeed finite. Let us consider, for simplicity, $\epsilon_{\mathbf{k}} = \epsilon(k)$, where $k = |\mathbf{k}|$. In this case, $\epsilon(k) = 0$ when $k = k_F$, the Fermi momentum. In units such that the volume of the unit cell $v = 1$,

$$S_{\mathbf{k}}(\epsilon_0, \epsilon_1, \omega_1) = \frac{1}{(2\pi)^{2d} \rho_0^3} \int d\mathbf{k}_0 \, d\mathbf{k}_1 \, d\mathbf{p}_1 \, \delta(\epsilon_0 - \epsilon(k_0)) \, \delta(\epsilon_1 - \epsilon(k_1)) \, \delta(\omega_1 - \epsilon(p_1))$$
$$\delta(\mathbf{k} + \mathbf{p}_1 - \mathbf{k}_0 - \mathbf{k}_1) \,.$$

Since $\epsilon_0 \sim \epsilon_1 \sim \omega_1 \sim 0$, then $k_0 \sim k_1 \sim p_1 \sim k_F$ and thus $|\mathbf{k}_0 + \mathbf{k}_1 - \mathbf{p}_1| \leq 3k_F$, so that momentum conservation $\mathbf{k} = \mathbf{k}_0 + \mathbf{k}_1 - \mathbf{p}_1$ implies that $S_{\mathbf{k}}(\epsilon_0, \epsilon_1, \omega_1)$ is finite only if $|\mathbf{k}| \leq 3k_F$, otherwise it vanishes. We note that $S_{\mathbf{k}}(\epsilon_0, \epsilon_1, \omega_1)$ does not depend on the direction of $\mathbf{k}$ but only on its magnitude k, so that, if $\mathbf{k} = k\mathbf{v}_1$ with $\mathbf{v}_1$ a unit vector, then

$$S_{\mathbf{k}}(\epsilon_0, \epsilon_1, \omega_1) = \int \frac{d\mathbf{v}_1}{\Omega_d} \, S_{\mathbf{k}}(\epsilon_0, \epsilon_1, \omega_1) \equiv S_k(\epsilon_0, \epsilon_1, \omega_1) \,, \qquad (4.63)$$

where Ω_d is the d-dimensional solid angle. Let us define $k \equiv \kappa_1 \, k_F$, and

$$
\begin{aligned}
& k_0 : \epsilon(k_0) = \epsilon_0 \,, && \mathbf{k}_0 \equiv -k_0 \, \mathbf{v}_2 \equiv -k_F \, \kappa_2 \, \mathbf{v}_2 \,, \\
& k_1 : \epsilon(k_1) = \epsilon_1 \,, && \mathbf{k}_1 \equiv -k_1 \, \mathbf{v}_3 \equiv -k_F \, \kappa_3 \, \mathbf{v}_3 \,, \\
& p_1 : \epsilon(p_1) = \omega_1 \,, && \mathbf{p}_1 \equiv p_1 \, \mathbf{v}_4 \equiv k_F \, \kappa_4 \, \mathbf{v}_4 \,,
\end{aligned}
$$

with all $\mathbf{v}_i$ unit vectors, through which we can write, assuming the density of states smooth around the chemical potential,

$$S_k(\epsilon_0, \epsilon_1, \omega_1) \simeq \frac{(2\pi)^d}{\Omega_d^4 \, k_F^d} \int \prod_{i=1}^{4} d\mathbf{v}_i \, \delta(\kappa_1 \, \mathbf{v}_1 + \kappa_2 \, \mathbf{v}_2 + \kappa_3 \, \mathbf{v}_3 + \kappa_4 \, \mathbf{v}_4)$$

$$= \frac{1}{\Omega_d^4 \, k_F^d} \int d\mathbf{r} \prod_{i=1}^{4} \int d\mathbf{v}_i \, e^{i\kappa_i \, \mathbf{r} \cdot \mathbf{v}_i}$$

$$\equiv \frac{1}{\Omega_d^3 \, k_F^d} \int_0^{\infty} r^{d-1} \, dr \prod_{i=1}^{4} I_d(\kappa_i \, r) \,.$$

The integral over r is convergent at $r = 0$ but may be divergent at $r \to \infty$. For that, we need to know how $I_d(\kappa_i \, r)$ behaves at $r \to \infty$. Noticing that in $d = 1$ the integral over a unit vector $\mathbf{v}_i$ is just the sum over $v_i = \pm 1$, we find

$$I_3(\kappa_i \, r) = \Omega_3 \, \frac{\sin(\kappa_i \, r)}{\kappa_i \, r} \,,$$

$$I_2(\kappa_i \, r) = \Omega_2 \, J_0(\kappa_i \, r) \xrightarrow[r \to \infty]{} \frac{1}{\sqrt{\kappa_i \, r}} \, \cos\left(\kappa_i \, r - \frac{\pi}{4}\right) \,,$$

$$I_1(\kappa_i \, r) = \Omega_1 \, \cos(\kappa_i \, r) \,.$$

In $d = 3$, $\prod_{i=1}^4 I_3(\kappa_i\, r) \sim 1/r^4$ and the integral converges even if we set all $\kappa_i = 1$, thus $\epsilon_0 = \epsilon_1 = \omega_1 = 0$. In other words, $S_k(0, 0, 0)$ in $d = 3$ is finite.

In $d < 3$, the integral with all $\kappa_i = 1$ diverges logarithmically in $d = 2$, where $I_2^4(r) \sim 1/r^2$, and linearly in $d = 1$, where $I_1^4(r) \sim$ constant. In both cases, the different $\kappa_i \neq 1$ play a crucial role in cutting off the singularity. Since close to the Fermi surface, e.g.,

$$\epsilon_0 = \epsilon(k_0) \simeq v_F \left(k_0 - k_F\right) = v_F\, k_F \left(\kappa_2 - 1\right) \;\Rightarrow\; \kappa_2 \simeq 1 + \frac{\epsilon_0}{v_F\, k_F}\,,$$

and similarly for all other momentum, taking, for simplicity, $k = k_F$, so that $\kappa_1 = 1$, in $d = 1, 2$ the product of the cosine functions yields, at leading order,

$$\prod_{i=1}^4 \cos\left(\kappa_i\, r - \frac{\pi}{4}\right) \simeq \frac{1}{8}\cos\left(\left(1 - \kappa_2 - \kappa_3 + \kappa_4\right)r\right) = \frac{1}{8}\cos\left(\frac{\epsilon_0 + \epsilon_1 - \omega_1}{v_F\, k_F}\, r\right)$$

$$= \frac{1}{8}\cos\left(\frac{\epsilon}{v_F\, k_F}\, r\right),$$

explicitly showing that the finite ϵ cuts off the large r singularity. As a result, in $d = 1$ and $d = 2$ the ϵ^2 behaviour in (4.62) is replaced, respectively, by $-\epsilon^2 \ln|\epsilon|$ and $|\epsilon|$.

In conclusion, we have shown that, order by order in perturbation theory and at $T = 0$,

$$\lim_{\epsilon \to 0} \operatorname{Im}\Sigma(\epsilon, \mathbf{k}) \sim \begin{cases} -\epsilon^2 & d = 3\,, \\ \epsilon^2 \ln|\epsilon| & d = 2\,, \\ -|\epsilon| & d = 1\,, \end{cases} \tag{4.64}$$

and thus vanishes. However, through (4.54), the one-dimensional behaviour $\operatorname{Im}\Sigma(\epsilon, \mathbf{k}) \sim -|\epsilon|$ implies that

$$\frac{\partial \operatorname{Re}\Sigma(\epsilon, \mathbf{k})}{\partial \epsilon} = \int \frac{d\omega}{\pi}\, \frac{\operatorname{Im}\Sigma(\omega, \mathbf{k})}{\left(\epsilon - \omega\right)^2}\,, \tag{4.65}$$

diverges as $\ln|\epsilon| \to -\infty$ as $\epsilon \to 0$, corresponding to a vanishing quasiparticle residue $Z(\epsilon \to 0, \mathbf{k})$, see (4.57). That is actually one of several singularities that appear in perturbation theory, symptoms that the latter does not converge in $d = 1$ however small the interaction strength is. Notice that, in $d = 2, 3$ the integral in (4.65) converges as $\epsilon \to 0$, implying a finite quasiparticle residue $Z(\epsilon \to 0, \mathbf{k})$.

Hereafter, we thus proceed assuming $d > 1$, so that perturbation theory is well behaved. However, even in that case, the observation that order by order in perturbation theory $\operatorname{Im}\Sigma(\epsilon \to 0, \mathbf{k}) \sim -\epsilon^2$, with log-corrections in $d = 2$, does not guarantee that the whole perturbation series converges and the result indeed vanishes like $-\epsilon^2$. For that reason, we make the assumption that in $d > 1$ the perturbation series has a finite convergence radius, and, further, that the interaction strength is within that radius. Later, when discussing Landau's Fermi liquid theory, we shall relax the latter assumption. Therefore, for now on we take for granted that, if each term in the perturbative expansion of $\text{Im}\,\Sigma(\epsilon, \mathbf{k})$ vanishes, e.g., as $-\epsilon^2$ at small ϵ, also the whole series vanishes similarly. Correspondingly, as discussed above the quasiparticle residue $Z(\epsilon, \mathbf{k})$ is finite as $\epsilon \to 0$ for any $\mathbf{k}$.

We conclude by highlighting another important side result that derives from (4.58) before taking the analytic continuation on the real axis. Indeed, if we consider the imaginary part of that equation for positive Matsubara frequency $\epsilon > 0$,

$$\text{Im}\,\Sigma(i\epsilon, \mathbf{k}) = -\frac{2U^2}{V^2} \sum_{\mathbf{k}_0 \mathbf{k}_1 \mathbf{p}_1} \delta_{\mathbf{k}+\mathbf{p}_1, \mathbf{k}_0+\mathbf{k}_1} \frac{\epsilon}{\left(\epsilon_{\mathbf{p}_1} - \epsilon_{\mathbf{k}_0} - \epsilon_{\mathbf{k}_1}\right)^2 + \epsilon^2}$$

$$\left[f(\epsilon_{\mathbf{p}_1}) \left(1 - f(\epsilon_{\mathbf{k}_0})\right) \left(1 - f(\epsilon_{\mathbf{k}_1})\right) + \left(1 - f(\epsilon_{\mathbf{p}_1})\right) f(\epsilon_{\mathbf{k}_0}) f(\epsilon_{\mathbf{k}_1}) \right],$$

and note that the Lorentzian function

$$\frac{1}{\pi} \frac{\epsilon}{\left(\epsilon_{\mathbf{p}_1} - \epsilon_{\mathbf{k}_0} - \epsilon_{\mathbf{k}_1}\right)^2 + \epsilon^2} \xrightarrow[\epsilon \to 0]{} \delta\left(\epsilon_{\mathbf{p}_1} - \epsilon_{\mathbf{k}_0} - \epsilon_{\mathbf{k}_1}\right),$$

so that

$$\lim_{\epsilon \to 0} \text{Im}\,\Sigma(i\epsilon, \mathbf{k}) = \lim_{\epsilon \to 0} \text{Im}\,\Sigma(\epsilon + i\eta, \mathbf{k}) = 0, \qquad (4.66)$$

and, in particular, $\text{Im}\,\Sigma(i\epsilon, \mathbf{k}) \sim -\epsilon$ for small ϵ. As before, this result remains true order by order in perturbation theory, so that, if the latter converges and the strength of interaction is within the convergence radius, we can conclude that (4.66) holds for the whole perturbation series.

4.3.5 Emergence of Quasiparticles

Assuming, e.g., $d = 3$, we just showed that $\text{Im}\,\Sigma(\epsilon \to 0, \mathbf{k}) \simeq -\gamma(\mathbf{k})\,\epsilon^2$, so that the single-particle density-of-states in (4.56)

$$A(\epsilon, \mathbf{k}) = \frac{1}{\pi} \frac{-\text{Im}\,\Sigma(\epsilon, \mathbf{k})}{\left(\epsilon - \epsilon_{\mathbf{k}} - \text{Re}\,\Sigma(\epsilon, \mathbf{k})\right)^2 + \text{Im}^2\,\Sigma(\epsilon, \mathbf{k})}$$

$$\xrightarrow[\epsilon \to 0]{} \frac{Z(0, \mathbf{k})}{\pi} \frac{\gamma_*(\mathbf{k})\,\epsilon^2}{\left(\epsilon - \epsilon_*(\mathbf{k})\right)^2 + \gamma_*(\mathbf{k})^2\,\epsilon^4} \equiv Z(0, \mathbf{k})\,A_{\text{qp}}(\epsilon, \mathbf{k}),$$

$$(4.67)$$

where $A_{\text{qp}}(\epsilon, \mathbf{k})$ will be denoted as the *quasiparticle* density of states, $Z(\epsilon, \mathbf{k})$ the quasiparticle residue in (4.57), and

$$\epsilon_*(\mathbf{k}) \equiv Z(0, \mathbf{k})\left(\epsilon_{\mathbf{k}} - \text{Re}\,\Sigma(0, \mathbf{k})\right), \qquad \gamma_*(\mathbf{k}) \equiv Z(0, \mathbf{k})\,\gamma(\mathbf{k}). \qquad (4.68)$$

$A_{qp}(\epsilon, \mathbf{k})$ in (4.67) typically vanishes at the chemical potential, i.e., $\epsilon = 0$, unless on the surface defined through $\epsilon_*(\mathbf{k}) = 0$. Since $Z(0, \mathbf{k})$ is finite, that occurs when

$$\epsilon_{\mathbf{k}_F} + \mathrm{Re}\,\Sigma(0, \mathbf{k}_F) = 0, \qquad (4.69)$$

which defines the interacting Fermi surface. As a consequence,

$$A_{qp}(\epsilon \to 0, \mathbf{k}) \simeq \frac{1}{\pi} \frac{\gamma_*(\mathbf{k})\,\epsilon^2}{\left(\epsilon - \epsilon_*(\mathbf{k})\right)^2 + \gamma_*(\mathbf{k})^2\,\epsilon^4} \xrightarrow[\mathbf{k} \to \mathbf{k}_F]{} \delta\big(\epsilon - \epsilon_*(\mathbf{k})\big), \quad (4.70)$$

looks like the density of states of non interacting particles, the *quasiparticles*, with dispersion $\epsilon_*(\mathbf{k})$ in momentum space. Strictly speaking, the quasiparticle at energy ϵ and momentum $\mathbf{k}$ has a finite decay rate $\gamma_*(\mathbf{k})\,\epsilon^2$, which however vanishes faster than its energy approaching the Fermi surface. We can thus state that the quasiparticles are coherent excitations, namely their lifetime diverges at the Fermi surface.

It follows that the physical electron DOS

$$A(\epsilon \to 0, \mathbf{k} \to \mathbf{k}_F) = Z(0, \mathbf{k})\,A_{qp}(\epsilon \to 0, \mathbf{k} \to \mathbf{k}_F) = Z(0, \mathbf{k}_F)\,\delta\big(\epsilon - \epsilon_*(\mathbf{k})\big),$$

also becomes a δ-function centred at the chemical potential for $\mathbf{k} \to \mathbf{k}_F$, though with finite weight $Z(0, \mathbf{k})$. Since the integral over ϵ of $A(\epsilon, \mathbf{k})$ is unity, $Z(0, \mathbf{k}_F)$ is positive and less than one. The rest $1 - Z(0, \mathbf{k}_F)$ of the spectral weight must be concentrated at finite energy, so that

$$A(\epsilon, \mathbf{k} \to \mathbf{k}_F) = Z(0, \mathbf{k}_F)\,\delta\big(\epsilon - \epsilon_*(\mathbf{k})\big) + A_{\mathrm{inc}}(\epsilon, \mathbf{k}), \qquad (4.71)$$

where $A_{\mathrm{inc}}(\epsilon, \mathbf{k})$ carries the rest of spectral weight and is commonly denoted as incoherent background to distinguish it from the quasiparticle coherent δ-peak. Correspondingly, through (4.25),

$$
\begin{aligned}
G(z, \mathbf{k}) = \int d\epsilon\, \frac{A(\epsilon, \mathbf{k})}{z - \epsilon} &\xrightarrow[\mathbf{k} \to \mathbf{k}_F]{} \frac{Z(0, \mathbf{k}_F)}{z - \epsilon_*(\mathbf{k})} + \int d\epsilon\, \frac{A_{\mathrm{inc}}(\epsilon, \mathbf{k})}{z - \epsilon} \\
&\equiv \frac{Z(0, \mathbf{k}_F)}{z - \epsilon_*(\mathbf{k})} + G_{\mathrm{inc}}(z, \mathbf{k}) \equiv G_{\mathrm{coh}}(z, \mathbf{k}) + G_{\mathrm{inc}}(z, \mathbf{k}).
\end{aligned}
$$
$$(4.72)$$

Therefore, the Green's function for $\mathbf{k}$ moving towards the Fermi surface develops a coherent part $G_{\mathrm{coh}}(z, \mathbf{k})$, i.e., a simple pole on the real axis at the quasiparticle energy $\epsilon_*(\mathbf{k})$ with finite quasiparticle residue, $Z(0, \mathbf{k}_F)$, thus its name, besides an incoherent component $G_{\mathrm{inc}}(z, \mathbf{k})$ that has instead a branch cut on the real axis.

4.4   Other Kinds of Perturbations

Before proceeding deep in analysing the perturbation theory in the electron-electron interaction, let us briefly discuss the diagrammatic techniques for other two kinds of perturbations: a scalar potential and the coupling to bosonic modes.

4.4.1   Scalar Potential

Let us suppose to have non-interacting electrons identified by a quantum label a, with non-interacting Green's functions

$$G_{0ab}(\tau; \mathbf{x}, \mathbf{y}) = \delta_{ab}\, G_{0a}(\tau; \mathbf{x}, \mathbf{y}),$$

in presence of the scalar potential

$$V = \sum_{ab} \int d\mathbf{x}\, \Psi_a^\dagger(\mathbf{x})\, V_{ab}(\mathbf{x})\, \Psi_b(\mathbf{x}). \tag{4.73}$$

The S-operator is now

$$S(\beta, 0) = T_\tau \left[\exp\left(-\sum_{ab} \int_0^\beta d\tau \int d\mathbf{x}\, \Psi_a^\dagger(\mathbf{x}, \tau)\, V_{ab}(\mathbf{x})\, \Psi_b(\mathbf{x}, \tau) \right) \right]$$

$$\equiv T_\tau \left[\exp\left(-\sum_{ab} \int dx\, \Psi_a^\dagger(x)\, V_{ab}(x)\, \Psi_b(x) \right) \right],$$

where, as before, we have introduced the four-dimensional coordinate $x = (\tau, \mathbf{x})$. Upon expanding $S(\beta, 0)$ up to first order and keeping only connected diagrams, one can readily obtain the Green's function expansion

$$G_{ab}(x, y) = \delta_{ab}\, G_{0a}(x, y) + \sum_{cd} \int dz\, V_{cd}(z)\, \langle T_\tau \left(\Psi_a(x)\, \Psi_b^\dagger(y)\, \Psi_c^\dagger(z)\, \Psi_d(z) \right) \rangle_{conn}$$

$$= \delta_{ab}\, G_{0a}(x, y) + \int dz\, V_{ab}(z)\, G_{0a}(x, z)\, G_{0b}(z, y). \tag{4.74}$$

If we represent graphically the perturbation as in Fig. 4.16a, then the Green's function (4.74) up to first order can be drawn as the first two diagrams in Fig. 4.16b. Higher order terms can be simply obtained by inserting other potential lines, as the second order term in Fig. 4.16b. All diagrams have a positive sign. The Dyson equation can be easily read out:

$$G_{ab}(x, y) = \delta_{ab}\, G_{0a}(x, y) + \sum_c \int dz\, G_{0a}(x, z)\, V_{ac}(z)\, G_{cb}(z, y), \tag{4.75}$$

and is drawn in Fig. 4.16b.

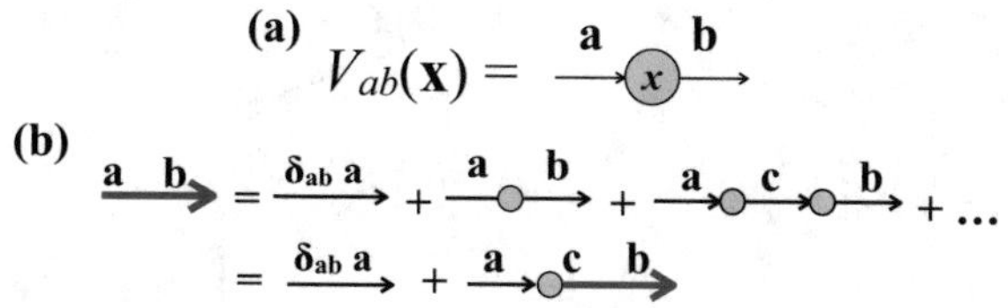

(a) $V_{ab}(\mathbf{x}) =$

(b)

Fig. 4.16 a Graphical representation of the scalar potential perturbation. b Perturbation expansion of the Green's function and Dyson equation. The bold blue line represents the Green's function at all orders in perturbation theory

4.4.2 Coupling to Bosonic Modes

Let us imagine now that our system of electrons is coupled to bosonic modes described by the free Hamiltonian

$$H_{bos} = \sum_{\mathbf{q}} \omega_{\mathbf{q}}\, a_{\mathbf{q}}^{\dagger} a_{\mathbf{q}}\,.$$

These modes could represent phonons, photons, or any other bosonic excitation. The coupling term is assumed to be

$$V = \sum_{\mathbf{q}} g(\mathbf{q})\, x_{\mathbf{q}}\, \rho(-\mathbf{q})\,, \tag{4.76}$$

where $g(\mathbf{q})^* = g(-\mathbf{q})$ is the coupling constant, the phonon coordinate is

$$x_{\mathbf{q}} = \sqrt{\frac{1}{2}}\left(a_{\mathbf{q}} + a_{-\mathbf{q}}^{\dagger}\right),$$

in terms of bosonic annihilation and creation operators, and

$$\rho(\mathbf{q}) = \sum_{\sigma}\sum_{\mathbf{k}} c_{\mathbf{k}\sigma}^{\dagger} c_{\mathbf{k}+\mathbf{q}\sigma}\,,$$

is the electron density operator.

We can perform perturbation theory in (4.76) using Wick's theorem both for the electrons and for the bosons. The latter amounts to contract an $x_{\mathbf{q}}$ with $x_{-\mathbf{q}}$ which leads to the free propagator $D_0(i\omega, \mathbf{q})$ of (4.33). Therefore, if we represent the electron-boson coupling as in Fig. 4.17, in which a dotted vertex line represents the boson-coordinate, by contracting two vertices one recovers an effective electron-electron interaction, see also Fig. 4.17, which is mediated by the bosons and given by

$$|g(\mathbf{q})|^2\, D_0(i\omega, \mathbf{q}) = -|g(\mathbf{q})|^2\, \frac{\omega_{\mathbf{q}}}{\omega^2 + \omega_{\mathbf{q}}^2}\,. \tag{4.77}$$

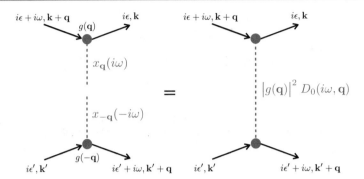

Fig. 4.17 On the left: electron-boson vertices at **q** and $-\mathbf{q}$. The external dotted vertex line represents boson coordinates. On the right: the effective electron-electron interaction after contracting the bosons

Fig. 4.18 Lowest orders in the diagrammatic perturbation expansion of $D(i\omega, \mathbf{q})$. The fully-interacting $D(i\omega, \mathbf{q})$ is represented by a bold dashed line, while $D_0(i\omega, \mathbf{q})$ by a thiner dashed line. The solid lines are electron Green's functions

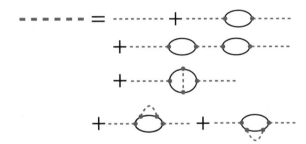

Such effective interaction is retarded, namely depends on the frequency, and attractive. Specifically, if we move on the real axis, then

$$\left|g(\mathbf{q})\right|^2 D_0(i\omega, \mathbf{q}) \xrightarrow[i\omega \to \omega + i\eta]{} -\left|g(\mathbf{q})\right|^2 \frac{\omega_{\mathbf{q}}}{\omega_{\mathbf{q}}^2 - \omega^2 - i\omega\eta} \, ,$$

where, as usual, $\eta > 0$ is infinitesimal, which is attractive if $\omega^2 < \omega_q^2$ and repulsive otherwise.

The rules for constructing diagrams are therefore the same as for the electron-electron interaction, with the additional complication that interaction is frequency dependent.

One may also investigate the effects of the electron-boson coupling on the boson Green's function $D(i\omega, \mathbf{q})$. The perturbation expansion is shown up to second order in Fig. 4.18. Two kinds of diagrams can be identified. The first class includes diagrams which can be divided into two lower order diagrams by cutting a $D_0(i\omega, \mathbf{q})$-line, like the first second order diagram shown in Fig. 4.18. These diagrams are called *reducible*. The other class includes all other diagrams which are therefore irreducible. If we define, similarly to Dyson's equation, a self-energy $\left|g(\mathbf{q})\right|^2 \Pi(i\omega, \mathbf{q})$ that includes all irreducible diagrams, shown up to second order in Fig. 4.19, we can

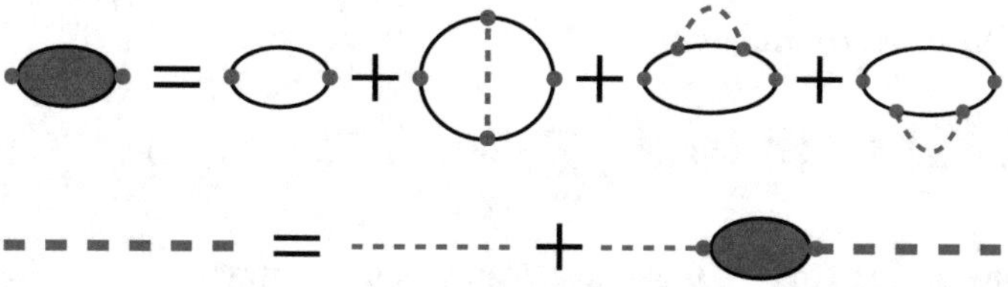

Fig. 4.19 Upper panel: Boson self-energy, represented by a filled circle, up to second order. Lower panel: Dyson equation for $D(q)$

easily derive the Dyson equation, also shown in the same figure, whose solution is

$$D(i\omega, \mathbf{q}) = \frac{D_0(i\omega, \mathbf{q})}{1 - |g(\mathbf{q})|^2 D_0(i\omega, \mathbf{q}) \Pi(i\omega, \mathbf{q})} = -\frac{\omega_{\mathbf{q}}}{\omega^2 + \omega_{\mathbf{q}}^2 - |g(\mathbf{q})|^2 \Pi(i\omega, \mathbf{q})}$$

$$\xrightarrow[i\omega \to \omega + i\eta]{} \frac{\omega_{\mathbf{q}}}{\omega^2 - \omega_{\mathbf{q}}^2 + |g(\mathbf{q})|^2 \Pi(\omega, \mathbf{q})}. \tag{4.78}$$

We note that, when continued on the real axis, the real solutions of

$$\omega_*^2(\mathbf{q}) = \omega_{\mathbf{q}}^2 - |g(\mathbf{q})|^2 \operatorname{Re} \Pi(\omega_*(\mathbf{q}), \mathbf{q}),$$

represent the boson dispersion relation modified by the coupling to electrons, which also provide a decay rate to the bosonic mode proportional to $|g(\mathbf{q})|^2 \operatorname{Im} \Pi(\omega_*(\mathbf{q}), \mathbf{q})$.

4.5 Two-Particle Green's Functions and Correlation Functions

Now we move back to the case of an instantaneous electron-electron interaction, and proceed in analysing the perturbative expansion.

In that case, the interacting Hamiltonian is

$$H = \sum_{\mathbf{k}\sigma} \epsilon_{\mathbf{k}} c_{\mathbf{k}\sigma}^\dagger c_{\mathbf{k}\sigma} + \frac{1}{2V} \sum_{\mathbf{kpq}} \sum_{\alpha\beta} U(\mathbf{q}) c_{\mathbf{p}\alpha}^\dagger c_{\mathbf{k}+\mathbf{q}\beta}^\dagger c_{\mathbf{k}\beta} c_{\mathbf{p}+\mathbf{q}\alpha}. \tag{4.79}$$

The Heisenberg imaginary time evolution of an annihilation operator is

$$-\frac{\partial c_{\mathbf{k}\sigma}(\tau)}{\partial \tau} = \left[c_{\mathbf{k}\sigma}(\tau), H(\tau) \right] = \epsilon_{\mathbf{k}} c_{\mathbf{k}\sigma}(\tau) + \frac{1}{V} \sum_{\mathbf{pq}\alpha} U(\mathbf{q}) c_{\mathbf{p}+\mathbf{q}\alpha}^\dagger(\tau) c_{\mathbf{p}\alpha}(\tau) c_{\mathbf{k}+\mathbf{q}\sigma}(\tau),$$

through which one readily find the equation of motion of the Green's function

$$-\frac{\partial}{\partial \tau} G(\tau, \mathbf{k}) = -\frac{\partial}{\partial \tau} \left(-\langle T_\tau \left(c_{\mathbf{k}\sigma}(\tau) c_{\mathbf{k}\sigma}^\dagger \right) \rangle \right)$$

$$= -\frac{\partial}{\partial \tau} \left(-\theta(\tau) \langle c_{\mathbf{k}\sigma}(\tau) c_{\mathbf{k}\sigma}^\dagger \rangle + \theta(-\tau) \langle c_{\mathbf{k}\sigma}^\dagger c_{\mathbf{k}\sigma}(\tau) \rangle \right) = \delta(\tau) - \langle T_\tau \left(\frac{\partial c_{\mathbf{k}\sigma}(\tau)}{\partial \tau} c_{\mathbf{k}\sigma}^\dagger \right) \rangle$$

$$= \delta(\tau) + \epsilon_{\mathbf{k}} G(\tau, \mathbf{k}) - \frac{1}{V} \sum_{\mathbf{pq}\alpha} U(\mathbf{q}) \langle T_\tau \left(c_{\mathbf{p}+\mathbf{q}\alpha}^\dagger(\tau) c_{\mathbf{p}\alpha}(\tau) c_{\mathbf{k}+\mathbf{q}\sigma}(\tau) c_{\mathbf{k}\sigma}^\dagger \right) \rangle.$$

This equation can be written as

$$\left(-\frac{\partial}{\partial\tau'} - \epsilon_{\mathbf{k}}\right) G(\tau', \mathbf{k}) = \delta(\tau') - \frac{1}{V}\sum_{\mathbf{pq}\alpha} U(\mathbf{q})\, \langle T_\tau \left(c_{\mathbf{p}+\mathbf{q}\alpha}^\dagger(\tau')\, c_{\mathbf{p}\alpha}(\tau')\, c_{\mathbf{k}+\mathbf{q}\sigma}(\tau')\, c_{\mathbf{k}\sigma}^\dagger\right)\rangle\,.$$

$$(4.80)$$

The non-interacting Green's function satisfies on the contrary

$$\left(-\frac{\partial}{\partial\tau'} - \epsilon_{\mathbf{k}}\right) G_0(\tau', \mathbf{k}) = \delta(\tau')\,.$$

If we multiply both sides of (4.80) by $G_0(\tau - \tau', \mathbf{k})$ and integrate over τ', we obtain

$$\int d\tau'\, G_0(\tau - \tau', \mathbf{k}) \left(-\frac{\partial}{\partial\tau'} - \epsilon_{\mathbf{k}}\right) G(\tau', \mathbf{k}) = G_0(\tau, \mathbf{k})$$

$$-\frac{1}{V}\sum_{\mathbf{pq}\alpha} U(\mathbf{q}) \int d\tau'\, G_0(\tau - \tau', \mathbf{k})\, \langle T_\tau \left(c_{\mathbf{p}+\mathbf{q}\alpha}^\dagger(\tau')\, c_{\mathbf{p}\alpha}(\tau')\, c_{\mathbf{k}+\mathbf{q}\sigma}(\tau')\, c_{\mathbf{k}\sigma}^\dagger\right)\rangle,$$

and upon integrating by part the left hand side, one obtains

$$G(\tau, \mathbf{k}) = G_0(\tau, \mathbf{k})$$

$$-\frac{1}{V}\sum_{\mathbf{pq}\alpha} U(\mathbf{q}) \int d\tau'\, G_0(\tau - \tau', \mathbf{k})\, \langle T_\tau \left(c_{\mathbf{p}+\mathbf{q}\alpha}^\dagger(\tau')\, c_{\mathbf{p}\alpha}(\tau')\, c_{\mathbf{k}+\mathbf{q}\sigma}(\tau')\, c_{\mathbf{k}\sigma}^\dagger\right)\rangle\,.$$

$$(4.81)$$

Therefore the single-particle Green's function can be expressed in terms of a two-particle Green's function. Namely, let us define the two-particle Green's function

$$K_{\sigma_1\sigma_2;\sigma_3\sigma_4}(\tau_1\,\mathbf{p}_1, \tau_2\,\mathbf{p}_2; \tau_3\,\mathbf{p}_3, \tau_4\,\mathbf{p}_4) = -\langle T_\tau \left(c_{\mathbf{p}_1\sigma_1}(\tau_1) c_{\mathbf{p}_2\sigma_2}(\tau_2)\, c_{\mathbf{p}_3\sigma_3}^\dagger(\tau_3) c_{\mathbf{p}_4\sigma_4}^\dagger(\tau_4)\right)\rangle\,, \quad (4.82)$$

where, by momentum conservation,

$$\mathbf{p}_1 + \mathbf{p}_2 = \mathbf{p}_3 + \mathbf{p}_4\,,$$

in terms of which

$$G(\mathbf{k}, \tau) = G_0(\mathbf{k}, \tau)$$

$$-\frac{1}{V}\sum_{\mathbf{pq}\alpha} U(\mathbf{q}) \int d\tau\, G_0(\tau - \tau', \mathbf{k})\, K_{\sigma\alpha;\alpha\sigma}(\tau'\,\mathbf{k} + \mathbf{q}, \tau'\,\mathbf{p}; \tau' + 0^+\,\mathbf{p} + \mathbf{q}, 0\,\mathbf{k})\,,$$

$$(4.83)$$

where, since in (4.81) the operator $c_{\mathbf{q}+\mathbf{q}\alpha}^\dagger(\tau')$ appears on the left, we have set its time to $\tau' + 0^+$ to enforce that.

4.5.1 Diagrammatic Representation of the Two-Particle Green's Function

In absence of interaction,

$$
K_{\sigma_1\sigma_2;\sigma_3\sigma_4}(\tau_1\,\mathbf{p}_1, \tau_2\,\mathbf{p}_2; \tau_3\,\mathbf{p}_3, \tau_4\,\mathbf{p}_4) = -\langle T_\tau \left(c_{\mathbf{p}_1\sigma_1}(\tau_1) c_{\mathbf{p}_2\sigma_2}(\tau_2)\, c_{\mathbf{p}_3\sigma_3}^\dagger(\tau_3) c_{\mathbf{p}_4\sigma_4}^\dagger(\tau_4) \right)\rangle_0
$$
$$
= -\delta_{\sigma_1\sigma_4}\delta_{\sigma_2\sigma_3}\delta_{\mathbf{p}_1\mathbf{p}_4}\delta_{\mathbf{p}_2\mathbf{p}_3}\, G_0(\tau_1 - \tau_4, \mathbf{p}_1)\, G_0(\tau_2 - \tau_3, \mathbf{p}_2)
$$
$$
+ \delta_{\sigma_1\sigma_3}\delta_{\sigma_2\sigma_4}\delta_{\mathbf{p}_1\mathbf{p}_3}\delta_{\mathbf{p}_2\mathbf{p}_4}\, G_0(\tau_1 - \tau_3, \mathbf{p}_1)\, G_0(\tau_2 - \tau_4, \mathbf{p}_2).
$$

The interaction in perturbation theory has two effects. First it turns each G_0 into the fully-interacting G, and next it couples together the two Green's functions. For that reason, we can formally write

$$
K_{\sigma_1\sigma_2;\sigma_3\sigma_4}(\tau_1\,\mathbf{p}_1, \tau_2\,\mathbf{p}_2; \tau_3\,\mathbf{p}_3, \tau_4\,\mathbf{p}_4) = -\delta_{\sigma_1\sigma_4}\delta_{\sigma_2\sigma_3}\delta_{\mathbf{p}_1\mathbf{p}_4}\delta_{\mathbf{p}_2\mathbf{p}_3}\, G(\tau_1 - \tau_4, \mathbf{p}_1)G(\tau_2 - \tau_3, \mathbf{p}_2)
$$
$$
+ \delta_{\sigma_1\sigma_3}\delta_{\sigma_2\sigma_4}\delta_{\mathbf{p}_1\mathbf{p}_3}\delta_{\mathbf{p}_2\mathbf{p}_4}\, G(\tau_1 - \tau_3, \mathbf{p}_1)G(\tau_2 - \tau_4, \mathbf{p}_2)
$$
$$
+ \int \prod_{i=1}^{4} d\tau_i'\, G(\tau_1 - \tau_1', \mathbf{p}_1)\, G(\tau_2 - \tau_2', \mathbf{p}_2)\, G(\tau_3' - \tau_3, \mathbf{p}_3)\, G(\tau_4' - \tau_4, \mathbf{p}_4)
$$
$$
\Gamma_{\sigma_1\sigma_2;\sigma_3\sigma_4}(\tau_1'\,\mathbf{p}_1, \tau_2'\,\mathbf{p}_2; \tau_3'\,\mathbf{p}_3, \tau_4'\,\mathbf{p}_4), \tag{4.84}
$$

which is represented graphically to Fig. 4.20. In Matsubara frequency, and taking into account spin, momentum and frequency conservation, we can write the four-leg interaction vertex Γ as

$$
\Gamma_{\sigma_1\sigma_2;\sigma_3\sigma_4}(i\epsilon + i\omega\,\mathbf{k} + \mathbf{q}, i\epsilon'\,\mathbf{k}'; i\epsilon' + i\omega\,\mathbf{k}' + \mathbf{q}, i\epsilon\,\mathbf{k}), \quad \sigma_1 + \sigma_2 = \sigma_3 + \sigma_4, \tag{4.85}
$$

where ϵ and ϵ' are fermionic Matsubara frequencies, ω a bosonic one, and the spin condition implies that the sum of the spin z-components of the incoming electrons, $\sigma_1 + \sigma_2$, must be equal to the sum of the outgoing electrons, $\sigma_3 + \sigma_4$. The skeleton diagrams for Γ up to second order are shown in Fig. 4.21, where, e.g., 1 stands for frequency $i\epsilon + i\omega$, momentum $\mathbf{k} + \mathbf{q}$ and spin σ_1.

In conclusion the (4.83) of the single particle Green's function in terms of the two-particle one can be expressed as function of the single particle Green's function itself and the interaction vertex as shown in Fig. 4.22. Notice that the interaction vertex acts as the bare interaction, so it carries a (-1) sign. Since there is a loop in

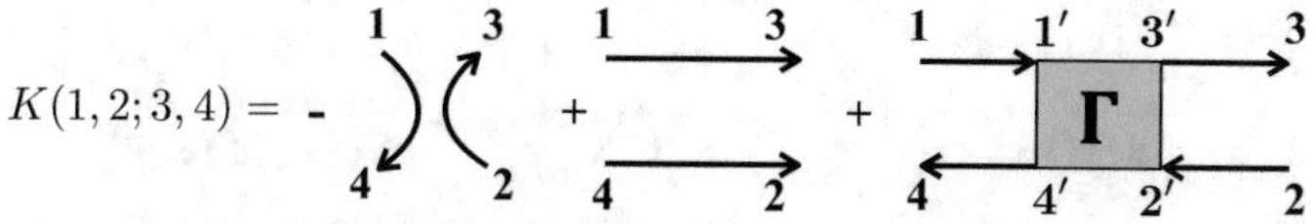

$$
K(1, 2; 3, 4) =
$$

Fig. 4.20 Graphical representation of the two-particle Green's function. Here, all numbers are short notation that include the imaginary time, momentum and spin of each vertex. The coloured box represent the four-leg vertex $\Gamma(1', 2'; 3', 4')$. Notice that all Green's functions are fully interacting ones

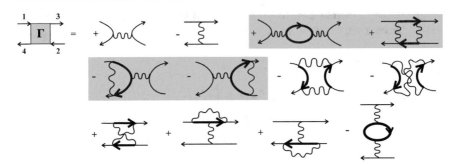

Fig. 4.21 Lowest order skeleton expansion of the interaction vertex $\Gamma(1, 2; 3, 4)$ in skeleton diagrams. The external vertices, 1, 2, 3 and 4, where each number stands for frequency, momentum and spin, are drawn as narrow arrow lines, while the internal interacting Green's functions as bold arrow lines. We explicitly show the sign of each diagram that can be found from the first two recalling that each interaction and loop bring an minus sign

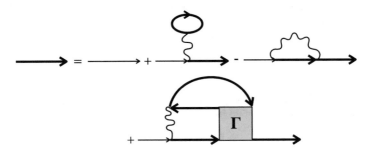

Fig. 4.22 Dyson equation in terms of the interaction vertex Γ

the third diagram of Fig. 4.22, the overall sign is plus. This result also provides an expression for the single-particle self-energy shown in Fig. 4.23. Figure 4.23 implies the following analytic equation

$$\Sigma(i\epsilon, \mathbf{k}) = \frac{1}{V} \sum_{\mathbf{k}'} \Big(2U(0) - U(\mathbf{k} - \mathbf{k}')\Big) n(\mathbf{k}')$$

$$+ \frac{1}{V^2} \sum_{\mathbf{k}'\mathbf{q}} \sum_{\sigma'} T^2 \sum_{\epsilon'\omega} U(\mathbf{q}) G(i\epsilon + i\omega, \mathbf{k} + \mathbf{q}) G(i\epsilon', \mathbf{k}') G(i\epsilon' + i\omega, \mathbf{k}' + \mathbf{q})$$

$$\Gamma_{\sigma,\sigma';\sigma',\sigma}\big(i\epsilon + i\omega \, \mathbf{k} + \mathbf{q}, i\epsilon' \, \mathbf{k}'; i\epsilon' + i\omega \, \mathbf{k}' + \mathbf{q}, i\epsilon \, \mathbf{k}\big), \qquad (4.86)$$

where

$$n(\mathbf{k}') \equiv \langle c^{\dagger}_{\mathbf{k}'\sigma} c_{\mathbf{k}'\sigma} \rangle = T \sum_{\epsilon'} e^{i\epsilon' 0^+} G(i\epsilon', \mathbf{k}'), \qquad (4.87)$$

is the interacting momentum distribution. Since $\Sigma = \Sigma[G, U]$, the above equation implies that also Γ can be expressed as a functional of G and U.

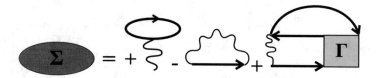

Fig. 4.23 Single-particle self-energy in terms of the interaction vertex Γ

4.5.2   Correlation Functions

Let us consider the density operators,

$$A(\mathbf{q}) = \sum_{\mathbf{k}\alpha\beta} \Lambda_0^A(\mathbf{k}\,\alpha, \mathbf{k} + \mathbf{q}\,\beta)\, c_{\mathbf{k}\alpha}^\dagger\, c_{\mathbf{k}+\mathbf{q}\beta}, \qquad B(\mathbf{q}) = \sum_{\mathbf{k}\alpha\beta} \Lambda_0^B(\mathbf{k}\,\alpha, \mathbf{k} + \mathbf{q}\,\beta)\, c_{\mathbf{k}\alpha}^\dagger\, c_{\mathbf{k}+\mathbf{q}\beta},$$

which have bosonic character, and the expectation value of their time-ordered product, which we shall denote as the correlation function χ_{AB} instead of the G_{AB} Green's function of (4.4),

$$\chi_{AB}(\tau, \mathbf{q}) \equiv -\frac{1}{V}\langle T_\tau\left(A(\mathbf{q}, \tau)\, B(-\mathbf{q})\right)\rangle$$

$$= -\frac{1}{V} \sum_{\mathbf{k}\alpha\beta} \sum_{\mathbf{p}\gamma\delta} \Lambda_0^A(\mathbf{k}\,\alpha, \mathbf{k} + \mathbf{q}\,\beta)\, \Lambda_0^B(\mathbf{p} + \mathbf{q}\,\gamma, \mathbf{p}\,\delta)\,\langle T_\tau\left(c_{\mathbf{k}\alpha}^\dagger(\tau) c_{\mathbf{k}+\mathbf{q}\beta}(\tau) c_{\mathbf{p}+\mathbf{q}\gamma}^\dagger c_{\mathbf{p}\delta}\right)\rangle$$

$$= \frac{1}{V} \sum_{\mathbf{k}\alpha\beta} \sum_{\mathbf{p}\gamma\delta} \Lambda_0^A(\mathbf{k}\,\alpha, \mathbf{k} + \mathbf{q}\,\beta)\, \Lambda_0^B(\mathbf{p} + \mathbf{q}\,\gamma, \mathbf{p}\,\delta)\, K_{\beta\delta;\gamma\alpha}\left(\tau\,\mathbf{k} + \mathbf{q}, 0\,\mathbf{p}; 0^+\,\mathbf{p} + \mathbf{q}, \tau^+\,\mathbf{k}\right),$$

where we use the definition (4.82) of the two-particle Green's function, and times 0^+ and τ^+ are assumed infinitesimally larger than 0 and τ, since in the definition of the density operators the creation operator is to the left of the annihilation one. Through the expression of the two-particle Green's function in terms of the interaction vertex, see (4.84) and also Fig. 4.20,

$$\chi_{AB}(\tau, \mathbf{q}) = -\frac{\delta_{\mathbf{q},0}}{V}\langle A(\mathbf{0})\rangle\langle B(\mathbf{0})\rangle$$

$$+ \frac{1}{V} \sum_{\mathbf{k}\alpha\beta} \Lambda_0^A(\mathbf{k}\,\alpha, \mathbf{k} + \mathbf{q}\,\beta)\, \Lambda_0^B(\mathbf{k} + \mathbf{q}\,\beta, \mathbf{k}\,\alpha)\, G(\tau, \mathbf{k} + \mathbf{q})\, G(-\tau, \mathbf{k})$$

$$+ \frac{1}{V} \sum_{\mathbf{kp}\alpha\beta\gamma\delta} \Lambda_0^A(\mathbf{k}\,\alpha, \mathbf{k} + \mathbf{q}\,\beta)\, \Lambda_0^B(\mathbf{p} + \mathbf{q}\,\gamma, \mathbf{p}\,\delta)$$

$$\int \prod_{i=1}^{4} d\tau_i\, G(\tau - \tau_1\,\mathbf{k} + \mathbf{q})\, G(-\tau_2, \mathbf{p})\, G(\tau_3, \mathbf{p} + \mathbf{q})\, G(\tau_4 - \tau, \mathbf{k})$$

$$\Gamma_{\beta,\delta;\gamma,\alpha}\left(\tau_1\,\mathbf{k} + \mathbf{q}, \tau_2\,\mathbf{p}; \tau_3\,\mathbf{p} + \mathbf{q}, \tau_4\,\mathbf{k}\right),$$

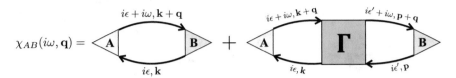

$$\chi_{AB}(i\omega, \mathbf{q}) =$$

Fig. 4.24 Graphical representation of the correlation function $\chi_{AB}(i\omega, \mathbf{q})$. All Green's functions are fully-interacting ones

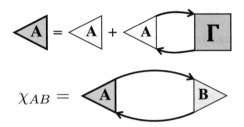

Fig. 4.25 Top panel: definition of interacting A-density vertex, in orange, as opposed to the non-interacting one, light yellow triangle. Bottom panel: diagrammatic representation of the correlation function χ_{AB} in terms of the interacting A-density vertex. Note that the B-density vertex is the non-interacting one. We could alternatively use the interacting B-density vertex, defined just by the same equation as in the top panel with A replaced by B, in which case, the A-density vertex would be the non-interacting one while the B-density vertex the interacting one

where, e.g.,

$$\langle A(\mathbf{0}) \rangle \equiv \sum_{\mathbf{k}\alpha} \Lambda_0^A(\mathbf{k}\,\alpha, \mathbf{k}\,\alpha)\, G(\mathbf{k}, 0^-)\,,$$

is just the expectation value of the density operator $A(\mathbf{q})$ at $\mathbf{q} = \mathbf{0}$. The first term contributes only at $\mathbf{q} = \mathbf{0}$, and can be generally absorbed into the definition of $\chi_{AB}(\tau, \mathbf{0})$. With that prescription, the Fourier transform in Matsubara frequencies of the correlation function has the diagrammatic representation of Fig. 4.24.

In the figure, the triangular vertices represent the matrix elements Λ_0^A and Λ_0^B of the corresponding density operators, which we shall denote as non-interacting A- and B-density vertices. We can correspondingly define interacting A- and B-density vertices as shown in Fig. 4.25. The explicit dependence upon frequency, momentum and spin variables of the density-vertex is shown in Fig. 4.26.

In the specific case of the charge-density,

$$\rho(\mathbf{q}) = \sum_{\mathbf{k}\sigma} c^\dagger_{\mathbf{k}\sigma}\, c_{\mathbf{k}+\mathbf{q}\sigma}\,,$$

the non-interacting vertex, which we hereafter denote simply as Λ_0 without any superscript, is simply unity, i.e., $\Lambda_0 = 1$, while the interacting one, consistently denoted as Λ, is graphically shown in Fig. 4.26. Comparing the graphical repre-

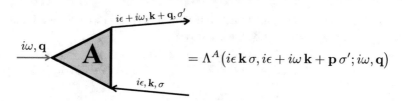

$$= \Lambda^A \left(i\epsilon\, \mathbf{k}\, \sigma, i\epsilon + i\omega\, \mathbf{k} + \mathbf{p}\, \sigma'; i\omega, \mathbf{q} \right)$$

Fig. 4.26 Explicit dependence upon frequency, momentum and spin variables of the, e.g., A-density-vertex. The lines are just external legs carrying the shown frequency, momentum and, eventually, spin

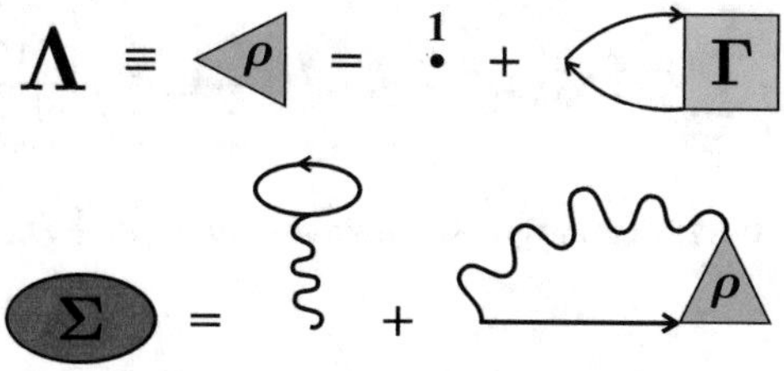

Fig. 4.27 Top panel: graphical representation of the interacting charge-density vertex Λ in terms of interacting Green's function G and the four-leg vertex Γ. Bottom panel: graphical representation of the self-energy Σ in terms of G and Λ, cf. with Fig. 4.23

sentation of Λ with that of the self-energy in Fig. 4.23, we find the alternative representation shown also in Fig. 4.27, whose analytic expression, cf. (4.86), reads

$$\Sigma(i\epsilon, \mathbf{k}) = U(0)\, n$$
$$- \frac{1}{V} \sum_{\mathbf{q}\sigma} T \sum_\omega G(i\epsilon - i\omega, \mathbf{k} - \mathbf{q})\, U(\mathbf{q})\, \Lambda\left(i\epsilon - i\omega\, \mathbf{k} - \mathbf{q}\, \sigma, i\epsilon\, \mathbf{k}\, \sigma; i\omega, \mathbf{q} \right),$$

$$(4.88)$$

where, see (4.87),

$$n = \frac{1}{V} \sum_{\mathbf{k}\sigma} n(\mathbf{k}) = \frac{1}{V} \sum_{\mathbf{k}\sigma} T \sum_\epsilon e^{i\epsilon 0^+}\, G(i\epsilon, \mathbf{k}),$$

is the total electron density.

4.5.2.1 Non-interacting Values

In the absence of interaction only the first term in Fig. 4.24 survives with the Green's function lines being the non-interacting ones. Since there is a loop, according to what we said before, the sign is $(-1)^{L-1} = (-1)^{1-1} = 1$ hence

$$\chi_{0\,AB}(i\omega, \mathbf{q}) = \frac{1}{V} \sum_{\mathbf{k}\alpha\beta} T \sum_\epsilon \Lambda_0^A(\mathbf{k}\,\alpha, \mathbf{k} + \mathbf{q}\,\beta)\, \Lambda_0^B(\mathbf{k} + \mathbf{q}\,\beta, \mathbf{k}\,\alpha)$$

$$G_0(i\epsilon + i\omega, \mathbf{k} + \mathbf{q})\, G_0(i\epsilon, \mathbf{k})$$

$$= \frac{1}{V} \sum_{\mathbf{k}} T \sum_n \mathrm{Tr}\left(\Lambda_0^A(\mathbf{k}, \mathbf{k} + \mathbf{q})\, \Lambda_0^B(\mathbf{k} + \mathbf{q}, \mathbf{k}) \right) \frac{1}{i\epsilon + i\omega - \epsilon_{\mathbf{k}+\mathbf{q}}} \frac{1}{i\epsilon - \epsilon_{\mathbf{k}}},$$

$$(4.89)$$

where the trace is over the spin variables. By means of (4.11) we can easily perform the sum over frequencies in (4.89), which gives

$$T \sum_\epsilon \frac{1}{i\epsilon + i\omega - \epsilon_{\mathbf{k}+\mathbf{q}}} \frac{1}{i\epsilon - \epsilon_{\mathbf{k}}} = f(\epsilon_{\mathbf{k}+\mathbf{q}} - i\omega) \frac{1}{\epsilon_{\mathbf{k}+\mathbf{q}} - i\omega - \epsilon_{\mathbf{k}}} + f(\epsilon_{\mathbf{k}}) \frac{1}{i\omega - \epsilon_{\mathbf{k}+\mathbf{q}} + \epsilon_{\mathbf{k}}}$$

$$= \frac{f(\epsilon_{\mathbf{k}}) - f(\epsilon_{\mathbf{k}+\mathbf{q}})}{i\omega - \epsilon_{\mathbf{k}+\mathbf{q}} + \epsilon_{\mathbf{k}}} ,$$

where we used the fact that, if $i\omega$ is bosonic, then $f(\epsilon \pm i\omega) = f(\epsilon)$. The final result is therefore

$$\chi_{0\,AB}(i\omega, \mathbf{q}) = \frac{1}{V} \sum_{\mathbf{k}} \mathrm{Tr}\left(\Lambda_0^A(\mathbf{k}, \mathbf{k}+\mathbf{q}) \Lambda_0^B(\mathbf{k}+\mathbf{q}, \mathbf{k})\right) \frac{f(\epsilon_{\mathbf{k}}) - f(\epsilon_{\mathbf{k}+\mathbf{q}})}{i\omega - \epsilon_{\mathbf{k}+\mathbf{q}} + \epsilon_{\mathbf{k}}} .$$

$$(4.90)$$

The linear response function is simply obtained by $i\omega \to \omega + i\eta$, with η an infinitesimal positive number.

4.6 Coulomb Interaction and Proper and Improper Response Functions

Let us now consider the case in which the electron-electron interaction is the Coulomb repulsion

$$U(\mathbf{q}) = \frac{4\pi e^2}{q^2} ,$$

which is singular for $q \to 0$. Strictly at $\mathbf{q} = \mathbf{0}$, the repulsion is cancelled by the neutralising ionic charge, which implies, for instance, that the Hartree term in (4.88) vanishes.

In order to cure that singularity, it is convenient to recast perturbation theory in a different manner, which naturally brings to identify the proper and improper response functions we earlier introduced. In Fig. 4.28 we draw the diagrammatic skeleton expansion up to first order of the density-density correlation function $\chi(i\omega, \mathbf{q})$, which is also the improper response function once analytically continued on the real frequency axis from above.

Already at first order we can distinguish two kinds of diagrams: those which can be cut into two by cutting an interaction line, as the last diagram, and those which cannot, all the others, which we define as irreducible with respect to the interaction.

$$\chi(i\omega, \mathbf{q}) =$$

Fig. 4.28 Diagrammatic skeleton expansion of the density-density correlation function up to first order. The black dots represent the non-interacting charge density vertex, Λ_0, which is just the identity

We define $\chi_*(i\omega, \mathbf{q})$ the sum of all irreducible diagrams, in the above sense. We can formally write $\chi_*(i\omega, \mathbf{q})$ as

$$\chi_*(i\omega, \mathbf{q}) = \frac{1}{V} \sum_{\mathbf{k}\sigma} T \sum_\epsilon G(i\epsilon, \mathbf{k}) \Lambda_*\big(i\epsilon\,\mathbf{k}\,\sigma, i\epsilon + i\omega\,\mathbf{k} + \mathbf{q}\,\sigma; i\omega, \mathbf{q}\big) G(i\epsilon + i\omega, \mathbf{k} + \mathbf{q}),$$

(4.91)

which defines the proper charge density-vertex $\Lambda_*\big(i\epsilon\,\mathbf{k}\,\sigma, i\epsilon + i\omega\,\mathbf{k} + \mathbf{q}\,\sigma; i\omega, \mathbf{q}\big)$, shown graphically in the upper panel of Fig. 4.29. In terms of $\chi_*(i\omega, \mathbf{q})$ the perturbation expansion of $\chi(i\omega, \mathbf{q})$ can be recast as in the lower panel of Fig. 4.29, which has the formal solution[3]

$$\chi(i\omega, \mathbf{q}) = \frac{\chi_*(i\omega, \mathbf{q})}{1 - \dfrac{4\pi e^2}{q^2} \chi_*(i\omega, \mathbf{q})} \equiv \frac{\chi_*(i\omega, \mathbf{q})}{\epsilon_\|(i\omega, \mathbf{q})}$$

$$= \frac{1}{V} \sum_{\mathbf{k}\sigma} T \sum_\epsilon G(i\epsilon, \mathbf{k}) \frac{\Lambda_*\big(i\epsilon\,\mathbf{k}\,\sigma, i\epsilon + i\omega\,\mathbf{k} + \mathbf{q}\,\sigma; i\omega, \mathbf{q}\big)}{\epsilon_\|(i\omega, \mathbf{q})} G(i\epsilon + i\omega, \mathbf{k} + \mathbf{q}),$$

(4.92)

proving that $\chi_*(i\omega, \mathbf{q})$ is actually the proper density-density response function that we have previously introduced, which defines the longitudinal dielectric constant $\epsilon_\|(i\omega, \mathbf{q})$. Since

$$\chi(i\omega, \mathbf{q}) = \frac{1}{V} \sum_{\mathbf{k}\sigma} T \sum_\epsilon G(i\epsilon, \mathbf{k}) \Lambda\big(i\epsilon\,\mathbf{k}\,\sigma, i\epsilon + i\omega\,\mathbf{k} + \mathbf{q}\,\sigma; i\omega, \mathbf{q}\big) G(i\epsilon + i\omega, \mathbf{k} + \mathbf{q}),$$

in terms of the interacting improper charge density-vertex, comparing the above equation with the last equation in (4.92), we find a relationship between improper and proper charge density-vertices

$$\Lambda\big(i\epsilon\,\mathbf{k}\,\sigma, i\epsilon + i\omega\,\mathbf{k} + \mathbf{q}\,\sigma; i\omega, \mathbf{q}\big) = \frac{1}{\epsilon_\|(i\omega, \mathbf{q})} \Lambda_*\big(i\epsilon\,\mathbf{k}\,\sigma, i\epsilon + i\omega\,\mathbf{k} + \mathbf{q}\,\sigma; i\omega, \mathbf{q}\big).$$

(4.93)

It follows that the expression (4.88) of the self-energy in terms of the improper charge density-vertex Λ changes into

$$\Sigma(i\epsilon, \mathbf{k}) = \Sigma_{HF} - U(0) \rho_{\text{ion}}$$

$$- \frac{1}{V} \sum_{\mathbf{q}\sigma} T \sum_\omega G(i\epsilon - i\omega, \mathbf{k} - \mathbf{q}) W(i\omega, \mathbf{q}) \Lambda_*\big(i\epsilon - i\omega\,\mathbf{k} - \mathbf{q}\,\sigma, i\epsilon\,\mathbf{k}\,\sigma; i\omega, \mathbf{q}\big)$$

$$\equiv \Sigma_{HF} - U(0) \rho_{\text{ion}} + \Sigma^{\text{xc}}(i\epsilon, \mathbf{k}),$$

(4.94)

---

[3] Notice both the interaction and $\chi_*(i\omega, \mathbf{q})$, which is a loop, bring a minus sign, hence the sign is plus.

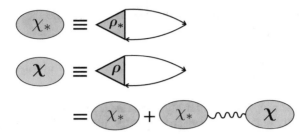

Fig. 4.29 Upper panel: graphical representation of $\chi_*(i\omega, \mathbf{q})$ in terms of the proper charge density-vertex $\Lambda_*(i\epsilon\,\mathbf{k}\,\sigma, i\epsilon + i\omega\,\mathbf{k} + \mathbf{q}\,\sigma; i\omega, \mathbf{q})$, blue triangle with label ρ_*. Note that the non-interacting charge density-vertex $\Lambda_0 = 1$. Lower panel: graphical representation of the equation that relates $\chi(i\omega, \mathbf{q})$ with $\chi_*(i\omega, \mathbf{q})$. We also show in the first line the representation of $\chi(i\omega, \mathbf{q})$ in terms of the interacting charge density-vertex Λ, green triangle with label ρ_*

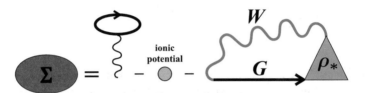

Fig. 4.30 Graphical representation of the self-energy in terms of the interacting Green's function G, the screened Coulomb interaction W and the proper charge density-vertex Λ_*

where the Hartree term cancels the contribution from the interaction with the positive ionic charges, and

$$W(i\omega, \mathbf{q}) = \frac{4\pi e^2}{q^2\,\epsilon(i\omega, \mathbf{q})} = \frac{4\pi e^2}{q^2 - 4\pi e^2\,\chi_*(i\omega, \mathbf{q})}, \qquad (4.95)$$

is the screened Coulomb interaction.

The graphical representation of (4.94) is shown in Fig. 4.30. The last contribution in (4.94) is named exchange-correlation self-energy $\Sigma^{xc}(i\epsilon, \mathbf{k})$, and includes all corrections beyond the simple Hartree potential.

4.7   Irreducible Vertices and the Bethe-Salpeter Equations

Let us consider again the interaction vertex $\Gamma(1, 2; 3, 4)$, whose lower order skeleton diagrams are shown in Fig. 4.21. Here we assume that $1 \equiv (i\epsilon + i\omega, \mathbf{k} + \mathbf{q}, \sigma_1)$, $2 \equiv (-i\epsilon - i\omega + i\Omega, -\mathbf{k} - \mathbf{q} + \mathbf{Q}, \sigma_2)$, $3 \equiv (-i\epsilon + i\Omega, -\mathbf{k} + \mathbf{Q}, \sigma_3)$ and $4 \equiv (i\epsilon, \mathbf{k}, \sigma_4)$, with the constraint $\sigma_1 + \sigma_2 - \sigma_3 - \sigma_4 = 0$ by conservation of the z-component of the total spin. We denote the vertices 1 and 2 that correspond to creation operators as particles, while 3 and 4 that correspond to annihilation operators as holes. It follows that $\Gamma(1, 2; 3, 4)$ can be regarded as a *particle-hole scattering amplitude*, namely the sum of all interaction processes that annihilate the particle-hole (p-h) pair

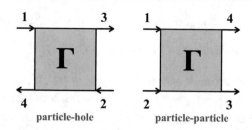

Fig. 4.31 Interaction vertex $\Gamma(1, 2; 3, 4)$ represented in the particle-hole channel, left, and particle-particle one, right

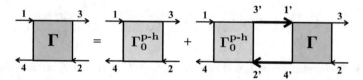

Fig. 4.32 Bethe-Salpeter equation for the interaction vertex Γ in the particle-hole channel

3&2 and create the pair 1&4, see left picture in Fig. 4.31. Each pair is characterised by the frequency transfer $i\omega$, momentum transfer $\mathbf{q}$ and z-component of the spin $\sigma_1 - \sigma_4 = \sigma_3 - \sigma_2$, in short $1 - 4 = 3 - 2$. Equivalently, $\Gamma(1, 2; 3, 4)$ can be seen as a *particle-particle scattering amplitude*, where the particle-particle (p-p) pair 4&3 is annihilated and the pair 1&2 created, see right picture in Fig. 4.31. Each pair, also known as Cooper pair, is characterised by the total frequency $i\Omega$, total momentum $\mathbf{Q}$ and z-component of the spin $\sigma_1 + \sigma_2 = \sigma_4 + \sigma_3$, in short $1 + 2 = 4 + 3$. We emphasise that, even though the two scattering process are different, $\Gamma(1, 2; 3, 4)$ is the same.

4.7.1 Particle-Hole Channel

The lowest order skeleton expansion of $\Gamma(1, 2; 3, 4)$ regarded as a p-h scattering amplitude was already shown in Fig. 4.21. In that figure we can distinguish two class of diagrams. Indeed, we note that the diagrams highlighted in red can be cut into two by cutting a particle line, from left to right, and a hole one, from right to left. In that case the frequency, momentum and z-component of the spin differences between the two lines is exactly $1 - 4 = 3 - 2$. We denote such class of diagrams as *reducible in the particle-hole channel*. All other diagrams are instead *irreducible in the particle-hole channel*. We name the sum of the irreducible diagrams at all orders as $\Gamma_0^{p\text{-}h}$, in terms of which Γ can be obtained by solving the equation shown in Fig. 4.32, which we shortly write as

$$\Gamma(1, 2; 3, 4) = \Gamma_0^{p\text{-}h}(1, 2; 3, 4) + \sum_{1',2',3',4'} \Gamma_0^{p\text{-}h}(1, 2'; 3', 4) \, R(1', 2'; 3', 4') \, \Gamma(1', 2; 3, 4')$$

$$\equiv \Gamma_0^{p\text{-}h} + \Gamma_0^{p\text{-}h} \odot R \odot \Gamma, \tag{4.96}$$

where, regarding Γ, $\Gamma_0^{p\text{-}h}$ and R as tensors, $\odot$ indicates the tensor product. We define

$$R(1', 2'; 3', 4') \equiv \delta_{1'3'}\,\delta_{2'4'}\,G(1')\,G(4')\,, \tag{4.97}$$

where, since by construction $1' - 4' = 1 - 4$, if $4' = (i\epsilon', \mathbf{k}', \sigma_{4'})$, then $1' = (i\epsilon' + i\omega, \mathbf{k}' + \mathbf{q}, \sigma_{1'})$, with $\sigma_{1'} - \sigma_{4'} = \sigma_1 - \sigma_4$. The lowest order skeleton expansion of $\Gamma_0^{p\text{-}h}$ contains the diagrams in Fig. 4.21 excluded the four highlighted reducible ones. Equation (4.96) is the Bethe-Salpeter equation in the particle-hole channel.

We can also exploit spin $SU(2)$ symmetry. We introduce the Pauli matrices in the spin space,

$$\hat{\sigma}_1 = \begin{pmatrix} 0 & 1 \\ 1 & 0 \end{pmatrix}, \qquad \hat{\sigma}_2 = \begin{pmatrix} 0 & -i \\ i & 0 \end{pmatrix}, \qquad \hat{\sigma}_3 = \begin{pmatrix} 1 & 0 \\ 0 & -1 \end{pmatrix},$$

including the identity $\hat{\sigma}_0 = \hat{I}$, and note that spin $SU(2)$ symmetry implies that

$$\frac{1}{4} \sum_{\sigma_1 \sigma_2 \sigma_3 \sigma_4} \left(\hat{\sigma}_a\right)_{\sigma_1 \sigma_4} \left(\hat{\sigma}_b\right)_{\sigma_2 \sigma_3} \Gamma_{\sigma_1,\sigma_2;\sigma_3,\sigma_4} = \delta_{ab}\,\Gamma_a\,, \quad a,b = 0,\ldots,3\,, \tag{4.98}$$

as well as $\Gamma_1 = \Gamma_2 = \Gamma_3$. We therefore define the scattering amplitude in the $S = 0$ p-h scattering channel through

$$\begin{aligned}
\Gamma_S &\equiv \frac{1}{4} \sum_{\sigma_1 \sigma_2 \sigma_3 \sigma_4} \left(\hat{\sigma}_0\right)_{\sigma_1 \sigma_4} \left(\hat{\sigma}_0\right)_{\sigma_2 \sigma_3} \Gamma_{\sigma_1,\sigma_2;\sigma_3,\sigma_4} \\
&= \frac{1}{4} \sum_{\sigma_1 \sigma_2 \sigma_3 \sigma_4} \delta_{\sigma_1 \sigma_4}\,\delta_{\sigma_2 \sigma_3}\,\Gamma_{\sigma_1,\sigma_2;\sigma_3,\sigma_4} = \frac{1}{4} \left(\Gamma_{\uparrow,\uparrow;\uparrow,\uparrow} + \Gamma_{\downarrow,\downarrow;\downarrow,\downarrow} + \Gamma_{\uparrow,\downarrow;\downarrow,\uparrow} + \Gamma_{\downarrow,\uparrow;\uparrow,\downarrow}\right) \\
&= \frac{1}{2} \left(\Gamma_{\uparrow,\uparrow;\uparrow,\uparrow} + \Gamma_{\uparrow,\downarrow;\downarrow,\uparrow}\right),
\end{aligned} \tag{4.99}$$

since $\Gamma_{\uparrow,\uparrow;\uparrow,\uparrow} = \Gamma_{\downarrow,\downarrow;\downarrow,\downarrow}$ and $\Gamma_{\uparrow,\downarrow;\downarrow,\uparrow} = \Gamma_{\downarrow,\uparrow;\uparrow,\downarrow}$ again by spin-$SU(2)$.

Similarly, we define the scattering amplitudes in the $S = 1$ p-h scattering channel Γ_a, $a = 1, 2, 3$, where, e.g.,

$$\Gamma_3 \equiv \frac{1}{4} \sum_{\sigma_1 \sigma_2 \sigma_3 \sigma_4} \left(\hat{\sigma}_3\right)_{\sigma_1 \sigma_4} \left(\hat{\sigma}_3\right)_{\sigma_2 \sigma_3} \Gamma_{\sigma_1,\sigma_2;\sigma_3,\sigma_4} = \frac{1}{2} \left(\Gamma_{\uparrow,\uparrow;\uparrow,\uparrow} - \Gamma_{\uparrow,\downarrow;\downarrow,\uparrow}\right). \tag{4.100}$$

Since, as we mentioned, Γ_a, $a = 1, 2, 3$, are all equal, then

$$\Gamma_1 = \Gamma_2 = \Gamma_3 \equiv \Gamma_A\,, \tag{4.101}$$

where $\Gamma_A \neq \Gamma_S$. We note that Γ_3 corresponds to the $S_z = 0$ component of the p-h spin triplet, while $\Gamma_+ \equiv (\Gamma_1 + i\,\Gamma_2)/2$ and $\Gamma_- \equiv (\Gamma_1 - i\,\Gamma_2)/2$ represent the $S_z = 1$ and $S_z = -1$ components, respectively. We emphasise that

$$\Gamma_{S/A} \equiv \Gamma_{S/A}\big(i\epsilon + i\omega\,\mathbf{k} + \mathbf{q}, -i\epsilon - i\omega + i\Omega\, -\mathbf{k} - \mathbf{q} + \mathbf{Q};\, -i\epsilon + i\Omega\, -\mathbf{k} + \mathbf{Q}, i\epsilon\,\mathbf{k}\big),$$

do not depend anymore on the spin variables.

We can similarly define spin-singlet and spin-triplet combinations of the irreducible vertex, respectively, $\Gamma_{0S}^{\text{p-h}}$ and $\Gamma_{0A}^{\text{p-h}}$, and note that, since R in (4.97) does not depend on spin, the Bethe-Salpeter equations can be shortly written as

$$\Gamma_{\uparrow,\uparrow;\uparrow,\uparrow} = \Gamma_{0\uparrow,\uparrow;\uparrow,\uparrow}^{\text{p-h}} + \sum_\sigma \Gamma_{0\uparrow,\sigma;\sigma,\uparrow}^{\text{p-h}} \odot R \odot \Gamma_{\sigma,\uparrow;\uparrow,\sigma}$$

$$= \Gamma_{0\uparrow,\uparrow;\uparrow,\uparrow}^{\text{p-h}} + \Gamma_{0\uparrow,\uparrow;\uparrow,\uparrow}^{\text{p-h}} \odot R \odot \Gamma_{\uparrow,\uparrow;\uparrow,\uparrow} + \Gamma_{0\uparrow,\downarrow;\downarrow,\uparrow}^{\text{p-h}} \odot R \odot \Gamma_{\downarrow,\uparrow;\uparrow,\downarrow},$$

$$\Gamma_{\uparrow,\downarrow;\downarrow,\uparrow} = \Gamma_{0\uparrow,\downarrow;\downarrow,\uparrow}^{\text{p-h}} + \sum_\sigma \Gamma_{0\uparrow,\sigma;\sigma,\uparrow}^{\text{p-h}} \odot R \odot \Gamma_{\sigma,\downarrow;\downarrow,\sigma}$$

$$= \Gamma_{0\uparrow,\downarrow;\downarrow,\uparrow}^{\text{p-h}} + \Gamma_{0\uparrow,\uparrow;\uparrow,\uparrow}^{\text{p-h}} \odot R \odot \Gamma_{\uparrow,\downarrow;\downarrow,\uparrow} + \Gamma_{0\uparrow,\downarrow;\downarrow,\uparrow}^{\text{p-h}} \odot R \odot \Gamma_{\downarrow,\downarrow;\downarrow,\downarrow}$$

$$= \Gamma_{0\uparrow,\downarrow;\downarrow,\uparrow}^{\text{p-h}} + \Gamma_{0\uparrow,\uparrow;\uparrow,\uparrow}^{\text{p-h}} \odot R \odot \Gamma_{\uparrow,\downarrow;\downarrow,\uparrow} + \Gamma_{0\uparrow,\downarrow;\downarrow,\uparrow}^{\text{p-h}} \odot R \odot \Gamma_{\uparrow,\uparrow;\uparrow,\uparrow},$$

where now the tensor product $\odot$ does not include the spin labels. One can easily show that

$$\frac{1}{2}\left(\Gamma_{\uparrow,\uparrow;\uparrow,\uparrow} + \Gamma_{\uparrow,\downarrow;\downarrow,\uparrow}\right) = \Gamma_S = \Gamma_{0S}^{\text{p-h}} + 2\Gamma_{0S}^{\text{p-h}} \odot R \odot \Gamma_S,$$

$$\frac{1}{2}\left(\Gamma_{\uparrow,\uparrow;\uparrow,\uparrow} - \Gamma_{\uparrow,\downarrow;\downarrow,\uparrow}\right) = \Gamma_A = \Gamma_{0A}^{\text{p-h}} + 2\Gamma_{0A}^{\text{p-h}} \odot R \odot \Gamma_A,$$

$$\tag{4.102}$$

namely that, unsurprisingly, the Bethe-Salpeter equations decouple into the two symmetry channels.

We can proceed further and, using the same notations, calculate the charge density-density correlation function of Fig. 4.24 with the non-interacting vertices replaced the charge-density ones, i.e., the identity, and find

$$\chi = \sum_\sigma \text{Tr}(R) + \sum_{\sigma\sigma'} \text{Tr}\left(R \odot \Gamma_{\sigma,\sigma';\sigma',\sigma} \odot R\right) = 2\,\text{Tr}(R) + 4\,\text{Tr}\left(R \odot \Gamma_S \odot R\right).$$

$$\tag{4.103}$$

Similarly, if we define the spin density operators as

$$\sigma_a(\mathbf{q}) \equiv \sum_{\mathbf{k}\sigma\sigma'} c_{\mathbf{k}\sigma}^\dagger (\hat{\sigma}_a)_{\sigma\sigma'} c_{\mathbf{k}+\mathbf{q}\sigma'}, \qquad a = 1, 2, 3,$$

the spin density-density correlation functions

$$\chi_{ab}(\tau, \mathbf{q}) \equiv -\langle T_\tau\left(\sigma_a(\tau, \mathbf{q})\,\sigma_b(0, -\mathbf{q})\right)\rangle = \delta_{ab}\,\chi_\sigma(\tau, \mathbf{q}),\tag{4.104}$$

the last equation following from spin $SU(2)$ symmetry, the Fourier transform in Matsubara frequency can be readily calculated and reads

$$\chi_\sigma = 2\,\text{Tr}(R) + 4\,\text{Tr}\left(R \odot \Gamma_A \odot R\right).\tag{4.105}$$

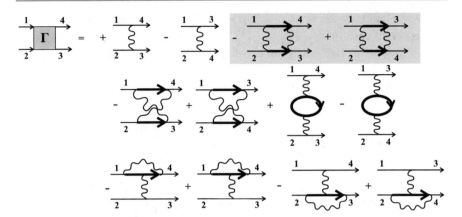

Fig. 4.33 Skeleton expansion up to second order of Γ in the particle-particle scattering channel

4.7.2 Particle-Particle Channel

In Fig. 4.33 we show the skeleton expansion up to second order of Γ in the particle-particle scattering channel. Note that the diagrams are equal to those in Fig. 4.21, but are simply drawn differently. Furthermore, one easily realises that Γ is odd interchanging 3 with 4, or 1 with 2. That is simply consequence of the Pauli exclusion principle, and leads to the following symmetry properties of the interaction vertex:

$$\Gamma(1, 2; 3, 4) = \Gamma(2, 1; 4, 3) = -\Gamma(1, 2; 4, 3) = -\Gamma(2, 1; 3, 4). \qquad (4.106)$$

As before, we can distinguish diagrams *reducible in the particle-particle channel*, namely that can be cut into two lower-order diagrams cutting two particle lines with total frequency, momentum and spin equal to $1 + 2 = 3 + 4$. Those are the two diagrams highlighted in red. All other diagrams are *irreducible in the particle-particle channel*. We define the sum of all order irreducible skeleton diagrams as $\Gamma_0^{p\text{-}p}$, through which Γ satisfies the following equation, graphically shown in Fig. 4.34,

$$\Gamma(1, 2; 3, 4) = \Gamma_0^{p\text{-}p}(1, 2; 3, 4) + \sum_{1', 2', 3', 4'} \Gamma_0^{p\text{-}p}(1, 2; 3', 4') \, S(1', 2'; 3', 4') \, \Gamma(1', 2'; 3, 4)$$

$$\equiv \Gamma_0^{p\text{-}p} + \Gamma_0^{p\text{-}p} \odot S \odot, \qquad (4.107)$$

with the meaning of $\odot$ the same as before, and

$$S(1', 2'; 3', 4') \equiv \delta_{1'\,4'} \, \delta_{2'\,3'} \, G(1') \, G(2'), \qquad (4.108)$$

where, since $1' + 2' = 1 + 2$, if $1' = \left(i\epsilon' + i\Omega, \mathbf{k}' + \mathbf{Q}, \sigma_{1'}\right)$, then $2' = \left(-i\epsilon', -\mathbf{k}', \sigma_{2'}\right)$, with $\sigma_{1'} + \sigma_{2'} = \sigma_1 + \sigma_2$. Equation (4.107) is the Bethe-Salpeter equation in the particle-particle channel. We emphasise that, even though Γ is the same, the irreducible vertices $\Gamma_0^{p\text{-}h}$ and $\Gamma_0^{p\text{-}p}$ are different.

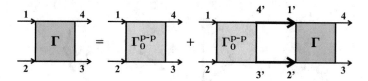

Fig. 4.34 Bethe-Salpeter equation for the interaction vertex Γ in the particle-particle channel

Also a Cooper pair can be a spin-singlet or a spin-triplet. Accordingly, if we redefine, as before, $1 \equiv (i\epsilon + i\omega, \mathbf{k} + \mathbf{q})$ without the spin quantum number, and similarly for all other variables 2, 3, 4, and show the spins as subscripts, the interaction vertex in the spin-singlet channel reads

$$
\begin{aligned}
\Gamma_{S=0}(1, 2; 3, 4) = \frac{1}{2} \Big(&\Gamma_{\uparrow,\downarrow;\downarrow,\uparrow}(1, 2; 3, 4) - \Gamma_{\uparrow,\downarrow;\uparrow,\downarrow}(1, 2; 3, 4) \\
&+ \Gamma_{\downarrow,\uparrow;\uparrow,\downarrow}(1, 2; 3, 4) - \Gamma_{\downarrow,\uparrow;\downarrow,\uparrow}(1, 2; 3, 4) \Big),
\end{aligned}
\tag{4.109}
$$

while those in the $S = 1$ channel with $S_z = 0, \pm 1$ as

$$
\begin{aligned}
\Gamma_{S=1,S_z=0}(1, 2; 3, 4) = \frac{1}{2} \Big(&\Gamma_{\uparrow,\downarrow;\downarrow,\uparrow}(1, 2; 3, 4) + \Gamma_{\uparrow,\downarrow;\uparrow,\downarrow}(1, 2; 3, 4) \\
&+ \Gamma_{\downarrow,\uparrow;\uparrow,\downarrow}(1, 2; 3, 4) + \Gamma_{\downarrow,\uparrow;\downarrow,\uparrow}(1, 2; 3, 4) \Big), \\
\Gamma_{S=1,S_z=+1}(1, 2; 3, 4) = \ &\Gamma_{\uparrow,\uparrow;\uparrow,\uparrow}(1, 2; 3, 4), \\
\Gamma_{S=1,S_z=-1}(1, 2; 3, 4) = \ &\Gamma_{\downarrow,\downarrow;\downarrow,\downarrow}(1, 2; 3, 4),
\end{aligned}
\tag{4.110}
$$

and they must be all equal by spin $SU(2)$ symmetry and different from $\Gamma_{S=0}(1, 2; 3, 4)$. We can similarly define the irreducible vertex in the spin-singlet and spin-triplet channels, and, as before, the Bethe-Salpeter equations decouple into different equations for each distinct channel. With this notation, (4.106) becomes

$$
\begin{aligned}
\Gamma_{\sigma_1,\sigma_2;\sigma_3,\sigma_4}(1, 2; 3, 4) = \Gamma_{\sigma_2,\sigma_1;\sigma_4,\sigma_3}(2, 1; 4, 3) &= -\Gamma_{\sigma_1,\sigma_2;\sigma_4,\sigma_3}(1, 2; 4, 3) \\
&= -\Gamma_{\sigma_2,\sigma_1;\sigma_3,\sigma_4}(2, 1; 3, 4),
\end{aligned}
$$

and implies that $\Gamma_{S=0}(1, 2; 3, 4)$ is even under $3 \leftrightarrow 4$ or $1 \leftrightarrow 2$, while $\Gamma_{S=1,S_z}(1, 2; 3, 4)$ odd, as expected.

4.7.3   Self-energy and Irreducible Vertices

In Fig. 4.35 we show the same skeleton diagrams of the self-energy as in Fig. 4.13, but rotated and with explicitly shown external legs. Each diagram is a functional of the interacting Green's function. Suppose we calculate the variation of the diagrams with respect to a Green's function variation. That amounts to consider an internal Green's function, label its initial and final points, e.g., as 3 and 2, respectively, cut the

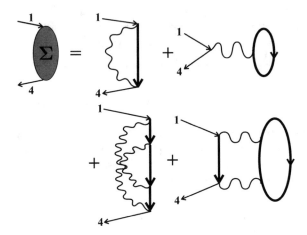

Fig. 4.35 Skeleton diagrams of the self-energy up to second order rotated by 90° with respect to those of Fig. 4.13, with the external legs explicitly indicated

line and multiply by $\delta G(3, 2)$, i.e., the Green's function variation. That procedure is shown explicitly for two diagrams in Fig. 4.36. It is easy to realise that, repeating this procedure for all diagrams and all internal Green's functions, we obtain all skeleton diagrams for the irreducible vertex $\Gamma_0^{\text{p-h}}$ in the particle-hole channel, also shown in Fig. 4.36 up to second order. We remark that, while in the self-energy diagrams 1 and 4 have same frequency, momentum and spin, and so 2 and 3 have, the variation $\delta G(3, 2)$ is generally off diagonal in those variables, and so the external legs 1 and 4 become. In other words, under the transformation $G(3, 2) \rightarrow G(3, 2) + \delta G(3, 2)$, with

$$\delta G(3, 2) \equiv \delta G\big(i\epsilon' + i\omega, \mathbf{k}' + \mathbf{q}, \sigma_3; i\epsilon', \mathbf{k}', \sigma_2\big),$$

the self-energy ceases to be diagonal in frequency, momentum and spin,

$$\Sigma(1, 4) \equiv \delta_{1,4}\, \Sigma(i\epsilon, \mathbf{k}) \rightarrow \Sigma(1, 4) + \delta\Sigma(1, 4)$$
$$\equiv \delta_{1,4}\, \Sigma(i\epsilon, \mathbf{k}) + \delta\Sigma\big(i\epsilon + i\omega, \mathbf{k} + \mathbf{q}, \sigma_1; i\epsilon, \mathbf{k}, \sigma_4\big),$$

and thus

$$\frac{\delta\Sigma(1, 4)}{\delta G(3, 2)} = \frac{\delta\Sigma\big(i\epsilon + i\omega, \mathbf{k} + \mathbf{q}, \sigma_1; i\epsilon, \mathbf{k}, \sigma_4\big)}{\delta G\big(i\epsilon' + i\omega, \mathbf{k}' + \mathbf{q}, \sigma_3; i\epsilon', \mathbf{k}', \sigma_2\big)} \equiv \Gamma_0^{p-h}(1, 2; 3, 4)$$
$$= \Gamma_0^{p-h}(i\epsilon + i\omega\,\mathbf{k} + \mathbf{q}\,\sigma_1, i\epsilon'\,\mathbf{k}'\,\sigma_2; i\epsilon' + i\omega\,\mathbf{k}' + \mathbf{q}\,\sigma_3, i\epsilon\,\mathbf{k}\,\sigma_4).$$
$$(4.111)$$

We have shown such equality up to second order, but it is actually true at any order in perturbation theory. Figure 4.37 is a graphical representation of the functional derivative of the self-energy with respect to the Green's function. We must imagine to take any of the internal Green's functions, extract out of the box that identifies

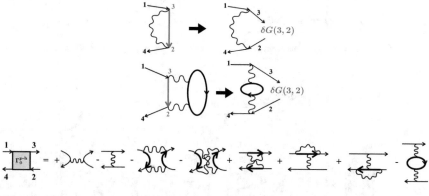

Fig. 4.36 Top panel: graphical representation of the functional derivative of the self-energy skeleton diagrams with respect to the Green's function. Bottom panel: skeleton diagrams of the irreducible vertex $\Gamma_0^{\text{p-h}}$ up to second order

Fig. 4.37 Graphical representation of the functional derivative of the self-energy with respect to the Green's function

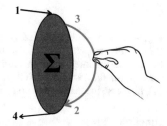

the self-energy and cut it. Evidently, what remains is a four leg vertex. However, since self-energy insertions do not appear by definition in the skeleton diagrams, that vertex must be irreducible in the particle-hole channel $1 - 4 = 3 - 2$, and thus is just $\Gamma_0^{\text{p-h}}$.

We mentioned that $\delta G(3, 2)$ needs not to satisfy the symmetries of the Hamiltonian. Therefore, we can also imagine a variation of Σ with respect to a Green's function that does not conserve the electron number. For that we define the anomalous Green's functions

$$\mathcal{F}(1, 2) \equiv -\langle T_\tau \big(c(1) \, c(2) \big) \rangle = -\mathcal{F}(2, 1)$$

$$= -\theta(\tau_1 - \tau_2) \langle c_{\mathbf{k}_1\sigma_1}(\tau_1) \, c_{\mathbf{k}_2\sigma_2}(\tau_2) \rangle + \theta(\tau_2 - \tau_1) \langle c_{\mathbf{k}_2\sigma_2}(\tau_2) \, c_{\mathbf{k}_1\sigma_1}(\tau_1) \rangle,$$

$$\overline{\mathcal{F}}(1, 2) \equiv -\langle T_\tau \big(c^\dagger(1) \, c^\dagger(2) \big) \rangle = -\overline{\mathcal{F}}(2, 1)$$

$$= -\theta(\tau_1 - \tau_2) \langle c_{\mathbf{k}_1\sigma_1}^\dagger(\tau_1) \, c_{\mathbf{k}_2\sigma_2}^\dagger(\tau_2) \rangle + \theta(\tau_2 - \tau_1) \langle c_{\mathbf{k}_2\sigma_2}^\dagger(\tau_2) \, c_{\mathbf{k}_1\sigma_1}^\dagger(\tau_1) \rangle$$

$$= \mathcal{F}(1, 2)^* - \langle \big[c_{\mathbf{k}_1\sigma_1}^\dagger(\tau_1), \, c_{\mathbf{k}_2\sigma_2}^\dagger(\tau_2) \big] \rangle, \tag{4.112}$$

as opposed to the normal one $G(1, 2)$, and the corresponding anomalous self-energies, both shown in the top of Fig. 4.38. Those allow us to introduce a Green's

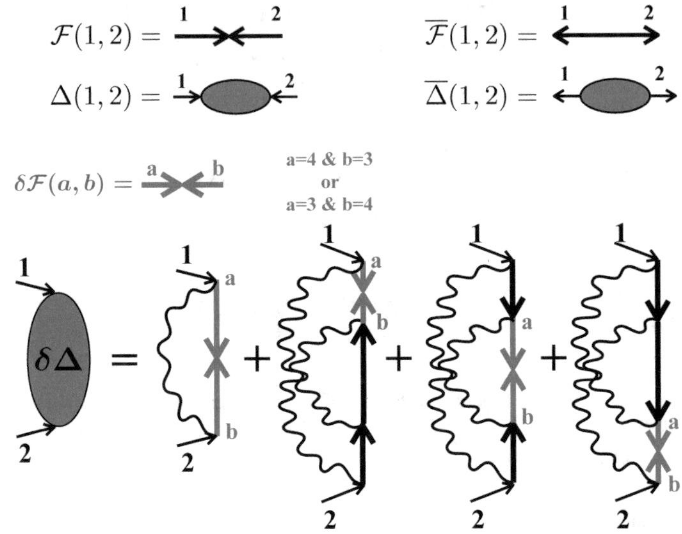

$$\mathcal{F}(1,2) = \begin{array}{c} 1 \quad 2 \\ \longrightarrow\!\!\times\!\!\longleftarrow \end{array} \qquad\qquad \overline{\mathcal{F}}(1,2) = \begin{array}{c} 1 \quad 2 \\ \longleftarrow\quad\longrightarrow \end{array}$$

$$\Delta(1,2) = \begin{array}{c} 1 \qquad\qquad 2 \\ \rightarrow\!\!\bullet\!\!\leftarrow \end{array} \qquad\qquad \overline{\Delta}(1,2) = \begin{array}{c} 1 \qquad\qquad 2 \\ \leftarrow\!\!\bullet\!\!\rightarrow \end{array}$$

Fig. 4.38 Top panel: graphical representation of the anomalous Green's functions $\mathcal{F}(3, 4)$ and $\overline{\mathcal{F}}(3, 4)$. Bottom panel: the skeleton diagrams up to second order of the self-energy variation obtained by changing an internal Green's function in the diagrams of Fig. 4.35 with $\delta\mathcal{F}(a, b)$. That can be done only in the first and third diagram. The internal vertices can be $a = 4$ and $b = 3$, or viceversa, in which case the diagram gets an additional minus sign since $\delta\mathcal{F}(3, 4) = -\delta\mathcal{F}(4, 3)$

function matrix

$$\hat{G}(1,2) \equiv \begin{pmatrix} -\langle T_\tau\big(c(1)\,c^\dagger(2)\big)\rangle & -\langle T_\tau\big(c(1)\,c(2)\big)\rangle \\ -\langle T_\tau\big(c^\dagger(1)\,c^\dagger(2)\big)\rangle & -\langle T_\tau\big(c^\dagger(1)\,c(2)\big)\rangle \end{pmatrix} = \begin{pmatrix} G(1,2) & \mathcal{F}(1,2) \\ \overline{\mathcal{F}}(1,2) & -G(2,1) \end{pmatrix}, \tag{4.113}$$

and a self-energy matrix

$$\hat{\Sigma}(1,2) \equiv \begin{pmatrix} \Sigma(1,2) & \Delta(1,2) \\ \overline{\Delta}(1,2) & -\Sigma(2,1) \end{pmatrix}, \tag{4.114}$$

functional of $\hat{G}$, which are related to each other also through Dyson's equation

$$\hat{G}(1,2) = \hat{G}_0(1,2) + \sum_{3,4} \hat{G}_0(1,3)\,\hat{\Sigma}(3,4)\,\hat{G}(4,2), \tag{4.115}$$

where

$$\hat{G}_0(1,2) = \begin{pmatrix} G_0(1,2) & 0 \\ 0 & -G_0(2,1) \end{pmatrix} = \delta_{\mathbf{k}_1,\mathbf{k}_2}\,\delta_{\sigma_1,\sigma_2} \begin{pmatrix} G_0(\tau_1 - \tau_2, \mathbf{k}_1) & 0 \\ 0 & -G_0(\tau_2 - \tau_1, \mathbf{k}_1) \end{pmatrix},$$

is the non-interacting Green's function.

With those notations, we consider the variation of the anomalous self-energy obtained upon replacing an internal normal Green's function of the diagrams in Fig. 4.35 with an anomalous Green's function variation $\delta\mathcal{F}(a, b) = \epsilon\,\mathcal{F}(a, b)$, with infinitesimal ϵ, and changing accordingly the directions of the normal Green's functions. The endpoints a and b can be either $a = 4$ and $b = 3$ or viceversa. Since $\delta\mathcal{F}(a, b) = -\delta\mathcal{F}(b, a)$, the two cases belong to the single variation with respect to $\delta\mathcal{F}(4, 3)$, though with opposite signs. Cutting in two $\delta\mathcal{F}(a, b)$ we recover all diagrams in Fig. 4.33 irreducible in the p-p channel, i.e., all diagrams but the highlighted two. The same holds at any order in the skeleton expansion. Therefore, similarly to (4.111),

$$\frac{\delta\Delta(1, 2)}{\delta\mathcal{F}(4, 3)}\bigg|_G = \Gamma_0^{\text{p-p}}(1, 2; 3, 4)\,, \tag{4.116}$$

where the functional derivative on the left-hand-side must be evaluated with all internal lines being normal, consistently with our original assumption that the global charge $U(1)$ symmetry is unbroken.

Since the irreducible vertex is unique, unlike the irreducible ones, hereafter we shall only deal with $\Gamma_0^{\text{p-h}}$ in the p-h channel. For that reason, we will refer to it simply at Γ_0, without any superscript.

4.7.3.1 Proper Interaction Vertex

In the case in which the interaction is the Coulomb repulsion, we better use the formalism described in Sect. 4.6. Therefore, given the self-energy (4.94), whose diagrammatic representation is shown in Fig. 4.30, the irreducible interaction vertex is defined through

$$\Gamma_0 = \frac{\delta\Sigma}{\delta G} = U(\mathbf{q}) + \frac{\delta\Sigma^{\text{xc}}}{\delta G}\,.$$

It is easy to realise that the second terms includes only skeleton diagrams that are irreducible with respect to cutting not only a particle-hole pair of lines with frequency and momentum transferred $i\omega$ and $\mathbf{q}$, respectively, but also an interaction line. Therefore,

$$\frac{\delta\Sigma^{\text{xc}}}{\delta G} \equiv \Gamma_{0*}\,, \tag{4.117}$$

defines the proper interaction vertex, which is related to the proper density-vertex (4.93) as shown in Fig. 4.39.

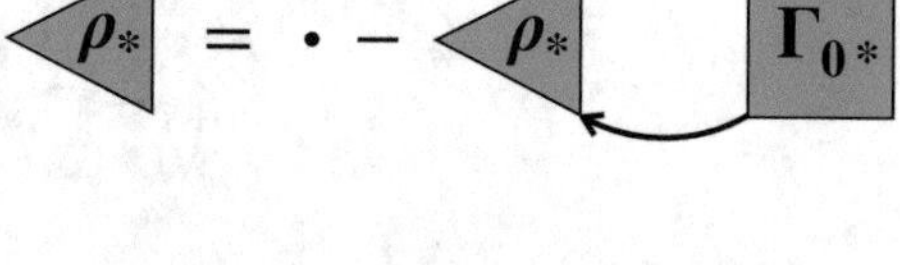

Fig. 4.39 Bethe-Salpeter equation relating the proper interaction vertex Γ_{0*} to the proper density-vertex Λ_*

4.8   The Luttinger-Ward Functional

The skeleton expansion shows that the self-energy $\Sigma[G, U]$ is functional of the Green's function as well as of the specific form of interaction, in short U. We previously demonstrated that

$$\frac{\delta\Sigma(1, 4)}{\delta G(3, 2)} = \Gamma_0(1, 2; 3, 4),$$

exists and is symmetric, i.e.,

$$\frac{\delta\Sigma(2, 3)}{\delta G(4, 1)} = \Gamma_0(2, 1; 4, 3) = \Gamma_0(1, 2; 3, 4).$$

Those conditions guarantee the existence of a functional $\Phi[G, U]$ such that

$$\frac{\delta\Phi[G, U]}{\delta G(4, 1)} = \Sigma(1, 4), \qquad \frac{\delta^2\Phi[G, U]}{\delta G(4, 1)\,\delta G(3, 2)} = \Gamma_0(1, 2; 3, 4) = \Gamma_0(2, 1; 4, 3).$$

$$(4.118)$$

This functional, known as Luttinger-Ward functional, has the simple expression in perturbation theory

$$\Phi[G, U] = \sum_{k\sigma} T \sum_{\epsilon} e^{i\epsilon\eta} \sum_{n\geq 1} \frac{1}{2n}\, G(i\epsilon, \mathbf{k})\, \Sigma^{(n)}(i\epsilon, \mathbf{k})$$

$$\equiv \sum_{k\sigma} T \sum_{\epsilon} e^{i\epsilon\eta}\, \Phi(i\epsilon, \mathbf{k}),$$

$$(4.119)$$

where $\Sigma^{(n)}(i\epsilon, \mathbf{k})$ includes all nth order skeleton diagrams, and satisfies

$$\delta\Phi[G, U] = \sum_{k\sigma} T \sum_{\epsilon} e^{i\epsilon\eta}\, \Sigma(i\epsilon, \mathbf{k})\,\delta G(i\epsilon, \mathbf{k}).$$

$$(4.120)$$

The parameter $\eta > 0$ is infinitesimal and can be set equal to zero only after performing the summation over Matsubara frequencies. The reason is that $e^{i\epsilon\eta}$ regularises that summation since $\Phi(i\epsilon, \mathbf{k})$ include terms that do not decay faster than $1/\epsilon$ for $|\epsilon| \to \infty$. Indeed, while $G(i\epsilon, \mathbf{k}) \sim 1/i\epsilon$, $\mathrm{Re}\,\Sigma^{(n)}(i\epsilon, \mathbf{k})$ is generally finite as $|\epsilon| \to \infty$. Therefore the sum over ϵ does not converge unless we explicitly take $e^{i\epsilon\eta}$ into account.[4] In Fig. 4.40 we show the skeleton expansion up to second order

---

[4] More rigorously,

$$T \sum_{\epsilon} e^{i\epsilon\eta}\, \Phi(i\epsilon, \mathbf{k}) = T \sum_{\epsilon>0} \cos(\epsilon\eta)\left(\Phi(i\epsilon, \mathbf{k}) + \Phi(-i\epsilon, \mathbf{k})\right)$$

$$+ i\, T \sum_{\epsilon>0} \sin(\epsilon\eta)\left(\Phi(i\epsilon, \mathbf{k}) - \Phi(-i\epsilon, \mathbf{k})\right).$$

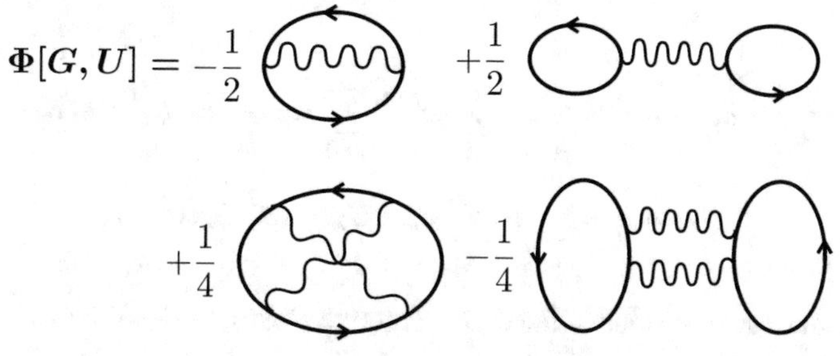

$$\Phi[G, U] = -\frac{1}{2} \quad \text{(diagram)} \quad +\frac{1}{2} \quad \text{(diagram)}$$

$$+\frac{1}{4} \quad \text{(diagram)} \quad -\frac{1}{4} \quad \text{(diagram)}$$

Fig. 4.40 Skeleton expansion up to second order of the Luttinger-Ward functional $\Phi[G, U]$. Note the pre-factors of each diagram

of the Luttinger-Ward functional, explicitly indicating sign and pre factor. We can easily realise that, since $2n$ is the number of Green's function of an nth order skeleton diagram, and the functional derivatives with respect to each of the $2n$ Green's functions yield topologically equivalent skeleton diagrams, that cancels the $1/2n$ and reproduces the skeleton expansion of the self-energy.

4.8.1 Thermodynamic Potential

Suppose we multiply the interaction U by a parameter $\lambda \in [0, 1]$, so that $\Phi[G, U] \to \Phi[G_\lambda, \lambda U]$, where G_λ is the Green's function parametrised by λ. Since the skeleton diagrams that define $\Sigma_\lambda^{(n)}$ contain n interaction lines, and

$$\frac{\partial \lambda^n}{\partial \lambda} = n \lambda^{n-1} = \frac{n}{\lambda} \lambda^n ,$$

The first term converges, since it involves either $\operatorname{Re} G \operatorname{Re} \Sigma^{(n)}$ or $\operatorname{Im} G \operatorname{Im} \Sigma^{(n)}$, both of which are even in ϵ, see (4.22) and (4.51), and decay at least as $1/\epsilon^2$, see (4.23) and (4.52). Therefore, in that term we can safely set $\eta = 0$ even before performing the sum. For $\eta \to 0$, the second term is dominated by the odd function $\operatorname{Im} G \operatorname{Re} \Sigma^{(n)}$, which decays as $1/\epsilon$ as $|\epsilon| \to \infty$, see (4.23) and (4.52), so that

$$i T \sum_{\epsilon>0} \sin(\epsilon \eta) \left(\Phi(i\epsilon, \mathbf{k}) - \Phi(-i\epsilon, \mathbf{k}) \right) \xrightarrow[\eta \simeq 0]{} 2 \left(\sum_{n \geq 0} \frac{1}{2n} \operatorname{Re} \Sigma^{(n)}(i\infty, \mathbf{k}) \right) T \sum_{\epsilon>0} \frac{\sin(\epsilon \eta)}{\epsilon} ,$$

where

$$T \sum_{\epsilon>0} \frac{\sin(\epsilon \eta)}{\epsilon} = \sum_{k \geq 0} \frac{\sin\left((2k+1)\pi T \eta\right)}{(2k+1)\pi} = \frac{1}{4} ,$$

is finite, and independent of η, only if the sum is done before sending $\eta \to 0$.

then

$$
\begin{aligned}
\frac{\partial \Phi[G_\lambda, \lambda U]}{\partial \lambda}\Big|_{G_\lambda} &= \frac{1}{2\lambda} \sum_{\mathbf{k}\sigma} T \sum_\epsilon e^{i\epsilon\eta} \sum_{n\geq 1} G_\lambda(i\epsilon, \mathbf{k}) \Sigma_\lambda^{(n)}(i\epsilon, \mathbf{k}) \\
&= \frac{1}{2\lambda} \sum_{\mathbf{k}\sigma} T \sum_\epsilon e^{i\epsilon\eta} G_\lambda(i\epsilon, \mathbf{k}) \Sigma_\lambda(i\epsilon, \mathbf{k}),
\end{aligned}
\tag{4.121}
$$

where the derivative is done at fixed G_λ. Through (4.80),

$$
\begin{aligned}
\frac{1}{2} \sum_{\mathbf{k}\sigma} \left(-\frac{\partial}{\partial \tau} - \epsilon_\mathbf{k} \right) G_\lambda(\tau, \mathbf{k}) &= \frac{T}{2} \sum_\epsilon \sum_{\mathbf{k}\sigma} e^{-i\epsilon\tau} G_0(i\epsilon, \mathbf{k})^{-1} G_\lambda(i\epsilon, \mathbf{k}) \\
&= \frac{\delta(\tau)}{2} + \frac{\lambda}{2V} \sum_{\mathbf{kpq}} \sum_{\sigma\sigma'} U(\mathbf{q}) \langle T_\tau \left(c_{\mathbf{k}\sigma}^\dagger(0) c_{\mathbf{p}+\mathbf{q}\sigma'}^\dagger(\tau + 0^+) c_{\mathbf{p}\sigma'}(\tau) c_{\mathbf{k}+\mathbf{q}\sigma}(\tau) \right) \rangle,
\end{aligned}
$$

so that, for $\tau = -\eta \to -0^+$,

$$
\begin{aligned}
\frac{T}{2} \sum_\epsilon \sum_{\mathbf{k}\sigma} e^{i\epsilon\eta} G_0(i\epsilon, \mathbf{k})^{-1} G_\lambda(i\epsilon, \mathbf{k}) &= \frac{T}{2} \sum_\epsilon \sum_{\mathbf{k}\sigma} e^{i\epsilon\eta} \left(G_\lambda(i\epsilon, \mathbf{k})^{-1} + \Sigma_\lambda(i\epsilon, \mathbf{k}) \right) G_\lambda(i\epsilon, \mathbf{k}) \\
&= \frac{T}{2} \sum_\epsilon \sum_{\mathbf{k}\sigma} e^{i\epsilon\eta} \Sigma_\lambda(i\epsilon, \mathbf{k}) G_\lambda(i\epsilon, \mathbf{k}) \\
&= \frac{\lambda}{2V} \sum_{\mathbf{kpq}} \sum_{\sigma\sigma'} U(\mathbf{q}) \langle c_{\mathbf{k}\sigma}^\dagger c_{\mathbf{p}+\mathbf{q}\sigma'}^\dagger c_{\mathbf{p}\sigma'} c_{\mathbf{k}+\mathbf{q}\sigma} \rangle \equiv \langle H_{int}(\lambda) \rangle.
\end{aligned}
$$

Comparing the above equation with (4.121) and using the Hellman-Feynman theorem we conclude that

$$
\begin{aligned}
\frac{\partial \Phi[G_\lambda, \lambda U]}{\partial \lambda}\Big|_{G_\lambda} &= \frac{1}{2\lambda} \sum_{\mathbf{k}\sigma} T \sum_\epsilon e^{i\epsilon\eta} G_\lambda(i\epsilon, \mathbf{k}) \Sigma_\lambda(i\epsilon, \mathbf{k}) = \frac{1}{\lambda} \langle H_{int}(\lambda) \rangle \\
&= \langle \frac{\partial H(\lambda)}{\partial \lambda} \rangle = \frac{\partial F(\lambda)}{\partial \lambda},
\end{aligned}
\tag{4.122}
$$

where $F(\lambda)$ is the free-energy of the interacting system as fixed λ.

We define a functional

$$
\overline{F}[G_\lambda, \lambda U] \equiv \sum_{\mathbf{k}\sigma} T \sum_\epsilon e^{i\epsilon\eta} \left(\ln G_\lambda(i\epsilon, \mathbf{k}) - G_0(i\epsilon, \mathbf{k})^{-1} G_\lambda(i\epsilon, \mathbf{k}) \right) + \Phi[G_\lambda, \lambda U].
$$

If G_λ is the solution of Dyson's equation

$$
G_\lambda(i\epsilon, \mathbf{k})^{-1} = G_0(i\epsilon, \mathbf{k})^{-1} - \Sigma_\lambda(i\epsilon, \mathbf{k}),
$$

then

$$
\frac{\partial \overline{F}[G_\lambda, \lambda U]}{\partial G_\lambda}\Big|_{\lambda U} = G_\lambda^{-1} - G_0^{-1} + \frac{\delta \Phi}{\delta G_\lambda} = G_\lambda^{-1} - G_0^{-1} + \Sigma_\lambda = 0.
$$

In other words, the Green's function solution of Dyson's equation is the saddle point of $\overline{F}[G, U]$.

It follows that

$$\frac{d\overline{F}[G_\lambda, \lambda U]}{d\lambda} = \frac{\partial \overline{F}[G_\lambda, \lambda U]}{\partial G_\lambda}\bigg|_{\lambda U} \frac{\partial G_\lambda}{\partial \lambda} + \frac{\partial \Phi[G_\lambda, \lambda U]}{\partial \lambda}\bigg|_{G_\lambda} = \frac{\partial \Phi[G_\lambda, \lambda U]}{\partial \lambda}\bigg|_{G_\lambda} = \frac{\partial F(\lambda)}{\partial \lambda},$$

and thus, upon integration over λ from 0 to 1,

$$\overline{F}[G, U] - \overline{F}[G_0, 0] = F(1) - F(0) = F - F_0, \qquad (4.123)$$

where $G = G_{\lambda=1}$ is the interacting Green's function, $F(\lambda = 1) = F$ the interacting free-energy, and $F(\lambda = 0) = F_0$ the non-interacting one. Since[5]

$$\overline{F}[G_0, 0] \equiv \sum_{k\sigma} T \sum_\epsilon e^{i\epsilon\eta} \ln G_0(i\epsilon, \mathbf{k}) = F_0,$$

we finally obtain that the free-energy of the interacting system

$$F = \overline{F}[G, U] = \sum_{k\sigma} T \sum_\epsilon e^{i\epsilon\eta} \left(\ln G(i\epsilon, \mathbf{k}) - G_0(i\epsilon, \mathbf{k})^{-1} G(i\epsilon, \mathbf{k}) \right) + \Phi[G, U], \quad (4.124)$$

with G solution of the Dyson equation

$$G^{-1} = G_0^{-1} - \frac{\delta\Phi[G, U]}{\delta G}.$$

---

[5] The free-energy of the non-interacting system is

$$F_0 = -T \ln Z_0 = -T \ln \prod_{k\sigma} \left(1 + e^{-\beta\epsilon_k}\right) = -T \sum_{k\sigma} \ln\left(1 + e^{-\beta\epsilon_k}\right).$$

On the other hand, using the standard tricks to perform the sum over Matsubara frequencies, and since

$$\ln G_0(\epsilon \pm i0^+, \mathbf{k}) = -\ln\left(\epsilon - \epsilon_k \pm i0^+\right) = -\ln|\epsilon - \epsilon_k| \mp i\pi\,\theta(\epsilon_k - \epsilon).$$

$$\sum_{k\sigma} T \sum_\epsilon e^{i\epsilon\eta} \ln G_0(i\epsilon, \mathbf{k}) = -\sum_{k\sigma} \int_{-\infty}^\infty \frac{d\epsilon}{2\pi i} f(\epsilon) e^{\epsilon\eta} \ln \frac{G_0(\epsilon + i0^+, \mathbf{k})}{G_0(\epsilon - i0^+, \mathbf{k})} = \sum_{k\sigma} \int_{-\infty}^{\epsilon_k} d\epsilon\, f(\epsilon) e^{\epsilon\eta}$$

$$= -T \sum_{k\sigma} \int_{-\infty}^{\epsilon_k} d\ln\left(1 + e^{-\beta\epsilon}\right) e^{\epsilon\eta} = -T \sum_{k\sigma} \ln\left(1 + e^{-\beta\epsilon_k}\right) = F_0,$$

which is desired result.

Fig. 4.41 Graphical
representation of (4.127)

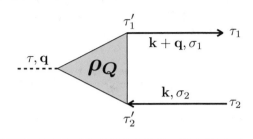

4.9   Ward-Takahashi Identities

Let us suppose that our model has a set of conserved quantities

$$Q = \int d\mathbf{r}\, \rho_Q(\mathbf{r})\,,$$

with the Fourier transform of the density having the general expression

$$\rho_Q(\mathbf{q}) = \sum_{\mathbf{k}} \sum_{\sigma\sigma'} c_{\mathbf{k}\sigma}^\dagger \, \Lambda_0^Q(\mathbf{k}\,\sigma, \mathbf{k}+\mathbf{q}\,\sigma')\, c_{\mathbf{k}+\mathbf{q}\sigma'}\,. \qquad (4.125)$$

The conservation of Q entails the existence of a continuity equation

$$-\frac{\partial \rho_Q(\mathbf{r})}{\partial \tau} = \big[\rho_Q(\mathbf{r}),\, H\big] \equiv -i\,\nabla\cdot\boldsymbol{J}_Q(\mathbf{r})\,, \qquad -\frac{\partial \rho_Q(\mathbf{q})}{\partial \tau} = \big[\rho_Q(\mathbf{q}),\, H\big] \equiv \mathbf{q}\cdot\boldsymbol{J}_Q(\mathbf{q})\,, \qquad (4.126)$$

which defines the corresponding current $\boldsymbol{J}_Q(\mathbf{r})$ or its Fourier transform $\boldsymbol{J}_Q(\mathbf{q})$.

Let us consider the following expectation value, graphically shown in Fig. 4.41,

$$-\,\big\langle T_\tau\big(\rho_Q(\tau,\mathbf{q})\, c_{\mathbf{k}+\mathbf{q}\sigma_1}^\dagger(\tau_1)\, c_{\mathbf{k}\sigma_2}(\tau_2)\big)\big\rangle$$

$$= \int d\tau_1'\, d\tau_2'\, \Lambda^Q\big(\tau_2'\,\mathbf{k}\,\sigma_2, \tau_1'\,\mathbf{k}+\mathbf{q}\,\sigma_1; \tau,\mathbf{q}\big)\, G(\tau_2-\tau_2',\mathbf{k})\, G(\tau_1'-\tau_1,\mathbf{k}+\mathbf{q})\,, \qquad (4.127)$$

where Λ^Q is the interacting Q-density vertex, and apply $-\partial/\partial\tau$. The derivative acts either directly on the charge density operator, in which case we can use the continuity equation (4.126), or on the θ-functions that define the time-ordered product. The latter is, dropping for simplicity all unnecessary indices,

$$\langle \rho_Q(\tau)\, c^\dagger(\tau_1)\, c(\tau_2)\rangle\, \partial_\tau\big(\theta(\tau-\tau_1)\theta(\tau_1-\tau_2)\big) + \langle c^\dagger(\tau_1)\,\rho_Q(\tau)\, c(\tau_2)\rangle\, \partial_\tau\big(\theta(\tau_1-\tau)\theta(\tau-\tau_2)\big)$$

$$+ \langle c^\dagger(\tau_1)\, c(\tau_2)\,\rho_Q(\tau)\rangle\, \partial_\tau\big(\theta(\tau_1-\tau_2)\theta(\tau_2-\tau)\big) - \langle \rho_Q(\tau)\, c(\tau_2)\, c^\dagger(\tau_1)\rangle\, \partial_\tau\big(\theta(\tau-\tau_2)\theta(\tau_2-\tau_1)\big)$$

$$- \langle c(\tau_2)\,\rho_Q(\tau)\, c^\dagger(\tau_1)\rangle\, \partial_\tau\big(\theta(\tau_2-\tau)\theta(\tau-\tau_1)\big) - \langle c(\tau_2)\, c^\dagger(\tau_1)\,\rho_Q(\tau)\rangle\, \partial_\tau\big(\theta(\tau_2-\tau_1)\theta(\tau_1-\tau)\big)$$

$$= \delta(\tau-\tau_1)\,\theta(\tau-\tau_2)\,\langle\big[\rho_Q(\tau),\, c^\dagger(\tau)\big]\, c(\tau_2)\rangle + \theta(\tau_1-\tau)\,\delta(\tau-\tau_2)\,\langle c^\dagger(\tau_1)\big[\rho_Q(\tau),\, c(\tau)\big]\rangle$$

$$- \delta(\tau-\tau_2)\,\theta(\tau-\tau_1)\,\langle\big[\rho_Q(\tau),\, c(\tau)\big]\, c^\dagger(\tau_1)\rangle - \theta(\tau_2-\tau)\,\delta(\tau-\tau_1)\,\langle c(\tau_2)\big[\rho_Q(\tau),\, c^\dagger(\tau)\big]\rangle$$

$$= \delta(\tau-\tau_1)\,\langle T_\tau\big(\big[\rho_Q(\tau),\, c^\dagger(\tau)\big]\, c(\tau_2)\big)\rangle + \delta(\tau-\tau_2)\,\langle T_\tau\big(c^\dagger(\tau_1)\big[\rho_Q(\tau),\, c(\tau)\big]\big)\rangle\,.$$

It follows, through (4.126), that

$$
-\frac{\partial}{\partial \tau} \int d\tau_1' \, d\tau_2' \, \Lambda^Q\big(\tau_2' \, \mathbf{k}\, \sigma_2, \tau_1' \, \mathbf{k}+\mathbf{q}\, \sigma_1; \tau, \mathbf{q}\big) \, G(\tau_2 - \tau_2', \mathbf{k}) \, G(\tau_1' - \tau_1, \mathbf{k}+\mathbf{q})
$$

$$
= \mathbf{q} \cdot \int d\tau_1' \, d\tau_2' \, \Lambda^Q\big(\tau_2' \, \mathbf{k}\, \sigma_2, \tau_1' \, \mathbf{k}+\mathbf{q}\, \sigma_1; \tau, \mathbf{q}\big) \, G(\tau_2 - \tau_2', \mathbf{k}) \, G(\tau_1' - \tau_1, \mathbf{k}+\mathbf{q})
$$

$$
+ \delta(\tau - \tau_1) \, \langle T_\tau\big(\big[\rho_Q(\tau, \mathbf{q}), \, c_{\mathbf{k}+\mathbf{q}\sigma_1}^\dagger(\tau)\big] c_{\mathbf{k}\sigma_2}(\tau_2) \big)\rangle
$$

$$
+ \delta(\tau - \tau_2) \, \langle T_\tau\big(c_{\mathbf{k}+\mathbf{q}\sigma_1}^\dagger(\tau_1) \big[\rho_Q(\tau, \mathbf{q}), \, c_{\mathbf{k}\sigma_2}(\tau)\big] \big)\rangle, \tag{4.128}
$$

where Λ^Q is the interacting Q-current density vertex. One can readily find that the equal-time commutators

$$
\big[\rho_Q(\tau, \mathbf{q}), \, c_{\mathbf{k}+\mathbf{q}\sigma_1}^\dagger(\tau)\big] = \sum_\sigma c_{\mathbf{k}\sigma}^\dagger(\tau) \, \Lambda_0^Q\big(\mathbf{k}\,\sigma, \mathbf{k}+\mathbf{q}\,\sigma_1\big),
$$

$$
\big[\rho_Q(\tau, \mathbf{q}), \, c_{\mathbf{k}\sigma_2}(\tau)\big] = -\sum_\sigma \Lambda_0^Q\big(\mathbf{k}\,\sigma_2, \mathbf{k}+\mathbf{q}\,\sigma\big) c_{\mathbf{k}+\mathbf{q}\sigma}(\tau),
$$

so that (4.128) becomes

$$
-\frac{\partial}{\partial \tau} \int d\tau_1' \, d\tau_2' \, \Lambda^Q\big(\tau_2' \, \mathbf{k}\, \sigma_2, \tau_1' \, \mathbf{k}+\mathbf{q}\, \sigma_1; \tau, \mathbf{q}\big) \, G(\tau_2 - \tau_2', \mathbf{k}) \, G(\tau_1' - \tau_1, \mathbf{k}+\mathbf{q})
$$

$$
= \mathbf{q} \cdot \int d\tau_1' \, d\tau_2' \, \Lambda^Q\big(\tau_2' \, \mathbf{k}\, \sigma_2, \tau_1' \, \mathbf{k}+\mathbf{q}\, \sigma_1; \tau, \mathbf{q}\big) \, G(\tau_2 - \tau_2', \mathbf{k}) \, G(\tau_1' - \tau_1, \mathbf{k}+\mathbf{q})
$$

$$
+ \delta(\tau - \tau_1) \, G(\tau_2 - \tau, \mathbf{k}) \, \Lambda_0^Q\big(\mathbf{k}\,\sigma_2, \mathbf{k}+\mathbf{q}\,\sigma_1\big)
$$

$$
- \delta(\tau - \tau_2) \, \Lambda_0^Q\big(\mathbf{k}\,\sigma_2, \mathbf{k}+\mathbf{q}\,\sigma_1\big) \, G\big(\tau - \tau_1, \mathbf{k}+\mathbf{q}\big),
$$

or, in Matsubara frequencies,

$$
i\omega \, \Lambda^Q\big(i\epsilon \, \mathbf{k}\, \sigma_2, i\epsilon + i\omega \, \mathbf{k}+\mathbf{q}\, \sigma_1; i\omega, \mathbf{q}\big) \, G(i\epsilon, \mathbf{k}) \, G(i\epsilon + i\omega, \mathbf{k}+\mathbf{q})
$$

$$
= \mathbf{q} \cdot \Lambda^Q\big(i\epsilon \, \mathbf{k}\, \sigma_2, i\epsilon + i\omega \, \mathbf{k}+\mathbf{q}\, \sigma_1; i\omega, \mathbf{q}\big) \, G(i\epsilon, \mathbf{k}) \, G(i\epsilon + i\omega, \mathbf{k}+\mathbf{q})
$$

$$
- \Lambda_0^Q\big(\mathbf{k}\,\sigma_2, \mathbf{k}+\mathbf{q}\,\sigma_1\big) \big(G(i\epsilon + i\omega, \mathbf{k}+\mathbf{q}) - G(i\epsilon, \mathbf{k}) \big). \tag{4.129}
$$

Dividing both sides by the product of the two Green's function, (4.129) is also equivalent to

$$
i\omega \, \Lambda^Q\big(i\epsilon \, \mathbf{k}\, \sigma_2, i\epsilon + i\omega \, \mathbf{k}+\mathbf{q}\, \sigma_1; i\omega, \mathbf{q}\big) - \mathbf{q} \cdot \Lambda^Q\big(i\epsilon \, \mathbf{k}\, \sigma_2, i\epsilon + i\omega \, \mathbf{k}+\mathbf{q}\, \sigma_1; i\omega, \mathbf{q}\big)
$$

$$
= \Lambda_0^Q\big(\mathbf{k}\,\sigma_2, \mathbf{k}+\mathbf{q}\,\sigma_1\big) \big(G(i\epsilon + i\omega, \mathbf{k}+\mathbf{q})^{-1} - G(i\epsilon, \mathbf{k})^{-1} \big)
$$

$$
= \Lambda_0^Q\big(\mathbf{k}\,\sigma_2, \mathbf{k}+\mathbf{q}\,\sigma_1\big) \big(G_0(i\epsilon + i\omega, \mathbf{k}+\mathbf{q})^{-1} - G_0(i\epsilon, \mathbf{k})^{-1} \big)
$$

$$
- \Lambda_0^Q\big(\mathbf{k}\,\sigma_2, \mathbf{k}+\mathbf{q}\,\sigma_1\big) \big(\Sigma(i\epsilon + i\omega, \mathbf{k}+\mathbf{q}) - \Sigma(i\epsilon, \mathbf{k}) \big). \tag{4.130}
$$

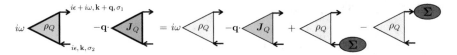

Fig. 4.42 Graphical representation of the Ward-Takahashi identity (4.131). The darker triangles with bold contours are fully interacting density vertices, while the other non-interacting ones. The incoming and outgoing arrow lines just indicate the external vertices. All incoming lines carry the same frequency $i\epsilon + i\omega$, momentum $\mathbf{k} + \mathbf{q}$, and spin σ_1, the reason why we show them only in the first vertex. Similarly, all outgoing lines carry the same frequency $i\epsilon$, momentum $\mathbf{k}$, and spin σ_2, only shown in the first vertex. The self-energies, red ellipses, carry the same frequency, momentum and spin of the line they are attached to

In absence of interaction, $\Lambda^Q = \Lambda_0^Q$, $\boldsymbol{\Lambda}^Q = \boldsymbol{\Lambda}_0^Q$, $\Sigma = 0$, so that

$$i\omega \Lambda_0^Q\left(\mathbf{k}\,\sigma_2, \mathbf{k} + \mathbf{q}\,\sigma_1\right) - \mathbf{q} \cdot \boldsymbol{\Lambda}_0^Q\left(\mathbf{k}\,\sigma_2, \mathbf{k} + \mathbf{q}\,\sigma_1\right)$$

$$= \Lambda_0^Q\left(\mathbf{k}\,\sigma_2, \mathbf{k} + \mathbf{q}\,\sigma_1\right)\left(G_0(i\epsilon + i\omega, \mathbf{k} + \mathbf{q})^{-1} - G_0(i\epsilon, \mathbf{k})^{-1}\right),$$

which allows us to rewrite (4.130) as

$$i\omega \Lambda^Q\left(i\epsilon\,\mathbf{k}\,\sigma_2, i\epsilon + i\omega\,\mathbf{k}{+}\mathbf{q}\,\sigma_1; i\omega, \mathbf{q}\right) - \mathbf{q} \cdot \boldsymbol{\Lambda}^Q\left(i\epsilon\,\mathbf{k}\,\sigma_2, i\epsilon + i\omega\,\mathbf{k} + \mathbf{q}\,\sigma_1; i\omega, \mathbf{q}\right)$$

$$= i\omega \Lambda_0^Q\left(\mathbf{k}\,\sigma_2, \mathbf{k} + \mathbf{q}\,\sigma_1\right) - \mathbf{q} \cdot \boldsymbol{\Lambda}_0^Q\left(\mathbf{k}\,\sigma_2, \mathbf{k} + \mathbf{q}\,\sigma_1\right)$$

$$- \Lambda_0^Q\left(\mathbf{k}\,\sigma_2, \mathbf{k} + \mathbf{q}\,\sigma_1\right)\left(\Sigma(i\epsilon + i\omega, \mathbf{k} + \mathbf{q}) - \Sigma(i\epsilon, \mathbf{k})\right),$$

$$(4.131)$$

which is the Ward-Takahashi identity, graphically shown in Fig. 4.42.

Let us study the consequences the Ward-Takahashi identity considering for simplicity the case $Q = N$, the number of electrons, so that $\rho_Q = \rho$ the charge density, Λ_0^Q the identity, and $\boldsymbol{\Lambda}_0^Q = \partial\epsilon_\mathbf{k}/\partial\mathbf{k}$ for small $\mathbf{q}$ times the identity in spin. In this case, the interacting density-vertices are also diagonal in spin, and (4.131) reads, for small ω and $\mathbf{q}$,

$$i\omega \Lambda\left(i\epsilon\,\mathbf{k}\,\sigma, i\epsilon + i\omega\,\mathbf{k} + \mathbf{q}\,\sigma; i\omega, \mathbf{q}\right) - \mathbf{q} \cdot \boldsymbol{\Lambda}\left(i\epsilon\,\mathbf{k}\,\sigma, i\epsilon + i\omega\,\mathbf{k} + \mathbf{q}\,\sigma; i\omega, \mathbf{q}\right)$$

$$\simeq i\omega \Lambda\left(i\epsilon\,\mathbf{k}\,\sigma, i\epsilon\,\mathbf{k}\,\sigma; i\omega, \mathbf{q}\right) - \mathbf{q} \cdot \boldsymbol{\Lambda}\left(i\epsilon\,\mathbf{k}\,\sigma, i\epsilon\,\mathbf{k}\,\sigma; i\omega, \mathbf{q}\right)$$

$$= \left(i\omega - \mathbf{q} \cdot \frac{\partial\epsilon_\mathbf{k}}{\partial\mathbf{k}}\right) - \left(\Sigma(i\epsilon + i\omega, \mathbf{k} + \mathbf{q}) - \Sigma(i\epsilon, \mathbf{k})\right)$$

$$\simeq i\omega\left(1 - \frac{\partial\Sigma(i\epsilon, \mathbf{k})}{\partial i\epsilon}\right) - \mathbf{q} \cdot \frac{\partial}{\partial\mathbf{k}}\left(\epsilon_\mathbf{k} + \Sigma(i\epsilon, \mathbf{k})\right),$$

$$(4.132)$$

It is common to denote the ω-limit as the limit sending first $q = |\mathbf{q}|$ to zero, and next ω to zero. The reverse case is accordingly denoted as q-limit. We thus find

$$\Lambda^\omega(i\epsilon, \mathbf{k}, \sigma) = \lim_{\omega \to 0} \lim_{q \to 0} \Lambda(i\epsilon\,\mathbf{k}\,\sigma, i\epsilon + i\omega\,\mathbf{k} + \mathbf{q}\,\sigma; i\omega, \mathbf{q}) = 1 - \frac{\partial \Sigma(i\epsilon, \mathbf{k})}{\partial i\epsilon} \, ,$$

$$\Lambda^q(i\epsilon, \mathbf{k}, \sigma) = \lim_{\omega \to 0} \lim_{q \to 0} \mathbf{\Lambda}(i\epsilon\,\mathbf{k}\,\sigma, i\epsilon + i\omega\,\mathbf{k} + \mathbf{q}\,\sigma; i\omega, \mathbf{q}) = \frac{\partial}{\partial \mathbf{k}} \left(\epsilon_{\mathbf{k}} + \Sigma(i\epsilon, \mathbf{k}) \right).$$

$$(4.133)$$

Those equations play an important role in Landau's Fermi liquid theory, as we shall see.

4.9.1   Ward-Takahashi Identity for the Heat Density

The Hamiltonian H is trivially associate to a conserved quantity, the total energy E. However, H is not a one-body operator, and therefore the previous derivation of the Ward-Takahashi identity cannot be used in this case.

Let us assume that the Hamiltonian is local, namely can be written as

$$H = \int d\mathbf{r}\, \rho_E(\mathbf{r}) \, .$$

It follows that, by definition,

$$-\frac{\partial \Psi_\sigma(\mathbf{r})}{\partial \tau} = \left[\Psi_\sigma(\mathbf{r}), H \right] = \left[\Psi_\sigma(\mathbf{r}), \rho_E(\mathbf{r}) \right],$$

$$-\frac{\partial \Psi_\sigma^\dagger(\mathbf{r})}{\partial \tau} = \left[\Psi_\sigma^\dagger(\mathbf{r}), H \right] = \left[\Psi_\sigma^\dagger(\mathbf{r}), \rho_E(\mathbf{r}) \right],$$

and, therefore,

$$\left[\rho_E(\mathbf{r}), \Psi_\sigma(\mathbf{r}') \right] = \delta(\mathbf{r} - \mathbf{r}') \frac{\partial \Psi_\sigma(\mathbf{r})}{\partial \tau} \, , \quad \left[\rho_E(\mathbf{r}), \Psi_\sigma^\dagger(\mathbf{r}') \right] = \delta(\mathbf{r} - \mathbf{r}') \frac{\partial \Psi_\sigma^\dagger(\mathbf{r})}{\partial \tau} \, .$$

$$(4.134)$$

Those equations can actually be used to find the heat density operator $\rho_E(\mathbf{r})$. Since the total energy is conserved, there exists a continuity equation

$$-\frac{\partial \rho_E(\mathbf{r})}{\partial \tau} + i\,\nabla \cdot \mathbf{J}_E(\mathbf{r}) = 0 \, ,$$

with $\mathbf{J}_E(\mathbf{r})$ a multi-particle operator just like $\rho_E(\mathbf{r})$ is. We consider the same function as in (4.127), but now in real space and imaginary time,

$$I_E(x; y, z) = -\langle T_\tau \left(\rho_E(x)\, \Psi_\sigma^\dagger(y)\, \Psi_\sigma(z) \right) \rangle \, , \qquad (4.135)$$

where $x = (\tau, \mathbf{r}) \equiv (x_0, \mathbf{x})$, and similarly for y and z. Proceeding as before, and using (4.134) we find

$$-\frac{\partial}{\partial x_0} I_E(x; y, z) = -\langle T_\tau \Big(\big(-i \, \boldsymbol{\nabla} \cdot \boldsymbol{J}_E(x) \big) \Psi_\sigma^\dagger(y) \, \Psi_\sigma(z) \Big) \rangle$$

$$- \delta(x_0 - y_0) \langle T_\tau \Big(\Psi_\sigma(z) \big[\rho_E(x), \Psi_\sigma^\dagger(y) \big] \Big) \rangle$$

$$- \delta(x_0 - z_0) \langle T_\tau \Big(\big[\rho_E(x), \Psi_\sigma(z) \big] \Psi_\sigma^\dagger(y) \Big) \rangle$$

$$= -i \, \boldsymbol{\nabla} \cdot \boldsymbol{I}_E(x; y, z) - \delta(x_0 - y_0) \, \delta(\mathbf{x} - \mathbf{y}) \langle T_\tau \Big(\Psi_\sigma(z) \frac{\partial \Psi_\sigma^\dagger(x)}{\partial x_0} \Big) \rangle$$

$$- \delta(x_0 - z_0) \, \delta(\mathbf{x} - \mathbf{z}) \langle T_\tau \Big(\frac{\partial \Psi_\sigma(x)}{\partial x_0} \Psi_\sigma^\dagger(y) \Big) \rangle$$

$$= -i \, \boldsymbol{\nabla} \cdot \boldsymbol{I}_E(x; y, z) + \delta(x - y) \frac{\partial}{\partial y_0} G(z, y) + \delta(x - z) \frac{\partial}{\partial z_0} G(z, y),$$

since the imaginary-time derivatives acting on the θ-functions that define G cancels out, and thus

$$-\frac{\partial}{\partial x_0} I_E(x; y, z) + i \, \boldsymbol{\nabla} \cdot \boldsymbol{I}_E(x; y, z) = \delta(x - y) \frac{\partial}{\partial y_0} G(z, y) + \delta(x - z) \frac{\partial}{\partial z_0} G(z, y).$$

Introducing the heat and heat-current density-vertices, the above equation becomes in frequency and momentum space

$$\Big(i\omega \, \Lambda^E(i\epsilon \, \mathbf{k} \, \sigma, i\epsilon + i\omega \, \mathbf{k} + \mathbf{q} \, \sigma; i\omega, \mathbf{q}) - \mathbf{q} \cdot \boldsymbol{\Lambda}^E(i\epsilon \, \mathbf{k} \, \sigma, i\epsilon + i\omega \, \mathbf{k} + \mathbf{q} \, \sigma; i\omega, \mathbf{q}) \Big)$$

$$G(i\epsilon, \mathbf{k}) \, G(i\epsilon + i\omega, \mathbf{k} + \mathbf{q})$$

$$= i\epsilon \, G(i\epsilon, \mathbf{k}) - i\big(\epsilon + \omega\big) G(i\epsilon + i\omega, \mathbf{k} + \mathbf{q}),$$

which, proceeding as before, leads to

$$i\omega \, \Lambda^E(i\epsilon \, \mathbf{k} \, \sigma, i\epsilon + i\omega \, \mathbf{k} + \mathbf{q} \, \sigma; i\omega, \mathbf{q}) - \mathbf{q} \cdot \boldsymbol{\Lambda}^E(i\epsilon \, \mathbf{k} \, \sigma, i\epsilon + i\omega \, \mathbf{k} + \mathbf{q} \, \sigma; i\omega, \mathbf{q})$$

$$= i\epsilon \, G(i\epsilon + i\omega, \mathbf{k} + \mathbf{q})^{-1} - \big(i\epsilon + i\omega\big) G(i\epsilon, \mathbf{k})^{-1}$$

$$= -i\epsilon \, \big(\epsilon(\mathbf{k} + \mathbf{q}) - \epsilon(\mathbf{k}) \big) + i\omega \, \epsilon(\mathbf{k})$$

$$- i\epsilon \, \Big(\Sigma(i\epsilon + i\omega, \mathbf{k} + \mathbf{q}) - \Sigma(i\epsilon, \mathbf{k}) \Big) + i\omega \, \Sigma(i\epsilon, \mathbf{k}). \tag{4.136}$$

Equation (4.136) is the Ward-Takahashi identity for the heat density. Expanding for small ω and $\mathbf{q}$, we get

$$i\omega \, \Lambda^E(i\epsilon \, \mathbf{k} \, \sigma, i\epsilon \, \mathbf{k} \, \sigma; i\omega \, \mathbf{q}) - \mathbf{q} \cdot \boldsymbol{\Lambda}^E(i\epsilon \, \mathbf{k} \, \sigma, i\epsilon \, \mathbf{k} \, \sigma; i\omega \, \mathbf{q})$$

$$= i\omega \left[\epsilon(\mathbf{k}) + \Sigma(i\epsilon, \mathbf{k}) - i\epsilon \, \frac{\partial \Sigma(i\epsilon, \mathbf{k})}{\partial i\epsilon} \right] - i\epsilon \, \mathbf{q} \cdot \left[\frac{\partial \epsilon_\mathbf{k}}{\partial \mathbf{k}} + \frac{\partial \Sigma(i\epsilon, \mathbf{k})}{\partial \mathbf{k}} \right],$$

so that

$$\left(\Lambda^E(i\epsilon, \mathbf{k}, \sigma)\right)^\omega = \epsilon(\mathbf{k}) + \Sigma(i\epsilon, \mathbf{k}) - i\epsilon \frac{\partial \Sigma(i\epsilon, \mathbf{k})}{\partial i\epsilon},$$

$$\left(\Lambda^E(i\epsilon, \mathbf{k}, \sigma)\right)^q = i\epsilon \left[\frac{\partial \epsilon_\mathbf{k}}{\partial \mathbf{k}} + \frac{\partial \Sigma(i\epsilon, \mathbf{k})}{\partial \mathbf{k}} \right]. \tag{4.137}$$

Like (4.133), those equations are important in Landau's Fermi liquid theory.

In particular, (4.137) allows calculating the interacting specific heat. We note that the internal energy is defined through

$$U(\beta) = -\frac{\partial \ln Z(\beta)}{\partial \beta} = \frac{\partial}{\partial \beta}\left(\beta F(\beta)\right).$$

Suppose we change $H \to \lambda H$, then it follows that $Z(\beta) \to Z(\lambda\beta)$, and thus

$$\frac{\delta F(\lambda\beta)}{\delta\lambda} = -T \frac{\delta \ln Z}{\delta\lambda} = -\frac{T}{Z}\frac{\partial}{\partial\lambda}\text{Tr}\left(e^{-\lambda\beta H}\right) = U(\lambda\beta).$$

The specific heat at fixed λ is defined through

$$c_V = \frac{\partial U(\lambda\beta)}{\partial T} = -\beta^2 \frac{\partial U(\lambda\beta)}{\partial\beta} = -\beta^2 \frac{\partial U(\lambda\beta)}{\partial\beta} = -\lambda\beta \frac{\partial U(\lambda\beta)}{\partial\lambda} = -\lambda\beta \frac{\partial^2 F(\lambda\beta)}{\partial\lambda^2},$$

so that the actual value at $\lambda = 1$ reads

$$c_V = -\frac{1}{T}\frac{\partial^2 F(\lambda\beta)}{\partial\lambda^2}\Big|_{\lambda=1}.$$

The second derivative of the free energy with respect to λ is just the q-limit of the heat density-heat density response function $\chi_E(i\omega, \mathbf{q})$, so that

$$c_V = -\frac{1}{T}\lim_{q\to 0}\lim_{\omega\to 0}\chi_E(0, \mathbf{q}). \tag{4.138}$$

We shall use both (4.137) and (4.138) to calculate the specific heat of a Landau's Fermi liquid.

4.10 Conserving Approximation Schemes

We already mentioned that, since the exact self-energy functional $\Sigma[G, U]$ is generally unknown, the common approach is to use an approximate expression $\Sigma_{approx}[G, U]$, and then solve the self-consistency Dyson's equation

$$G^{-1} = G_0^{-1} - \Sigma_{approx}[G, U],$$

to get the corresponding approximation G_{approx} of the actual Green's function. The question we here address is how to extend that approximation scheme to calculate linear response functions without spoiling conservation laws.

Let us go back to (4.131), which, in shorthand notations, can be written as

$$i\omega \Lambda^Q - \mathbf{q} \cdot \Lambda^Q = i\omega \Lambda_0^Q - \mathbf{q} \cdot \Lambda_0^Q - \Lambda_0^Q \odot \Sigma + \Sigma \odot \Lambda_0^Q, \quad (4.139)$$

while (4.129) as, see (4.97),

$$i\omega \Lambda^Q \odot R - \mathbf{q} \Lambda^Q \odot R = -\Lambda_0^Q \odot G + G \odot \Lambda_0^Q. \quad (4.140)$$

In this representation, both equations also account for the case in which there are additional quantum numbers, e.g., band, orbital or reciprocal lattice vector indices, and the Green's functions becomes matrices with finite off-diagonal elements. By definition, see Fig. 4.25, and using the Bethe-Salpeter equation in the particle-hole channel, see Fig. 4.32,

$$\Lambda^Q \equiv \Lambda_0^Q + \Lambda_0^Q \odot R \odot \Gamma = \Lambda_0^Q + \Lambda_0^Q \odot R \odot \Gamma_0 + \Lambda_0^Q \odot R \odot \Gamma \odot R \odot \Gamma_0$$
$$= \Lambda_0^Q + \Lambda^Q \odot R \odot \Gamma_0, \quad (4.141)$$

and similarly for the current density-vertex, which are just the Bethe-Salpeter equation for the density-vertices. Through (4.141) we can rewrite (4.139) as

$$i\omega \left(\Lambda_0^Q + \Lambda^Q \odot R \odot \Gamma_0 \right) - \mathbf{q} \cdot \left(\Lambda_0^Q + \Lambda^Q \odot R \odot \Gamma_0 \right) = i\omega \Lambda_0^Q - \mathbf{q} \cdot \Lambda_0^Q$$
$$- \Lambda_0^Q \odot \Sigma + \Sigma \odot \Lambda_0^Q,$$

namely, through (4.141),

$$\left(i\omega \Lambda^Q \odot R - \mathbf{q} \cdot \Lambda^Q \odot R \right) \odot \Gamma_0 = \left(-\Lambda_0^Q \odot G + G \odot \Lambda_0^Q \right) \odot \Gamma_0$$
$$= -\Lambda_0^Q \odot \Sigma + \Sigma \odot \Lambda_0^Q.$$

That equation corresponds analytically to

$$\frac{1}{V} \sum_{\mathbf{k}'} \sum_{\sigma_1' \sigma_2'} T \sum_{\epsilon'} \Lambda_0^Q (\mathbf{k}' \sigma_2', \mathbf{k}' + \mathbf{q} \sigma_1') \left(G(i\epsilon' + i\omega, \mathbf{k}' + \mathbf{q}) - G(i\epsilon', \mathbf{k}') \right)$$
$$\Gamma_0 (i\epsilon' + i\omega \mathbf{k}' + \mathbf{q} \sigma_1', i\epsilon \mathbf{k} \sigma_2; i\epsilon + i\omega \mathbf{k} + \mathbf{q} \sigma_1, i\epsilon' \mathbf{k}' \sigma_2')$$
$$= \Lambda_0^Q (\mathbf{k} \sigma_2, \mathbf{k} + \mathbf{q} \sigma_1) \left(\Sigma(i\epsilon + i\omega, \mathbf{k} + \mathbf{q}) - \Sigma(i\epsilon, \mathbf{k}) \right). \quad (4.142)$$

If we regard

$$\Lambda_0^Q (\mathbf{k}' \sigma_2', \mathbf{k}' + \mathbf{q} \sigma_1') \left(G(i\epsilon' + i\omega, \mathbf{k}' + \mathbf{q}) - G(i\epsilon', \mathbf{k}') \right) \equiv \delta G(i\epsilon' + i\omega \mathbf{k}' + \mathbf{q} \sigma_1', i\epsilon' \mathbf{k}' \sigma_2'),$$

$$\Lambda_0^Q (\mathbf{k} \sigma_2, \mathbf{k} + \mathbf{q} \sigma_1) \left(\Sigma(i\epsilon + i\omega, \mathbf{k} + \mathbf{q}) - \Sigma(i\epsilon, \mathbf{k}) \right) \equiv \delta \Sigma(i\epsilon + i\omega \mathbf{k} + \mathbf{q} \sigma_1, i\epsilon \mathbf{k} \sigma_2),$$

as variations of G and Σ, which, as we mentioned, are generally non diagonal in frequency, momentum and spin, then (4.142)

$$\frac{1}{V} \sum_{\mathbf{k}'} \sum_{\sigma_1' \sigma_2'} T \sum_{\epsilon'} \delta G\left(i\epsilon' + i\omega\,\mathbf{k}' + \mathbf{q}\,\sigma_1', i\epsilon'\,\mathbf{k}'\,\sigma_2'\right)$$

$$\Gamma_0\left(i\epsilon' + i\omega\,\mathbf{k}' + \mathbf{q}\,\sigma_1', i\epsilon\,\mathbf{k}\,\sigma_2; i\epsilon + i\omega\,\mathbf{k} + \mathbf{q}\,\sigma_1, i\epsilon'\,\mathbf{k}'\,\sigma_2'\right)$$

$$= \delta\Sigma\left(i\epsilon + i\omega\,\mathbf{k} + \mathbf{q}\,\sigma_1, i\epsilon\,\mathbf{k}\,\sigma_2\right),$$

is just the definition of Γ_0 as functional derivative of Σ with respect to G. In turn, that result implies that an irreducible vertex obtained as the functional derivative of Σ with respect to G preserves the Ward-Takahashi identities, hence does not spoil conservation laws.

Therefore, if we approximate $\Sigma[G, U]$ with $\Sigma_{approx}[G, U]$, then the irreducible vertex is accordingly

$$\Gamma_{0\,approx} = \frac{\delta\Sigma_{approx}[G, U]}{\delta G}. \tag{4.143}$$

A *conserving* approximation scheme consistent with all conservation laws consists in solving the Bethe-Salpeter for the reducible vertex

$$\Gamma = \Gamma_0 + \Gamma_0 \odot R \odot \Gamma,$$

with $\Gamma_0 = \Gamma_{0\,approx}$ and $R = R_{approx} = G_{approx}\,G_{approx}$, and thus obtain an approximation Γ_{approx} for the interaction vertex. A generic response function must be calculated through

$$\chi_{AB} = \text{Tr}\left(\Lambda_0^A \odot R_{approx} \odot \Lambda_0^B\right) + \text{Tr}\left(\Lambda_0^A \odot R_{approx} \odot \Gamma_{approx} \odot R_{approx} \odot \Lambda_0^B\right). \tag{4.144}$$

In that way, we are guaranteed that the linear response theory is consistent with all conservation laws.

4.10.1 Conserving Hartree-Fock Approximation

As a first example, we shall consider the Hartree-Fock approximation within the diagrammatic technique.

For that, we consider an interacting Hamiltonian

$$H = \sum_{ab} t_{ab}\,c_a^\dagger c_b + \frac{1}{2} \sum_{abcd} U_{acdb}\,c_a^\dagger c_c^\dagger c_d c_b,$$

Fig. 4.43 Self-energy
functional in the
Hartree-Fock approximation

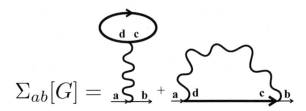

$$\Sigma_{ab}[G] =$$

written in a generic single-particle basis of wavefunctions. Let us approximate the skeleton expansion of the self-energy with the two first-order diagrams, as shown in Fig. 4.43. That amounts to the following the self-energy functional

$$\Sigma_{ab} = \sum_{cd} U_{acdb}\, T \sum_{\epsilon} e^{i\epsilon\, 0^+} G_{dc}(i\epsilon) - \sum_{cd} U_{acbd}\, T \sum_{\epsilon} e^{i\epsilon\, 0^+} G_{dc}(i\epsilon)\,,$$

where

$$T \sum_{\epsilon} e^{i\epsilon\, 0^+} G_{dc}(i\epsilon) = \langle c_c^\dagger c_d \rangle \equiv \Delta_{cd}. \tag{4.145}$$

The Green's function is a matrix and satisfies the Dyson equation

$$\left(\hat{G}(i\epsilon)^{-1}\right)_{ab} = i\epsilon - t_{ab} - \Sigma_{ab} = i\epsilon - t_{ab} - \sum_{cd} \left(U_{acdb} - U_{acbd}\right)\Delta_{cd}\,,$$

which is diagonalised by diagonalising

$$t_{ab} + \sum_{cd} \Delta_{cd}\,(U_{acdb} - U_{acbd}) = \left(\hat{H}_{HF}\right)_{ab}\,,$$

where $\hat{H}_{HF}$ is nothing but the matrix representation of the Hartree-Fock Hamiltonian. Namely, the self-consistency requirement (4.145) is just the Hartree-Fock self-consistency equation.

In order for the Hartree-Fock approximation to be a consistent scheme we need to approximate Γ_0 as

$$\Gamma_0(a, c; d, b) = \frac{\delta \Sigma_{ab}}{\delta G_{dc}} = U_{acdb} - U_{acbd}\,,$$

and use it to solve the Bethe-Salpeter equation for the reducible vertex Γ, see Fig. 4.44. One can readily verify that the correlation functions calculated with the above Γ coincide with the time-dependent Hartree-Fock, which is therefore a consistent approximation.

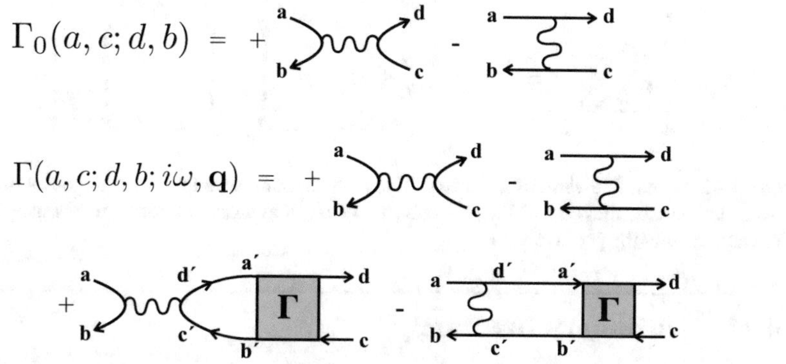

$$\Gamma_0(a, c; d, b) =$$

$$\Gamma(a, c; d, b; i\omega, \mathbf{q}) =$$

Fig. 4.44 Top panel: irreducible vertex Γ_0 in the particle-hole channel within the Hartree-Fock approximation. Bottom panel: Bethe-Salpeter equation for the reducible interaction vertex Γ in the Hartree-Fock approximation. In this approximation the interaction vertex only depends on the frequency $i\omega$ and momentum $\mathbf{q}$ transferred, namely on the frequency and momentum difference between the two internal lines

Fig. 4.45 Top panel: self-energy functional in the GW approximation. Bottom panel: Dyson's equation for the screened interaction $W(\mathbf{q})$

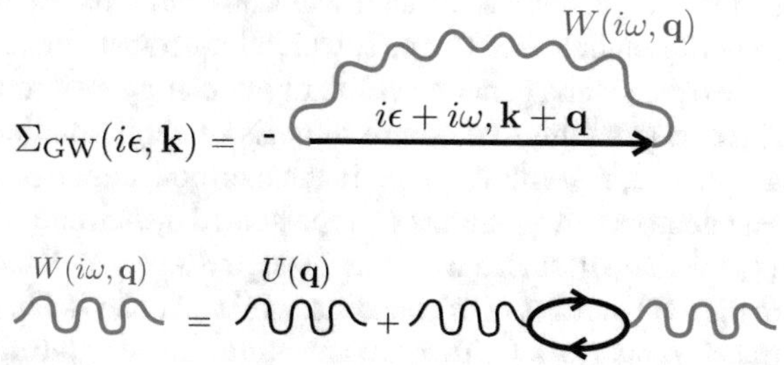

$$\Sigma_{GW}(i\epsilon, \mathbf{k}) = -$$

4.10.2 Conserving GW Approximation

Figure 4.30 is an exact representation of the self-energy in terms of the Green's function, the screened Coulomb interaction $W(i\omega, \mathbf{q})$, and the proper charge density-vertex Λ_*. An approximation that allows solving the self-consistent Dyson equation and yet taking into account the screening of the Coulomb interaction that is missing in the above Hartree-Fock approximation, is to assume consistently $\Lambda_* = \Lambda_0 = 1$. This is the so-called GW approximation, graphically shown in Fig. 4.45.

In the GW approximation one has to solve two coupled equations, since $W(i\omega, \mathbf{q})$ is also functional of G, specifically

$$W(i\omega, \mathbf{q}) = \frac{U(\mathbf{q})}{\epsilon_{\parallel}(i\omega, \mathbf{q})}, \quad \epsilon_{\parallel}(i\omega, \mathbf{q}) = 1 - \frac{U(\mathbf{q})}{V} \sum_{\mathbf{k}\sigma} T \sum_{\epsilon} G(i\epsilon + i\omega, \mathbf{k} + \mathbf{q}) G(i\epsilon, \mathbf{k}).$$

In order to make the GW approximation a conserving scheme, the irreducible vertex Γ_0 in the particle-hole channel must be defined through the functional derivative of Σ_{GW} with respect to G, which we show in Fig. 4.46, and used to solve the Bethe-Salpeter equations and calculate linear response functions.

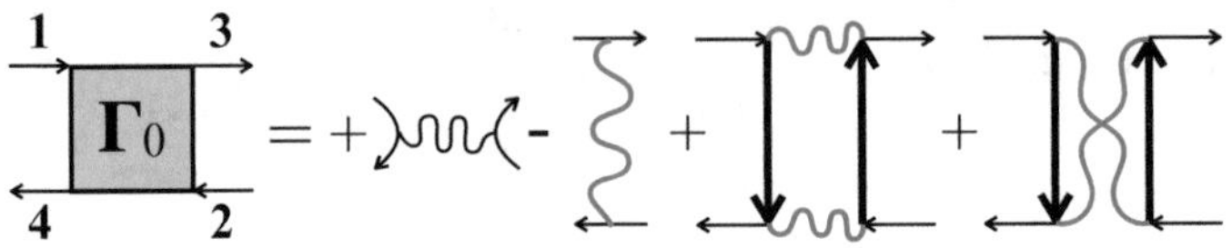

Fig. 4.46 Irreducible vertex Γ_0 in the GW approximation. Note that the first contribution comes from the Hartree diagram in the self-energy, which does exists although it is cancelled by the interaction with the positive ions

4.11 Luttinger's Theorem

We end this chapter presenting the so-called Luttinger theorem [4], which is often invoked in the context of strongly-correlated materials.

Luttinger's theorem refers to any conserved quantity, though it is commonly discussed for the total electron number and assuming translational symmetry. Since the latter symmetry is not necessary, we here present a derivation [5] that does not assume it since we shall later use Luttinger's theorem in the context of Anderson impurity models, see Chap. 7, which lack translational invariance.

We consider a generic system of interacting electrons with annihilation operators c_α corresponding to a complete basis of single-particle wavefunctions labelled by $\alpha = 1, \ldots, K$, with $K \to \infty$ in the thermodynamic limit. The Hamiltonian admits a set of conserved quantities Q, represented by hermitian matrices $\hat{Q}$ with components $Q_{\alpha\beta}$ defined in such a way that the eigenvalues are integers. $Q_{\alpha\beta} = \delta_{\alpha\beta}$ corresponds to the total number N of electrons, while all other independent Q's are represented by traceless matrices $\hat{Q}$. To avoid any issue related to the discontinuity at zero imaginary time of the Green's functions, we use instead of N the deviation $N - K/2$ of the electron number with respect to half-filling, so that we can write the expectation value of any conserved quantity as

$$
\begin{aligned}
Q &= \frac{1}{2} \sum_{\alpha\beta} Q_{\beta\alpha} \left(\langle c_\beta^\dagger c_\alpha \rangle - \langle c_\alpha c_\beta^\dagger \rangle \right) = \frac{1}{2} \left[\mathrm{Tr}\left(\hat{G}(\tau = 0^-) \, \hat{Q} \right) + \mathrm{Tr}\left(\hat{G}(\tau = 0^+) \, \hat{Q} \right) \right] \\
&= \frac{T}{2} \sum_n \left(e^{i\epsilon_n 0^+} + e^{-i\epsilon_n 0^+} \right) \mathrm{Tr}\left(\hat{G}(i\epsilon_n) \, \hat{Q} \right) = T \sum_n \cos\left(\epsilon_n 0^+ \right) \mathrm{Tr}\left(\hat{G}(i\epsilon_n) \, \hat{Q} \right) \\
&= T \sum_n \mathrm{Tr}\left(\hat{G}(i\epsilon_n) \, \hat{Q} \right),
\end{aligned}
\tag{4.146}
$$

where $\hat{G}(i\epsilon_n) = \hat{G}(-i\epsilon_n)^\dagger$ is the Green's function matrix in the Matsubara frequencies $\epsilon_n = (2n + 1)\pi T$, and the last equality is valid since the sum involves only the component of the trace even in ϵ_n, and since that decays at least as $1/\epsilon_n^2$ we can safely take $\cos\left(\epsilon_n 0^+ \right) = 1$.

According to Dyson's equation,

$$
\hat{G}^{-1}(i\epsilon_n) = i\epsilon_n \, \hat{I} - \hat{H}_0 - \hat{\Sigma}(i\epsilon_n),
\tag{4.147}
$$

with $\hat{I}$ the identity matrix, and $\hat{H}_0$ the non-interacting Hamiltonian, including the chemical potential term, represented in the chosen basis. $\hat{\Sigma}(i\epsilon_n) = \hat{\Sigma}(-i\epsilon_n)^\dagger$ is the self-energy matrix that accounts for all interaction effects. We can thus write (4.146) as

$$Q = -T \sum_n \frac{\partial}{\partial i\epsilon_n} \text{Tr}\left(\ln \hat{G}(i\epsilon_n) \, \hat{Q} \right) + I_L(Q), \tag{4.148}$$

where

$$I_L(Q) = T \sum_n \text{Tr}\left(\hat{G}(i\epsilon_n) \frac{\partial \hat{\Sigma}(i\epsilon_n)}{\partial i\epsilon_n} \, \hat{Q} \right), \tag{4.149}$$

is the Luttinger integral for the conserved quantity Q. Hereafter, we use simply I_L for the case $\hat{Q} = \hat{I}$, i.e., for the number of electrons.

We note that at particle-hole symmetry[6] $I_L(Q)$ vanishes identically for all Q's that are odd under a particle-hole transformation, thus also the electron number operator. Seemingly, $I_L(Q) = 0$ in absence of interaction, where $\hat{\Sigma}(i\epsilon_n) = 0$.

Let us instead analyse $I_L(Q)$ in more general circumstances. By definition, if $\Phi[G]$ is the Luttinger-Ward functional, then, see (4.120) and (4.119),

$$\delta\Phi[G] = T \sum_n e^{i\epsilon_n \eta} \text{Tr}\left(\hat{\Sigma}(i\epsilon_n) \, \delta\hat{G}(i\epsilon_n) \right), \tag{4.150}$$

with $\eta > 0$ that must be sent to zero after performing the summation, and where

$$\Phi[G] = T \sum_n e^{i\epsilon_n \eta} \sum_{m \geq 1} \frac{1}{2m} \text{Tr}\left(\hat{G}(i\epsilon_n) \, \hat{\Sigma}^{(m)}(i\epsilon_n) \right) \equiv T \sum_n e^{i\epsilon_n \eta} \, \Phi(i\epsilon_n), \tag{4.151}$$

---

[6] The Hamiltonian has particle-hole symmetry if there is a one-to-one correspondence between the fermionic operators with quantum number, e.g., $\alpha = 1, \ldots, K/2$ and those with quantum number $\alpha + K/2$, assuming even K, such that the Hamiltonian remains invariant under the transformation $P := c_\alpha \leftrightarrow c_{\alpha+K/2}^\dagger$. It follows that, for generic $\alpha, \beta = 1, \ldots, K$ defined modulus K, i.e., $\ell + K \equiv \ell$ for $1 \leq \ell \leq K$,

$$G_{\alpha\beta}(\tau - \tau') = -\langle T\left(c_\alpha(\tau) c_\beta^\dagger(\tau') \right) \rangle \rightarrow -\langle T\left(c_{\alpha+K/2}^\dagger(\tau) c_{\beta+K/2}(\tau') \right) \rangle = -G_{\beta+K/2\,\alpha+K/2}(\tau' - \tau),$$

so that $G_{\alpha\beta}(i\epsilon) = -G_{\beta+K/2\,\alpha+K/2}(-i\epsilon)$. Similarly, $\Sigma_{\alpha\beta}(i\epsilon) = -\Sigma_{\beta+K/2\,\alpha+K/2}(-i\epsilon)$. Therefore, $I_L(Q)$ in (4.149), for instance with $\hat{Q} = \hat{I}$, transforms as

$$I_L = T \sum_n \sum_{\alpha\beta} G_{\alpha\beta}(i\epsilon) \frac{\partial \Sigma_{\beta\alpha}(i\epsilon_n)}{\partial i\epsilon_n} \rightarrow T \sum_n \sum_{\alpha\beta} G_{\beta+K/2\,\alpha+K/2}(-i\epsilon) \frac{\partial \Sigma_{\alpha+K/2\,\beta+K/2}(-i\epsilon_n)}{\partial i\epsilon_n} = -I_L,$$

thus $I_L = 0$. One can readily prove that the same holds for all Q's odd under particle-hole transformation, namely such that $Q_{\alpha\beta} = Q_{\beta-K/2\,\alpha-K/2}$, but not for those even, $Q_{\alpha\beta} = -Q_{\beta-K/2\,\alpha-K/2}$.

where $\hat{\Sigma}^{(m)}(i\epsilon_n)$ is the sum of all mth order skeleton diagrams. Through (4.150) and (4.151) it readily follows that

$$\frac{\delta\Phi[G]}{\delta i\epsilon} \equiv T\sum_n \text{Tr}\left(\hat{\Sigma}(i\epsilon_n)\frac{\partial\hat{G}(i\epsilon_n)}{\partial i\epsilon_n}\right) = T\sum_n \frac{\partial\Phi(i\epsilon_n)}{\partial i\epsilon_n}, \quad (4.152)$$

where we set $\eta = 0$ before performing the sum since the function decays faster than $1/\epsilon_n$ for $\epsilon_n \to \pm\infty$. Equation (4.152) allows us to rewrite $I_L(Q)$ of (4.149) for $\hat{Q} = \hat{I}$ simply as

$$I_L = T\sum_n \frac{\partial I_L(i\epsilon_n)}{\partial i\epsilon_n}, \quad (4.153)$$

where

$$I_L(i\epsilon_n) = \text{Tr}\left(\hat{\Sigma}(i\epsilon_n)\,\hat{G}(i\epsilon_n)\right) - \Phi(i\epsilon_n). \quad (4.154)$$

In other words, it is always possible to represent the Luttinger integral as the summation over Matsubara frequencies of the derivative of a function. It follows that the total number of electrons can be written as[7]

$$N = \frac{K}{2} - T\sum_n \frac{\partial}{\partial i\epsilon_n}\text{Tr}\left(\ln\hat{G}(i\epsilon_n)\right) + T\sum_n \text{Tr}\left(\hat{G}(i\epsilon_n)\frac{\partial\hat{\Sigma}(i\epsilon_n)}{\partial i\epsilon_n}\right)$$

$$= \frac{K}{2} - T\sum_n \frac{\partial}{\partial i\epsilon_n}\text{Tr}\left(\ln\hat{G}(i\epsilon_n)\right) + T\sum_n \frac{\partial I_L(i\epsilon_n)}{\partial i\epsilon_n}$$

$$\xrightarrow[T\to 0]{} \frac{K}{2} - \int_{-\infty}^{\infty}\frac{d\epsilon}{2\pi}\frac{\partial}{\partial i\epsilon}\text{Tr}\left(\ln\hat{G}(i\epsilon)\right) + \int_{-\infty}^{\infty}\frac{d\epsilon}{2\pi}\frac{\partial I_L(i\epsilon)}{\partial i\epsilon}.$$

Since $\hat{G}(-i\epsilon) = \hat{G}(i\epsilon)^\dagger$ and, similarly, $I_L(-i\epsilon) = I_L(i\epsilon)^*$, if we define, through the polar decomposition of $\hat{G}(i\epsilon)$, the matrix

$$\hat{\delta}(\epsilon) \equiv \arg\left(\hat{G}(i\epsilon)\right) = \text{Im}\ln\left(\hat{G}(i\epsilon)\right), \quad (4.155)$$

then, for $T \to 0$,

$$N = \frac{K}{2} - \int_{0^+}^{\infty}\frac{d\epsilon}{\pi}\frac{\partial}{\partial\epsilon}\text{Tr}\left(\hat{\delta}(\epsilon)\right) + \int_{0^+}^{\infty}\frac{d\epsilon}{\pi}\frac{\partial\text{Im}\,I_L(i\epsilon)}{\partial\epsilon}.$$

---

[7] Since $\epsilon_n = (2n+1)\pi T$, it becomes a continuous variable for $T \to 0$ such that

$$T\sum_n F(i\epsilon_n) \xrightarrow[T\to 0]{} \int_{-\infty}^{\infty}\frac{d\epsilon}{2\pi}F(i\epsilon).$$

Since $\hat{G}(i\epsilon) \to \hat{I}/i\epsilon$ for $\epsilon \to \infty$, then

$$\hat{\delta}(\epsilon) \xrightarrow[\epsilon \to \infty]{} -\frac{\pi}{2}\,\hat{I}\,, \qquad\qquad \mathrm{Tr}\left(\hat{\delta}(\epsilon)\right) \xrightarrow[\epsilon \to \infty]{} -\frac{\pi}{2}\,K\,,$$

while, through its definition (4.154), we realise that $\mathrm{Im}\,I_L(i\epsilon) \to 0$ for $\epsilon \to \infty$, so that we finally find that

$$
\begin{aligned}
N &= \frac{K}{2} + \int_{-\infty}^{\infty} \frac{d\epsilon}{2\pi}\,\mathrm{Tr}\left(\hat{G}(i\epsilon)\right) \\
&= \frac{K}{2} - \int_{-\infty}^{\infty} \frac{d\epsilon}{2\pi}\,\frac{\partial}{\partial i\epsilon}\,\mathrm{Tr}\left(\ln\hat{G}(i\epsilon)\right) + \int_{-\infty}^{\infty} \frac{d\epsilon}{2\pi}\,\frac{\partial I_L(i\epsilon)}{\partial i\epsilon} \qquad (4.156) \\
&= K + \frac{1}{\pi}\,\mathrm{Tr}\left(\hat{\delta}(0^+)\right) - \frac{1}{\pi}\,\mathrm{Im}\,I_L(i0^+)\,.
\end{aligned}
$$

This expression is exact. It is still not Luttinger's theorem but a kind of generalisation of it, and it is remarkable as it shows that a quantity requiring integration over all frequencies can be alternatively calculated by just knowing the Green's function at the single point $\epsilon = 0^+$, namely, just a boundary term. In reality, Luttinger's theorem statement is that $\mathrm{Im}\,I_L(i0^+) = 0$ in (4.156), which is not to be expected a priori.

Hereafter, we analyse in detail the proof of that theorem, and highlight under which conditions it is valid, and what are its physical consequences.

4.11.1 Validity Conditions for Luttinger's Theorem

The Luttinger-Ward functional $\Phi[G]$ is invariant if the Matsubara frequency of each internal Green's function is replaced, see (4.147), by $i\epsilon_n\,\hat{I} + i\omega\,\hat{Q}$ for any conserved Q, where $\omega = 2\pi\,T$ is the first bosonic Matsubara frequency.[8] Therefore,

$$0 = \frac{\Delta^Q \Phi[G]}{i\omega} = T\sum_n \mathrm{Tr}\left(\hat{\Sigma}(i\epsilon_n)\,\frac{\Delta^Q \hat{G}(i\epsilon)}{i\omega}\right),$$

---

[8] Suppose we change the interacting Hamiltonian $H = H_0 + H_{int}$ into the non-hermitian one $H - i\omega\,Q = (H_0 - i\omega\,Q) + H_{int}$. It follows that

$$S(\beta, 0) = \int_0^\beta d\tau\,e^{H_0\tau}\,H_{int}\,e^{-H_0\tau} \to \int_0^\beta d\tau\,e^{(H_0 - i\omega_0\,Q)\tau}\,H_{int}\,e^{-(H_0 - i\omega_0\,Q)\tau} = S(\beta, 0)\,,$$

is invariant since both H_0 and H_{int} commute with Q. Therefore also $\langle H_{int}\rangle = \langle S(\beta, 0)\,H_{int}\rangle/\langle S(\beta, 0)\rangle$ is invariant under $H_0 \to H_0 - i\omega_0\,Q$. Through (4.122), we conclude that the Luttinger-Ward functional

$$\Phi[G] = \int_0^1 \frac{d\lambda}{\lambda}\,\langle H_{int}(\lambda)\rangle$$

is also invariant under $H_0 \to H_0 - i\omega_0\,Q$, which however transforms (4.147) into

$$\hat{G}^{-1}(i\epsilon_n) = i\epsilon_n\,\hat{I} - \hat{H}_0 + i\,\omega\,\hat{Q} - \hat{\Sigma}(i\epsilon_n) = \hat{G}^{-1}\left(i\epsilon_n + i\,\omega\,\hat{Q}\right)\,.$$

Therefore, $\Phi[G]$ must be invariant under such Green's function transformation.

with

$$\frac{\Delta^Q \hat{G}(i\epsilon)}{i\omega} \equiv \frac{\hat{G}(i\epsilon_n + i\omega\,\hat{Q}) - \hat{G}(i\epsilon_n)}{i\omega},$$

the finite difference of $\hat{G}(i\epsilon)$. For $\hat{Q} = \hat{I}$, we thus find

$$\begin{aligned}
0 &= T \sum_n \text{Tr}\left(\hat{\Sigma}(i\epsilon_n) \frac{\hat{G}(i\epsilon_n + i\omega) - \hat{G}(i\epsilon_n)}{i\omega} \right) \\
&= -T \sum_n \text{Tr}\left(\hat{G}(i\epsilon_n) \frac{\hat{\Sigma}(i\epsilon_n + i\omega) - \hat{\Sigma}(i\epsilon_n)}{i\omega} \right) \qquad (4.157) \\
&\equiv T \sum_n \frac{I_L(i\epsilon_n + i\omega) - I_L(i\epsilon_n)}{i\omega} \equiv I_L^{\Delta},
\end{aligned}$$

which just means that the convergence of the series allows the change of variable $i\epsilon_n + i\omega \to i\epsilon_n$ that makes I_L^{Δ} trivially vanish. It is tempting to assume that I_L^{Δ}, i.e., the sum over ϵ_n of the finite difference, coincides with I_L in (4.153), i.e., the sum over ϵ_n of the derivative, in the limit $T \to 0$, thus $\omega \to 0$. That is actually what Luttinger assumed, in which case $I_L = 0$ follows, and thus $\text{Im}\,I_L(i0^+) = 0$ in (4.156). However, that apparently reasonable assumption is not at all guaranteed, as we now discuss.

For that, let us analyse in detail the differences between the two series

$$I_L^{\Delta} \equiv T \sum_n \text{Tr}\left(\hat{G}(i\epsilon_n) \frac{\hat{\Sigma}(i\epsilon_n + i\omega) - \hat{\Sigma}(i\epsilon_n - i\omega)}{2i\omega} \right) = 0,$$

$$I_L \equiv T \sum_n \text{Tr}\left(\hat{G}(i\epsilon_n) \frac{\partial \hat{\Sigma}(i\epsilon_n)}{\partial i\epsilon_n} \right),$$

in the limit of zero temperature, thus also $\omega \to 0$, noticing that the invariance of the Luttinger-Ward functional implies $I_L^{\Delta} = 0$. Let us start from the latter. Since $\hat{G}(i\epsilon)$ and $\hat{\Sigma}(i\epsilon)$ have discontinuous antihermitian components at $\epsilon = 0$, I_L^{Δ} must be dealt with care in the $T \to 0$ limit, since the functions that are summed may be on different sides of the imaginary axis. Therefore, we can write

$$\begin{aligned}
I_L^{\Delta} = T &\sum_{n \geq 1 \vee n \leq -2} \text{Tr}\left(\hat{G}(i\epsilon_n) \frac{\hat{\Sigma}(i\epsilon_n + i\omega) - \hat{\Sigma}(i\epsilon_n - i\omega)}{2i\omega} \right) \\
&+ \frac{1}{4\pi i} \text{Tr}\left[\hat{G}(i\pi T) \left(\hat{\Sigma}(3i\pi T) - \hat{\Sigma}(-i\pi T) \right) \right] \\
&+ \frac{1}{4\pi i} \text{Tr}\left[\hat{G}(-i\pi T) \left(\hat{\Sigma}(i\pi T) - \hat{\Sigma}(-3i\pi T) \right) \right],
\end{aligned}$$

so that the two summations $n \geq 1$ and $n \leq -2$ only involve functions on the same side of the imaginary axis. At this stage, it is tempting to straight take the $T \to 0$ limit and conclude that

$$
I_L^{\Delta} \xrightarrow[T \to 0]{} \int_{0^+}^{\infty} \frac{d\epsilon}{2\pi} \operatorname{Tr}\left(\hat{G}(i\epsilon) \, \frac{\partial \hat{\Sigma}(i\epsilon)}{\partial i\epsilon} \right) + \int_{-\infty}^{0^-} \frac{d\epsilon}{2\pi} \operatorname{Tr}\left(\hat{G}(i\epsilon) \, \frac{\partial \hat{\Sigma}(i\epsilon)}{\partial i\epsilon} \right)
$$
$$
+ \frac{1}{4\pi i} \operatorname{Tr}\left[\left(\hat{G}(i0^+) + \hat{G}(i0^+)^{\dagger} \right) \left(\hat{\Sigma}(i0^+) - \hat{\Sigma}(i0^+)^{\dagger} \right) \right],
$$

which is however correct only at leading order in T. At the same leading order, I_L simply reads

$$
I_L \xrightarrow[T \to 0]{} \int_{0^+}^{\infty} \frac{d\epsilon}{2\pi} \operatorname{Tr}\left(\hat{G}(i\epsilon) \, \frac{\partial \hat{\Sigma}(i\epsilon)}{\partial i\epsilon} \right) + \int_{-\infty}^{0^-} \frac{d\epsilon}{2\pi} \operatorname{Tr}\left(\hat{G}(i\epsilon) \, \frac{\partial \hat{\Sigma}(i\epsilon)}{\partial i\epsilon} \right).
$$

Since $I_L^{\Delta} = 0$, it follows that

$$
I_L = -\frac{1}{\pi} \operatorname{Im} I_L(i0^+) \simeq -\frac{1}{4\pi i} \operatorname{Tr}\left[\left(\hat{G}(i0^+) + \hat{G}(i0^+)^{\dagger} \right) \left(\hat{\Sigma}(i0^+) - \hat{\Sigma}(i0^+)^{\dagger} \right) \right],
$$
$$
\text{(4.158)}
$$

which, we stress again, is an equality only at leading order in T. Therefore, if

$$
S(i\epsilon) \equiv \operatorname{Tr}\left[\left(\hat{G}(i\epsilon) + \hat{G}(i\epsilon)^{\dagger} \right) \left(\hat{\Sigma}(i\epsilon) - \hat{\Sigma}(i\epsilon)^{\dagger} \right) \right], \qquad \text{(4.159)}
$$

is finite as $\epsilon \to 0^+$, we can definitely conclude that $I_L \neq 0$, and therefore that Luttinger's theorem is not valid.

On the contrary, one can readily prove that $S(i\epsilon \to 0) = 0$ in the perturbative regime. Indeed, if we define

$$
\sqrt{ \hat{Z}(i\epsilon)^{\dagger -1} \, \hat{Z}(i\epsilon)^{-1} } \equiv \hat{I} - \frac{\hat{\Sigma}(i\epsilon) - \hat{\Sigma}(i\epsilon)^{\dagger}}{2i\epsilon} = \hat{I} - \frac{\hat{\Sigma}(-i\epsilon) - \hat{\Sigma}(-i\epsilon)^{\dagger}}{-2i\epsilon},
$$
$$
\text{(4.160)}
$$

where $\hat{Z}(i\epsilon) = \hat{Z}(-i\epsilon)$ and $\hat{Z}(i\epsilon \to 0)$ is the quasiparticle residue (4.57) at zero energy, we do know that perturbatively $\hat{Z}(0) = \hat{Z}(0)^{\dagger}$ is positive definite, so that

$$
\hat{\Sigma}(i\epsilon) - \hat{\Sigma}(i\epsilon)^{\dagger} \xrightarrow[\epsilon \to 0]{} 2 \left(\hat{I} - \hat{Z}(0)^{-1} \right) i\epsilon,
$$

and thus $S(i\epsilon)$ vanishes as $\epsilon \to 0$. However, $S(i\epsilon \to 0) = 0$, though necessary for $I_L = 0$, is not a sufficient condition. The reason is that the right hand side of (4.158) is just the leading term of an expansion in T. Its vanishing means that each term of the expansion goes to zero as $T \to 0$, which does not guarantee that the whole series vanishes, too. In other words, while we can safely state that, in the regime where perturbation theory is valid, $S(i\epsilon \to 0) = 0$ does imply that $I_L = 0$ hence Luttinger's theorem is valid, we cannot exclude that the theorem is violated when perturbation theory breaks down.

As an example explicitly showing why $S(i\epsilon \to 0) = 0$ is a necessary but not sufficient condition for Luttinger's theorem to hold, let us assume the hypothetical case in which

$$I_L(i\epsilon) = \Lambda \ln\left(1 - \Omega(i\epsilon)\right), \qquad (4.161)$$

where $\Lambda \in \mathbb{R}$ and $\Omega(i\epsilon) = \Omega(-i\epsilon)^*$ is perturbative, i.e., it can be expanded in powers of the interaction and vanishes when the latter does, correctly yielding $I_L = 0$ in the non-interacting case. Accordingly, $\Omega(i\epsilon)$ must be analytic and thus admit a Taylor expansion in powers of ϵ,

$$\Omega(i\epsilon) \simeq \Omega(0) + \Omega'(0)\, i\epsilon + O(\epsilon^2).$$

Expanding $\mathrm{Im}\, I_L(i\pi T)$ of (4.161) at leading order in T, we get

$$I_L = -\frac{1}{\pi}\, \mathrm{Im}\, I_L(i\pi T) \simeq \frac{\Lambda\, \Omega'(0)}{1 - \Omega(0)}\, T \xrightarrow[T \to 0]{} 0.$$

Similarly, all higher order terms vanish, too. Since $\Omega(i\epsilon)$ is perturbative in the interaction, we come to the conclusion that order by order in perturbation theory $I_L \to 0$ for $T \to 0$, which is what happens in reality, as earlier discussed. This result remains valid so long as $\Omega(0) < 1$, which represents the convergence radius of the series. When $\Omega(0) > 1$, perturbation theory breaks down and

$$\mathrm{Im}\, I_L(i\epsilon) = \Lambda\, \mathrm{Im} \ln\left(1 - \Omega(i\epsilon)\right) \xrightarrow[\epsilon \to 0^+]{} -\pi\, \Lambda\, \mathrm{sign}\left(\Omega'(0)\right),$$

jumps to a finite value.

4.11.1.1 Quasiparticle Hamiltonian
Even though $S(i\epsilon \to 0) = 0$, see (4.159), does not guarantee the validity of Luttinger's theorem, let us try to uncover its physical meaning.

By definition, the single-particle density of states at the chemical potential A is

$$A = -\lim_{\epsilon \to 0^+} \frac{1}{2\pi i}\, \mathrm{Tr}\left(\hat{G}(i\epsilon) - \hat{G}(i\epsilon)^\dagger\right),$$

so that, if

$$\hat{A}(i\epsilon) \equiv -\frac{1}{2\pi i}\left(\hat{G}(i\epsilon) - \hat{G}(i\epsilon)^\dagger\right) = \hat{A}(i\epsilon)^\dagger = -\hat{A}(-i\epsilon),$$

then

$$A = \lim_{\epsilon \to 0^+} \text{Tr}\left(\hat{A}(i\epsilon)\right).$$

Through $\hat{A}(i\epsilon)$, we can write

$$\hat{\Sigma}(i\epsilon) - \hat{\Sigma}(i\epsilon)^\dagger = 2i\epsilon + \hat{G}(i\epsilon)^{-1}\left(\hat{G}(i\epsilon) - \hat{G}(i\epsilon)^\dagger\right)\hat{G}(i\epsilon)^{\dagger-1}$$

$$= 2i\epsilon - 2\pi i\,\hat{G}(i\epsilon)^{-1}\,\hat{A}(i\epsilon)\,\hat{G}(i\epsilon)^{\dagger-1},$$

and thus $S(i\epsilon)$ in (4.159) becomes

$$S(i\epsilon) = 2i\epsilon\,\text{Tr}\left(\hat{G}(i\epsilon) + \hat{G}(-i\epsilon)\right) - 2\pi i\,\text{Tr}\left[\left(\hat{G}(i\epsilon)^{-1} + \hat{G}(i\epsilon)^{\dagger-1}\right)\hat{A}(i\epsilon)\right].$$

We define, through (4.160),

$$\hat{G}_{\text{qp}}(i\epsilon)^{-1} \equiv \sqrt{\hat{Z}(i\epsilon)^\dagger}\,\hat{G}(i\epsilon)^{-1}\sqrt{\hat{Z}(i\epsilon)} = i\epsilon\,\hat{I} - \hat{\Xi}(i\epsilon), \tag{4.162}$$

where

$$\hat{\Xi}(i\epsilon) = \hat{\Xi}(i\epsilon)^\dagger = \hat{\Xi}(-i\epsilon) \equiv -\frac{1}{2}\sqrt{\hat{Z}(i\epsilon)^\dagger}\left(\hat{G}(i\epsilon)^{-1} + \hat{G}(i\epsilon)^{\dagger-1}\right)\sqrt{\hat{Z}(i\epsilon)}$$

$$= \frac{1}{2}\sqrt{\hat{Z}(i\epsilon)^\dagger}\left(2\hat{H}_0 + \hat{\Sigma}(i\epsilon) + \hat{\Sigma}(i\epsilon)^\dagger\right)\sqrt{\hat{Z}(i\epsilon)}, \tag{4.163}$$

is a $K \times K$ hermitian matrix, and thus has real eigenvalues $\epsilon_{*\ell}(\epsilon) = \epsilon_{*\ell}(-\epsilon)$, $\ell = 1, \ldots, K$. Therefore, if we further define

$$\hat{A}_{\text{qp}}(i\epsilon) \equiv -\frac{1}{2\pi i}\left(\hat{G}_{\text{qp}}(i\epsilon) - \hat{G}_{\text{qp}}(i\epsilon)^\dagger\right) = \frac{\epsilon}{\pi}\,\hat{G}_{\text{qp}}(i\epsilon)\,\hat{G}_{\text{qp}}(i\epsilon)^\dagger$$

$$= \frac{\epsilon}{\pi}\,\frac{1}{\epsilon^2 + \hat{\Xi}(i\epsilon)^2} = \sqrt{\hat{Z}(i\epsilon)^{-1}}\,\hat{A}(i\epsilon)\sqrt{\hat{Z}(i\epsilon)^{\dagger-1}}, \tag{4.164}$$

which is diagonal in the basis that diagonalises $\hat{\Xi}(i\epsilon)$ with elements

$$A_{\text{qp}\,\ell}(i\epsilon) = \frac{1}{\pi}\,\frac{\epsilon}{\epsilon^2 + \epsilon_{*\ell}(\epsilon)^2},$$

then

$$S(i\epsilon) = 2i\epsilon \operatorname{Tr}\left(\hat{G}(i\epsilon) + \hat{G}(-i\epsilon)\right) + 4\pi i \operatorname{Tr}\left[\hat{\Xi}(i\epsilon)\,\hat{A}_{\mathrm{qp}}(i\epsilon)\right]$$

$$= 2i\epsilon \operatorname{Tr}\left(\hat{G}(i\epsilon) + \hat{G}(-i\epsilon)\right) + 4\pi i \sum_{\ell=1}^{K} \frac{\epsilon_{*\ell}(\epsilon)}{\pi} \frac{\epsilon}{\epsilon^2 + \epsilon_{*\ell}(\epsilon)^2} . \tag{4.165}$$

Since the first term on the right hand side of (4.165) vanishes for $\epsilon \to 0$, the necessary condition for Luttinger's theorem to hold is therefore

$$\lim_{\epsilon \to 0^+} \operatorname{Tr}\left[\hat{\Xi}(i\epsilon)\,\hat{A}_{\mathrm{qp}}(i\epsilon)\right] = \lim_{\epsilon \to 0^+} \sum_{\ell=1}^{K} \frac{\epsilon_{*\ell}(\epsilon)}{\pi} \frac{\epsilon}{\epsilon^2 + \epsilon_{*\ell}(\epsilon)^2} = 0 . \tag{4.166}$$

In the thermodynamic limit, $K \to \infty$, $\epsilon_{*\ell}(\epsilon)$ defines a continuous spectrum where ℓ runs in a d-dimensional space, with d the spatial dimension of the system times the number of internal degrees of freedom. For instance, in the periodic case, ℓ labels the momentum within the Brillouin zone, the band index and the spin. The condition (4.166) depends on the analytic properties of the matrix function $\hat{\Xi}(i\epsilon)$.

> **Analytic assumption**

For instance, if we assume that

$$\boxed{\hat{\Xi}(i\epsilon) \text{ is analytic at } \epsilon = 0, \text{ at least to leading order}} , \tag{4.167}$$

then the necessary condition for Luttinger's theorem to hold is satisfied.

In this case, $\epsilon_{*\ell}(\epsilon \to 0) \simeq \epsilon_{*\ell}(0) + O(\epsilon^2)$, where $\epsilon_{*\ell}(0) \equiv \epsilon_{*\ell}$ are the eigenvalues of

$$\hat{H}_* \equiv \sqrt{\hat{Z}(0)^\dagger}\left(\hat{H}_0 + \hat{\Sigma}(0)\right)\sqrt{\hat{Z}(0)} . \tag{4.168}$$

Accordingly, the quasiparticle Green's function and density of states at the chemical potential are

$$\hat{G}_{\mathrm{qp}}(i\epsilon) \xrightarrow[\epsilon \to 0]{} \frac{1}{i\epsilon\,\hat{I} - \hat{H}_*} ,$$

$$A_{\mathrm{qp}} = \lim_{\epsilon \to 0^+} \operatorname{Tr}\left(\hat{A}_{\mathrm{qp}}(i\epsilon)\right) = \operatorname{Tr}\left(\delta(\hat{H}_*)\right) = \sum_{\ell} \delta\left(\epsilon_{*\ell}\right) , \tag{4.169}$$

and correspond to those of free particles, the 'quasiparticles', described by the 'quasiparticle' Hamiltonian $\hat{H}_*$ with eigenvalues $\epsilon_{*\ell}$. We will see in Sect. 5.1 that (4.169) is the starting point of Landau's Fermi liquid theory.

We remark that also a non-analytic $\hat{\Xi}(i\epsilon)$ may satisfy (4.166), as it happens for interacting electrons in one dimension, which we discuss in Chap. 6. Those systems do not sustain quasiparticles in the sense of (4.169), and yet Luttinger's theorem is valid. Conversely, since $S(i\epsilon \to i0^+) = 0$ is not sufficient for Luttinger's theorem to hold, we must also conclude that 'quasiparticles' may exist even when Luttinger's theorem is violated.

Let us assume the existence of quasiparticles, i.e., of (4.167). Through (4.155), and since

$$\text{Im} \ln \det\left(\sqrt{\hat{Z}(i\epsilon)^\dagger \, \hat{Z}(i\epsilon)}\,\right) = 0,$$

we readily find that

$$\hat{\delta}(\epsilon) \xrightarrow[\epsilon \to 0^+]{} \tan^{-1} \frac{-\epsilon}{-\hat{H}_*},$$

is diagonal with elements $-\pi\,\theta(\epsilon_{*\ell}) = -\pi + \pi\,\theta(-\epsilon_{*\ell})$ in the basis that diagonalises $\hat{H}_*$. It follows that (4.156) becomes

$$N = K + \frac{1}{\pi} \text{Tr}\left(\hat{\delta}(0^+)\right) - \frac{1}{\pi} \text{Im} \, I_L(i0^+)$$

$$= K + \frac{1}{\pi} \sum_{\ell=1}^{K} \left(-\pi + \pi\,\theta(-\epsilon_{*\ell})\right) - \frac{1}{\pi} \text{Im} \, I_L(i0^+) = \sum_{\ell=1}^{K} \theta(-\epsilon_{*\ell}) - \frac{1}{\pi} \text{Im} \, I_L(i0^+),$$

(4.170)

which represents the general statement (4.156) of Luttinger's theorem when quasiparticles exist. We note that the first term of the last equation is integer, as N is integer at $T = 0$, which implies that

$$-\frac{1}{\pi} \text{Im} \, I_L(i0^+) = \mathcal{L} \in \mathbb{Z},$$

in other words that the Luttinger integral (4.149) for $\hat{Q} = \hat{I}$ is quantised in integer values. We note that this condition is compatible with the example (4.161) provided Λ is integer.

Therefore, when perturbation theory is valid, and thus quasiparticles exist, then

$$N = \sum_{\ell=1}^{K} \theta(-\epsilon_{*\ell}),$$

(4.171)

which is the conventional statement of Luttinger's theorem, while, when perturbation theory breaks down and yet quasiparticles still exist, then

$$N = \sum_{\ell=1}^{K} \theta(-\epsilon_{*\ell}) + \mathcal{L}, \quad \mathcal{L} \qquad\qquad \in \mathbb{Z}, \qquad (4.172)$$

namely, Luttinger's counting formula (4.171) misses an integer number of electrons.

4.11.2 Luttinger's Theorem in Presence of Quasiparticles and in Periodic Systems

In a periodic system invariant under spin $SU(2)$ symmetry, we have the possibility to further elaborate on the meaning of 'quasiparticle'. In this case, $\hat{G}(i\epsilon)$ is diagonal in momentum and spin with elements $G(i\epsilon, \mathbf{k})$ independent of spin, and thus $\hat{\Xi}(i\epsilon)$ is diagonal, too, with elements $\epsilon_*(\epsilon, \mathbf{k})$ equal for spin $\sigma = \uparrow$ and $\downarrow$, now defined, see (4.163), as

$$\epsilon_*(\epsilon, \mathbf{k}) = \epsilon_*(\epsilon, \mathbf{k})^* = \Xi(i\epsilon, \mathbf{k}) = \frac{1}{2} \sqrt{Z(i\epsilon, \mathbf{k})^* \, Z(i\epsilon, \mathbf{k})} \left(2\epsilon(\mathbf{k}) + \Sigma(i\epsilon, \mathbf{k}) + \Sigma(i\epsilon, \mathbf{k})^*\right). \tag{4.173}$$

Correspondingly, the quasiparticle, A_{qp}, and physical electron, A, density of states at the chemical potential are, in units of the volume V, see (4.169),

$$A_{\mathrm{qp}} = \frac{1}{V} \sum_{\mathbf{k}\sigma} \delta\big(\epsilon_*(\mathbf{k})\big), \qquad A = \frac{1}{V} \sum_{\mathbf{k}\sigma} Z(0, \mathbf{k}) \, \delta\big(\epsilon_*(\mathbf{k})\big), \qquad (4.174)$$

where $\epsilon_*(\mathbf{k}) = \epsilon_*(\epsilon \to 0, \mathbf{k})$.

We already know that (4.166) and (4.167) imply that, if a manifold $\mathbf{k} = \mathbf{k}_{FL}$ exists such that $\epsilon_*(\epsilon \to 0, \mathbf{k}_{FL}) = 0$, then $\epsilon_*(\epsilon \to 0, \mathbf{k}_{FL}) \simeq \epsilon^2$. We observe that $\epsilon_*(\epsilon \to 0, \mathbf{k}_{FL}) = 0$ occurs

---

Fermi Surface if $\epsilon(\mathbf{k}_F) + \Sigma(0, \mathbf{k}_F) = 0$ while $0 < Z(0, \mathbf{k}_F) < 1$, which defines a conventional Fermi surface $\mathbf{k} = \mathbf{k}_F$ through the roots of $G(0, \mathbf{k})^{-1}$ in momentum space. The Fermi surface contribution to the physical electron DOS (4.174) is finite since $Z(0, \mathbf{k}_F) \neq 0$.

Luttinger Surface if $\epsilon(\mathbf{k}_F) + \Sigma(0, \mathbf{k}_L) \neq 0$ but

$$\lim_{\epsilon \to 0^+} \sqrt{Z(i\epsilon, \mathbf{k}_L)^* \, Z(i\epsilon, \mathbf{k}_L)} = \lim_{\epsilon \to 0^+} \frac{2i\epsilon}{2i\epsilon - \Sigma(i\epsilon, \mathbf{k}_L) + \Sigma(i\epsilon, \mathbf{k}_L)^*}$$

$$\sim \lim_{\epsilon \to 0} \epsilon^2 = 0, \qquad (4.175)$$

which implies $\Sigma(i\epsilon, \mathbf{k}_L) \sim 1/i\epsilon$ and, correspondingly, $G(i\epsilon, \mathbf{k}_L) \to 0$ as $\epsilon \to 0$. Therefore, (4.175) defines the so-called Luttinger surface, i.e., the manifold of roots $\mathbf{k} = \mathbf{k}_L$ of $G(0, \mathbf{k})$ in momentum space, whose existence is due to a singular self-energy and thus signals the breakdown of perturbation theory. Remarkably, even though the Luttinger surface contribution to the quasiparticle DOS, A_{qp} in (4.174), is finite, its contribution to the physical electron DOS vanishes.

Therefore, under the analyticity assumption (4.167), Fermi and Luttinger surfaces are both described by the single equation $\epsilon_*(\epsilon \to 0, \mathbf{k}_{FL}) = 0$. Moreover, if perturbation theory is valid, there are quasiparticles, only Fermi surfaces exist within the Brillouin zone, and, see (4.171),

$$N = \sum_{\mathbf{k}\sigma} \theta\left(-\epsilon_*(\mathbf{k})\right), \tag{4.176}$$

which is the standard perturbative statement that the volume fraction of the quasi-particle Fermi volume, i.e., the manifold of $\mathbf{k} : \epsilon_*(\mathbf{k}) < 0$, with respect to the whole Brillouin zone is equal to the electron filling fraction.

When perturbation theory breaks down without breaking translational and spin $SU(2)$ symmetries, and Luttinger surfaces appear inside the Brillouin zone, the more general formula must be used, i.e.,

$$N = \sum_{\mathbf{k}\sigma} \theta\left(-\epsilon_*(\mathbf{k})\right) + \mathcal{L}, \quad \mathcal{L} \in \mathbb{Z}, \tag{4.177}$$

and thus the volume fraction of the quasiparticle Fermi volume no more accounts for the electron filling fraction.

Problems

4.1 Electron decay rate due to phonons—Consider the Hamiltonian of free electrons coupled to phonons

$$H = \sum_{\mathbf{k}\sigma} \epsilon_\mathbf{k} \, c_{\mathbf{k}\sigma}^\dagger \, c_{\mathbf{k}\sigma} + \sum_{\mathbf{q}\mathbf{k}\sigma} g(\mathbf{q}) \, x(\mathbf{q}) \, c_{\mathbf{k}+\mathbf{q}\sigma}^\dagger \, c_{\mathbf{k}\sigma} + \sum_{\mathbf{q}} \frac{\omega_\mathbf{q}}{2} \left(x^\dagger(\mathbf{q}) \, x(\mathbf{q}) + p^\dagger(\mathbf{q}) \, p(\mathbf{q}) \right),$$

where $x(\mathbf{q}) = x^\dagger(-\mathbf{q})$ and $p(\mathbf{q}) = p^\dagger(-\mathbf{q})$ are the Fourier transforms of the phonon coordinate and its conjugate momentum, satisfying

$$\left[x(\mathbf{q}), \, p^\dagger(\mathbf{q}') \right] = i \, \delta_{\mathbf{q},\mathbf{q}'}.$$

Consider the Fock diagram $\Sigma_F(i\epsilon, \mathbf{k})$ for the self-energy in Fig. 4.47. Find the expression of the electron decay rate due to phonons contributed by this diagram, i.e., $-\text{Im} \, \Sigma_F(\epsilon + i0^+, \mathbf{k})$, at finite temperature.

Fig. 4.47 Fock contribution
to the self-energy due to the
exchange of phonons. The
effective interaction is the
phonon propagator times
$|g(\mathbf{q})|^2$

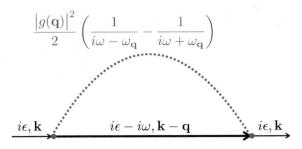

$$\frac{|g(\mathbf{q})|^2}{2}\left(\frac{1}{i\omega-\omega_{\mathbf{q}}}-\frac{1}{i\omega+\omega_{\mathbf{q}}}\right)$$

$i\epsilon,\mathbf{k} \qquad\qquad i\epsilon-i\omega,\mathbf{k}-\mathbf{q} \qquad\qquad i\epsilon,\mathbf{k}$

4.2 Conserving Hartree-Fock approximation—Consider the Hamiltonian

$$H = \sum_{\mathbf{k}\sigma}\left(\epsilon_{\mathbf{k}}-\mu\right)c^{\dagger}_{\mathbf{k}\sigma}c_{\mathbf{k}\sigma} + \frac{U}{2V}\sum_{\mathbf{kpq}}\sum_{\sigma\sigma'}c^{\dagger}_{\mathbf{k}\sigma}c^{\dagger}_{\mathbf{p}+\mathbf{q}\sigma'}c_{\mathbf{p}\sigma'}c_{\mathbf{k}+\mathbf{q}\sigma}, \qquad (4.178)$$

with chemical potential $\mu \neq 0$, and $U > 0$. Approximate the self-energy functional
with the Hartree-Fock skeleton diagrams, see Fig. 4.43, forcing translation and spin
$SU(2)$ symmetries.

- Write, but not solve, the Hartree-Fock self-consistency equation.
- Use the conserving Hartree-Fock approximation to calculate the irreducible interaction vertex in the particle-hole channel and, through the Bethe-Salpeter equation, the reducible interaction vertices in the $S = 0$ and $S = 1$ particle-hole channels. Show that the latter are only function of the transferred frequency $i\omega$ and
  momentum $\mathbf{q}$, thus $\Gamma_{S=0,1} = \Gamma_{S=0,1}(i\omega, \mathbf{q})$.
- After the analytic continuation $i\omega \to \omega + i0^+$, search for possible poles on the
  real frequency axis of the reducible interaction vertices $\Gamma_{S=0,1}(\omega, \mathbf{q})$ in the limit
  $q \ll k_F$, where, e.g.,

$$\epsilon_{\mathbf{k}+\mathbf{q}} - \epsilon_{\mathbf{k}} \simeq \frac{\partial\epsilon_{\mathbf{k}}}{\partial\mathbf{k}}\cdot\mathbf{q} \equiv v_{\mathbf{k}}\cdot\mathbf{q}, \quad f(\epsilon_{\mathbf{k}}) - f(\epsilon_{\mathbf{k}+\mathbf{q}}) \simeq -\frac{\partial f(\epsilon_{\mathbf{k}})}{\partial\epsilon_{\mathbf{k}}}\,v_{\mathbf{k}}\cdot\mathbf{q}.$$

  You should find that only in the spin-singlet channel there is a pole on the real
  frequency axis, which corresponds to a collective particle-hole excitation whose
  spectrum is that of an acoustic mode, so called Landau's zero sound.
- Using the reducible interaction vertices $\Gamma_{S=0,1}(i\omega, \mathbf{q})$ calculate the charge compressibility and spin susceptibility through the corresponding linear response
  function at $\omega = 0$ in the limit $q \to 0$. You will notice that the spin susceptibility
  diverges at a finite value of U_* and change sign. Discuss its meaning.

References

1. A. Abrikosov, L. Gorkov, I. Dzyaloshinskii, *Methods of Quantum Field Theory in Statistical Physics* (Dover, New York, 1975). See Sect. 19.4
2. P. Noziéres, *Theory of Interacting Fermi Systems* (CRC Press, Boca Raton, 1998)
3. J.M. Luttinger, J.C. Ward, Phys. Rev. **118**, 1417 (1960). https://doi.org/10.1103/PhysRev.118.1417
4. J.M. Luttinger, Phys. Rev. **119**, 1153 (1960). https://doi.org/10.1103/PhysRev.119.1153
5. J. Skolimowski, M. Fabrizio, Phys. Rev. B **106**, 045109 (2022). https://doi.org/10.1103/PhysRevB.106.045109

Landau's Fermi Liquid Theory

5

Landau's Fermi liquid theory is by now a paradigm of interacting fermions, even strongly interacting ones as ^{3}He or heavy fermions. For instance, it explains why the Drude-Sommerfeld theory, which essentially deals with non-interacting electrons that scatter off randomly distributed impurities, and the semiclassical Boltzmann equations for Bloch wave-packets describe so well the thermodynamic and transport properties of metals.

Originally, Landau [1,2] derived his celebrated theory of Fermi liquids by making the audacious assumption that the many-body excited states of non-interacting electrons evolve smoothly into the excited states of interacting electrons when the interaction is slowly turned on, and that despite the absence of any energy scale that could justify an adiabatic switching.

The non-interacting excited states $\mid \{\delta n_{\mathbf{k}\sigma}\}\rangle_0$ are uniquely identified by the variations $\delta n_{\mathbf{k}\sigma}$ of the occupation number of Bloch states with momentum $\mathbf{k}$ and spin σ with respect to the equilibrium values $f(\epsilon_{\mathbf{k}})$, the Fermi distribution function, and have excitation energy

$$\delta E_0\big[\{\delta n_{\mathbf{k}\sigma}\}\big] = \sum_{\mathbf{k}\sigma} \epsilon_{\mathbf{k}}\, \delta n_{\mathbf{k}\sigma}\,.$$

Landau's hypothesis implies that each $\mid \{\delta n_{\mathbf{k}\sigma}\}\rangle_0$ is in one-to-one correspondence with an interacting excited state, which allows labelling the latter as its non-interacting counterpart, thus $\mid \{\delta n_{\mathbf{k}\sigma}\}\rangle$. However, $\delta n_{\mathbf{k}\sigma}$ cannot refer anymore to the occupation numbers of the physical electrons, but to different entities that Landau dubbed *quasiparticles*.

The excitation energy $\delta E\big[\{\delta n_{\mathbf{k}\sigma}\}\big]$ of the interacting excited state $\mid \{\delta n_{\mathbf{k}\sigma}\}\rangle$ is evidently different from the non-interacting one. However, for small deviations from equilibrium, i.e. $\delta n_{\mathbf{k}\sigma} \ll 1$, one can expand $\delta E\big[\{\delta n_{\mathbf{k}\sigma}\}\big]$ in Taylor series,

© The Author(s), under exclusive license to Springer Nature Switzerland AG 2022
M. Fabrizio, *A Course in Quantum Many-Body Theory*, Graduate Texts in Physics,
https://doi.org/10.1007/978-3-031-16305-0_5

$$\delta E\big[\{\delta n_{\mathbf{k}\sigma}\}\big] \simeq \sum_{\mathbf{k}\sigma} \epsilon_*(\mathbf{k})\,\delta n_{\mathbf{k}\sigma} + \frac{1}{2} \sum_{\mathbf{k}\mathbf{k}'} \sum_{\sigma\sigma'} f_{\mathbf{k}\,\sigma,\mathbf{k}'\,\sigma'}\,\delta n_{\mathbf{k}\sigma}\,\delta n_{\mathbf{k}'\sigma'} + \cdots , \qquad (5.1)$$

and stop the expansion at second order. The reason why one has to keep the second order term is because it has essentially the magnitude of the first order one, since, for small deviations from equilibrium, $\mathbf{k}$ is close to the Fermi surface and thus $\epsilon_*(\mathbf{k})$ is as small as $\delta n_{\mathbf{k}\sigma}$. Starting from the energy functional (5.1), Landau was able to derive many properties, specifically the low temperature linear response functions in the long wavelength and small frequency limits, and from those the thermodynamic susceptibilities, the transport properties, and the collective modes, which compare extremely well with those observed in experiments. We emphasise that, since adiabatic evolution may only occur in same-symmetry subspaces—different symmetry states are allowed to cross in energy—Landau's Fermi liquid theory gives only access to linear response functions of density operators associated with conserved quantities. For instance, the non-interacting state $|\,\{\delta n_{\mathbf{k}\sigma}\}\rangle_0$ is characterised by a deviation of the total physical electron number

$$\delta N = \sum_{\mathbf{k}\sigma} \delta n_{\mathbf{k}\sigma} ,$$

and, by definition, that must be the same as the deviation of the total quasiparticle number in the interacting state $|\,\{\delta n_{\mathbf{k}\sigma}\}\rangle$. In other words, the quasiparticle carries the same conserved quantum numbers of the physical electron, i.e., the same charge and spin.

In reality, Landau's adiabatic hypothesis is nothing but the assumption that perturbation theory is valid. Indeed, right after the development of diagrammatic many-body perturbation theory, Landau's Fermi liquid theory was given a microscopic justification [3–7] just assuming the perturbative result (4.64). However, the discovery of strongly correlated metals which, at the verge of a Mott transition, display anomalous properties, partly at odds with conventional Fermi liquids, concurrently with striking phenomena, like high-temperature superconductivity in doped Mott insulator copper-oxides, elicited a closer inspection of Landau's Fermi liquid theory. That resulted mostly in searching for new states of matter alternative to Fermi liquids, see, e.g., [8–14], though that is not an exhaustive list, but, partly, also in reexamining [15, 16] the basic assumption that, for Landau's Fermi liquid theory to hold true, perturbation theory must be valid, which is evidently not the case in those materials. Indeed, we already observed, when discussing Luttinger's theorem in Sect. 4.11, that quasiparticles exist also at Luttinger's surfaces, see Sect. 4.11.2, which appear only when perturbation theory has broken down. That already gives a strong indication that Landau's Fermi liquid theory remains valid beyond the perturbative regime.

In this chapter, we derive microscopically Landau's Fermi liquid theory through the diagrammatic technique under a more general assumption that includes the validity of perturbation theory as a special case. The derivation closely follows the conventional one [3,4] with some new developments [15,16]. For simplicity, throughout this chapter, we deal with interacting electrons in three dimensions, although the final results can be straightforwardly extended in two dimensions. Moreover, we assume that none of the symmetries of the Hamiltonian is broken, namely, translation, charge $U(1)$, and spin $SU(2)$ symmetries, so that the interacting Green's function $G(i\epsilon, \mathbf{k})$ is diagonal in momentum and spin, and independent of the latter.

5.1 Emergence of Quasiparticles Reexamined

In Sect. 4.3.4, we have demonstrated, see (4.64), that order by order in perturbation theory

$$\operatorname{Im} \Sigma(\epsilon, \mathbf{k}) \xrightarrow[\epsilon \to 00]{} -\gamma(\mathbf{k})\,\epsilon^2 \,, \tag{5.2}$$

where $\Sigma(\epsilon, \mathbf{k}) \equiv \Sigma(\epsilon + i0^+, \mathbf{k})$ is the retarded self-energy, i.e., the causal analytic continuation of $\Sigma(i\epsilon, \mathbf{k})$ in the complex frequency plane and calculated on the real frequency axis.

Moreover, in Sect. 4.3.5, we have shown that, under the assumption that the perturbative result (5.2) holds true for the whole perturbation series, one can prove the existence of 'quasiparticles', namely of coherent fermionic excitations with dispersion $\epsilon_*(\mathbf{k}) \neq \epsilon_{\mathbf{k}}$ in momentum space and whose decay rate $\Gamma(\epsilon, \mathbf{k})$ vanishes at the chemical potential and on the Fermi surface $\mathbf{k} = \mathbf{k}_F$. Specifically, the quasiparticle dispersion and decay rate are defined by the equations

$$\epsilon_*(\mathbf{k}) = Z\big(\epsilon_*(\mathbf{k}), \mathbf{k}\big)\left(\epsilon_{\mathbf{k}} + \operatorname{Re} \Sigma\big(\epsilon_*(\mathbf{k}), \mathbf{k}\big)\right), \qquad \mathbf{k}_F : \epsilon_*(\mathbf{k}_F) = 0\,,$$
$$\Gamma(\epsilon, \mathbf{k}) = -Z(\epsilon, \mathbf{k}) \operatorname{Im} \Sigma(\epsilon, \mathbf{k}) \simeq Z(\epsilon, \mathbf{k})\,\gamma(\mathbf{k})\,\epsilon^2\,, \tag{5.3}$$

where, see (4.57),

$$Z(\epsilon, \mathbf{k}) = \left(1 - \frac{\partial \operatorname{Re} \Sigma(\epsilon, \mathbf{k})}{\partial \epsilon}\right)^{-1}, \tag{5.4}$$

is the quasiparticle residue. Specifically, $Z(0, \mathbf{k}_F) < 1$ is the weight of a quasiparticle in the physical electron at the Fermi surface and at the chemical potential.

However, the simple requirement that a coherent 'quasiparticle' does exist, namely that its lifetime grows indefinitely as $\epsilon \to 0$, just implies

$$\Gamma(\epsilon, \mathbf{k}) = -Z(\epsilon, \mathbf{k}) \operatorname{Im} \Sigma(\epsilon, \mathbf{k}) \xrightarrow[\epsilon \to 0]{} \gamma_*(\mathbf{k})\,\epsilon^2 \to 0\,, \tag{5.5}$$

which also occurs if, e.g., $\operatorname{Im} \Sigma(0, \mathbf{k})$ is finite but $Z(\epsilon, \mathbf{k}) \sim \epsilon^2$, completely at odds with the perturbative result (4.64). In other words, the condition (5.5) is actually more general than (5.2), which is just a special case. We emphasise that the continuously vanishing $\Gamma(\epsilon, \mathbf{k})$ implies a finite and smooth single-particle density of states (DOS) around $\epsilon = 0$, see (4.67), as long as $\gamma_*(\mathbf{k}) > 0$, while a gapped DOS if $\gamma_*(\mathbf{k}) = 0$, signals of an interaction-driven non-symmetry breaking Mott insulator, which may occur only at half-filling in the present single-band model. To maintain full generality, in what follows we keep using $\gamma_*(\mathbf{k})$ without specifying whether it is finite or zero, with the prescription that, if $\gamma_*(\mathbf{k}) = 0$, then $\gamma_*(\mathbf{k})\,\epsilon^2$ in (5.5) is replaced by an infinitesimal constant $\eta > 0$.

In principle, one could even imagine that $\Gamma(\epsilon, \mathbf{k})$ vanishes not analytically as $\epsilon \to 0$. However, recalling that $\mathrm{Re}\,\Sigma(\epsilon, \mathbf{k})$ and $\mathrm{Im}\,\Sigma(\epsilon, \mathbf{k})$ are related to each other by the Kramers-Krönig equations, that would most likely correspond to a perturbation theory ill-defined already at weak coupling,[1] as it happens in one dimension, see Chap. 6.

> **Analytic assumption**

Hereafter, we exclude that possibility, although it might be realised even beyond one dimension, and assume that

$\boxed{\Sigma(\epsilon, \mathbf{k})\ \text{is everywhere analytic at small}\ \epsilon\ \text{but, eventually, right at}\ \epsilon = 0}$.   (5.6)

As a matter of fact, the above statement is the real frequency counterpart of the statement (4.167) that we already showed in Sect. 4.11.2 to yield well-defined quasiparticles. Indeed, what follows is exactly the same we discussed in Sect. 4.11.2 but now seen on the real axis.

We thus assume the analytic conjecture (5.6) and the analytic behaviour (5.5), at least to leading order in ϵ, which does include the weak coupling perturbative regime as a special case, and try to draw all its consequences, which coincide with the results of Sect. 4.11.2. In terms of $\Gamma(\epsilon, \mathbf{k})$ and $Z(\epsilon, \mathbf{k})$, we can write the single-particle DOS as, cf. (4.67),

$$
\begin{aligned}
A(\epsilon, \mathbf{k}) &= \frac{1}{\pi}\,\frac{-\mathrm{Im}\,\Sigma(\epsilon, \mathbf{k})}{\left(\epsilon - \epsilon_{\mathbf{k}} - \mathrm{Re}\,\Sigma(\epsilon, \mathbf{k})\right)^2 + \left(\mathrm{Im}\,\Sigma(\epsilon, \mathbf{k})\right)^2} \\
&= \frac{Z(\epsilon, \mathbf{k})}{\pi}\,\frac{\Gamma(\epsilon, \mathbf{k})}{\Xi(\epsilon, \mathbf{k})^2 + \Gamma(\epsilon, \mathbf{k})^2} \equiv Z(\epsilon, \mathbf{k})\,A_{\mathrm{qp}}(\epsilon, \mathbf{k}),
\end{aligned}
$$ (5.7)

where $A_{\mathrm{qp}}(\epsilon, \mathbf{k})$, see also (4.164), is the 'quasiparticle' DOS as opposed to the physical electron one, $A(\epsilon, \mathbf{k})$, and, see also (4.173),

$$
\Xi(\epsilon, \mathbf{k}) = \Xi(i\epsilon \to \epsilon + i0^+, \mathbf{k}) = Z(\epsilon, \mathbf{k})\left(\epsilon_{\mathbf{k}} + \mathrm{Re}\,\Sigma(\epsilon, \mathbf{k}) - \epsilon\right).
$$ (5.8)

---

[1] In two dimensions, see (4.64), $\mathrm{Im}\,\Sigma(\epsilon \to 0, \mathbf{k}) \sim \epsilon^2\ \ln|\epsilon|$, which, despite being weakly not analytic, does not yield singularities in perturbation theory.

Let us at first concentrate on $A_{qp}(\epsilon, \mathbf{k})$. We note that, because of (5.5),

$$A_{qp}(\epsilon, \mathbf{k}) = \frac{1}{\pi} \frac{\Gamma(\epsilon, \mathbf{k})}{\Xi(\epsilon, \mathbf{k})^2 + \Gamma(\epsilon, \mathbf{k})^2} \xrightarrow[\epsilon \to 0]{} \frac{1}{\pi} \frac{\gamma_*(\mathbf{k})\, \epsilon^2}{\Xi(\epsilon \to 0, \mathbf{k})^2 + \gamma_*(\mathbf{k})^2\, \epsilon^4}\,,$$
(5.9)

typically vanishes unless $\Xi(\epsilon \to 0, \mathbf{k}) = 0$, in which case $A_{qp}(\epsilon \to 0, \mathbf{k})$ diverges. Let us, therefore, define the 'quasiparticle' energy dispersion through

$$\epsilon_*(\mathbf{k}): \quad \Xi\big(\epsilon_*(\mathbf{k}), \mathbf{k}\big) = Z\big(\epsilon_*(\mathbf{k}), \mathbf{k}\big) \Big(\epsilon_{\mathbf{k}} + \mathrm{Re}\,\Sigma\big(\epsilon_*(\mathbf{k}), \mathbf{k}\big) - \epsilon_*(\mathbf{k})\Big) = 0\,.$$
(5.10)

The equation $\epsilon_*(\mathbf{k}) = 0$, or, equivalently, $\Xi(0, \mathbf{k}) = 0$, defines surfaces $\mathbf{k} = \mathbf{k}_{FL}$ within the Brillouin zone, exactly the Fermi, $\mathbf{k} = \mathbf{k}_F$, and Luttinger, $\mathbf{k} = \mathbf{k}_L$, surfaces of Sect. 4.11.2. Moreover, at $\mathbf{k} = \mathbf{k}_{FL}$, (5.8) as well as our analytical conjecture (5.6) suggest that $\Xi(\epsilon, \mathbf{k}_{FL})$ vanishes linearly in ϵ. Indeed, we can envisage two different analytic scenarios:

Fermi Surface: $\epsilon_{\mathbf{k}_{FL}} + \mathrm{Re}\,\Sigma(0, \mathbf{k}_{FL}) = 0$, so that

$$\mathrm{Re}\, G\big(\epsilon \to 0, \mathbf{k}_{FL}\big)^{-1} = Z(0, \mathbf{k}_{FL})^{-1}\, \epsilon, \quad \Xi(\epsilon \to 0, \mathbf{k}_{FL}) = -\epsilon.$$

This is the case of conventional Fermi liquids, hereafter denoted as (F), where $\mathbf{k}_{FL} \equiv \mathbf{k}_F$ belongs to the Fermi surface. At the Fermi surface, the Green's function has a simple pole at $\epsilon = 0$ with residue $Z(0, \mathbf{k}_F)$, or, equivalently, $G(0, \mathbf{k}_F)^{-1} = 0$.

Luttinger Surface: The quasiparticle residue $Z(\epsilon \to 0, \mathbf{k}_{FL})$ vanishes faster than $\mathrm{Re}\,\Sigma(\epsilon \to 0, \mathbf{k}_{FL})$ diverges, which implies

$$\mathrm{Re}\,\Sigma\big(\epsilon \to 0, \mathbf{k}_{FL}\big) \simeq \frac{\Delta(\mathbf{k}_{FL})^2}{\epsilon}, \quad Z\big(\epsilon \to 0, \mathbf{k}_{FL}\big) \simeq \frac{\epsilon^2}{\epsilon^2 + \Delta(\mathbf{k}_{FL})^2}\,,$$

so that

$$\lim_{\epsilon \to 0} \Xi(\epsilon, \mathbf{k}_{FL}) = \epsilon\,.$$

In this case, hereafter denoted as (L), $\mathbf{k}_{FL} = \mathbf{k}_L$ defines a Luttinger surface [17], i.e., the location within the Brillouin zone of the zeros of the Green's function at $\epsilon = 0$, $G(0, \mathbf{k}_L) = 0$.

By our main assumption (5.6), $\Xi(\epsilon, \mathbf{k})$ has a regular Taylor expansion at small ϵ and $\mathbf{k} \simeq \mathbf{k}_{FL}$, at least to leading order, in which case

$$\Xi\big(\epsilon \to 0, \mathbf{k} \simeq \mathbf{k}_{FL}\big) \simeq \Xi(0, \mathbf{k}) \mp \epsilon \equiv \mp\big(\epsilon - \epsilon_*(\mathbf{k})\big)\,,$$
(5.11)

where the minus and plus signs refer, respectively, to the cases (F) and (L) above. It follows that, for small ϵ and close to the Fermi or Luttinger surfaces, the quasiparticle DOS (5.9) as function of a small ϵ becomes

$$A_{qp}(\epsilon, \mathbf{k}) \simeq \frac{1}{\pi} \frac{\gamma_*(\mathbf{k})\epsilon^2}{\left(\epsilon - \epsilon_*(\mathbf{k})\right)^2 + \gamma_*(\mathbf{k})^2\epsilon^4} \simeq \delta\left(\epsilon - \epsilon_*(\mathbf{k})\right), \qquad (5.12)$$

both close to a Luttinger surface as well as a Fermi surface. The precise meaning of the last equality in (5.12) is that as $\epsilon \to 0$ the quasiparticle DOS always vanishes unless $\epsilon = \epsilon_*(\mathbf{k})$ and $\mathbf{k}$ moves towards the Fermi or Luttinger surfaces. In the sense of a distribution, the above statement corresponds to the following equivalence at small temperature T,

$$\int d\epsilon \left(-\frac{\partial f(\epsilon)}{\partial \epsilon}\right) A_{qp}(\epsilon, \mathbf{k}) F(\epsilon) \equiv -\frac{\partial f\left(\epsilon_*(\mathbf{k})\right)}{\partial \epsilon_*(\mathbf{k})} F\left(\epsilon_*(\mathbf{k})\right). \qquad (5.13)$$

However, the two cases (F) and (L) differ substantially from the point of view of the the physical electron DOS (5.7), which for $\mathbf{k}$ close to $\mathbf{k}_{FL}$ can be written as, cf. (4.71),

$$A(\epsilon, \mathbf{k}) \simeq \frac{Z(\epsilon, \mathbf{k})}{\pi} \frac{\gamma_*(\mathbf{k})\epsilon^2}{\left(\epsilon - \epsilon_*(\mathbf{k})\right)^2 + \gamma_*(\mathbf{k})^2\epsilon^4} + A_{inc}(\epsilon, \mathbf{k}), \qquad (5.14)$$

where $A_{inc}(\epsilon, \mathbf{k})$ is the finite energy incoherent background that carries the rest of spectral weight. Indeed, in case (F), the quasiparticle residue $Z(0, \mathbf{k}_F)$ is finite and thus the physical electron DOS is also characterised by a peak at the quasiparticle energy $\epsilon_*(\mathbf{k})$, which becomes a δ-function with weight $Z(0, \mathbf{k}_F)$ for $\mathbf{k}$ moving towards the Fermi surface. On the contrary, in case (L)

$$Z(\epsilon, \mathbf{k}_L) \simeq \frac{\epsilon^2}{\Delta(\mathbf{k}_L)^2 + \epsilon^2},$$

vanishes quadratically as $\epsilon \to 0$, and so does $A(\epsilon, \mathbf{k})$. In other words, the physical electron DOS vanishes at the Luttinger surface, and yet coherent quasiparticles exist. The fact that $Z(\epsilon, \mathbf{k}) \to 0$ just means that the quasiparticles, though they do exist, have zero weight in the physical electron excitation.

We remark that, if the system is a symmetry-invariant Mott insulator, then only a Luttinger surface can exist. The physical electron DOS $A(\epsilon, \mathbf{k}) = 0$ for all ϵ within the insulating gap, and yet $A_{qp}(\epsilon, \mathbf{k})$ is δ-peaked at the quasiparticle energy $\epsilon_*(\mathbf{k})$, a rather striking physical scenario. In that case, which must correspond to half-filled density $N = V$, with V the number of sites, the generalised Luttinger's theorem (4.177) determines unequivocally the value of $\mathcal{L}$, and suggests that a Luttinger surface can only accommodate $N = V$ electrons, whatever is the enclosed volume. As a consequence, the Fermi pockets that must appear upon doping the Mott insulator only count for the doping away from half-filling. In other words, if v_{EP} and v_{HP}, respectively, are the volume fractions of electron and hole Fermi pockets with respect to the Brillouin zone volume, then the electron filling fraction $\nu = N/2V$ in the absence of Luttinger surfaces, i.e., when perturbation theory is valid and thus Luttinger's theorem holds, reads

$$\nu = v_{EP} + \left(1 - v_{HP}\right) = 1 + \left(v_{EP} - v_{HP}\right), \tag{5.15}$$

while, when a Luttinger surface is present, thus perturbation theory breaks down and Luttinger's theorem is violated,

$$\nu = \frac{1}{2} + \left(v_{EP} - v_{HP}\right). \tag{5.16}$$

5.2 Manipulating the Bethe-Salpeter Equation

We know that the Bethe-Salpeter equation is important to calculate the interaction vertex and, from that, all linear response functions. In the particle-hole channel, that equation reads explicitly, see (4.96),

$$\Gamma_{\sigma_1,\sigma_2;\sigma_3,\sigma_4}\left(i\epsilon_1 + i\omega\, \mathbf{k}_1 + \mathbf{q}, i\epsilon_2\, \mathbf{k}_2; i\epsilon_2 + i\omega\, \mathbf{k}_2 + \mathbf{q}, i\epsilon_1\, \mathbf{k}_1\right)$$
$$= \Gamma_{0\,\sigma_1,\sigma_2;\sigma_3,\sigma_4}\left(i\epsilon_1 + i\omega\, \mathbf{k}_1 + \mathbf{q}, i\epsilon_2\, \mathbf{k}_2; i\epsilon_2 + i\omega\, \mathbf{k}_2 + \mathbf{q}, i\epsilon_1\, \mathbf{k}_1\right)$$
$$+ \frac{1}{V} \sum_{\mathbf{k}_1\,\sigma\sigma'} T \sum_{i\epsilon_1} \Gamma_{0\,\sigma_1,\sigma';\sigma,\sigma_4}\left(i\epsilon_1 + i\omega\, \mathbf{k}_1 + \mathbf{q}, i\epsilon\, \mathbf{k}; i\epsilon + i\omega\, \mathbf{k} + \mathbf{q}, i\epsilon_1\, \mathbf{k}_1\right)$$
$$G(i\epsilon + i\omega, \mathbf{k} + \mathbf{q})\, G(i\epsilon, \mathbf{k})$$
$$\Gamma_{\sigma,\sigma_2;\sigma_3,\sigma'}\left(i\epsilon + i\omega\, \mathbf{k} + \mathbf{q}, i\epsilon_2\, \mathbf{k}_2; i\epsilon_2 + i\omega\, \mathbf{k}_2 + \mathbf{q}, i\epsilon\, \mathbf{k}\right),$$

which, as in Sect. 4.7, we shortly write as

$$\Gamma = \Gamma_0 + \Gamma_0 \odot R \odot \Gamma, \tag{5.17}$$

with $\odot$ implying the sum of all internal variables, i.e., frequencies, spins and momenta, and the kernel

$$R(i\epsilon + i\omega\, \mathbf{k} + \mathbf{q}, i\epsilon\, \mathbf{k}) \equiv G(i\epsilon + i\omega, \mathbf{k} + \mathbf{q})\, G(i\epsilon, \mathbf{k}). \tag{5.18}$$

We recall that in each interaction vertex, the z-component of the spin is conserved, which implies that $\sigma_1 - \sigma_4 = \sigma_3 - \sigma_2 = \sigma - \sigma'$.

5.2.1   A Lengthy but Necessary Preliminary Calculation

Since our goal is to calculate low temperature linear response functions at long wavelength and small frequency, we hereafter analyse the behaviour of Γ at a very small frequency, ω, and momentum, $q = |\mathbf{q}|$, transferred, as well as at very low temperature T.

Let us start by studying, as representative of the Bethe-Salpeter equation, the Matsubara frequency summation

$$
\begin{aligned}
C_{\mathbf{k}}(i\omega, \mathbf{q}) &\equiv T \sum_{\epsilon} R(i\epsilon + i\omega\,\mathbf{k} + \mathbf{q}, i\epsilon\,\mathbf{k})\, F(i\epsilon) \\
&= T \sum_{\epsilon} G(i\epsilon + i\omega, \mathbf{k} + \mathbf{q})\, G(i\epsilon, \mathbf{k})\, F(i\epsilon)\,,
\end{aligned}
\tag{5.19}
$$

with the test function $F(i\epsilon) = F(-i\epsilon)^*$ smooth around $\epsilon = 0$ and not singular for $\epsilon \to \pm\infty$. Our scope here is to understand whether the kernel R is analytic at $\omega = q = 0$ so that the value of $C_{\mathbf{k}}(i\omega, \mathbf{q})$ for $q \to 0$ and $\omega \to 0$ is independent of the order of the limits.

We recall that in a system with gapless single-particle excitations, $G(i\epsilon, \mathbf{k})$ has generally a discontinuous imaginary part crossing $\epsilon = 0$, see Sect. 4.1.3, but otherwise is a non-singular function. Taking that property into account, we write, assuming the bosonic Matsubara frequency $\omega > 0$,

$$
\begin{aligned}
C_{\mathbf{k}}(i\omega, \mathbf{q}) = \;& T \sum_{\epsilon > 0} G(i\epsilon + i\omega, \mathbf{k} + \mathbf{q})\, G(i\epsilon, \mathbf{k})\, F(i\epsilon) \\
& + T \sum_{\epsilon < -\omega} G(i\epsilon + i\omega, \mathbf{k} + \mathbf{q})\, G(i\epsilon, \mathbf{k})\, F(i\epsilon) \\
& + T \sum_{-\omega < \epsilon < 0} G(i\epsilon + i\omega, \mathbf{k} + \mathbf{q})\, G(i\epsilon, \mathbf{k})\, F(i\epsilon) \\
\equiv \;& C_{\mathbf{k}}^{(1)}(i\omega, \mathbf{q}) + C_{\mathbf{k}}^{(2)}(i\omega, \mathbf{q}) + C_{\mathbf{k}}^{(3)}(i\omega, \mathbf{q})\,,
\end{aligned}
\tag{5.20}
$$

so that the functions in each separate sum have no discontinuity in the range of summation. If we take $\omega = q = 0$ from the start

$$
\begin{aligned}
C_{\mathbf{k}}(0, 0) &= T \sum_{\epsilon > 0} G(i\epsilon, \mathbf{k})\, G(i\epsilon, \mathbf{k})\, F(i\epsilon) + T \sum_{\epsilon < 0} G(i\epsilon, \mathbf{k})\, G(i\epsilon, \mathbf{k})\, F(i\epsilon) \\
&= C_{\mathbf{k}}^{(1)}(0, 0) + C_{\mathbf{k}}^{(2)}(0, 0)\,,
\end{aligned}
$$

while, if we first send $\omega \to 0$ and then send $q \to 0$, so-called q-limit,

$$
\begin{aligned}
C_{\mathbf{k}}^{q} &\equiv \lim_{\mathbf{q} \to 0} C_{\mathbf{k}}(0, \mathbf{q}) \\
&= \lim_{\mathbf{q} \to 0} \left[T \sum_{\epsilon > 0} G(i\epsilon, \mathbf{k} + \mathbf{q})\, G(i\epsilon, \mathbf{k})\, F(i\epsilon) + T \sum_{\epsilon < 0} G(i\epsilon, \mathbf{k} + \mathbf{q})\, G(i\epsilon, \mathbf{k})\, F(i\epsilon) \right] \\
&= C_{\mathbf{k}}^{(1)}(0, 0) + C_{\mathbf{k}}^{(2)}(0, 0)\,.
\end{aligned}
$$

In the opposite ω-limit, i.e., sending first $q \to 0$ and then $\omega \to 0$

$$C_{\mathbf{k}}^{\omega} \equiv \lim_{\omega \to 0} C_{\mathbf{k}}(i\omega, \mathbf{0})$$

$$= \lim_{\omega \to 0} \left[T \sum_{\epsilon > 0} G(i\epsilon + i\omega, \mathbf{k}) G(i\epsilon, \mathbf{k}) F(i\epsilon) + T \sum_{\epsilon < -\omega} G(i\epsilon + i\omega, \mathbf{k}) G(i\epsilon, \mathbf{k}) F(i\epsilon) \right.$$

$$\left. + T \sum_{-\omega < \epsilon < 0} G(i\epsilon + i\omega, \mathbf{k}) G(i\epsilon, \mathbf{k}) F(i\epsilon) \right]$$

$$= C_{\mathbf{k}}^{(1)}(0, \mathbf{0}) + C_{\mathbf{k}}^{(2)}(0, \mathbf{0}) + \lim_{\omega \to 0} C_{\mathbf{k}}^{(3)}(i\omega, \mathbf{0}) .$$

$$(5.21)$$

Therefore, a non-analytic behaviour may arise only because $C_{\mathbf{k}}^{(3)}(i\omega, \mathbf{q})$ could remain finite in the ω-limit. Naïvely, having assumed $F(i\epsilon)$ smooth at small ϵ, $C_{\mathbf{k}}^{(3)}(i\omega, \mathbf{q})$, being limited to frequencies $-\omega < \epsilon < 0$, would seem to vanish as $\omega \to 0$. That is surely the case of insulators with gaps to all kinds of excitations, where the Green's functions are also smooth. However, in a system with gapless excitations, the product of the two Green's functions might have a singularity at $\omega = 0$ that compensates the vanishing frequency range of the summation, thus yielding a finite $C_{\mathbf{k}}^{(3)}(i\omega, \mathbf{0})$ as $\omega \to 0$. In other words, at small ω, q and T, the kernel R for $\epsilon \in [-\omega, 0]$, which we hereafter name $\widetilde{\Delta}$, might be singular and non analytic at $\omega = q = 0$, as opposed to R for $\epsilon \notin [-\omega, 0]$, which we denote as $\widetilde{R}_{inc}$ and is instead analytic at $\omega = q = 0$. Our aim here is to explicitly calculate $\widetilde{\Delta}$.

By standard tricks, we can rewrite (5.20) as

$$C_{\mathbf{k}}(i\omega, \mathbf{q}) = -\oint \frac{dz}{2\pi i} f(z) G(z + i\omega, \mathbf{k} + \mathbf{q}) G(z, \mathbf{k}) F(z) ,$$

with the contour shown in Fig. 5.1 that encircles clockwise the two branch cuts $z = \epsilon$ and $z = \epsilon - i\omega$, with real $\epsilon \in [-\infty, \infty]$.

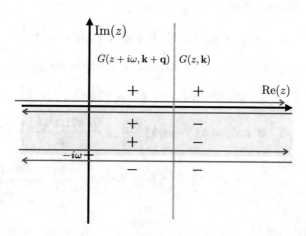

Fig. 5.1 Contour used in the integration (5.22), red and blue arrow lines. The contour goes along two branch cuts at $z = \epsilon$ and $z = -i\omega + \epsilon$ with real $\epsilon \in [-\infty, \infty]$. For each Green's function, $G(z + i\omega, \mathbf{k})$ and $G(z, \mathbf{k})$, we indicate its retarded (+) or advanced (−) character above or below the branch cuts

The singular contribution to R may derive from the green region in that figure where $\mathrm{Im}\, z \in [-\omega, 0]$, thus the integral over the two lines in blue that reads

$$C_{\mathbf{k}}^{(3)}(i\omega, \mathbf{q}) = -\int_{-\infty}^{\infty} \frac{d\epsilon}{2\pi i}\, f(\epsilon)\, F(\epsilon) \Big[G_+(\epsilon, \mathbf{k}+\mathbf{q})\, G_-(\epsilon - i\omega, \mathbf{k}) - G_+(\epsilon + i\omega, \mathbf{k}+\mathbf{q})\, G_-(\epsilon, \mathbf{k}) \Big],$$

where $G_+(\epsilon, \mathbf{k}) = G(\epsilon + i0^+, \mathbf{k})$ and $G_-(\epsilon, \mathbf{k}) = G(\epsilon - i0^+, \mathbf{k}) = G_+(\epsilon, \mathbf{k})^*$ are, respectively, the retarded and advanced components of the Green's functions, and similar definitions and properties hold for the self-energy as well. Since ω is small, we can write the term in square brackets as

$$G_+(\epsilon, \mathbf{k}+\mathbf{q})\, G_-(\epsilon - i\omega, \mathbf{k}) - G_+(\epsilon + i\omega, \mathbf{k}+\mathbf{q})\, G_-(\epsilon, \mathbf{k})$$

$$= G_+(\epsilon, \mathbf{k}+\mathbf{q})\, G_-(\epsilon - i\omega, \mathbf{k}) - \Big(G_+(\epsilon, \mathbf{k}+\mathbf{q})\, G_-(\epsilon - i\omega, \mathbf{k}) \Big)\Big|_{\epsilon \to \epsilon + i\omega}$$

$$\simeq -i\omega\, \frac{\partial}{\partial \epsilon} \Big(G_+(\epsilon, \mathbf{k}+\mathbf{q})\, G_-(\epsilon - i\omega, \mathbf{k}) \Big),$$

where

$$G_+(\epsilon, \mathbf{k}+\mathbf{q})\, G_-(\epsilon - i\omega, \mathbf{k}) = \Big(G_-(\epsilon - i\omega, \mathbf{k}) - G_+(\epsilon, \mathbf{k}+\mathbf{q}) \Big)$$

$$\frac{1}{i\omega - \epsilon_{\mathbf{k}+\mathbf{q}} + \epsilon_{\mathbf{k}} - \Sigma_+(\epsilon, \mathbf{k}+\mathbf{q}) + \Sigma_-(\epsilon - i\omega, \mathbf{k})}\, .$$

At leading order in ω and $\mathbf{q}$,

$$G_-(\epsilon - i\omega, \mathbf{k}) - G_+(\epsilon, \mathbf{k}+\mathbf{q}) \simeq 2\pi i\, A(\epsilon, \mathbf{k}),$$

while, recalling our definitions of $\Gamma(\epsilon, \mathbf{k})$ in (5.5), $\Xi(\epsilon, \mathbf{k})$ in (5.8) and $Z(\epsilon, \mathbf{k})$ in (5.4), and, expanding in ω and $\mathbf{q}$,

$$i\omega - \epsilon_{\mathbf{k}+\mathbf{q}} + \epsilon_{\mathbf{k}} - \Sigma_+(\epsilon, \mathbf{k}+\mathbf{q}) + \Sigma_-(\epsilon - i\omega, \mathbf{k})$$

$$= i\omega - \epsilon_{\mathbf{k}+\mathbf{q}} + \epsilon_{\mathbf{k}} - \mathrm{Re}\, \Sigma_+(\epsilon, \mathbf{k}+\mathbf{q}) + \mathrm{Re}\, \Sigma_+(\epsilon - i\omega, \mathbf{k})$$

$$- i\, \mathrm{Im}\, \Sigma_+(\epsilon, \mathbf{k}+\mathbf{q}) - i\, \mathrm{Im}\, \Sigma_+(\epsilon - i\omega, \mathbf{k})$$

$$\simeq i\omega\, Z(\epsilon, \mathbf{k})^{-1} - \mathbf{q}\cdot \left(\frac{\partial \epsilon_{\mathbf{k}}}{\partial \mathbf{k}} + \frac{\partial \mathrm{Re}\, \Sigma_+(\epsilon, \mathbf{k})}{\partial \mathbf{k}} \right) - 2i\, \mathrm{Im}\, \Sigma_+(\epsilon, \mathbf{k})$$

$$\simeq Z(\epsilon, \mathbf{k})^{-1} \left(i\omega - \mathbf{q}\cdot Z(\epsilon, \mathbf{k}) \left(\frac{\partial \epsilon_{\mathbf{k}}}{\partial \mathbf{k}} + \frac{\partial \mathrm{Re}\, \Sigma_+(\epsilon, \mathbf{k})}{\partial \mathbf{k}} \right) + 2i\, \Gamma(\epsilon, \mathbf{k}) \right).$$

Therefore

$$G_+(\epsilon, \mathbf{k} + \mathbf{q})\, G_-(\epsilon - i\omega, \mathbf{k}) - G_+(\epsilon + i\omega, \mathbf{k} + \mathbf{q})\, G_-(\epsilon, \mathbf{k})$$

$$\simeq -2\pi i\, \frac{\partial}{\partial \epsilon} \left(A(\epsilon, \mathbf{k})\, \frac{i\omega\, Z(\epsilon, \mathbf{k})}{i\omega - \mathbf{q} \cdot Z(\epsilon, \mathbf{k}) \left(\dfrac{\partial \epsilon_\mathbf{k}}{\partial \mathbf{k}} + \dfrac{\partial \mathrm{Re}\, \Sigma_+(\epsilon, \mathbf{k})}{\partial \mathbf{k}} \right) + 2i\, \Gamma(\epsilon, \mathbf{k})} \right)$$

so that, upon integration by part, and noting that $\partial F(\epsilon)/\partial \epsilon$ can be neglected with respect to the singular $\partial f(\epsilon)/\partial \epsilon$ at low T, we finally find through (5.13) that

$$C_\mathbf{k}^{(3)}(i\omega, \mathbf{q}) \simeq \int_{-\infty}^{\infty} d\epsilon \left(-\frac{\partial f(\epsilon)}{\partial \epsilon} \right) A(\epsilon, \mathbf{k})\, F(\epsilon)$$

$$\frac{i\omega\, Z(\epsilon, \mathbf{k})}{i\omega - \mathbf{q} \cdot Z(\epsilon, \mathbf{k}) \left(\dfrac{\partial \epsilon_\mathbf{k}}{\partial \mathbf{k}} + \dfrac{\partial \mathrm{Re}\, \Sigma_+(\epsilon, \mathbf{k})}{\partial \mathbf{k}} \right) + 2i\, \Gamma(\epsilon, \mathbf{k})}$$

$$= \int_{-\infty}^{\infty} d\epsilon \left(-\frac{\partial f(\epsilon)}{\partial \epsilon} \right) A_{\mathrm{qp}}(\epsilon, \mathbf{k})\, F(\epsilon)\, Z(\epsilon, \mathbf{k})^2$$

$$\frac{i\omega\, Z(\epsilon, \mathbf{k})}{i\omega - \mathbf{q} \cdot Z(\epsilon, \mathbf{k}) \left(\dfrac{\partial \epsilon_\mathbf{k}}{\partial \mathbf{k}} + \dfrac{\partial \mathrm{Re}\, \Sigma_+(\epsilon, \mathbf{k})}{\partial \mathbf{k}} \right) + 2i\, \Gamma(\epsilon, \mathbf{k})}$$

$$\simeq \int_{-\infty}^{\infty} d\epsilon \left(-\frac{\partial f(\epsilon)}{\partial \epsilon} \right) \delta\big(\epsilon - \epsilon_*(\mathbf{k})\big)\, F(\epsilon)\, Z(\epsilon, \mathbf{k})^2$$

$$\frac{i\omega\, Z(\epsilon, \mathbf{k})}{i\omega - \mathbf{q} \cdot Z(\epsilon, \mathbf{k}) \left(\dfrac{\partial \epsilon_\mathbf{k}}{\partial \mathbf{k}} + \dfrac{\partial \mathrm{Re}\, \Sigma_+(\epsilon, \mathbf{k})}{\partial \mathbf{k}} \right) + 2i\, \Gamma(\epsilon, \mathbf{k})} \tag{5.22}$$

$$\simeq -\frac{\partial f\big(\epsilon_*(\mathbf{k})\big)}{\partial \epsilon_*(\mathbf{k})}\, Z\big(\epsilon_*(\mathbf{k}), \mathbf{k}\big)^2\, \frac{i\omega}{i\omega - \mathbf{q} \cdot v_\mathbf{k}}\, F\big(\epsilon_*(\mathbf{k})\big)$$

$$\simeq -\frac{\partial f\big(\epsilon_*(\mathbf{k})\big)}{\partial \epsilon_*(\mathbf{k})}\, Z\big(\epsilon_*(\mathbf{k}), \mathbf{k}\big)^2\, \frac{i\omega}{i\omega - \mathbf{q} \cdot v_\mathbf{k}}\, F(0)\,,$$

the last almost equalities deriving from $-\partial f(\epsilon)/\partial \epsilon \simeq \delta(\epsilon)$ at low T, which picks out only the small ϵ component of the quasiparticle DOS, thus $A_{\mathrm{qp}}(\epsilon, \mathbf{k}) \simeq \delta\big(\epsilon - \epsilon_*(\mathbf{k})\big)$, and further implies that $\Gamma(\epsilon \to 0, \mathbf{k}) \simeq 0$ because of (5.5). Moreover, through the definition (5.10) we can readily show that

$$0 = \frac{d\, \Xi\big(\epsilon_*(\mathbf{k}), \mathbf{k}\big)}{d\mathbf{k}} = \frac{\partial \Xi(\epsilon, \mathbf{k})}{\partial \epsilon}\bigg|_{\epsilon = \epsilon_*(\mathbf{k})} \frac{\partial \epsilon_*(\mathbf{k})}{\partial \mathbf{k}} + \frac{\partial \Xi\big(\epsilon_*(\mathbf{k}), \mathbf{k}\big)}{\partial \mathbf{k}}$$

$$= -\frac{\partial \epsilon_*(\mathbf{k})}{\partial \mathbf{k}} + Z\big(\epsilon_*(\mathbf{k}), \mathbf{k}\big) \left(\frac{\partial \epsilon_\mathbf{k}}{\partial \mathbf{k}} + \frac{\partial \mathrm{Re}\, \Sigma_+(\epsilon, \mathbf{k})}{\partial \mathbf{k}} \bigg|_{\epsilon = \epsilon_*(\mathbf{k})} \right),$$

namely

$$Z\big(\epsilon_*(\mathbf{k}), \mathbf{k}\big) \left(\frac{\partial \epsilon_\mathbf{k}}{\partial \mathbf{k}} + \frac{\partial \operatorname{Re} \Sigma_+(\epsilon, \mathbf{k})}{\partial \mathbf{k}} \bigg|_{\epsilon = \epsilon_*(\mathbf{k})} \right) = \frac{\partial \epsilon_*(\mathbf{k})}{\partial \mathbf{k}} \equiv v_\mathbf{k} \, , \qquad (5.23)$$

which is the quasiparticle group velocity.

In conclusion, coming back to our original task (5.19), we find that a singular part $\widetilde{\Delta}$ of the kernel R, namely that in the Matsubara frequency interval $-\omega < \epsilon < 0$, does exist and can be written, in the sense of distribution, as

$$\widetilde{\Delta}(i\epsilon + i\omega \, \mathbf{k} + \mathbf{q}, i\epsilon \, \mathbf{k}) \equiv -\frac{\delta_{\epsilon,0}}{T} \frac{\partial f\big(\epsilon_*(\mathbf{k})\big)}{\partial \epsilon_*(\mathbf{k})} Z(\mathbf{k})^2 \, \frac{i\omega}{i\omega - \mathbf{q} \cdot v_\mathbf{k}} \, , \qquad (5.24)$$

having defined

$$Z(\mathbf{k}) \equiv Z\big(\epsilon_*(\mathbf{k}), \mathbf{k}\big) \, . \qquad (5.25)$$

Indeed, we note that in the ω-limit

$$\widetilde{\Delta}^\omega(i\epsilon, \mathbf{k}) \equiv \lim_{\omega \to 0} \lim_{q \to 0} \widetilde{\Delta}(i\epsilon + i\omega \, \mathbf{k} + \mathbf{q}, i\epsilon \, \mathbf{k}) = -\frac{\delta_{\epsilon,0}}{T} \frac{\partial f\big(\epsilon_*(\mathbf{k})\big)}{\partial \epsilon_*(\mathbf{k})} Z(\mathbf{k})^2 \, , \qquad (5.26)$$

yields a finite contribution to $C_\mathbf{k}^{sing}(i\omega \to 0, \mathbf{0})$ despite the vanishing range of the sum. On the contrary, in the opposite q-limit,

$$\widetilde{\Delta}^q(i\epsilon, \mathbf{k}) \equiv \lim_{q \to 0} \lim_{\omega \to 0} \widetilde{\Delta}(i\epsilon + i\omega \, \mathbf{k} + \mathbf{q}, i\epsilon \, \mathbf{k}) = 0 \, , \qquad (5.27)$$

as expected, which explicitly shows that $\widetilde{\Delta}$ is non analytic at $\omega = q = 0$. The component $\widetilde{R}_{inc}$ of the kernel R is instead, by definition, analytic at $\omega = q = 0$, thus $\widetilde{R}_{inc}^\omega = \widetilde{R}_{inc}^q = \widetilde{R}_{inc}$, the last equality due to our assumption of small q and ω. It thus follows that

$$R^\omega = \widetilde{\Delta}^\omega + \widetilde{R}_{inc}^\omega = \widetilde{\Delta}^\omega + \widetilde{R}_{inc} \, , \quad R^q = \widetilde{\Delta}^q + \widetilde{R}_{inc}^q = \widetilde{R}_{inc} \, . \qquad (5.28)$$

For later convenience, we define

$$\begin{aligned}\Delta(i\epsilon + i\omega \, \mathbf{k} + \mathbf{q}, i\epsilon \, \mathbf{k}) &\equiv \widetilde{\Delta}(i\epsilon + i\omega \, \mathbf{k} + \mathbf{q}, i\epsilon \, \mathbf{k}) - \widetilde{\Delta}^\omega(i\epsilon, \mathbf{k}) \\ &= -\frac{\delta_{\epsilon,0}}{T} \frac{\partial f\big(\epsilon_*(\mathbf{k})\big)}{\partial \epsilon_*(\mathbf{k})} Z(\mathbf{k})^2 \, \frac{\mathbf{q} \cdot v_\mathbf{k}}{i\omega - \mathbf{q} \cdot v_\mathbf{k}} \, , \end{aligned} \qquad (5.29)$$

and consistently

$$R = \widetilde{\Delta} + \widetilde{R}_{inc} = \big(\widetilde{\Delta} - \widetilde{\Delta}^\omega\big) + \big(\widetilde{R}_{inc} + \widetilde{\Delta}^\omega\big) \equiv \Delta + R_{inc} \, . \qquad (5.30)$$

With those notations

$$\Delta^q(i\epsilon, \mathbf{k}) = -\tilde{\Delta}^\omega(i\epsilon, \mathbf{k}), \quad \Delta^\omega(i\epsilon, \mathbf{k}) = 0, \tag{5.31}$$

so that

$$R^\omega = \Delta^\omega + R^\omega_{inc} = R_{inc}, \quad R^q = \Delta^q + R^q_{inc} = \Delta^q + R_{inc}. \tag{5.32}$$

We emphasise that, unlike the complex kernel R, Δ only depends on two unknown quantities: the quasiparticle dispersion $\epsilon_*(\mathbf{k})$ and the quasiparticle residue $Z(\mathbf{k})$. In the case of a Luttinger surface, $Z(\mathbf{k} \to \mathbf{k}_L) \to 0$ and thus $\Delta \to 0$, too, which, one might erroneously conclude, should put an end to the story, including the circumstance that the Luttinger surface appears in a Mott insulator. However, the key to the derivation of Landau's Fermi liquid theory is the ability to absorb $Z(\mathbf{k})$ into effective quantities that are well behaved even though $Z(\mathbf{k} \to \mathbf{k}_L) \to 0$.

5.2.2 Interaction Vertex and Density-Vertices

The Bethe-Salpeter equation (5.17) can, therefore, be written as

$$\begin{aligned} \Gamma &= \Gamma_0 + \Gamma_0 \odot R \odot \Gamma = \Gamma_0 + \Gamma_0 \odot \left(\Delta + R_{inc}\right) \odot \Gamma \\ &= \Gamma_0 + \Gamma \odot R \odot \Gamma_0 = \Gamma_0 + \Gamma \odot \left(\Delta + R_{inc}\right) \odot \Gamma_0, \end{aligned} \tag{5.33}$$

so that, since Γ_0 is analytic at $\omega = q = 0$,[2] through (5.32),

$$\Gamma^\omega = \Gamma_0 + \Gamma_0 \odot R^\omega \odot \Gamma^\omega = \Gamma_0 + \Gamma_0 \odot R_{inc} \odot \Gamma^\omega = \Gamma_0 + \Gamma^\omega \odot R_{inc} \odot \Gamma_0.$$

We can express Γ_0 in terms of Γ

$$\Gamma_0 = \Gamma \odot \left(1 + R \odot \Gamma\right)^{-1} = \left(1 + \Gamma \odot R\right)^{-1} \odot \Gamma,$$

or in terms of Γ^ω, i.e.,

$$\Gamma_0 = \Gamma^\omega \odot \left(1 + R_{inc} \odot \Gamma^\omega\right)^{-1} = \left(1 + \Gamma^\omega \odot R_{inc}\right)^{-1} \odot \Gamma^\omega,$$

which implies, e.g., that

$$\Gamma \odot \left(1 + R \odot \Gamma\right)^{-1} = \left(1 + \Gamma^\omega \odot R_{inc}\right)^{-1} \odot \Gamma^\omega$$

---

[2] By definition, Γ_0 does not include the kernel R, i.e., a particle line and a hole one differing by frequency $i\omega$ and momentum $\mathbf{q}$, and therefore, cannot have any singularity in that channel. However, the analyticity at $\omega = q = 0$ does not exclude singularities in the other variables.

namely,

$$\left(1 + \Gamma^\omega \odot R_{inc}\right) \odot \Gamma = \Gamma^\omega \odot \left(1 + R \odot \Gamma\right),$$

which corresponds to a new definition of the Bethe-Salpeter equation

$$\Gamma = \Gamma^\omega + \Gamma^\omega \odot \left(R - R_{inc}\right) \odot \Gamma = \Gamma^\omega + \Gamma^\omega \odot \Delta \odot \Gamma. \tag{5.34}$$

We note that we have been able to absorb the unknown Γ_0 and R_{inc} into a single unknown quantity, the interaction vertex Γ^ω.

Through (5.34) we find that

$$\Gamma^\omega = \Gamma \odot \left(1 + \Delta \odot \Gamma\right)^{-1},$$

which, since $R^\omega + \Delta = R_{inc} + \Delta = R$, also implies

$$1 + R^\omega \odot \Gamma^\omega = \left(1 + \Delta \odot \Gamma\right) \odot \left(1 + \Delta \odot \Gamma\right)^{-1} + R^\omega \odot \Gamma \odot \left(1 + \Delta \odot \Gamma\right)^{-1}$$

$$= \left(1 + R \odot \Gamma\right) \odot \left(1 + \Delta \odot \Gamma\right)^{-1}.$$

$$\tag{5.35}$$

Let us consider now the interacting A density-vertex Λ^A that is defined, see Fig. 4.25, by the equation

$$\Lambda^A = \Lambda_0^A + \Lambda_0^A \odot R \odot \Gamma, \tag{5.36}$$

where Λ_0^A is the non-interacting value. As before, we can express the latter in terms of the interacting Λ^A,

$$\Lambda_0^A = \Lambda^A \odot \left(1 + R \odot \Gamma\right)^{-1} = \left(1 + \Gamma \odot R\right)^{-1} \odot \Lambda^A. \tag{5.37}$$

Alternatively, we can take the ω-limit

$$\Lambda^{A\omega} = \Lambda_0^A + \Lambda_0^A \odot R^\omega \odot \Gamma^\omega,$$

and again solve for Λ_0^A and get, making use of (5.35),

$$\Lambda_0^A = \Lambda^{A\omega} \odot \left(1 + R^\omega \odot \Gamma^\omega\right)^{-1} = \Lambda^{A\omega} \odot \left(1 + \Delta \odot \Gamma\right) \odot \left(1 + R \odot \Gamma\right)^{-1}$$

$$= \left(1 + \Gamma \odot R\right)^{-1} \odot \left(1 + \Gamma \odot \Delta\right) \odot \Lambda^{A\omega}.$$

$$\tag{5.38}$$

Comparing (5.37) with (5.38) we find

$$\Lambda^A \odot \left(1 + R \odot \Gamma\right)^{-1} = \Lambda^{A\omega} \odot \left(1 + \Delta \odot \Gamma\right) \odot \left(1 + R \odot \Gamma\right)^{-1},$$

namely

$$\Lambda^A = \Lambda^{A\omega} + \Lambda^{A\omega} \odot \Delta \odot \Gamma = \Lambda^{A\omega} \odot \left(1 + \Delta \odot \Gamma\right). \tag{5.39}$$

Just like for the interacting vertex, we have been able to absorb R_{inc} into the ω-limit of the vertex.

We can straightforwardly repeat all the above calculations focusing on the q-limit, i.e., using Γ^q in place of Γ^ω and Λ^{Aq} in place of $\Lambda^{A\omega}$. In that case, the (5.34), (5.38) and (5.39) are replaced, respectively, by

$$\Gamma = \Gamma^q + \Gamma^q \odot \tilde{\Delta} \odot \Gamma,$$

$$\Lambda_0^A = \Lambda^{Aq} \odot \left(1 + \tilde{\Delta} \odot \Gamma\right) \odot \left(1 + R \odot \Gamma\right)^{-1} = \left(1 + \Gamma \odot R\right)^{-1} \odot \left(1 + \Gamma \odot \tilde{\Delta}\right) \odot \Lambda^{Aq},$$

$$\Lambda^A = \Lambda^{Aq} + \Lambda^{Aq} \odot \tilde{\Delta} \odot \Gamma = \Lambda^{Aq} \odot \left(1 + \tilde{\Delta} \odot \Gamma\right).$$

$$\tag{5.40}$$

5.3 Linear Response Functions

We can now represent a generic linear response function in terms of the above quantities. Let us consider the response function $\chi_{AB}(\omega + i\eta, \mathbf{q})$, with $\eta > 0$ infinitesimal, which we shortly write as

$$\chi_{AB} \equiv \mathrm{Tr}\left(\Lambda^A \odot R \odot \Lambda_0^B\right) = \mathrm{Tr}\left(\Lambda_0^A \odot R \odot \Lambda^B\right). \tag{5.41}$$

Using (5.39), the last equation in (5.38), and recalling that $R = \Delta + R^\omega$, we can write

$$\begin{aligned}
\chi_{AB} &= \mathrm{Tr}\left[\Lambda^{A\omega} \odot \left(1 + \Delta \odot \Gamma\right) \odot R \odot \Lambda_0^B\right] \\
&= \mathrm{Tr}\left(\Lambda^{A\omega} \odot R^\omega \odot \Lambda_0^B\right) + \mathrm{Tr}\left(\Lambda^{A\omega} \odot \Delta \odot \Lambda_0^B\right) + \mathrm{Tr}\left(\Lambda^{A\omega} \odot \Delta \odot \Gamma \odot R \odot \Lambda_0^B\right) \\
&= \chi_{AB}^\omega + \mathrm{Tr}\left[\Lambda^{A\omega} \odot \Delta \odot \left(1 + \Gamma \odot R\right) \odot \Lambda_0^B\right] \\
&= \chi_{AB}^\omega + \mathrm{Tr}\left[\Lambda^{A\omega} \odot \Delta \odot \left(1 + \Gamma \odot R\right) \odot \left(1 + \Gamma \odot R\right)^{-1} \odot \left(1 + \Gamma \odot \Delta\right) \odot \Lambda^{B\omega}\right] \\
&= \chi_{AB}^\omega + \mathrm{Tr}\left[\Lambda^{A\omega} \odot \Delta \odot \left(1 + \Gamma \odot \Delta\right) \odot \Lambda^{B\omega}\right] \\
&= \chi_{AB}^\omega + \mathrm{Tr}\left(\Lambda^{A\omega} \odot \Delta \odot \Lambda^{B\omega}\right) + \mathrm{Tr}\left(\Lambda^{A\omega} \odot \Delta \odot \Gamma \odot \Delta \odot \Lambda^{B\omega}\right),
\end{aligned}$$

$$\tag{5.42}$$

where

$$\chi_{AB}^{\omega} = \text{Tr}\left(\Lambda^{A\omega} \odot R^{\omega} \odot \Lambda_0^{B}\right) = \text{Tr}\left(\Lambda^{A\omega} \odot R_{inc} \odot \Lambda_0^{B}\right), \tag{5.43}$$

is the ω-limit of the response function.

We define

$$\Delta(i\epsilon + i\omega \, \mathbf{k} + \mathbf{q}, i\epsilon \, \mathbf{k}) = -Z(\mathbf{k})^2 \frac{\delta_{\epsilon,0}}{T} \frac{\partial f(\epsilon_*(\mathbf{k}))}{\partial \epsilon_*(\mathbf{k})} \frac{\mathbf{q} \cdot v_{\mathbf{k}}}{i\omega - \mathbf{q} \cdot v_{\mathbf{k}}} \tag{5.44}$$

$$\equiv Z(\mathbf{k})^2 \, R_{\text{qp}}(i\epsilon + i\omega \, \mathbf{k} + \mathbf{q}, i\epsilon \, \mathbf{k}),$$

and note that, for small q

$$T \sum_{\epsilon} R_{\text{qp}}(i\epsilon + i\omega \, \mathbf{k} + \mathbf{q}, i\epsilon \, \mathbf{k}) \simeq T \sum_{\epsilon} \frac{1}{i\epsilon + i\omega - \epsilon_*(\mathbf{k} + \mathbf{q})} \frac{1}{i\epsilon - \epsilon_*(\mathbf{k})}$$

$$\equiv T \sum_{\epsilon} G_{\text{qp}}(i\epsilon + i\omega, \mathbf{k} + \mathbf{q}) \, G_{\text{qp}}(i\epsilon, \mathbf{k}),$$

$$\tag{5.45}$$

is just the sum over Matsubara frequencies of the product of two non-interacting Green's functions with $\epsilon_{\mathbf{k}} \to \epsilon_*(\mathbf{k})$, thus the Green's functions G_{qp} (4.169) of the quasiparticles.

In addition, we define the quasiparticle interaction vertex

$$A_{\sigma_1,\sigma_2;\sigma_3,\sigma_4}(\mathbf{k}, \mathbf{k}'; i\omega, \mathbf{q}) \equiv Z(\mathbf{k}) \, Z(\mathbf{k}') \, \Gamma_{\sigma_1,\sigma_2;\sigma_3,\sigma_4}(i\epsilon + i\omega \, \mathbf{k} + \mathbf{q}, i\epsilon' \, \mathbf{k}'; i\epsilon' + i\omega \, \mathbf{k}' + \mathbf{q}, i\epsilon \, \mathbf{k}), \tag{5.46}$$

where $i\epsilon \to \epsilon_*(\mathbf{k}) \simeq 0$ and $i\epsilon' \to \epsilon_*(\mathbf{k}') \simeq 0$, and the quasiparticle density-vertices for small q

$$\lambda^{A}\left(i\epsilon \, \mathbf{k} \, \sigma, i\epsilon + i\omega \, \mathbf{k} + \mathbf{q} \, \sigma'; i\omega, \mathbf{q}\right) = Z(\mathbf{k}) \, \Lambda^{A}\left(i\epsilon \, \mathbf{k} \, \sigma, i\epsilon + i\omega \, \mathbf{k} + \mathbf{q} \, \sigma'; i\omega, \mathbf{q}\right), \tag{5.47}$$

both shown in Fig. 5.2, through which we can write (5.42) as

$$\chi_{AB} = \chi_{AB}^{\omega} + \text{Tr}\left(\lambda^{A\omega} \odot R_{\text{qp}} \odot \lambda^{B\omega}\right) + \text{Tr}\left(\lambda^{A\omega} \odot R_{\text{qp}} \odot A \odot R_{\text{qp}} \odot \lambda^{B\omega}\right). \tag{5.48}$$

We can repeat step-by-step the above derivation using (5.40) instead of (5.39) and (5.38), in which case we find

$$\chi_{AB} = \chi_{AB}^{q} + \text{Tr}\left(\lambda^{Aq} \odot \tilde{R}_{\text{qp}} \odot \lambda^{Bq}\right) + \text{Tr}\left(\lambda^{Aq} \odot \tilde{R}_{\text{qp}} \odot A \odot \tilde{R}_{\text{qp}} \odot \lambda^{Bq}\right), \tag{5.49}$$

where

$$\tilde{R}_{\text{qp}}(i\epsilon + i\omega \, \mathbf{k} + \mathbf{q}, i\epsilon \, \mathbf{k}) = \frac{1}{Z(\mathbf{k})^2} \tilde{\Delta}(i\epsilon + i\omega \, \mathbf{k} + \mathbf{q}, i\epsilon \, \mathbf{k})$$

$$= -\frac{\delta_{\epsilon,0}}{T} \frac{\partial f(\epsilon_*(\mathbf{k}))}{\partial \epsilon_*(\mathbf{k})} \frac{i\omega}{i\omega - \mathbf{q} \cdot v_{\mathbf{k}}}. \tag{5.50}$$

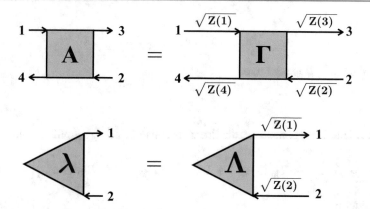

Fig. 5.2 Top panel: definition of the quasiparticle interaction vertex A in terms of the physical electron one Γ and the quasiparticle residue Z. The numerals indicate the frequency, momentum, and spin of each external leg. Bottom panel: same as top one but for the quasiparticle triangular density-vertex λ in terms of the physical electron one Λ

In both cases (5.48) and (5.49), all the complexity of the interacting linear response function has been notably reduced at small ω, q and T. That simplification has been achieved by simply exploiting the analytic properties of the kernel R, which is remarkable.

5.3.1   Response Functions of Densities Associated to Conserved Quantities

If A or B are conserved quantities, so that $\rho_A(\mathbf{0})$ or $\rho_B(\mathbf{0})$ are just numbers, then in (5.48) $\chi^{\omega}_{AB} = 0$, and

$$\chi_{AB} = \mathrm{Tr}\left(\lambda^{A\,\omega} \odot R_{\mathrm{qp}} \odot \lambda^{B\,\omega}\right) + \mathrm{Tr}\left(\lambda^{A\,\omega} \odot R_{\mathrm{qp}} \odot A \odot R_{\mathrm{qp}} \odot \lambda^{B\,\omega}\right). \tag{5.51}$$

We note that R_{qp}, see (5.44), forces all internal frequencies equal to the quasiparticle energies, which, in turn, are vanishingly small because of the Fermi distribution derivative. Therefore, the Ward-Takahashi identity (4.131) in the ω-limit implies, if A is conserved, that[3]

$$\Lambda^{A\,\omega}(\epsilon, \mathbf{k}) = \Lambda^A_0\left(\mathbf{k}\,\sigma_2, \mathbf{k}\,\sigma_1\right)\left(1 - \frac{\partial \Sigma(\epsilon, \mathbf{k})}{\partial \epsilon}\bigg|_{\epsilon=\epsilon_*(\mathbf{k})}\right) = \Lambda^A_0\left(\mathbf{k}\,\sigma_2, \mathbf{k}\,\sigma_1\right) Z(\mathbf{k})^{-1}.$$

---

[3] Since $\epsilon_*(\mathbf{k}) \to 0$, namely $\mathbf{k}$ is close to the Fermi or Luttinger surface, then Im Σ is either vanishing in case (F), or negligible with respect to Re Σ in case (L). Therefore, in both case the derivative of the self-energy can be replaced by the derivative of its real part.

namely

$$\Lambda^{A\omega} = \Lambda_0^A \, Z^{-1}, \quad \lambda^{A\omega} = Z\,\Lambda^{A\omega} = \Lambda_0^A, \tag{5.52}$$

and, similarly, if B is conserved

$$\lambda^{B\omega} = \Lambda_0^B. \tag{5.53}$$

If both A and B are conserved, the linear response function simplifies even more

$$\chi_{AB} = \mathrm{Tr}\left(\Lambda_0^A \odot R_{\mathrm{qp}} \odot \Lambda_0^B\right) + \mathrm{Tr}\left(\Lambda_0^A \odot R_{\mathrm{qp}} \odot A \odot R_{\mathrm{qp}} \odot \Lambda_0^B\right), \tag{5.54}$$

and explicitly reads

$$
\begin{aligned}
\chi_{AB}(\omega, \mathbf{q}) = & -\frac{1}{V} \sum_{\mathbf{k}\sigma_1\sigma_2} \frac{\partial f(\epsilon_*(\mathbf{k}))}{\partial \epsilon_*(\mathbf{k})} \frac{\mathbf{q}\cdot \mathbf{v_k}}{i\omega - \mathbf{q}\cdot \mathbf{v_k}} \Lambda_0^A(\mathbf{k}\,\sigma_2, \mathbf{k}+\mathbf{q}\,\sigma_1)\, \Lambda_0^B(\mathbf{k}+\mathbf{q}\,\sigma_1, \mathbf{k}\,\sigma_2) \\
& + \frac{1}{V^2} \sum_{\mathbf{k}\mathbf{k}'} \frac{\partial f(\epsilon_*(\mathbf{k}))}{\partial \epsilon_*(\mathbf{k})} \frac{\partial f(\epsilon_*(\mathbf{k}'))}{\partial \epsilon_*(\mathbf{k}')} \frac{\mathbf{q}\cdot \mathbf{v_k}}{i\omega - \mathbf{q}\cdot \mathbf{v_k}} \frac{\mathbf{q}\cdot \mathbf{v_{k'}}}{i\omega - \mathbf{q}\cdot \mathbf{v_{k'}}} \\
& \quad \sum_{\sigma_1\sigma_2\sigma_3\sigma_4} \Lambda_0^A(\mathbf{k}\,\sigma_4, \mathbf{k}+\mathbf{q}\,\sigma_1)\, A_{\sigma_1,\sigma_2;\sigma_3,\sigma_4}(\mathbf{k}, \mathbf{k}'; \omega, \mathbf{q})\,\Lambda_0^B(\mathbf{k}'+\mathbf{q}\,\sigma_3, \mathbf{k}'\,\sigma_2) \\
= & \, \frac{T}{V} \sum_{\mathbf{k}\sigma_1\sigma_2} \sum_{\epsilon} G_{\mathrm{qp}}(i\epsilon + i\omega, \mathbf{k}+\mathbf{q})\, G_{\mathrm{qp}}(i\epsilon, \mathbf{k})\, \Lambda_0^A(\mathbf{k}\,\sigma_2, \mathbf{k}+\mathbf{q}\,\sigma_1)\, \Lambda_0^B(\mathbf{k}+\mathbf{q}\,\sigma_1, \mathbf{k}\,\sigma_2) \\
& + \frac{T^2}{V^2} \sum_{\mathbf{k}\mathbf{k}'} \sum_{\epsilon\epsilon'} G_{\mathrm{qp}}(i\epsilon + i\omega, \mathbf{k}+\mathbf{q})\, G_{\mathrm{qp}}(i\epsilon, \mathbf{k})\, G_{\mathrm{qp}}(i\epsilon' + i\omega, \mathbf{k}'+\mathbf{q})\, G_{\mathrm{qp}}(i\epsilon', \mathbf{k}') \\
& \quad \sum_{\sigma_1\sigma_2\sigma_3\sigma_4} \Lambda_0^A(\mathbf{k}\,\sigma_4, \mathbf{k}+\mathbf{q}\,\sigma_1)\, A_{\sigma_1,\sigma_2;\sigma_3,\sigma_4}(\mathbf{k}, \mathbf{k}'; \omega, \mathbf{q})\,\Lambda_0^B(\mathbf{k}'+\mathbf{q}\,\sigma_3, \mathbf{k}'\,\sigma_2).
\end{aligned}
$$
$$\tag{5.55}$$

Remarkably, the linear response functions in the case of conserved quantities just depend on the quasiparticle interaction vertex A and quasiparticle energy $\epsilon_*(\mathbf{k})$, and not on $Z(\mathbf{k})$; the complexity of the interacting response function has been reduced just to two unknown functions at the Fermi or Luttinger surfaces.

Similar to (4.99)–(4.101), we can define the interaction vertex A in the $S = 0$ and $S = 1$ quasiparticle-quasihole channels, respectively,

$$
\begin{aligned}
A_S(\mathbf{k}, \mathbf{k}'; \omega, \mathbf{q}) &= \frac{1}{2}\left(A_{\uparrow,\uparrow;\uparrow,\uparrow}(\mathbf{k}, \mathbf{k}'; \omega, \mathbf{q}) + A_{\uparrow,\downarrow;\downarrow,\uparrow}(\mathbf{k}, \mathbf{k}'; \omega, \mathbf{q})\right), \\
A_A(\mathbf{k}, \mathbf{k}'; \omega, \mathbf{q}) &= \frac{1}{2}\left(A_{\uparrow,\uparrow;\uparrow,\uparrow}(\mathbf{k}, \mathbf{k}'; \omega, \mathbf{q}) - A_{\uparrow,\downarrow;\downarrow,\uparrow}(\mathbf{k}, \mathbf{k}'; \omega, \mathbf{q})\right).
\end{aligned}
\tag{5.56}
$$

We finally observe that in the case of a Luttinger surface, finite quasiparticle interaction vertex and density-vertices despite the vanishing quasiparticle residue imply diverging physical electron interaction vertex and interacting density-vertices. That is certainly the case of quasiparticle density-vertices corresponding to conserved quantities, which through (5.52) are equal to the corresponding non-interacting vertices of the physical electrons. It is, therefore, remarkable that a theory full of singularities can nonetheless yield a regular quasiparticle description.

5.4 Thermodynamic Susceptibilities

If A and B are both conserved quantities, through (5.55) the thermodynamic susceptibility κ_{AB} reads

$$
\begin{aligned}
\kappa_{AB} \equiv -\chi_{AB}^q = -\frac{1}{V} \sum_{\mathbf{k}\sigma_1\sigma_2} & \frac{\partial f\left(\epsilon_*(\mathbf{k})\right)}{\partial\epsilon_*(\mathbf{k})} \Lambda_0^A(\mathbf{k}\,\sigma_2, \mathbf{k}\,\sigma_1) \Lambda_0^B(\mathbf{k}\,\sigma_1, \mathbf{k}\,\sigma_2) \\
& -\frac{1}{V^2} \sum_{\mathbf{k}\mathbf{k}'} \frac{\partial f\left(\epsilon_*(\mathbf{k})\right)}{\partial\epsilon_*(\mathbf{k})} \frac{\partial f\left(\epsilon_*(\mathbf{k}')\right)}{\partial\epsilon_*(\mathbf{k}')} \\
& \sum_{\sigma_1\sigma_2\sigma_3\sigma_4} \Lambda_0^A(\mathbf{k}\,\sigma_4, \mathbf{k}\,\sigma_1) A_{\sigma_1,\sigma_2;\sigma_3,\sigma_4}^q(\mathbf{k}, \mathbf{k}') \Lambda_0^B(\mathbf{k}'\,\sigma_3, \mathbf{k}'\,\sigma_2),
\end{aligned}
\tag{5.57}
$$

where

$$
A_{\sigma_1,\sigma_2;\sigma_3,\sigma_4}^q(\mathbf{k}, \mathbf{k}') = \lim_{q\to 0} \lim_{\omega\to 0} A_{\sigma_1,\sigma_2;\sigma_3,\sigma_4}(\mathbf{k}, \mathbf{k}'; \omega, \mathbf{q}),
$$

is the q-limit of the quasiparticle interaction vertex. In what follows we derive the explicit expressions of the most relevant susceptibilities characterising Landau's Fermi liquid.

5.4.1 Charge Compressibility

If $A = B = N$, the number of particles, the non-interacting vertex $\Lambda_0(\mathbf{k}\,\sigma, \mathbf{k}\,\sigma') = \delta_{\sigma\sigma'}$, and the corresponding susceptibility, i.e., the charge compressibility κ, is, see (5.56),

$$
\begin{aligned}
\kappa = & -\frac{1}{V} \sum_{\mathbf{k}\sigma} \frac{\partial f\left(\epsilon_*(\mathbf{k})\right)}{\partial\epsilon_*(\mathbf{k})} \\
& -\frac{1}{V^2} \sum_{\mathbf{k}\mathbf{k}'} \sum_{\sigma\sigma'} \frac{\partial f\left(\epsilon_*(\mathbf{k})\right)}{\partial\epsilon_*(\mathbf{k})} \frac{\partial f\left(\epsilon_*(\mathbf{k}')\right)}{\partial\epsilon_*(\mathbf{k}')} A_{\sigma,\sigma';\sigma',\sigma}^q(\mathbf{k}, \mathbf{k}') \\
= & -\frac{2}{V} \sum_{\mathbf{k}} \frac{\partial f\left(\epsilon_*(\mathbf{k})\right)}{\partial\epsilon_*(\mathbf{k})} - \frac{4}{V^2} \sum_{\mathbf{k}\mathbf{k}'} \frac{\partial f\left(\epsilon_*(\mathbf{k})\right)}{\partial\epsilon_*(\mathbf{k})} \frac{\partial f\left(\epsilon_*(\mathbf{k}')\right)}{\partial\epsilon_*(\mathbf{k}')} A_S^q(\mathbf{k}, \mathbf{k}') \equiv 2\rho_{qp}\left(1 - A_S\right),
\end{aligned}
\tag{5.58}
$$

where $\rho_{qp} \equiv A_{qp}(0)$ is the quasiparticle local density of states at the chemical potential as opposed to the non-interacting one ρ_0, with, see also (4.169),

$$A_{qp}(\epsilon) = -\frac{1}{V} \sum_{\mathbf{k}} \frac{\partial f\left(\epsilon_*(\mathbf{k})\right)}{\partial \epsilon_*(\mathbf{k})} \simeq \frac{1}{V} \sum_{\mathbf{k}} \delta\left(\epsilon - \epsilon_*(\mathbf{k})\right), \tag{5.59}$$

and

$$\begin{aligned} A_S &\equiv \rho_{qp}^{-1} \frac{2}{V^2} \sum_{\mathbf{k}\mathbf{k}'} \frac{\partial f\left(\epsilon_*(\mathbf{k})\right)}{\partial \epsilon_*(\mathbf{k})} \frac{\partial f\left(\epsilon_*(\mathbf{k}')\right)}{\partial \epsilon_*(\mathbf{k}')} A_S^q(\mathbf{k}, \mathbf{k}') \\ &\simeq \rho_{qp}^{-1} \frac{2}{V^2} \sum_{\mathbf{k}\mathbf{k}'} \delta\left(\epsilon_*(\mathbf{k})\right) \delta\left(\epsilon_*(\mathbf{k}')\right) A_S^q(\mathbf{k}, \mathbf{k}'), \end{aligned} \tag{5.60}$$

is the average over the Fermi or Luttinger surface of $A_S^q(\mathbf{k}, \mathbf{k}')$. We remark that the compressibility κ_0 of non-interacting electrons is just $2\rho_0$, so that

$$\frac{\kappa}{\kappa_0} = \frac{\rho_{qp}}{\rho_0} \left(1 - A_S\right). \tag{5.61}$$

We expect that a repulsive interaction slows down quasiparticle motion yielding a reduced quasiparticle bandwidth, or, equivalently, increased quasiparticle effective mass, and thus $\rho_{qp} > \rho_0$. In addition, we also expect that the interacting compressibility κ is smaller than the non-interacting one κ_0. It follows that

$$A_S > 1 - \frac{\rho_0}{\rho_{qp}} > 0.$$

As long as $A_S < 1$, the system is compressible and hence metallic. The repulsion driven Mott metal-to-insulator transition is, therefore, signalled by

$$A_S = \frac{1}{2} \left(A_{\uparrow,\uparrow;\uparrow,\uparrow} + A_{\uparrow,\downarrow;\downarrow,\uparrow}\right) \to 1.$$

Since the more localised the charge, the smaller $A_{\uparrow,\uparrow;\uparrow,\uparrow}$ is because of Pauli exclusion principle, the approach to a Mott transition actually corresponds to $A_{\uparrow,\uparrow;\uparrow,\uparrow} \to 0$ and $A_{\uparrow,\downarrow;\downarrow,\uparrow} \to 2$. Therefore, in a Mott insulator that has a Luttinger surface, thus a finite ρ_{qp}, $A_{\uparrow,\uparrow;\uparrow,\uparrow} = 0$, $A_{\uparrow,\downarrow;\downarrow,\uparrow} = 2$, and so A_S is strictly equal to one.

5.4.2 Spin Susceptibility

If we take, instead

$$A = B = S_z = \frac{1}{2} \sum_{\mathbf{k}} \left(c_{\mathbf{k}\uparrow}^\dagger c_{\mathbf{k}\uparrow} - c_{\mathbf{k}\downarrow}^\dagger c_{\mathbf{k}\downarrow}\right),$$

i.e., the z-component of the total spin, the non-interacting vertex $\Lambda_0^{S_z}(\mathbf{k}\,\sigma, \mathbf{k}\,\sigma') = \delta_{\sigma\sigma'}(\delta_{\sigma\uparrow} - \delta_{\sigma\downarrow})$, and simply following the derivation above, the corresponding spin susceptibility χ is finite at low temperature and reads

$$\chi = \rho_{\text{qp}}\left(1 - A_A\right), \tag{5.62}$$

where A_A is now the average over the Fermi or Luttinger surface of $A_A^q(\mathbf{k}, \mathbf{k}')$. Evidently, because of spin $SU(2)$ symmetry, the spin susceptibilities to fields along any other direction different from z is exactly the same as (5.62). It follows that

$$\frac{\chi}{\chi_0} = \frac{\rho_{\text{qp}}}{\rho_0}\left(1 - A_A\right), \tag{5.63}$$

where $\chi_0 = \rho_0$ is the non-interacting spin susceptibility.

Contrary to the charge compressibility, we expect that $\chi > \chi_0$ when the interaction is repulsive. Moreover, approaching a Mott transition, and in light of the previous discussion

$$A_A = \frac{1}{2}\left(A_{\uparrow,\uparrow;\uparrow,\uparrow} - A_{\uparrow,\downarrow;\downarrow,\uparrow}\right) \simeq -\frac{1}{2}\,A_{\uparrow,\downarrow;\downarrow,\uparrow} \to -1,$$

so that $\chi \to 2\rho_{\text{qp}}$, i.e., twice the value expected for non-interacting quasiparticles. Strikingly, this result remains true even inside a Mott insulator that sustains a Luttinger surface, and which, despite the charge gap, should display a conventional Pauli paramagnetic susceptibility; what is commonly known as a spin-liquid insulator.

5.4.3 Specific Heat

In Sect. 4.9.1, we derived the Ward-Takahashi identity (4.137) for the heat density, which, in the ω-limit and, as discussed earlier, for real and small frequency ϵ reads

$$
\begin{aligned}
\Lambda^{E\,\omega}(\epsilon, \mathbf{k}, \sigma) &= \epsilon(\mathbf{k}) + \Sigma(\epsilon, \mathbf{k}) - \epsilon\,\frac{\partial\Sigma(\epsilon, \mathbf{k})}{\partial\epsilon} = \left(\epsilon(\mathbf{k}) + \text{Re}\,\Sigma(\epsilon, \mathbf{k}) - \epsilon\right) \\
&\quad + i\,\text{Im}\,\Sigma(\epsilon, \mathbf{k}) + \epsilon\left(1 - \frac{\partial\text{Re}\Sigma(\epsilon, \mathbf{k})}{\partial\epsilon}\right) - i\,\epsilon\,\frac{\partial\text{Im}\Sigma(\epsilon, \mathbf{k})}{\partial\epsilon} \\
&= Z(\epsilon, \mathbf{k})^{-1}\left(\Xi(\epsilon, \mathbf{k}) + \epsilon + i\,\Gamma(\epsilon, \mathbf{k}) - i\,Z(\epsilon, \mathbf{k})\,\epsilon\,\frac{\partial\text{Im}\Sigma(\epsilon, \mathbf{k})}{\partial\epsilon}\right) \\
&\simeq Z(\epsilon, \mathbf{k})^{-1}\left(\Xi(\epsilon, \mathbf{k}) + \epsilon\right) \equiv Z(\epsilon, \mathbf{k})^{-1}\,\lambda^{E\,\omega}(\epsilon, \mathbf{k}, \sigma),
\end{aligned}
\tag{5.64}
$$

the last almost equality deriving from the fact that both in case (F) and (L) the neglected imaginary terms vanishes at least as ϵ^2. Since R_{qp} in (5.54) implies $\epsilon = \epsilon_*(\mathbf{k})$ and $\Xi(\epsilon_*(\mathbf{k}), \mathbf{k}) = 0$, then the specific heat through (5.57) becomes

$$
c_V = -\frac{1}{T} \chi_{EE}^q(0, \mathbf{q}) = -\frac{2}{T V} \sum_{\mathbf{k}} \epsilon_*(\mathbf{k})^2 \frac{\partial f(\epsilon_*(\mathbf{k}))}{\partial \epsilon_*(\mathbf{k})}
$$

$$
- \frac{4}{T V^2} \sum_{\mathbf{kk'}} \epsilon_*(\mathbf{k}) \epsilon_*(\mathbf{k'}) \frac{\partial f(\epsilon_*(\mathbf{k}))}{\partial \epsilon_*(\mathbf{k})} \frac{\partial f(\epsilon_*(\mathbf{k'}))}{\partial \epsilon_*(\mathbf{k'})} A_S^q(\mathbf{k}, \mathbf{k'})
$$

$$
\equiv c_V^{(1)} + c_V^{(2)}.
$$

$$(5.65)$$

The first term $c_V^{(1)}$ can be written as

$$
c_V^{(1)} = \frac{2}{T} \int_{-\infty}^{\infty} d\epsilon \, A_*(\epsilon) \, \epsilon^2 \left(-\frac{\partial f(\epsilon)}{\partial \epsilon} \right) \simeq \frac{2\pi^2}{3} T \, \rho_{qp} . \tag{5.66}
$$

The second contribution $c_V^{(2)}$ requires expanding $A_S^q(\mathbf{k}, \mathbf{k'})$ at first order in $\epsilon_*(\mathbf{k}) \, \epsilon_*(\mathbf{k'})$, i.e., $A_S^q(\mathbf{k}, \mathbf{k'}) - A_S^q(\mathbf{k}_{FL}, \mathbf{k'}_{FL}) \sim \epsilon_*(\mathbf{k}) \epsilon_*(\mathbf{k'})$ times known, see, e.g., [18], logarithmic corrections that yield $c_V^{(2)} \sim T^3 \ln T$, subleading with respect to $c_V^{(1)} \sim T$.

Therefore, at leading order in T,

$$
c_V = \frac{2\pi^2}{3} T \, \rho_{qp} , \tag{5.67}
$$

which is the same expression of the specific heat c_{0V} of non-interacting electrons, with the non-interacting DOS ρ_0 replaced by the quasiparticle one ρ_{qp}. It is common to quantify the strength of correlations in a Fermi liquid through the so-called Wilson ratio, defined as

$$
R_W \equiv \frac{\chi}{\chi_0} \frac{c_{0V}}{c_V} = 1 - A_A . \tag{5.68}
$$

The Wilson ratio $R_W = 1$ in absence of interaction, and grows with the strength of interaction reaching $R_W = 2$ at the Mott transition, and even inside the Mott insulator if it has a Luttinger surface. That is also striking since it implies that such Mott insulator has a Pauli-like paramagnetic susceptibility, linear in temperature specific heat, and all that despite the charge gap.

5.5   Current-Current Response Functions

The densities ρ_A of conserved quantities are associated via the continuity equation to the current densities $\mathbf{J}_A$. Although the latter are not densities of conserved quantities, we can still derive a simple expression of their response functions, which, in presence of Coulomb repulsion, correspond to the proper ones, i.e., the response to the internal fields

$$\chi_{iA\,jB}(\tau, \mathbf{q}) = -\frac{1}{V} \langle T_\tau \Big(J_{iA}(\tau, \mathbf{q}) \, J_{jB}(0, -\mathbf{q}) \Big) \rangle,$$

where J_{iA} and J_{iB}, $i = x, y, z$, are the components of $\mathbf{J}_A$ and $\mathbf{J}_B$, respectively. We remark that, since we are considering small q and normal, i.e., non superconducting, Fermi liquids, we can safely assume that the proper longitudinal and transverse current-current response functions are coincident, and thus focus just on the proper longitudinal response.

For that, we use the alternative expression (5.49) of a generic response function, namely

$$\chi_{iA\,jB} = \chi_{iA\,jB}^q + \text{Tr}\Big(\lambda_i^{A\,q} \odot \tilde{R}_{\text{qp}} \odot \lambda_j^{B\,q} \Big) + \text{Tr}\Big(\lambda_i^{A\,q} \odot \tilde{R}_{\text{qp}} \odot A \odot \tilde{R}_{\text{qp}} \odot \lambda_j^{B\,q} \Big), \tag{5.69}$$

where now, through (4.131), the interacting current density-vertices satisfy in the q-limit the equation

$$\lambda^{Q\,q} = Z(\epsilon, \mathbf{k}) \left(\Lambda_0^{Q\,q} + \Lambda_0^{Q\,q} \, \frac{\partial \Sigma(\epsilon, \mathbf{k})}{\partial \mathbf{k}} \right),$$

for $Q = A, B$. We note that, in the response function (5.69), the kernel $\tilde{R}_{\text{qp}}$ imply that $\epsilon = \epsilon_*(\mathbf{k})$. In our specific case, the only conserved quantities are the total charge, spin, and energy. Therefore, for the charge current vertex, we find

$$\lambda^q = Z\big(\epsilon_*(\mathbf{k}), \mathbf{k}\big) \left(\frac{\partial \epsilon_{\mathbf{k}}}{\partial \mathbf{k}} + \frac{\partial \Sigma(\epsilon, \mathbf{k})}{\partial \mathbf{k}} \Big|_{\epsilon = \epsilon_*(\mathbf{k})} \right) = \frac{\partial \epsilon_*(\mathbf{k})}{\partial \mathbf{k}} = v_{\mathbf{k}}, \tag{5.70}$$

for the spin, e.g., the component $a = x, y, z$,

$$\lambda_a^q = \frac{\sigma_a}{2} \frac{\partial \epsilon_*(\mathbf{k})}{\partial \mathbf{k}} = \frac{\sigma_a}{2} v_{\mathbf{k}}, \tag{5.71}$$

with σ_a the Pauli matrices, and finally, for the heat current, through (4.137)

$$\lambda^{E\,q} = \epsilon_*(\mathbf{k}) \, Z\big(\epsilon_*(\mathbf{k}), \mathbf{k}\big) \left(\frac{\partial \epsilon_{\mathbf{k}}}{\partial \mathbf{k}} + \frac{\partial \Sigma(i\epsilon, \mathbf{k})}{\partial \mathbf{k}} \right) = \epsilon_*(\mathbf{k}) \frac{\partial \epsilon_*(\mathbf{k})}{\partial \mathbf{k}} = \epsilon_*(\mathbf{k}) \, v_{\mathbf{k}}. \tag{5.72}$$

The next question concerns the q-limit of the current-current response functions (5.69). In the case of the charge current that is not a real issue, since gauge invariance implies that the q-limit of the current-current response function cancels the diamagnetic term. In that way, we can readily derive the expression of the ω-limit of the conductivity, which, assuming space isotropy or cubic symmetry, and real frequency, $i\omega \to \omega + i\eta$ with $\eta > 0$ infinitesimal, is simply

$$
\begin{aligned}
\sigma_\parallel(\omega) = \sigma_\perp(\omega) &= i\,\frac{e^2}{\omega + i\eta}\left[\mathrm{Tr}\!\left(\lambda_i^q \odot \widetilde{R}_*^\omega \odot \lambda_i^q\right) + \mathrm{Tr}\!\left(\lambda_i^q \odot \widetilde{R}_*^\omega \odot A^\omega \odot \widetilde{R}_*^\omega \odot \lambda_i^q\right)\right] \\
&= i\,\frac{e^2}{\omega + i\eta}\left[\frac{2}{3V}\sum_{\mathbf{k}} v_{\mathbf{k}} \cdot v_{\mathbf{k}} \left(-\frac{\partial f\big(\epsilon_*(\mathbf{k})\big)}{\partial\epsilon_*(\mathbf{k})}\right)\right.\\
&\qquad\left. + \frac{4}{3V^2}\sum_{\mathbf{k}\mathbf{k}'} v_{\mathbf{k}} \cdot v_{\mathbf{k}'} \frac{\partial f\big(\epsilon_*(\mathbf{k})\big)}{\partial\epsilon_*(\mathbf{k})}\frac{\partial f\big(\epsilon_*(\mathbf{k}')\big)}{\partial\epsilon_*(\mathbf{k}')} A_S^\omega(\mathbf{k},\mathbf{k}')\right]\\
&= D_*\left(\delta(\omega) + \frac{i}{\omega}\right),
\end{aligned}
\tag{5.73}
$$

with the Drude weight

$$
\begin{aligned}
D_* &= \frac{2e^2}{3V}\sum_{\mathbf{k}} v_{\mathbf{k}} \cdot v_{\mathbf{k}} \left(-\frac{\partial f\big(\epsilon_*(\mathbf{k})\big)}{\partial\epsilon_*(\mathbf{k})}\right)\\
&\quad + \frac{4e^2}{3V^2}\sum_{\mathbf{k}\mathbf{k}'} v_{\mathbf{k}} \cdot v_{\mathbf{k}'} \frac{\partial f\big(\epsilon_*(\mathbf{k})\big)}{\partial\epsilon_*(\mathbf{k})}\frac{\partial f\big(\epsilon_*(\mathbf{k}')\big)}{\partial\epsilon_*(\mathbf{k}')} A_S^\omega(\mathbf{k},\mathbf{k}')\\
&= \frac{2e^2}{3V}\sum_{\mathbf{k}} v_{\mathbf{k}} \cdot v_{\mathbf{k}} \left(-\frac{\partial f\big(\epsilon_*(\mathbf{k})\big)}{\partial\epsilon_*(\mathbf{k})}\right)\\
&\quad + \frac{4e^2}{3V^2}\sum_{\mathbf{k}\mathbf{k}'} v_{\mathbf{k}} \cdot v_{\mathbf{k}'} \frac{\partial f\big(\epsilon_*(\mathbf{k})\big)}{\partial\epsilon_*(\mathbf{k})}\frac{\partial f\big(\epsilon_*(\mathbf{k}')\big)}{\partial\epsilon_*(\mathbf{k}')} f_S(\mathbf{k},\mathbf{k}'),
\end{aligned}
\tag{5.74}
$$

having defined

$$
A_S^\omega(\mathbf{k},\mathbf{k}') \equiv f_S(\mathbf{k},\mathbf{k}'),\qquad A_A^\omega(\mathbf{k},\mathbf{k}') \equiv f_A(\mathbf{k},\mathbf{k}').
\tag{5.75}
$$

Assuming instead a fictitious longitudinal and static vector potential opposite for the two spin directions, we could, similarly as in the charge case, gauge it away through a unitary transformation, and thus conclude that also the q-limit of the spin current-spin current response function vanishes, which allows calculating the latter with ease.

5.5.1   Thermal Response

The thermal response is less simple to analyse, since the conserved quantity, the total energy, is a many-body operator. First, let us discuss how thermal response may arise. We shall do that in an unusual way. Specifically, we assume the system is initially at equilibrium with a bath at temperature T, but later, the bath changes and thus its temperature $T \to T + \delta T(t, \mathbf{r})$, with $\left|\delta T(t, \mathbf{r})\right| \ll T$ and so slowly varying in space and time that it is still possible to describe the system density matrix just replacing T with $T + \delta T(t, \mathbf{r})$, thus

$$\hat{\rho}(t) = \frac{1}{Z} \exp\left[-\int d\mathbf{r} \, \frac{\rho_E(\mathbf{r})}{T + \delta T(t, \mathbf{r})} \right].$$

Therefore

$$\frac{\rho_E(\mathbf{r})}{T + \delta T(t, \mathbf{r})} \simeq \frac{\rho_E(\mathbf{r})}{T} - \rho_E(\mathbf{r}) \frac{\delta T(t, \mathbf{r})}{T^2} = \frac{1}{T} \left(\rho_E(\mathbf{r}) + \phi_E(t, \mathbf{r}) \rho_E(\mathbf{r}) \right),$$

with the dimensionless field

$$\phi_E(t, \mathbf{r}) \equiv -\frac{\delta T(t, \mathbf{r})}{T}, \qquad (5.76)$$

so that the density matrix

$$\hat{\rho}(t) \simeq \frac{1}{Z} \exp\left[-\frac{1}{T} \int d\mathbf{r} \left(\rho_E(\mathbf{r}) + \phi_E(t, \mathbf{r}) \rho_E(\mathbf{r}) \right) \right] = \frac{e^{-\beta\left(H + \delta H(t)\right)}}{\mathrm{Tr}\left(e^{-\beta\left(H + \delta H(t)\right)}\right)},$$

with the perturbation

$$\delta H = \int d\mathbf{r} \, \phi_E(t, \mathbf{r}) \rho_E(\mathbf{r}) = \frac{1}{V} \sum_{\mathbf{q}} \phi_E(t, \mathbf{q}) \rho_E(-\mathbf{q}),$$

with the convention

$$A(\mathbf{q}) = \int d\mathbf{r} \, e^{-i\mathbf{q}\cdot\mathbf{r}} A(\mathbf{r}),$$

for both $\phi_E(t, \mathbf{q})$ and $\rho_E(\mathbf{q})$. At $\phi_E \neq 0$, the heat density at finite $\mathbf{q}$ acquires a finite expectation value that, in linear response, reads

$$\rho_E(\omega, \mathbf{q}) = \chi_{EE}(\omega, \mathbf{q}) \, \phi_E(\omega, \mathbf{q}),$$

where $\chi_{EE}(\omega, \mathbf{q})$ is the Fourier transform of

$$\chi_{EE}(t, \mathbf{q}) = -\frac{i}{V} \theta(t) \left\langle \left[\rho_E(t, \mathbf{q}), \rho_E(0, -\mathbf{q}) \right] \right\rangle,$$

calculated on the initial state at temperature T. We observe that the change in total energy δE due to a static $\delta T(\mathbf{r})$ in the uniform limit $\delta T(\mathbf{r}) \to \delta T$ can be calculated through

$$\delta E = \lim_{\mathbf{q}\to 0} \lim_{\omega\to 0} \rho_E(\omega, \mathbf{q}) = \lim_{\mathbf{q}\to 0} \lim_{\omega\to 0} \chi_{EE}(\omega, \mathbf{q}) \, \phi_E(\omega, \mathbf{q}) = -V \frac{\delta T}{T} \chi_{EE}^q,$$

yielding the known expression of the specific heat

$$c_V = \frac{1}{V} \frac{\delta E}{\delta T} = -\frac{1}{T} \chi_{EE}^q .$$

On the other end, by the continuity equation

$$i \dot\rho_E(t, \mathbf{q}) = \Big[\rho_E(t, \mathbf{q}), H\Big] = \mathbf{q} \cdot \mathbf{J}_E(t, \mathbf{q}), \quad \omega \rho_E(\omega, \mathbf{q}) = \mathbf{q} \cdot \mathbf{J}_E(\omega, \mathbf{q}).$$

so that

$$\mathbf{J}_{\parallel E}(\omega, \mathbf{q}) = \frac{\omega}{q^2} \chi_{EE}(\omega, \mathbf{q}) \; \mathbf{q} \; \phi_E(\omega, \mathbf{q}). \tag{5.77}$$

We note that since $\chi_{EE}(0, \mathbf{q})$ is a thermodynamic susceptibility, and thus finite, the heat current vanishes at $\omega = 0$, and therefore, the heat current-heat current response function must vanish too, exactly like for the charge and spin. That includes evidently also the q-limit.

However, we still need to relate (5.77) to the current-current response. For that, we can readily show that

$$
\begin{aligned}
-i \frac{\partial}{\partial t'} \, i \frac{\partial}{\partial t} \, \chi_{EE}(t - t', \mathbf{q}) &= -\frac{1}{V} \, i \frac{\partial}{\partial t'} \, i \frac{\partial}{\partial t} \left(-i \, \theta(t - t') \langle \Big[\rho_E(t, \mathbf{q}), \rho_E(t', -\mathbf{q})\Big] \rangle \right) \\
&= \frac{1}{V} \Big(\delta(t - t') \langle \Big[\mathbf{q} \cdot \mathbf{J}_E(\mathbf{q}), \rho_E(-\mathbf{q})\Big] \rangle \\
&\quad - i \, \theta(t - t') \langle \Big[\mathbf{q} \cdot \mathbf{J}_E(t, \mathbf{q}), \mathbf{q} \cdot \mathbf{J}_E(t', -\mathbf{q})\Big] \rangle \Big) \\
&= \delta(t - t') \frac{1}{V} \langle \Big[\mathbf{q} \cdot \mathbf{J}_E(\mathbf{q}), \rho_E(-\mathbf{q})\Big] \rangle + q^2 \chi_{\parallel EE}(t - t', \mathbf{q}),
\end{aligned}
$$

$$\tag{5.78}$$

where $\chi_{\parallel EE}$ is the longitudinal component of the heat current-heat current response function. We further note that, at $T = 0$,

$$
\begin{aligned}
\frac{1}{V} \langle 0 | \Big[\mathbf{q} \cdot \mathbf{J}_E(\mathbf{q}), \rho_E(-\mathbf{q})\Big] | 0 \rangle &= \frac{1}{V} \langle 0 | \Big[\big[\rho_E(\mathbf{q}), H\big], \rho_E(-\mathbf{q})\Big] | 0 \rangle \\
&= \frac{1}{V} \langle 0 | \rho_E(\mathbf{q}) H \rho_E(-\mathbf{q}) | 0 \rangle - \frac{E_0}{V} \langle 0 | \rho_E(\mathbf{q}) \rho_E(-\mathbf{q}) | 0 \rangle \\
&\equiv \epsilon_E(\mathbf{q}) \, S_E(\mathbf{q}),
\end{aligned}
$$

where

$$\epsilon_E(\mathbf{q}) \equiv \frac{1}{V} \frac{\langle 0 | \rho_E(\mathbf{q}) H \rho_E(-\mathbf{q}) | 0 \rangle}{\langle 0 | \rho_E(\mathbf{q}) \rho_E(-\mathbf{q}) | 0 \rangle} - \frac{E_0}{V},$$

is the excitation energy per unit volume of an heat-density fluctuation at momentum $\mathbf{q}$ and, in a Fermi liquid, vanishes linearly in q as $\mathbf{q} \to \mathbf{0}$, while

$$S_E(\mathbf{q}) = \langle 0 | \rho_E(\mathbf{q}) \rho_E(-\mathbf{q}) | 0 \rangle,$$

is the heat-density structure factor, which also vanishes linearly in a Fermi liquid with gapless excitations. Therefore, at small q, $\epsilon_E(\mathbf{q}) S_E(\mathbf{q}) \sim q^2$, and we can write

$$\frac{1}{V} \langle 0 \mid \left[\mathbf{q} \cdot \mathbf{J}_E(\mathbf{q}), \, \rho_E(-\mathbf{q}) \right] \mid 0 \rangle = q^2 N_E(\mathbf{q}).$$

with $N_E(\mathbf{0})$ finite, so that, upon Fourier transform (5.78), we find

$$\omega^2 \chi_{EE}(\omega, \mathbf{q}) = q^2 \left(\chi_{\| EE}(\omega, \mathbf{q}) + N_E(\mathbf{q}) \right), \tag{5.79}$$

which is the counterpart of the similar equation for the charge density, in which case $N_E(\mathbf{q}) \to n/m$, the diamagnetic term. It thus follows that, through (5.76),

$$\mathbf{J}_{\| E}(\omega, \mathbf{q}) = \frac{\omega}{q^2} \chi_{EE}(\omega, \mathbf{q}) \; \mathbf{q} \, \phi_E(\omega, \mathbf{q}) = \left(\chi_{\| EE}(\omega, \mathbf{q}) + N_E(\mathbf{q}) \right) \frac{\mathbf{q} \, \phi_E(\omega, \mathbf{q})}{\omega}$$

$$= \frac{i}{\omega} \left(\chi_{\| EE}(\omega, \mathbf{q}) + N_E(\mathbf{q}) \right) \int d\mathbf{r} \, e^{-\mathbf{q} \cdot \mathbf{r}} \, \frac{\nabla \delta T(\omega, \mathbf{r})}{T}.$$

Let us take the limit $\mathbf{q} \to \mathbf{0}$ assuming that that $\nabla \delta T(\omega, \mathbf{r}) \to \nabla \delta T(\omega)$ remains finite and space independent, thus

$$\frac{1}{V} \mathbf{J}_{\| E}(\omega, \mathbf{0}) = -\frac{1}{i\omega T} \left(\chi_{\| EE}(\omega, \mathbf{0}) + N_E(\mathbf{0}) \right) \nabla \delta T(\omega) \equiv K(\omega) \, \nabla \delta T(\omega), \tag{5.80}$$

which defines the thermal conductivity $K(\omega \to 0)$. Since, as we earlier demonstrated, $\chi^q_{\| EE} + N_E(\mathbf{0}) = 0$, through (5.69), we readily find

$$K(\omega) = \frac{1}{T} \frac{i}{\omega + i\eta} \left[\frac{2}{3V} \sum_{\mathbf{k}} \epsilon_*(\mathbf{k})^2 \, v_{\mathbf{k}} \cdot v_{\mathbf{k}} \left(-\frac{\partial f(\epsilon_*(\mathbf{k}))}{\partial \epsilon_*(\mathbf{k})} \right) \right.$$

$$\left. + \frac{4}{3V^2} \sum_{\mathbf{k}\mathbf{k}'} \epsilon_*(\mathbf{k}) \, \epsilon_*(\mathbf{k}') \, v_{\mathbf{k}} \cdot v_{\mathbf{k}'} \frac{\partial f(\epsilon_*(\mathbf{k}))}{\partial \epsilon_*(\mathbf{k})} \frac{\partial f(\epsilon_*(\mathbf{k}'))}{\partial \epsilon_*(\mathbf{k}')} \, f_S(\mathbf{k}, \mathbf{k}') \right]$$

$$= K_* \left(\delta(\omega) + \frac{i}{\omega} \right), \tag{5.81}$$

thus a thermal Drude weight $K_* \sim T$.

We note that the sign in (5.80) is opposite with respect to the usual expression $\mathbf{J}_E = -V K \nabla T$. The reason lies in the way we have added the thermal perturbation. Namely, we have assumed that the system is initially in equilibrium at uniform temperature T, and later on, a temperature drop across the opposite sides of the sample is applied as a perturbation. On the contrary, the common way to measure heat conductivity is in a sample that, at equilibrium, is already in contact with two baths, one at higher temperature than the other. To appreciate the difference, we can consider the equivalent case of an electrochemical potential drop. If the system is initially at equilibrium with constant electrochemical potential $-e \mu$, thus Hamiltonian

$H - e\,\mu\,N$, and a $\delta\mu(t, \mathbf{r})$ is switched on, such that $\nabla\delta\mu(t) = \nabla\mu(t)$ is constant in space, then for $\nabla\mu(t) \to \nabla\mu$, namely, in the ω-limit, we would find

$$\boldsymbol{J} = V\,\sigma\,\nabla\mu. \tag{5.82}$$

Therefore, in this set up, electrons flow towards the side with larger chemical potential, and thus $\boldsymbol{J}$ is parallel to $\nabla\mu$. On the contrary, should we assume a sample that, at equilibrium, had an electrochemical potential drop, electrons would flow from the side with larger μ towards that with lower one, thus $\boldsymbol{J}$ antiparallel to $\nabla\mu$.

5.5.2 Coulomb Interaction

The whole derivation above outlined in Landau's Fermi liquid Theory is based on the non-analytic properties of the kernel R in (5.18). A long-range Coulomb interaction yields an additional singularity at small q of all improper vertices, both interaction and density ones. However, if we concentrate just on the proper vertices, we can repeat step-by-step the whole derivation, which implies that the density-density response functions (5.55) of densities associated with conserved quantities remain the same if we regard them as proper density-density response functions.

For instance, the proper density-density response function is still given by (5.55), and reads

$$
\begin{aligned}
\chi_*(\omega, \mathbf{q}) = &-\frac{2e^2}{V} \sum_{\mathbf{k}} \frac{\partial f\big(\epsilon_*(\mathbf{k})\big)}{\partial\epsilon_*(\mathbf{k})} \frac{\mathbf{q}\cdot v_{\mathbf{k}}}{\omega - \mathbf{q}\cdot v_{\mathbf{k}}} \\
&+ \frac{4e^2}{V^2} \sum_{\mathbf{k}\mathbf{k}'} \frac{\partial f\big(\epsilon_*(\mathbf{k})\big)}{\partial\epsilon_*(\mathbf{k})} \frac{\partial f\big(\epsilon_*(\mathbf{k}')\big)}{\partial\epsilon_*(\mathbf{k}')} \frac{\mathbf{q}\cdot v_{\mathbf{k}}}{\omega - \mathbf{q}\cdot v_{\mathbf{k}}} \frac{\mathbf{q}\cdot v_{\mathbf{k}'}}{\omega - \mathbf{q}\cdot v_{\mathbf{k}'}} A_S(\mathbf{k}, \mathbf{k}'; \omega, \mathbf{q}),
\end{aligned}
\tag{5.83}
$$

while the improper one is, as usual,

$$\chi(\omega, \mathbf{q}) = \frac{\chi_*(\omega, \mathbf{q})}{1 - \dfrac{4\pi e^2}{q^2} \chi_*(\omega, \mathbf{q})} \equiv \frac{\chi_*(\omega, \mathbf{q})}{\epsilon_\parallel(\omega, \mathbf{q})}.$$

5.6   Mott Insulators with a Luttinger Surface

With the inclusion of Coulomb repulsion, we can now discuss more realistically what a Mott insulator with a Luttinger surface implies for transport properties, and more importantly, whether the resulting behaviour is physically consistent, e.g., compatible with gauge invariance.

The longitudinal current-current proper response function can be written either using (5.48) or (5.49), thus

$$\chi_{*\|} = \chi_{*\|}^{\omega} + \text{Tr}\left(\lambda_{\|}^{\omega} \odot R_{\text{qp}} \odot \lambda_{\|}^{\omega}\right) + \text{Tr}\left(\lambda_{\|}^{\omega} \odot R_{\text{qp}} \odot A \odot R_{\text{qp}} \odot \lambda_{\|}^{\omega}\right)$$
$$= \chi_{*\|}^{q} + \text{Tr}\left(\lambda_{\|}^{q} \odot \tilde{R}_{\text{qp}} \odot \lambda_{\|}^{q}\right) + \text{Tr}\left(\lambda_{\|}^{q} \odot \tilde{R}_{\text{qp}} \odot A \odot \tilde{R}_{\text{qp}} \odot \lambda_{\|}^{q}\right),$$

(5.84)

where

$$\lambda_{\|}^{\omega} = \frac{\mathbf{q} \cdot \boldsymbol{\lambda}^{\omega}}{q}, \quad \lambda_{\|}^{q} = -e\,\frac{\mathbf{q} \cdot \boldsymbol{v}_{\mathbf{k}}}{q},$$

the last equation deriving from the Ward-Takahashi identity.

In a lattice model where a Mott transition at half-filling may occur, gauge invariance (2.79) implies, similar to (5.78), that, for small $\mathbf{q}$

$$\omega^2\,\chi_*(\omega, \mathbf{q}) = q^2\,\chi_{*\|}(\omega, \mathbf{q}) + \frac{1}{V}\,\langle\left[\mathbf{q} \cdot \boldsymbol{J}(\mathbf{q}), \rho(-\mathbf{q})\right]\rangle,$$

where

$$-e\sum_{\mathbf{k}\sigma}\left(\epsilon_{\mathbf{k+q}} - \epsilon_{\mathbf{k}}\right) c_{\mathbf{k}\sigma}^{\dagger} c_{\mathbf{k+q}\sigma} \xrightarrow[q\to0]{} \mathbf{q} \cdot \boldsymbol{J}(\mathbf{q}), \quad \rho(-\mathbf{q}) = -e\sum_{\mathbf{k}\sigma} c_{\mathbf{k+q}\sigma}^{\dagger} c_{\mathbf{k}\sigma}.$$

Therefore

$$\frac{e^2}{V}\sum_{\mathbf{k}\sigma}\left(\epsilon_{\mathbf{k+q}} + \epsilon_{\mathbf{k-q}} - 2\epsilon_{\mathbf{k}}\right)\langle c_{\mathbf{k}\sigma}^{\dagger} c_{\mathbf{k}\sigma}\rangle \xrightarrow[q\to0]{} \frac{1}{V}\,\langle\left[\mathbf{q} \cdot \boldsymbol{J}(\mathbf{q}), \rho(-\mathbf{q})\right]\rangle \equiv q^2\,T,$$

where

$$T \equiv \frac{e^2}{V}\sum_{\mathbf{k}\sigma}\sum_{i,j=1}^{3} \frac{q_i\,q_j}{q^2}\,\frac{\partial^2 \epsilon_{\mathbf{k}}}{\partial k_i \partial k_j}\,\langle c_{\mathbf{k}\sigma}^{\dagger} c_{\mathbf{k}\sigma}\rangle,$$

is the diamagnetic term of the lattice model. It follows that gauge invariance entails

$$\frac{q^2}{\omega^2}\left(\chi_{*\|}(\omega, \mathbf{q}) + T\right) = \chi_*(\omega, \mathbf{q}), \quad \chi_{*\|}^{q} = -T,$$

(5.85)

while, through (5.84), the optical conductivity read

$$\sigma_{\|}(\omega, \mathbf{q}) = e^2\,\frac{i}{\omega + i\eta}\left(\chi_{*\|}(\omega, \mathbf{q}) + T\right) \xrightarrow[\omega-\text{limit}]{} \sigma_{\|}^{\omega} = D_*\,\frac{i}{\omega + i\eta}.$$

In the Mott insulator, $D_* = 0$ and that implies $\chi_{*\|}^{\omega} = -T$, which is not surprising since the non-analyticity at $\omega = q = 0$ typical of metal must disappear in the insulator.

If we now take the ω-limit and use the second equation in (5.84), or the q-limit and use instead the first equation, we find that

$$\chi^\omega_{*\|} = -T + \mathrm{Tr}\left(\lambda^q_\| \odot \tilde{R}^\omega_{\mathrm{qp}} \odot \lambda^q_\|\right) + \mathrm{Tr}\left(\lambda^q_\| \odot \tilde{R}^\omega_{\mathrm{qp}} \odot A^\omega \odot \tilde{R}^\omega_{\mathrm{qp}} \odot \lambda^q_\|\right),$$

$$\chi^q_{*\|} = -T = \chi^\omega_{*\|} + \mathrm{Tr}\left(\lambda^\omega_\| \odot R^q_{\mathrm{qp}} \odot \lambda^\omega_\|\right) + \mathrm{Tr}\left(\lambda^\omega_\| \odot R^q_{\mathrm{qp}} \odot A^q \odot R^q_{\mathrm{qp}} \odot \lambda^\omega_\|\right).$$

Therefore, in a Mott insulator where $\chi^\omega_{*\|} = -T$, and recalling that

$$\tilde{R}^\omega_{\mathrm{qp}} = -R^q_{\mathrm{qp}} = -\frac{\partial f\left(\epsilon_*(\mathbf{k})\right)}{\partial \epsilon_*(\mathbf{k})} \equiv -f',$$

the following two equations are fulfilled:

$$\begin{aligned}
0 &= -\mathrm{Tr}\left(\lambda^q_\| \odot f' \odot \lambda^q_\|\right) + \mathrm{Tr}\left(\lambda^q_\| \odot f' \odot A^\omega \odot f' \odot \lambda^q_\|\right), \\
0 &= \mathrm{Tr}\left(\lambda^\omega_\| \odot f' \odot \lambda^\omega_\|\right) + \mathrm{Tr}\left(\lambda^\omega_\| \odot f' \odot A^q \odot f' \odot \lambda^\omega_\|\right).
\end{aligned} \tag{5.86}$$

On the other hand, the density-density proper response function reads

$$\chi_* = \mathrm{Tr}\left(\lambda^\omega \odot R_{\mathrm{qp}} \odot \lambda^\omega\right) + \mathrm{Tr}\left(\lambda^\omega \odot R_{\mathrm{qp}} \odot A \odot R_{\mathrm{qp}} \odot \lambda^\omega\right), \tag{5.87}$$

where $\lambda^\omega = -e$ by the Ward-Takahashi identity. The charge compressibility is, therefore

$$\begin{aligned}
\kappa = -\chi^q_* &= -\mathrm{Tr}\left(\lambda^\omega \odot R^q_{\mathrm{qp}} \odot \lambda^\omega\right) - \mathrm{Tr}\left(\lambda^\omega \odot R^q_{\mathrm{qp}} \odot A^q \odot R^q_{\mathrm{qp}} \odot \lambda^\omega\right) \\
&= -\mathrm{Tr}\left(\lambda^\omega \odot f' \odot \lambda^\omega\right) - \mathrm{Tr}\left(\lambda^\omega \odot f' \odot A^q \odot f' \odot \lambda^\omega\right),
\end{aligned}$$

and must vanish in the Mott insulator, yielding an additional equation

$$0 = \mathrm{Tr}\left(\lambda^\omega \odot f' \odot \lambda^\omega\right) + \mathrm{Tr}\left(\lambda^\omega \odot f' \odot A^q \odot f' \odot \lambda^\omega\right), \tag{5.88}$$

which, compared with the second equation in (5.86), and recalling (5.85), implies that

$$\lambda^\omega_\| = \frac{\omega}{q}\,\lambda^\omega = -e\,\frac{\omega}{q}\,, \tag{5.89}$$

in a Mott insulator, besides the already discussed property $A_S = 1$.

The first equation in (5.86) reads instead, since $A^\omega = f$, the Landau f-parameter,

$$
0 = \frac{1}{V} \sum_{\mathbf{k}} \left(-\frac{\partial f\big(\epsilon_*(\mathbf{k})\big)}{\partial \epsilon_*(\mathbf{k})} \right) \mathbf{q} \cdot \mathbf{v}_{\mathbf{k}} \left\{ \mathbf{q} \cdot \mathbf{v}_{\mathbf{k}} - \frac{2}{V} \sum_{\mathbf{k}'} \mathbf{q} \cdot \mathbf{v}_{\mathbf{k}'} \frac{\partial f\big(\epsilon_*(\mathbf{k}')\big)}{\partial \epsilon_*(\mathbf{k}')} f_S(\mathbf{k}, \mathbf{k}') \right\}
$$

$$
\Rightarrow 0 = \frac{1}{V} \sum_{\mathbf{k}} \left(-\frac{\partial f\big(\epsilon_*(\mathbf{k})\big)}{\partial \epsilon_*(\mathbf{k})} \right) \mathbf{v}_{\mathbf{k}} \cdot \left\{ \mathbf{v}_{\mathbf{k}} - \frac{2}{V} \sum_{\mathbf{k}'} \mathbf{v}_{\mathbf{k}'} \frac{\partial f\big(\epsilon_*(\mathbf{k}')\big)}{\partial \epsilon_*(\mathbf{k}')} f_S(\mathbf{k}, \mathbf{k}') \right\}
$$

$$
= \frac{1}{V} \sum_{\mathbf{k}} \left(-\frac{\partial f\big(\epsilon_*(\mathbf{k})\big)}{\partial \epsilon_*(\mathbf{k})} \right) \mathbf{v}_{\mathbf{k}} \cdot \bar{\mathbf{v}}_{\mathbf{k}} ,
$$

$$(5.90)$$

where the last two equations follow from the fact that in the ω-limit longitudinal and transverse response coincide, and thus the Ward-Takahashi identity actually implies $\boldsymbol{\lambda}^\omega = -e\, \mathbf{v}_{\mathbf{k}}$, thus the second and third equations. In particular, the latter equation states that the flux of $\bar{\mathbf{v}}_{\mathbf{k}}$ out of the Luttinger surface vanishes.

The density-density proper response function in (5.87) becomes in the ω-limit

$$
\chi_*^\omega \simeq -\frac{2e^2}{V} \sum_{\mathbf{k}} \frac{\partial f\big(\epsilon_*(\mathbf{k})\big)}{\partial \epsilon_*(\mathbf{k})} \left(\frac{\mathbf{q} \cdot \mathbf{v}_{\mathbf{k}}}{\omega} \right)^2
$$

$$
+ \frac{4e^2}{V^2} \sum_{\mathbf{k}\mathbf{k}'} \frac{\partial f\big(\epsilon_*(\mathbf{k})\big)}{\partial \epsilon_*(\mathbf{k})} \frac{\partial f\big(\epsilon_*(\mathbf{k}')\big)}{\partial \epsilon_*(\mathbf{k}')} \frac{\mathbf{q} \cdot \mathbf{v}_{\mathbf{k}}}{\omega} \frac{\mathbf{q} \cdot \mathbf{v}_{\mathbf{k}'}}{\omega} f_S(\mathbf{k}, \mathbf{k}) \equiv \frac{q^2}{\omega^2} T_{\text{qp}} ,
$$

$$(5.91)$$

where T_{qp} is actually the quasiparticle contribution to the diamagnetic term T, and must vanish in the Mott insulator. Indeed, in the same ω-limit, the longitudinal conductivity, which can be also written as

$$
\sigma_{||}(\omega, \mathbf{q}) = \frac{\omega}{4\pi i} \Big(\epsilon_{||}(\omega, \mathbf{q}) - 1 \Big) = -\frac{\omega}{i\, q^2} \chi_*(\omega, \mathbf{q}) ,
$$

becomes

$$
\sigma_{||}^\omega = T_{\text{qp}} \frac{i}{\omega + i\eta} \equiv D_* \frac{i}{\omega + i\eta} ,
$$

$$(5.92)$$

which does imply that $T_{\text{qp}} = 0$ in the Mott insulator. We observe that $T_{\text{qp}} = 0$ from (5.91) is equivalent to the first equation in (5.90).

Therefore, a Mott insulator with a Luttinger surface sustaining well-defined quasiparticles is perfectly consistent with gauge invariance, and just requires two simple conditions to be verified

$$
\frac{1}{V} \sum_{\mathbf{k}} \left(-\frac{\partial f\big(\epsilon_*(\mathbf{k})\big)}{\partial \epsilon_*(\mathbf{k})} \right) \mathbf{v}_{\mathbf{k}} \cdot \left\{ \mathbf{v}_{\mathbf{k}} - \frac{2}{V} \sum_{\mathbf{k}'} \mathbf{v}_{\mathbf{k}'} \frac{\partial f\big(\epsilon_*(\mathbf{k}')\big)}{\partial \epsilon_*(\mathbf{k}')} f_S(\mathbf{k}, \mathbf{k}') \right\} = 0 , \quad A_S = 1 .
$$

$$(5.93)$$

As earlier discussed, in the Mott insulator, we expect $A_{\uparrow,\uparrow;\uparrow,\uparrow} = 0$, and thus also $f_{\uparrow,\uparrow;\uparrow,\uparrow} = 0$. That implies $f_A = -f_S$, and therefore, if the charge Drude weight (5.74) vanishes because of (5.93), the spin Drude weight D_σ

$$
\begin{aligned}
D_\sigma &= \frac{2\mu_B^2}{3V} \sum_{\mathbf{k}} \left(-\frac{\partial f\big(\epsilon_*(\mathbf{k})\big)}{\partial \epsilon_*(\mathbf{k})} \right) \boldsymbol{v}_{\mathbf{k}} \cdot \left\{ \boldsymbol{v}_{\mathbf{k}} - \frac{2}{V} \sum_{\mathbf{k}'} \boldsymbol{v}_{\mathbf{k}'} \frac{\partial f\big(\epsilon_*(\mathbf{k}')\big)}{\partial \epsilon_*(\mathbf{k}')} f_A(\mathbf{k}, \mathbf{k}') \right\} \\
&\simeq \frac{2\mu_B^2}{3V} \sum_{\mathbf{k}} \left(-\frac{\partial f\big(\epsilon_*(\mathbf{k})\big)}{\partial \epsilon_*(\mathbf{k})} \right) \boldsymbol{v}_{\mathbf{k}} \cdot \left\{ \boldsymbol{v}_{\mathbf{k}} + \frac{2}{V} \sum_{\mathbf{k}'} \boldsymbol{v}_{\mathbf{k}'} \frac{\partial f\big(\epsilon_*(\mathbf{k}')\big)}{\partial \epsilon_*(\mathbf{k}')} f_S(\mathbf{k}, \mathbf{k}') \right\} \\
&= \frac{4\mu_B^2}{3V} \sum_{\mathbf{k}} \left(-\frac{\partial f\big(\epsilon_*(\mathbf{k})\big)}{\partial \epsilon_*(\mathbf{k})} \right) \boldsymbol{v}_{\mathbf{k}} \cdot \boldsymbol{v}_{\mathbf{k}} ,
\end{aligned}
$$

$$(5.94)$$

is finite and consistent with $\chi = 2\,\rho_{\mathrm{qp}}$, which is indeed remarkable.

Let us now discuss more in detail the thermal properties. We already showed that also in a Mott insulator quasiparticles at a Luttinger surface yield a leading linear in temperature specific heat

$$
c_V = \frac{2\pi^2}{3}\, T\, \rho_{\mathrm{qp}} . \tag{5.95}
$$

We here rewrite for convenience the thermal Drude weight of (5.81),

$$
\begin{aligned}
K_* = \frac{1}{T} \Bigg[&\frac{2}{3V} \sum_{\mathbf{k}} \epsilon_*(\mathbf{k})^2\, \boldsymbol{v}_{\mathbf{k}} \cdot \boldsymbol{v}_{\mathbf{k}} \left(-\frac{\partial f\big(\epsilon_*(\mathbf{k})\big)}{\partial \epsilon_*(\mathbf{k})} \right) \\
&+ \frac{4}{3V^2} \sum_{\mathbf{k}\mathbf{k}'} \epsilon_*(\mathbf{k})\,\epsilon_*(\mathbf{k}')\, \boldsymbol{v}_{\mathbf{k}} \cdot \boldsymbol{v}_{\mathbf{k}'} \frac{\partial f\big(\epsilon_*(\mathbf{k})\big)}{\partial \epsilon_*(\mathbf{k})} \frac{\partial f\big(\epsilon_*(\mathbf{k}')\big)}{\partial \epsilon_*(\mathbf{k}')} f_S(\mathbf{k}, \mathbf{k}') \Bigg],
\end{aligned}
$$

since it highlights that, just like in the case of specific heat, the leading in temperature contribution derives just from the first term in the square brackets. Therefore, in leading order

$$
K_* \simeq \frac{1}{T}\, \frac{2}{3V} \sum_{\mathbf{k}} \epsilon_*(\mathbf{k})^2\, \boldsymbol{v}_{\mathbf{k}} \cdot \boldsymbol{v}_{\mathbf{k}} \left(-\frac{\partial f\big(\epsilon_*(\mathbf{k})\big)}{\partial \epsilon_*(\mathbf{k})} \right), \tag{5.96}
$$

and again is consistent with the specific heat (5.95).

In conclusion, a Mott insulator that displays a Luttinger surface is expected to have gapless quasiparticles, which do not contribute to charge properties, thus the insulator is correctly incompressible and has vanishing Drude weight, and yet yield the spin and thermal properties expected in a conventional Fermi liquid. This striking scenario does not seem to contradict fundamental properties, like gauge invariance, and therefore, it may be in principle realised. The obvious caveat is that the existence of a Mott insulator that has a Luttinger surface and does not break any symmetry is highly hypothetical. However, the discussion of this section proves that at least such possibility is contemplated within Landau's Fermi liquid theory.

5.7 Luttinger's Theorem and Quasiparticle Distribution Function

Under the assumption (5.6), which corresponds to the assumption (4.167) in Sect. 4.11.1.1, the electron density is obtained by the generalised Luttinger's theorem

$$
\begin{aligned}
\frac{N}{V} &= -\frac{1}{V} \sum_{\mathbf{k}\sigma} T \sum_{i\epsilon} e^{i\epsilon 0^+} \frac{\partial \ln G(i\epsilon, \mathbf{k})}{\partial i\epsilon} + \frac{\mathcal{L}}{V} \\
&= \frac{1}{V} \sum_{\mathbf{k}\sigma} \int_{-\infty}^{\infty} \frac{d\epsilon}{\pi} f(\epsilon) \frac{\partial \operatorname{Im} \ln G(\epsilon + i0^+, \mathbf{k})}{\partial \epsilon} + \frac{\mathcal{L}}{V},
\end{aligned}
\tag{5.97}
$$

with integer $\mathcal{L}$ and for $T \to 0$. We define the real frequency counterpart of (4.155),

$$
\delta_\sigma(\epsilon, \mathbf{k}) \equiv \pi + \operatorname{Im} \ln G_\sigma(\epsilon + i0^+, \mathbf{k}) = \tan^{-1} \frac{-\operatorname{Im} G_\sigma(\epsilon + i0^+, \mathbf{k})}{-\operatorname{Re} G_\sigma(\epsilon + i0^+, \mathbf{k})} \in [0, \pi],
\tag{5.98}
$$

where $\delta_\sigma(-\infty, \mathbf{k}) = 0$, $\delta_\sigma(\infty, \mathbf{k}) = \pi$, and, for convenience, we have indicated the spin label though the Green's function is independent of it, so that

$$
\frac{N}{V} = \frac{1}{V} \sum_{\mathbf{k}\sigma} \int_{-\infty}^{\infty} \frac{d\epsilon}{\pi} f(\epsilon) \frac{\partial \delta_\sigma(\epsilon, \mathbf{k})}{\partial \epsilon} + \frac{\mathcal{L}}{V}.
\tag{5.99}
$$

In the perturbative regime $\mathcal{L} = 0$, and we can legitimately regard

$$
n_{\mathbf{k}\sigma} \equiv \int_{-\infty}^{\infty} \frac{d\epsilon}{\pi} f(\epsilon) \frac{\partial \delta_\sigma(\epsilon, \mathbf{k})}{\partial \epsilon} = -\int_{-\infty}^{\infty} \frac{d\epsilon}{\pi} \frac{\partial f(\epsilon)}{\partial \epsilon} \delta_\sigma(\epsilon, \mathbf{k}),
\tag{5.100}
$$

as the quasiparticle distribution function in momentum space, not to be confused with the electron distribution function $\langle c_{\mathbf{k}\sigma}^\dagger c_{\mathbf{k}\sigma} \rangle$. When perturbation theory breaks down, thus $\mathcal{L} \neq 0$, (5.100) is not valid anymore. However, we here show that any variation of the actual quasiparticle distribution function with respect to a thermodynamic variable λ_Q conjugate to a conserved quantity Q is given by

$$
\frac{\partial n_{\mathbf{k}\sigma}}{\partial \lambda_Q} = -\int_{-\infty}^{\infty} \frac{d\epsilon}{\pi} \frac{\partial}{\partial \lambda_Q} \left(\frac{\partial f(\epsilon)}{\partial \epsilon} \delta_\sigma(\epsilon, \mathbf{k}) \right).
\tag{5.101}
$$

This result can be anticipated without even proving it. Indeed, since $\mathcal{L}$ must be an integer, it cannot vary continuously upon varying λ_Q. Therefore, $\partial \mathcal{L}/\partial \lambda_Q = 0$, and thus (5.101) follows. However, the explicit proof is a useful exercise.

In our case of interest, the conserved quantities are the charge, the spin, and the total energy, thus the corresponding λ_Q are, respectively, the chemical potential μ, the Zeeman splitting field μ_Z, and $\beta = 1/T$. For simplicity, we hereafter consider the chemical potential case. Let us, therefore, calculate

$$\frac{\partial n_{\mathbf{k}\sigma}}{\partial \mu} = -\int_{-\infty}^{\infty} \frac{d\epsilon}{\pi} \frac{\partial f(\epsilon)}{\partial \epsilon} \frac{\partial \delta_\sigma(\epsilon, \mathbf{k})}{\partial \mu} = \int_{-\infty}^{\infty} \frac{d\epsilon}{\pi} \frac{\partial f(\epsilon)}{\partial \epsilon} \operatorname{Im} \left\{ G_\sigma(\epsilon, \mathbf{k}) \left(1 - \frac{\partial \Sigma_\sigma(\epsilon, \mathbf{k})}{\partial \mu} \right) \right\},$$

$$(5.102)$$

where $G(\epsilon, \mathbf{k}) \equiv G(\epsilon + i0^+, \mathbf{k})$ and similarly for $\Sigma(\epsilon, \mathbf{k})$, and we recall that, under changing $\mu \to \mu + \delta\mu$

$$G_\sigma(\epsilon, \mathbf{k})^{-1} \to \epsilon - \epsilon_{\mathbf{k}} + \delta\mu - \Sigma_\sigma(\epsilon, \mathbf{k}) - \frac{\partial \Sigma_\sigma(\epsilon, \mathbf{k})}{\partial \mu} \delta\mu.$$

On the other hand, by definition of the irreducible interaction vertex and moving back to Matsubara frequencies

$$\frac{\partial \Sigma_\sigma(i\epsilon, \mathbf{k})}{\partial \mu} = \frac{1}{V} \sum_{\mathbf{k}'\sigma'} T \sum_{\epsilon'} \Gamma_0 \big(i\epsilon\,\mathbf{k}\,\sigma, i\epsilon'\,\mathbf{k}'\,\sigma'; i\epsilon'\,\mathbf{k}'\,\sigma', i\epsilon\,\mathbf{k}\,\sigma \big) \frac{\partial G_\sigma(i\epsilon', \mathbf{k}')}{\partial \mu},$$

or, in shorthand notation,

$$\frac{\partial \Sigma_\sigma}{\partial \mu} = \Gamma_0 \odot \frac{\partial G_\sigma}{\partial \mu} = -\Gamma_0 \odot G^2 \odot \left(1 - \frac{\partial \Sigma_\sigma}{\partial \mu} \right).$$

Therefore, in sequence

$$\left(1 - \Gamma_0 \odot G^2 \right) \odot \frac{\partial \Sigma_\sigma}{\partial \mu} = -\Gamma_0 \odot G^2 \ \Rightarrow \ \frac{\partial \Sigma_\sigma}{\partial \mu} = -\left(1 - \Gamma_0 \odot G^2 \right)^{-1} \Gamma_0 \odot G^2$$

$$\Rightarrow \frac{\partial \Sigma_\sigma}{\partial \mu} = -\Gamma \odot G^2,$$

thus

$$1 - \frac{\partial \Sigma_\sigma(i\epsilon, \mathbf{k})}{\partial \mu} = 1 + \frac{1}{V} \sum_{\mathbf{k}'\sigma'} T \sum_{\epsilon'} \Gamma \big(i\epsilon\,\mathbf{k}\,\sigma, i\epsilon'\,\mathbf{k}'\,\sigma'; i\epsilon'\,\mathbf{k}'\,\sigma', i\epsilon\,\mathbf{k}\,\sigma \big) G_{\sigma'}(i\epsilon', \mathbf{k}')^2$$

$$= 1 + \Gamma^q \odot R^q = \Lambda^q,$$

$$(5.103)$$

where the last equation derives from the fact that adding a chemical potential is equivalent to adding a static and a long-wavelength potential and then take the limit of infinite wavelength, thus just the q-limit.

Let us now recall the Ward-Takahashi identity (4.131), see Fig. 4.42, which yields for the charge density-vertex and in the ω-limit

$$1 - \frac{\partial \Sigma_\sigma(i\epsilon, \mathbf{k})}{\partial i\epsilon} = \Lambda^\omega,$$

$$(5.104)$$

We recall (5.39) for the charge density-vertex

$$\Lambda = \Lambda^\omega + \Lambda^\omega \odot \Delta \odot \Gamma \,,$$

which reads in the q-limit

$$\Lambda^q = \left(1 + \Gamma^q \odot \Delta^q\right) \odot \Lambda^\omega = \left(1 + \Gamma^q \odot \Delta^q\right) \odot \left(1 - \frac{\partial \Sigma_\sigma}{\partial i\epsilon}\right)$$

$$= Z^{-1} \odot \left(1 + A^q \odot R^q_{\text{qp}}\right) = 1 - \frac{\partial \Sigma_\sigma}{\partial \mu} \,,$$

so that, see (5.102),

$$\frac{\partial \delta_\sigma(\epsilon, \mathbf{k})}{\partial \mu} = -\text{Im}\left\{ G_\sigma(\epsilon, \mathbf{k}) \left(1 - \frac{\partial \Sigma_\sigma(\epsilon, \mathbf{k})}{\partial \mu}\right)\right\}$$

$$= -\text{Im}\left\{ G_\sigma(\epsilon, \mathbf{k}) Z(\epsilon, \mathbf{k})^{-1} \left(1 + \frac{1}{V} \sum_{\mathbf{k}'\sigma'} A^q_{\sigma,\sigma';\sigma',\sigma}(\mathbf{k}, \mathbf{k}') \frac{\partial f\big(\epsilon_*(\mathbf{k}')\big)}{\partial \epsilon_*(\mathbf{k}')}\right)\right\}$$

$$= \pi A_{\text{qp}}(\epsilon, \mathbf{k}) \left(1 + \frac{1}{V} \sum_{\mathbf{k}'\sigma'} A^q_{\sigma,\sigma';\sigma',\sigma}(\mathbf{k}, \mathbf{k}') \frac{\partial f\big(\epsilon_*(\mathbf{k}')\big)}{\partial \epsilon_*(\mathbf{k}')}\right),$$

and thus, recalling (5.13),

$$\frac{1}{V} \sum_{\mathbf{k}\sigma} \frac{\partial n_{\mathbf{k}\sigma}}{\partial \mu} = -\frac{1}{V} \sum_{\mathbf{k}\sigma} \int_{-\infty}^{\infty} \frac{d\epsilon}{\pi} \frac{\partial f(\epsilon)}{\partial \epsilon} \frac{\partial \delta_\sigma(\epsilon, \mathbf{k})}{\partial \mu}$$

$$= -\frac{1}{V} \sum_{\mathbf{k}\sigma} \frac{\partial f\big(\epsilon_*(\mathbf{k})\big)}{\partial \epsilon_*(\mathbf{k})} \left(1 + \frac{1}{V} \sum_{\mathbf{k}'\sigma'} A^q_{\sigma,\sigma';\sigma',\sigma}(\mathbf{k}, \mathbf{k}') \frac{\partial f\big(\epsilon_*(\mathbf{k}')\big)}{\partial \epsilon_*(\mathbf{k}')}\right),$$

$$(5.105)$$

which is just the charge compressibility (5.58), and thus proves that (5.101) is true

for $\lambda_Q = \mu$. One can straightforwardly prove the same result also for a Zeeman field $\lambda_Q = \mu_Z$. The case $\lambda_Q = \beta$ is slightly more complicated but can be worked as well. For that, it is important to notice the following equation:

$$\frac{\partial \Sigma_\sigma(\epsilon, \mathbf{k})}{\partial T} = Z(\epsilon, \mathbf{k})^{-1} \frac{1}{V} \sum_{\mathbf{k}\sigma'} \int d\epsilon' \frac{\partial f\big(\epsilon_*(\mathbf{k}')\big)}{\partial T} A^q_{\sigma,\sigma';\sigma',\sigma}(\mathbf{k}, \mathbf{k}') \,, \qquad (5.106)$$

which can be derived by inspection of the perturbative expansion of the self-energy in terms of skeleton diagrams.

Therefore, even though in general

$$n_{\mathbf{k}\sigma} \neq \int_{-\infty}^{\infty} \frac{d\epsilon}{\pi} \, f(\epsilon) \, \frac{\partial \delta_\sigma(\epsilon, \mathbf{k})}{\partial \epsilon} \xrightarrow{T \to 0} \delta_\sigma(0^-, \mathbf{k}) \,, \qquad (5.107)$$

unless in the perturbative regime, upon varying the chemical potential of each spin species, $\mu_\sigma = \mu \to \mu + \delta\mu_\sigma$, or the temperature $T \simeq 0$,

$$\delta n_{\mathbf{k}\sigma} = \sum_{\sigma'} \frac{1}{\pi} \frac{\partial \delta_\sigma(0^-, \mathbf{k})}{\partial \mu_{\sigma'}} \, \delta\mu_{\sigma'} + \frac{1}{\pi} \frac{\partial \delta_\sigma(0^-, \mathbf{k})}{\partial T} \, \delta T \,, \qquad (5.108)$$

is instead true irrespective of whether perturbation theory is valid or breaks down, provided (5.6) holds. In other words, the variation of the quasiparticle distribution function is just the variation of the phase of the Green's function $G_\sigma(0^-, \mathbf{k})$ in units of π.

Let us unveil the meaning of the above result in the case of a Mott insulator of Sect. 5.6, in which case

$$\delta N = \sum_{\mathbf{k}\sigma} \frac{\delta n_{\mathbf{k}\sigma}}{\delta\mu} \, \delta\mu = \sum_{\mathbf{k}\sigma} \frac{1}{\pi} \frac{\partial \delta_\sigma(0^-, \mathbf{k})}{\partial \mu} \, \delta\mu = 0 \,,$$

$$\delta M = \sum_{\mathbf{k}} \left(\frac{\delta n_{\mathbf{k}\uparrow}}{\delta\mu_Z} - \frac{\delta n_{\mathbf{k}\downarrow}}{\delta\mu_Z} \right) \delta\mu_Z = \sum_{\mathbf{k}} \frac{1}{\pi} \left(\frac{\partial \delta_\uparrow(0^-, \mathbf{k})}{\partial \mu_Z} - \frac{\partial \delta_\downarrow(0^-, \mathbf{k})}{\partial \mu_Z} \right) \delta\mu_Z \neq 0 \,,$$

$$\delta E = \sum_{\mathbf{k}\sigma} \epsilon_*(\mathbf{k}) \frac{\delta n_{\mathbf{k}\sigma}}{\delta T} \, \delta T = \sum_{\mathbf{k}\sigma} \epsilon_*(\mathbf{k}) \frac{1}{\pi} \frac{\partial \delta_\sigma(0^-, \mathbf{k})}{\partial T} \, \delta T \neq 0 \,.$$

$$(5.109)$$

Since the system has a single-particle gap, $\delta_\sigma(0^-, \mathbf{k})$ is equal, e.g., to π inside the Luttinger surface of spin-σ quasiparticles and 0 outside, and that must remain true even under the perturbations $\delta\mu$, $\delta\mu_Z$ and δT. It follows that the vanishing compressibility, $\delta N/\delta\mu = 0$ in (5.109), implies that the volume enclosed by the Luttinger surface does not change upon small variations of the chemical potential. On the contrary, the relative volumes of spin-up and spin-down quasiparticles are free to change in a Zeeman field, provided the sum of the two volumes remains constant. Similarly, the finite specific heat, $\delta E/\delta T \neq 0$ in (5.109), implies that the shape changes at fixed volume upon varying temperature.

5.7.1 Oshikawa's Topological Derivation of Luttinger's Theorem

In periodic systems, one can derive Luttinger's theorem in a much simpler and physically more transparent way [19]. For simplicity, we here consider a two dimensional lattice model with a torus geometry, see Fig. 5.3, and, in order to avoid any complication related to the difference between internal and external electromagnetic

Fig. 5.3 A two-dimensional system with a torus geometry. A magnetic field, concentrated in the centre of the torus, is adiabatically switched on till the flux threading the torus reaches the flux quantum $\Phi_0 = 2\pi\hbar/q$

fields, we assume a fictitious flux $\Phi(t)$ that threads the torus and is coupled to fictitious spin-dependent charges $-q_\sigma$, with either $q_\uparrow = q_\downarrow = q$ or $q_\uparrow = -q_\downarrow = q$ to discuss separately spin and charge response. We further assume that $\Phi(t)$ slowly grows in time from 0 at $t = 0$ to a unit flux quantum $\Phi_0 = 2\pi\hbar/q$ at $t = \tau$.[4] That amounts to add a fictitious vector potential $A(t) = \Phi(t)/L_x$, where, see Fig. 5.3, $L_x = 2\pi R = N_x a$, with $N_{x(y)}$ the number of unit cells along $x(y)$, and a the lattice spacing. The corresponding Hamiltonian reads

$$H[\Phi(t)] = \frac{1}{2m} \sum_\sigma \int dx\, dy\, \Psi_\sigma^\dagger(x, y) \left(\left(-i\hbar\, \partial_x + q_\sigma\, A(t) \right) - \hbar^2\, \partial_y^2 \right) \Psi_\sigma(x, y)$$

$$+ \sum_\sigma \int dx\, dy\, \Psi_\sigma^\dagger(x, y)\, V(x, y)\, \Psi_\sigma(x, y) + H_{int} ,$$

(5.110)

where $V(x, y) = V(x + na, y + ma)$ is the periodic potential, and H_{int} the translationally invariant interaction. During the switching time, $\Phi(t)$ drives currents into the system, and the initial state, assumed to be the ground state $| \Psi_0 \rangle$ of $H[0]$, evolves according to

$$| \Psi(t) \rangle = e^{-\frac{i}{\hbar} \int_0^t dt'\, H[\Phi(t')]} | \Psi_0 \rangle \equiv U(t) | \Psi_0 \rangle .$$

After the flux has reached Φ_0, the system flows to a stationary state described by the wavefunction

$$| \Psi_*(t) \rangle = e^{-\frac{i}{\hbar} H[\Phi_0](t-\tau)} | \Psi(\tau) \rangle .$$

(5.111)

At $\Phi = 0$, the primitive lattice translation operator along x is defined through $T_x[0] = e^{-i P_x a/\hbar}$, where P_x is the x-component of the lattice momentum operator. Since $H[\Phi(t)]$ commutes with $T_x[0]$, if the initial ground-state wavefunction is eigenstate of P_x with eigenvalue P_{ix}, then

---

[4] In presence of a static and uniform flux Φ, the momentum quantisation in the gauge in which the vector potential $A \parallel x \perp B$ reads, cfr. (5.110) with $q_\sigma \to e/c$,

$$k_x = \frac{2\pi\hbar}{N_x a}\, n_x \to \frac{2\pi\hbar}{N_x a}\, n_x + \frac{e}{c}\, \frac{\Phi}{N_x a} , \qquad n_x \in \mathbb{Z} .$$

We note that, if $\Phi = \Phi_0 \equiv 2\pi\hbar c/e$ the magnetic flux quantum, the quantisation is invariant, simply $n_x \to n_x + 1$.

$$T_x(0) \mid \Psi_*(t)\rangle = e^{-\frac{i}{\hbar} H[\Phi_0](t-\tau)} \, U(\tau) \, T_x(0) \mid \Psi_0\rangle = e^{-i P_{ix} a/\hbar} \mid \Psi_*(t)\rangle \,,$$

thus $\mid \Psi_*(t)\rangle$ is also eigenstate with same eigenvalue. Note that lattice translation implies that

$$P_{ix} = 2\pi \frac{\hbar}{a} \frac{M_{ix}}{N_x} \,, \quad M_{ix} \in \mathbb{Z}, \tag{5.112}$$

is quantised. However, at finite Φ, the actual translation operator $T_x[\Phi] \neq T_x[0]$ is gauge dependent. Since a static flux Φ can be reabsorbed by a gauge transformation

$$H[\Phi] = U_G[\Phi]^\dagger \, H[0] \, U_G[\Phi], \quad U_G[\Phi] = \exp\left[i \frac{\Phi}{\hbar L_x} \sum_\sigma \int d\mathbf{r} \, q_\sigma \, x \, \rho_\sigma(\mathbf{r})\right],$$

then

$$T_x[\Phi_0] = U_G^\dagger[\Phi_0] \, T_x[0] \, U_G[\Phi_0] = T_x[0] \exp\left[-i \sum_\sigma \frac{a \, \Phi_0 \, q_\sigma}{\hbar L_x} N_\sigma\right]$$

$$= T_x[0] \exp\left[-i \frac{2\pi}{N_x} \sum_\sigma \frac{q_\sigma}{q} N_\sigma\right],$$

and thus

$$T_x[\Phi] \mid \Psi_*(t)\rangle \equiv e^{-i P_{fx} a/\hbar} \mid \Psi_*(t)\rangle = e^{-i \frac{2\pi}{N_x} \sum_\sigma \frac{q_\sigma}{q} N_\sigma} \, T_x[0] \mid \Psi_*(t)\rangle$$

$$= e^{-i \frac{2\pi}{N_x} \sum_\sigma \frac{q_\sigma}{q} N_\sigma} \, e^{-i P_{ix} a/\hbar} \mid \Psi_*(t)\rangle \,.$$

In conclusion

$$P_{fx} = P_{ix} + \frac{2\pi\hbar}{N_x a} \sum_\sigma \frac{q_\sigma}{q} N_\sigma \equiv P_{ix} + \Delta P_x \,, \tag{5.113}$$

and it is still quantised as in the absence of flux because Φ_0 is a unit flux quantum. In other words, the adiabatic threading of a unit flux quantum transforms the initial ground state with lattice momentum P_{ix} along x into a final state with different lattice momentum P_{fx}.

Let us now assume to be in the perturbative regime, and thus that Landau's adiabatic principle holds true. In that case, the above fictitious charges q_σ are inherited by quasiparticles, and the final state just corresponds to $\epsilon_*(\mathbf{k}) \to \epsilon_*(\mathbf{k}_\sigma)$, where

$$\mathbf{k}_\sigma = \mathbf{k} + \frac{q_\sigma}{q} \frac{2\pi\hbar}{N_x a} \,, \tag{5.114}$$

so that the lattice momentum, assumed to vanish in the ground state at $\Phi = 0$, changes for $q_\uparrow = q_\downarrow$ into

$$\Delta P_x = \frac{2\pi\hbar}{N_x a} N_{F\mathrm{qp}} \,, \tag{5.115}$$

where N_{Fqp} is twice the number of **k**-points within the Fermi volume. Therefore, comparing (5.113) for $q_\uparrow = q_\downarrow = q$, namely

$$\Delta P_x = \frac{2\pi\hbar}{N_x a} \left(N_\uparrow + N_\downarrow \right) = \frac{2\pi\hbar}{N_x a} N , \tag{5.116}$$

with (5.115), we conclude that $N = N_{Fqp}$, which is just the conventional statement of Luttinger's theorem.

Let us consider now the non-symmetry breaking half-filled Mott insulator with a Luttinger surface discussed in Sect. 5.6. Here, we cannot invoke Landau's adiabatic principle, so that we cannot anticipate what is the momentum acquired by the quasiparticles at the Luttinger surface when $q_\uparrow = q_\downarrow$, which we can write as

$$\Delta P_x = \frac{2\pi\hbar}{N_x a} N_{Lqp} , \tag{5.117}$$

in terms of an unknown N_{Lqp}. On the other hand, (5.116) implies, for $N = N_x N_y$ that

$$\Delta P_x = \frac{2\pi\hbar}{N_x a} N_x N_y , \tag{5.118}$$

and thus $N_{Lqp} = N_x N_y$. Moreover, since ΔP_x is a multiple of the primitive reciprocal lattice vector $2\pi\hbar/a$, it is equivalent to zero, which is consistent with the system being insulating, and thus having a charge gap. Indeed, since $H[0]$ and $H[\Phi_0]$ have the same spectrum, and if the ground state is unique, we can always design an adiabatic switching protocol such that the initial state evolves into itself, hence the acquired momentum must strictly vanish.

On the other hand, we could follow a different route and repeat the same hypothetical experiment taking, e.g. $q_\uparrow = q$ but $q_\downarrow = 0$, in which case we would find, since $N_\uparrow = N_x N_y/2$,

$$\Delta P_x = \frac{2\pi\hbar}{a} N_y \frac{N_\uparrow}{N_x N_y} = \frac{\pi\hbar}{a} N_y \equiv \begin{cases} 0 & N_y \text{ even}, \\ \frac{\pi\hbar}{a} & N_y \text{ odd}, \end{cases} \tag{5.119}$$

which, for odd N_y, may look in contradiction with the previous physical argument, but in reality it is not so. The reason is that the Luttinger surface entails the existence of quasiparticles that, as we showed in Sect. 5.6, do not contribute to the charge response but yields conventional thermal and magnetic properties. Therefore, while for $q_\uparrow = q_\downarrow$ the flux insertion affects the charge, whose fluctuations are suppressed in a Mott insulator, for $q_\uparrow = q$ and $q_\downarrow = 0$ the spin degrees of freedom are heavily involved and can freely absorb a $\pi\hbar/a$ momentum without any cost in energy.

Let us finally consider the metallic case in which Fermi pockets and Luttinger surfaces coexist. Here, we may reasonably argue that the quasiparticles at the Fermi surfaces carry the same quantum numbers of the physical particles. It follows that equating (5.116) with the sum of (5.115) and (5.117)

$$\Delta P_x = \frac{2\pi\hbar}{N_x a} N = \frac{4\pi\hbar}{N_x a} N_x N_y \nu \equiv \frac{2\pi\hbar}{N_x a} N_{Fqp} + \frac{2\pi\hbar}{N_x a} N_{Lqp}$$

$$= \frac{2\pi\hbar}{N_x a} N_{Fqp} + \frac{2\pi\hbar}{N_x a} N_x N_y ,$$

where the equivalence holds apart from a reciprocal lattice vector. Since the hole and electron Fermi pockets contain, respectively, $N_x N_y (1 - v_{HP})$ and $N_x N_y v_{EP}$ k-points, then

$$\frac{4\pi\hbar}{N_x a} N_x N_y \nu \equiv \frac{4\pi\hbar}{N_x a} N_x N_y \left(v_{EP} + 1 - v_{HP} + \frac{1}{2} \right) \equiv \frac{4\pi\hbar}{N_x a} N_x N_y \left(v_{EP} - v_{HP} + \frac{1}{2} \right),$$

which is just (5.16).

5.8   Quasiparticle Hamiltonian and Landau-Boltzmann Transport Equation

Since G_{qp} in (5.45) can be regarded as the *non-interacting* quasiparticle Green's function, (5.55) looks like the linear response function within the conserving Hartree-Fock approximation of Sect. 4.10.1. It follows that the quasiparticle interaction vertex A must satisfy a Bethe-Salpeter equation like that in Fig. 4.44, namely

$$A = f + f \odot R_{qp} \odot A , \tag{5.120}$$

with f the Hartree-Fock irreducible interaction vertex

$$f = f_{\sigma_1,\sigma_2;\sigma_3,\sigma_4}(\mathbf{k} + \mathbf{q}, \mathbf{k}'; \mathbf{k}' + \mathbf{q}, \mathbf{k}) . \tag{5.121}$$

Note that, because of spin $SU(2)$, $\sigma_1 + \sigma_2 = \sigma_3 + \sigma_4$ in (5.121). Therefore, if we consider the quasiparticle Hamiltonian

$$H_* \equiv \sum_{\mathbf{k}\sigma} \epsilon_{*\mathbf{k}} d_{\mathbf{k}\sigma}^\dagger d_{\mathbf{k}\sigma}$$

$$+ \frac{1}{2V} \sum_{\mathbf{k}\mathbf{k}'\mathbf{q}} \sum_{\sigma_1\sigma_2\sigma_3\sigma_4} U_{\sigma_1,\sigma_2;\sigma_3,\sigma_4}(\mathbf{k} + \mathbf{q}, \mathbf{k}'; \mathbf{k}' + \mathbf{q}, \mathbf{k}) \, d_{\mathbf{k}+\mathbf{q}\sigma_1}^\dagger d_{\mathbf{k}'\sigma_2}^\dagger d_{\mathbf{k}'+\mathbf{q}\sigma_3} d_{\mathbf{k}\sigma_4} ,$$

$$\tag{5.122}$$

where $d_{\mathbf{k}\sigma}$ and $d_{\mathbf{k}\sigma}^\dagger$ are the quasiparticle annihilation and creation operators, and the interaction U such that

$$f_{\sigma_1,\sigma_2;\sigma_3,\sigma_4}(\mathbf{k} + \mathbf{q}, \mathbf{k}'; \mathbf{k}' + \mathbf{q}, \mathbf{k}) = U_{\sigma_1,\sigma_2;\sigma_3,\sigma_4}(\mathbf{k} + \mathbf{q}, \mathbf{k}'; \mathbf{k}' + \mathbf{q}, \mathbf{k})$$

$$- U_{\sigma_1,\sigma_2;\sigma_4,\sigma_3}(\mathbf{k} + \mathbf{q}, \mathbf{k}'; \mathbf{k}, \mathbf{k}' + \mathbf{q}) ,$$

then all previous results imply that the conserving Hartree-Fock approximation yields reliable linear response functions of the physical electrons.

Within the Hartree-Fock approximation, the energy $\epsilon_*(\mathbf{k})$ of the quasiparticle at momentum $\mathbf{k}$ and spin σ must therefore be identified with

$$\epsilon_*(\mathbf{k}) = \epsilon_{*\mathbf{k}} + \frac{1}{V} \sum_{\mathbf{k}'\sigma'} f_{\sigma,\sigma';\sigma',\sigma}(\mathbf{k}, \mathbf{k}'; \mathbf{k}', \mathbf{k}) \, n^0_{\mathbf{k}'\sigma'}, \tag{5.123}$$

where

$$n^0_{\mathbf{k}'\sigma'} = \langle d^\dagger_{\mathbf{k}'\sigma'} d_{\mathbf{k}'\sigma'} \rangle_0 = f(\epsilon_*(\mathbf{k})), \tag{5.124}$$

is the equilibrium distribution function. The total energy thus reads

$$E_0 = \sum_{\mathbf{k}\sigma} \epsilon_{*\mathbf{k}} n^0_{\mathbf{k}\sigma} + \frac{1}{2V} \sum_{\mathbf{k}\mathbf{k}'} \sum_{\sigma\sigma'} f_{\sigma,\sigma';\sigma',\sigma}(\mathbf{k}, \mathbf{k}'; \mathbf{k}', \mathbf{k}) \, n^0_{\mathbf{k}\sigma} n^0_{\mathbf{k}'\sigma'}.$$

An excited state with quasiparticle distribution $n_{\mathbf{k}\sigma} = n^0_{\mathbf{k}\sigma} + \delta n_{\mathbf{k}\sigma}$ has instead total excitation energy

$$
\begin{aligned}
\delta E[\{\delta n_{\mathbf{k}\sigma}\}] = {} & \sum_{\mathbf{k}\sigma} \epsilon_{*\mathbf{k}} n_{\mathbf{k}\sigma} + \frac{1}{2V} \sum_{\mathbf{k}\mathbf{k}'} \sum_{\sigma\sigma'} f_{\sigma,\sigma';\sigma',\sigma}(\mathbf{k}, \mathbf{k}'; \mathbf{k}', \mathbf{k}) \, n_{\mathbf{k}\sigma} n_{\mathbf{k}'\sigma'} \\
& - \sum_{\mathbf{k}\sigma} \epsilon_{*\mathbf{k}} n^0_{\mathbf{k}\sigma} - \frac{1}{2V} \sum_{\mathbf{k}\mathbf{k}'} \sum_{\sigma\sigma'} f_{\sigma,\sigma';\sigma',\sigma}(\mathbf{k}, \mathbf{k}'; \mathbf{k}', \mathbf{k}) \, n^0_{\mathbf{k}\sigma} n^0_{\mathbf{k}'\sigma'} \\
= {} & \sum_{\mathbf{k}\sigma} \epsilon_*(\mathbf{k}) \delta n_{\mathbf{k}\sigma} + \frac{1}{V} \sum_{\mathbf{k}\mathbf{k}'} \sum_{\sigma\sigma'} f_{\sigma,\sigma';\sigma',\sigma}(\mathbf{k}, \mathbf{k}'; \mathbf{k}', \mathbf{k}) \, \delta n_{\mathbf{k}\sigma} \delta n_{\mathbf{k}'\sigma'},
\end{aligned}
\tag{5.125}
$$

which is just Landau's energy functional (5.1) after defining

$$f_{\sigma,\sigma';\sigma',\sigma}(\mathbf{k}, \mathbf{k}'; \mathbf{k}', \mathbf{k}) \equiv f_{\mathbf{k}\sigma,\mathbf{k}'\sigma'}.$$

The parameters f in (5.121) are just the Landau's f-parameters in a more general spin $SU(2)$ formulation. Since $R^\omega_* = 0$, see (5.44), it follows that

$$A^\omega = \lim_{\omega \to 0} \lim_{q \to 0} A_{\sigma_1,\sigma_2;\sigma_3,\sigma_4}(\mathbf{k}, \mathbf{k}'; \omega, \mathbf{q}) = f_{\sigma_1,\sigma_2;\sigma_3,\sigma_4}(\mathbf{k}, \mathbf{k}'; \mathbf{k}', \mathbf{k}), \tag{5.126}$$

namely, the f-parameter is just the ω-limit of the quasiparticle interaction vertex, related through (5.46), see also Fig. 5.2, to the physical electron vertex Γ. Similarly to (5.56) we can define f is the $S = 0$ and $S = 1$ channels, specifically

$$
\begin{aligned}
f_S(\mathbf{k}, \mathbf{k}') &= \frac{1}{2}\left(f_{\uparrow,\uparrow;\uparrow,\uparrow}(\mathbf{k}, \mathbf{k}'; \mathbf{k}', \mathbf{k}) + f_{\uparrow,\downarrow;\downarrow,\uparrow}(\mathbf{k}, \mathbf{k}'; \mathbf{k}', \mathbf{k})\right), \\
f_A(\mathbf{k}, \mathbf{k}') &= \frac{1}{2}\left(f_{\uparrow,\uparrow;\uparrow,\uparrow}(\mathbf{k}, \mathbf{k}'; \mathbf{k}', \mathbf{k}) - f_{\uparrow,\downarrow;\downarrow,\uparrow}(\mathbf{k}, \mathbf{k}'; \mathbf{k}', \mathbf{k})\right),
\end{aligned}
\tag{5.127}
$$

which are just the functions defined in (5.75). The Bethe-Salpeter equation (5.120) becomes evidently diagonal in the $S = 0$ and $S = 1$ channels.

We end remarking that, since $\delta n_{\mathbf{k}\sigma} = \delta \operatorname{Im} \ln G_\sigma(0^-, \mathbf{k})/\pi$, Landau's energy functional is actually a functional of the phases of the Green's functions at zero energy.

5.8.1   Landau-Boltzmann Transport Equation for Quasiparticles

The Ward-Takahashi identities imply that, when the physical electron Hamiltonian is perturbed, $H \to H + \delta H_A(t)$, with

$$\delta H^A(t) = \sum_{\mathbf{kq}} \sum_{\sigma_1 \sigma_2} V_A(t, \mathbf{q})\, c^\dagger_{\mathbf{k}+\mathbf{q}\sigma_1} \Lambda_0^A(\mathbf{k} + \mathbf{q}\,\sigma_1, \mathbf{k}\,\sigma_2)\, c_{\mathbf{k}\sigma_2} = \sum_{\mathbf{q}} V_A(t, \mathbf{q})\, \rho_A(-\mathbf{q}),$$

where, by definition,

$$V_A(t, \mathbf{q}) = \frac{1}{V} \int d\mathbf{r}\, e^{-i\mathbf{q}\cdot\mathbf{r}}\, V_A(t, \mathbf{r}), \quad V_A(t, \mathbf{r}) = \sum_{\mathbf{q}} e^{i\mathbf{q}\cdot\mathbf{r}}\, V_A(t, \mathbf{q}),$$

(5.128)

if $\rho_A(\mathbf{q})$ is associated to a conserved quantity, i.e., $\rho_A(\mathbf{0})$ commutes with H, then that corresponds to perturbing the quasiparticle Hamiltonian (5.122) with

$$\delta H_*^A(t) = \sum_{\mathbf{kq}} \sum_{\sigma_1 \sigma_2} V_A(t, \mathbf{q})\, d^\dagger_{\mathbf{k}+\mathbf{q}\sigma_1} \Lambda_0^A(\mathbf{k} + \mathbf{q}\,\sigma_1, \mathbf{k}\,\sigma_2)\, d_{\mathbf{k}\sigma_2}. \tag{5.129}$$

In other words, external fields couple to physical electron densities associated with conserved quantities as they do with the corresponding quasiparticles densities. Said differently, quasiparticles carry the same conserved quantum numbers as the physical electrons.

In our case, the only conserved quantities besides the total energy are the total electron number N, the total spin as well as its component S_a along an arbitrary quantisation axis $a = x, y, z$. The physical electron densities associated with N and $2S = 2(S_x, S_y, S_z)$ are

$$\rho(\mathbf{q}) = \sum_{\mathbf{k}\sigma} c^\dagger_{\mathbf{k}\sigma} c_{\mathbf{k}+\mathbf{q}\sigma}, \quad \boldsymbol{\sigma}(\mathbf{q}) = \sum_{\mathbf{k}\alpha\beta} c^\dagger_{\mathbf{k}\alpha} \boldsymbol{\sigma}_{\alpha\beta} c_{\mathbf{k}+\mathbf{q}\beta},$$

where $\boldsymbol{\sigma} = (\sigma_1, \sigma_2, \sigma_3)$, with σ_i, $i = 1, 2, 3$, the Pauli matrices. Without loss of generality, we can specify the quantisation axis as the z-one, and consider instead

$$\rho_\sigma(\mathbf{q}) = \sum_{\mathbf{k}} c^\dagger_{\mathbf{k}\sigma} c_{\mathbf{k}+\mathbf{q}\sigma}, \quad \sigma = \uparrow, \downarrow,$$

so that the perturbation

$$\delta H(t) = \sum_{\mathbf{q}\sigma} V_\sigma(t, \mathbf{q})\, \rho_\sigma(-\mathbf{q}) \xrightarrow[\text{quasiparticles}]{} \delta H_*(t) = \sum_{\mathbf{k}\mathbf{q}\sigma} V_\sigma(t, \mathbf{q})\, d^\dagger_{\mathbf{k}+\mathbf{q}\sigma} d_{\mathbf{k}\sigma} ,$$

which implies that $H_* \to H_*(t)$ with

$$H_*(t) = H_* + \sum_{\mathbf{k}\mathbf{q}\sigma} V_\sigma(t, \mathbf{q})\, d^\dagger_{\mathbf{k}+\mathbf{q}\sigma} d_{\mathbf{k}\sigma} . \tag{5.130}$$

We can now simply apply the time-dependent Hartree-Fock approximation of Sect. 3.3 to calculate the equation of motion of the expectation value $n_{\mathbf{k}\sigma}(t, \mathbf{q}) = \langle d^\dagger_{\mathbf{k}\sigma}(t)\, d_{\mathbf{k}+\mathbf{q}\sigma}(t) \rangle$ at linear order in the perturbation, which implies also at linear order in the deviation from equilibrium

$$\delta n_{\mathbf{k}\sigma}(t, \mathbf{q}) = n_{\mathbf{k}\sigma}(t, \mathbf{q}) - \delta_{\mathbf{q}0}\, n^0_{\mathbf{k}\sigma} ,$$

and get

$$\begin{aligned}
\frac{\partial \delta n_{\mathbf{k}\sigma}(t, \mathbf{q})}{\partial t} &= -i\left(\epsilon_*(\mathbf{k}+\mathbf{q}) - \epsilon_*(\mathbf{k})\right) \delta n_{\mathbf{k}\sigma}(t, \mathbf{q}) - i\left(n^0_{\mathbf{k}\sigma} - n^0_{\mathbf{k}+\mathbf{q}\sigma}\right) V_\sigma(t, \mathbf{q}) \\
&\quad - i\left(n^0_{\mathbf{k}\sigma} - n^0_{\mathbf{k}+\mathbf{q}\sigma}\right) \frac{1}{V} \sum_{\mathbf{k}'\sigma'} f_{\mathbf{k}\sigma,\mathbf{k}'\sigma'}\, \delta n_{\mathbf{k}'\sigma'}(t, \mathbf{q}) \\
&\simeq -i\, v_{\mathbf{k}} \cdot \mathbf{q}\, \delta n_{\mathbf{k}\sigma}(t, \mathbf{q}) + i\, \frac{\partial f(\epsilon_*(\mathbf{k}))}{\partial \epsilon_*(\mathbf{k})}\, v_{\mathbf{k}} \cdot \mathbf{q}\left[V_\sigma(t, \mathbf{q}) + \frac{1}{V} \sum_{\mathbf{k}'\sigma'} f_{\mathbf{k}\sigma,\mathbf{k}'\sigma'}\, \delta n_{\mathbf{k}'\sigma'}(t, \mathbf{q})\right],
\end{aligned} \tag{5.131}$$

the last equation valid at small q. Upon defining the Wigner distribution

$$n_{\mathbf{k}\sigma}(t, \mathbf{r}) = \frac{1}{V} \sum_{\mathbf{q}} e^{i\mathbf{q}\cdot\mathbf{r}}\, n_{\mathbf{k}\sigma}(t, \mathbf{q}), \quad n_{\mathbf{k}\sigma}(t, \mathbf{q}) = \int d\mathbf{r}\, e^{-i\mathbf{q}\cdot\mathbf{r}}\, n_{\mathbf{k}\sigma}(t, \mathbf{r}),$$

under the assumption that $q \ll k \simeq k_{FL}$, which also implies that, at equilibrium, $n^0_{\mathbf{k}\sigma}(t, \mathbf{r}) = f(\epsilon_*(\mathbf{k}))/V$, and, through (5.128), (5.131) becomes, in real space and time,

$$\begin{aligned}
\frac{\partial \delta n_{\mathbf{k}\sigma}(t, \mathbf{r})}{\partial t} &+ v_{\mathbf{k}} \cdot \nabla \delta n_{\mathbf{k}\sigma}(t, \mathbf{r}) \\
&- \frac{1}{V} \frac{\partial f(\epsilon_*(\mathbf{k}))}{\partial \epsilon_*(\mathbf{k})}\, v_{\mathbf{k}} \cdot \nabla\left[V_\sigma(t, \mathbf{r}) + \sum_{\mathbf{k}'\sigma'} f_{\mathbf{k}\sigma,\mathbf{k}'\sigma'}\, \delta n_{\mathbf{k}'\sigma'}(t, \mathbf{r})\right] = I_{\mathbf{k}\sigma}(t, \mathbf{r}),
\end{aligned} \tag{5.132}$$

where we also added a collision integral $I_{\mathbf{k}\sigma}(t, \mathbf{r})$ defined as the probability per unit time that a quasiparticle with momentum $\mathbf{k}$ and spin σ is created in an infinitesimal volume around $\mathbf{r}$ and infinitesimal time interval around t minus the probability that it is destroyed. The collision integral is due to scattering off impurities and lattice vibrations, as well as to interaction effects not accounted for by the Hartree-Fock approximation, specifically, the imaginary part of the quasiparticle self-energy yielding the quasiparticle decay rate.

We observe that (5.132) is the conventional Boltzmann transport equation

$$
\frac{\partial n_{\mathbf{k}\sigma}(t, \mathbf{r})}{\partial t} + \nabla n_{\mathbf{k}\sigma}(t, \mathbf{r}) \cdot \frac{\partial \bar{\epsilon}_{*\mathbf{k}\sigma}(t, \mathbf{r})}{\partial \mathbf{k}} - \frac{\partial n_{\mathbf{k}\sigma}(t, \mathbf{r})}{\partial \mathbf{k}} \cdot \nabla \epsilon_{*\mathbf{k}\sigma}(t, \mathbf{r}) = I_{\mathbf{k}\sigma}(t, \mathbf{r}),
$$
(5.133)

linearised in the deviation from equilibrium, where

$$
\epsilon_{*\mathbf{k}\sigma}(t, \mathbf{r}) = \epsilon_*(\mathbf{k}) + \sum_{\mathbf{k}'\sigma'} f_{\mathbf{k}\sigma, \mathbf{k}'\sigma'}\, \delta n_{\mathbf{k}'\sigma'}(t, \mathbf{r}) + V_\sigma(t, \mathbf{r}) \equiv \bar{\epsilon}_{*\mathbf{k}\sigma}(t, \mathbf{r}) + V_\sigma(t, \mathbf{r}),
$$

is the quasiparticle energy in presence of the perturbation and of other excited quasiparticles. Alternatively, we can define a deviation from local equilibrium

$$
\overline{\delta n}_{\mathbf{k}\sigma}(t, \mathbf{r}) = n_{\mathbf{k}\sigma}(t, \mathbf{r}) - \frac{1}{V}\, f\big(\bar{\epsilon}_{*\mathbf{k}\sigma}(t, \mathbf{r})\big) = \delta n_{\mathbf{k}\sigma}(t, \mathbf{r}) - \frac{1}{V}\left(f\big(\bar{\epsilon}_{*\mathbf{k}\sigma}(t, \mathbf{r})\big) - f\big(\epsilon_*(\mathbf{k})\big)\right)
$$

$$
\simeq \delta n_{\mathbf{k}\sigma}(t, \mathbf{r}) - \frac{1}{V}\, \frac{\partial f\big(\epsilon_*(\mathbf{k})\big)}{\partial \epsilon_*(\mathbf{k})} \sum_{\mathbf{k}'\sigma'} f_{\mathbf{k}\sigma, \mathbf{k}'\sigma'}\, \delta n_{\mathbf{k}'\sigma'}(t, \mathbf{r}),
$$
(5.134)

through which (5.132) acquires a more compact expression

$$
\frac{\partial \delta n_{\mathbf{k}\sigma}(t, \mathbf{r})}{\partial t} + \mathbf{v}_{\mathbf{k}} \cdot \nabla \overline{\delta n}_{\mathbf{k}\sigma}(t, \mathbf{r}) - \frac{1}{V}\, \frac{\partial f\big(\epsilon_*(\mathbf{k})\big)}{\partial \epsilon_*(\mathbf{k})}\, \mathbf{v}_{\mathbf{k}} \cdot \nabla V_\sigma(t, \mathbf{r}) = I_{\mathbf{k}\sigma}(t, \mathbf{r}).
$$
(5.135)

If the temperature $T(t, \mathbf{r})$ smoothly varies in space and time, with average T, the quasiparticles are coupled to $\delta T(t, \mathbf{r}) = T(t, \mathbf{r}) - T$ just like physical electrons are, which implies that

$$
\frac{\delta E}{\delta n_{\mathbf{k}\sigma}(t, \mathbf{r})} = \bar{\epsilon}_{*\mathbf{k}\sigma}(t, \mathbf{r}) \to \bar{\epsilon}_{*\mathbf{k}\sigma}(t, \mathbf{r})\left(1 - \frac{\delta T(t, \mathbf{r})}{T}\right),
$$

so that

$$
\nabla \bar{\epsilon}_{*\mathbf{k}\sigma}(t, \mathbf{r}) \to \nabla \bar{\epsilon}_{*\mathbf{k}\sigma}(t, \mathbf{r}) - \bar{\epsilon}_{*\mathbf{k}\sigma}(t, \mathbf{r})\, \frac{\nabla T(t, \mathbf{r})}{T} \simeq \nabla \bar{\epsilon}_{*\mathbf{k}\sigma}(t, \mathbf{r}) - \epsilon_*(\mathbf{k})\, \frac{\nabla T(t, \mathbf{r})}{T},
$$
(5.136)

where the last expression derives from the implicit assumption that $\big|\delta T(t, \mathbf{r})\big| \ll T$. Therefore, the linearised Boltzmann equation becomes

$$
\frac{\partial \delta n_{\mathbf{k}\sigma}(t, \mathbf{r})}{\partial t} + \mathbf{v}_{\mathbf{k}} \cdot \nabla \overline{\delta n}_{\mathbf{k}\sigma}(t, \mathbf{r}) + \frac{1}{V}\, \frac{\partial f\big(\epsilon_*(\mathbf{k})\big)}{\partial \epsilon_*(\mathbf{k})}\, \frac{\epsilon_*(\mathbf{k})}{T}\, \mathbf{v}_{\mathbf{k}} \cdot \nabla T(t, \mathbf{r})
$$

$$
- \frac{1}{V}\, \frac{\partial f\big(\epsilon_*(\mathbf{k})\big)}{\partial \epsilon_*(\mathbf{k})}\, \mathbf{v}_{\mathbf{k}} \cdot \nabla V_\sigma(t, \mathbf{r}) = I_{\mathbf{k}\sigma}(t, \mathbf{r}).
$$
(5.137)

Equation (5.137) with $I_{\mathbf{k}\sigma}(t, \mathbf{r}) = 0$ reproduces all thermal and transport properties we earlier calculated through the response functions.

By definition, the collision integral satisfies

$$\sum_{\mathbf{k}\sigma} I_{\mathbf{k}\sigma}(t, \mathbf{r}) = 0 \,, \quad \sum_{\mathbf{k}\sigma} \bar{\epsilon}_{*\mathbf{k}\sigma}(t, \mathbf{r}) \, I_{\mathbf{k}\sigma}(t, \mathbf{r}) = 0 \,,$$

where the last equation follows from the conservation of local quasiparticle energy. Therefore, in absence of any external perturbation and at linear order,

$$0 = \sum_{\mathbf{k}\sigma} I_{\mathbf{k}\sigma}(t, \mathbf{r}) = \frac{\partial}{\partial t} \sum_{\mathbf{k}\sigma} \delta n_{\mathbf{k}\sigma}(t, \mathbf{r}) + \nabla \cdot \sum_{\mathbf{k}\sigma} v_{\mathbf{k}} \, \overline{\delta n}_{\mathbf{k}\sigma}(t, \mathbf{r}) \,,$$

$$0 = \sum_{\mathbf{k}\sigma} \bar{\epsilon}_{*\mathbf{k}\sigma}(t, \mathbf{r}) \, I_{\mathbf{k}\sigma}(t, \mathbf{r}) \simeq \frac{\partial}{\partial t} \sum_{\mathbf{k}\sigma} \epsilon_*(\mathbf{k}) \, \delta n_{\mathbf{k}\sigma}(t, \mathbf{r}) + \nabla \cdot \sum_{\mathbf{k}\sigma} v_{\mathbf{k}} \, \epsilon_*(\mathbf{k}) \, \overline{\delta n}_{\mathbf{k}\sigma}(t, \mathbf{r}) \,,$$

which, according to the continuity equations for charge and heat, provide the definition of charge and heat currents

$$\boldsymbol{J}(t, \mathbf{r}) = \sum_{\mathbf{k}\sigma} v_{\mathbf{k}} \, \overline{\delta n}_{\mathbf{k}\sigma}(t, \mathbf{r}) = \sum_{\mathbf{k}\sigma} \bar{v}_{\mathbf{k}} \, \delta n_{\mathbf{k}\sigma}(t, \mathbf{r}) \,,$$

$$\boldsymbol{J}_E(t, \mathbf{r}) = \sum_{\mathbf{k}\sigma} v_{\mathbf{k}} \, \epsilon_*(\mathbf{k}) \, \overline{\delta n}_{\mathbf{k}\sigma}(t, \mathbf{r}) \,, \tag{5.138}$$

where

$$\bar{v}_{\mathbf{k}} \equiv v_{\mathbf{k}} - \frac{2}{V} \sum_{\mathbf{k}'} f_S(\mathbf{k}, \mathbf{k}') \, v_{\mathbf{k}'} \, \frac{\partial f(\epsilon_*(\mathbf{k}'))}{\partial \epsilon_*(\mathbf{k}')} \,. \tag{5.139}$$

5.8.2   Transport Equation in Presence of an Electromagnetic Field

We earlier mentioned that, by construction, Landau's Fermi liquid theory gives only access to the proper response functions when the electron-electron interaction is the Coulomb repulsion, i.e., the response to the internal field including the potential generated via Gauss's law by the same electrons. That simply amounts in (5.137) to the replacement

$$V_\sigma(t, \mathbf{r}) \rightarrow V_\sigma(t, \mathbf{r}) + \int d\mathbf{r}' \, \frac{e^2}{|\mathbf{r} - \mathbf{r}'|} \, \delta\rho(t, \mathbf{r}') \,, \quad \delta\rho(t, \mathbf{r}') = \sum_{\mathbf{k}'\sigma'} \delta n_{\mathbf{k}'\sigma'}(t, \mathbf{r}') \,, \tag{5.140}$$

after which we can deal with both proper and improper linear response to the longitudinal component of the electromagnetic field.

In order to extend the transport equation in the presence of a transverse electromagnetic field, we can exploit the semiclassical approximation, see, e.g., [7]. Let us consider the Boltzmann equation (5.133) before linearisation. In the presence of a vector potential $A(t, \mathbf{r})$, the conjugate momentum is not anymore $\mathbf{k}$ but

$$\mathbf{K} = \mathbf{k} - \frac{e}{c} A(t, \mathbf{r}), \qquad (5.141)$$

Therefore, the quasiparticle distribution function in the semiclassical limit can be still parametrised by $n_{\mathbf{K}\sigma}(t, \mathbf{r})$. Nevertheless, it is more convenient to define the occupation density in an equivalent way

$$n_{\mathbf{K}\sigma}(t, \mathbf{r}) \equiv n_{\mathbf{k}\sigma}(t, \mathbf{r}) = n_{\mathbf{K}+\frac{e}{c}A(t,\mathbf{r})\,\sigma}(t, \mathbf{r}), \qquad (5.142)$$

since, in presence of $\mathbf{A}$, the quasiparticle excitation energy, (5.133), changes simply into

$$\bar{\epsilon}_{*\mathbf{k}\sigma}(t, \mathbf{r}) \to \bar{\epsilon}_{*\mathbf{K}+\frac{e}{c}A(t,\mathbf{r})\,\sigma}(t, \mathbf{r}) = \bar{\epsilon}_{*\mathbf{k}\sigma}(t, \mathbf{r}).$$

From the above equation, it follows that

$$\frac{\partial \bar{\epsilon}_{*\mathbf{K}+\frac{e}{c}A(t,\mathbf{r})\,\sigma}(t, \mathbf{r})}{\partial \mathbf{K}} = \frac{\partial \bar{\epsilon}_{*\mathbf{k}\sigma}(t, \mathbf{r})}{\partial \mathbf{k}} = v_{\mathbf{k}\sigma}(t, \mathbf{r}),$$

is the quasiparticle group velocity in presence of the field, and

$$\frac{\partial \bar{\epsilon}_{*\mathbf{K}+\frac{e}{c}A(t,\mathbf{r})\,\sigma}(t, \mathbf{r})}{\partial \mathbf{r}} = \sum_i \frac{e}{c}\, v_{\mathbf{k}\sigma\,i}(t, \mathbf{r})\, \frac{\partial A_i(t, \mathbf{r})}{\partial \mathbf{r}} + \frac{\partial V_\sigma(t, \mathbf{r})}{\partial \mathbf{r}} + \sum_{\mathbf{k}'\sigma'} f_{\mathbf{k}\sigma,\mathbf{k}'\sigma'}\, \frac{\partial n_{\mathbf{k}'\sigma'}(t, \mathbf{r})}{\partial \mathbf{r}},$$

with $V_\sigma(t, \mathbf{r})$ defined in (5.140).

Through the definition (5.142), it follows that

$$\left(\frac{\partial n_{\mathbf{K}\sigma}(t, \mathbf{r})}{\partial t}\right)_{\mathbf{K}\mathbf{r}} = \left(\frac{\partial n_{\mathbf{k}\sigma}(t, \mathbf{r})}{\partial t}\right)_{\mathbf{k}\mathbf{r}} + \frac{e}{c}\, \frac{\partial n_{\mathbf{k}\sigma}(t, \mathbf{r})}{\partial \mathbf{k}} \cdot \frac{\partial A(t, \mathbf{r})}{\partial t},$$

$$\left(\frac{\partial n_{\mathbf{K}\sigma}(t, \mathbf{r})}{\partial \mathbf{r}}\right)_{\mathbf{K}} = \left(\frac{\partial n_{\mathbf{k}\sigma}(t, \mathbf{r})}{\partial \mathbf{r}}\right)_{\mathbf{k}} + \sum_i \frac{e}{c}\, \frac{\partial n_{\mathbf{k}\sigma}(t, \mathbf{r})}{\partial k_i}\, \frac{\partial A_i(t, \mathbf{r})}{\partial \mathbf{r}},$$

$$\left(\frac{\partial n_{\mathbf{K}\sigma}(t, \mathbf{r})}{\partial \mathbf{K}}\right)_{\mathbf{r}} = \left(\frac{\partial n_{\mathbf{k}\sigma}(t, \mathbf{r})}{\partial \mathbf{k}}\right)_{\mathbf{r}}.$$

Putting everything together, we find the following transport equation for charged electrons:

$$I_{\mathbf{k}\sigma}(t, \mathbf{r}) = \frac{\partial n_{\mathbf{k}\sigma}(t, \mathbf{r})}{\partial t} + \frac{e}{c}\, \frac{\partial n_{\mathbf{k}\sigma}(t, \mathbf{r})}{\partial \mathbf{k}} \cdot \frac{\partial A(t, \mathbf{r})}{\partial t}$$

$$+ v_{\mathbf{k}\sigma}(t, \mathbf{r}) \cdot \frac{\partial n_{\mathbf{k}\sigma}(t, \mathbf{r})}{\partial \mathbf{x}} + \sum_{ij} v_{\mathbf{k}\sigma\,j}(t, \mathbf{r})\, \frac{e}{c}\, \frac{\partial n_{\mathbf{k}\sigma}(t, \mathbf{r})}{\partial k_i}\, \frac{\partial A_i(t, \mathbf{r})}{\partial r_j}$$

$$-\frac{\partial n_{\mathbf{k}\sigma}(t, \mathbf{r})}{\partial \mathbf{k}} \cdot \left[\frac{\partial V_\sigma(t, \mathbf{r})}{\partial \mathbf{r}} + \sum_{\mathbf{k}'\sigma'} f_{\mathbf{k}\sigma\,\mathbf{k}'\sigma'} \frac{\partial n_{\mathbf{k}'\sigma'}(t, \mathbf{r})}{\partial \mathbf{r}} \right]$$

$$- \sum_{ij} \frac{e}{c} \frac{\partial n_{\mathbf{k}\sigma}(t, \mathbf{r})}{\partial k_j} v_{\mathbf{k}\sigma\,i}(t, \mathbf{r}) \frac{\partial A_i(t, \mathbf{r})}{\partial r_j} .$$

We note that the two terms with the $\sum_{ij}$ can be written as

$$\sum_{ij} \frac{\partial n_{\mathbf{k}\sigma}(t, \mathbf{r})}{\partial k_i} v_{\mathbf{k}\sigma\,j}(t, \mathbf{r}) \left(\frac{\partial A_i(t, \mathbf{r})}{\partial r_j} - \frac{\partial A_j(t, \mathbf{r})}{\partial r_i} \right) = -\left(v_{\mathbf{k}\sigma}(t, \mathbf{r}) \wedge B(t, \mathbf{r}) \right) \cdot \frac{\partial n_{\mathbf{k}\sigma}(t, \mathbf{r})}{\partial \mathbf{k}},$$

where $B(t, \mathbf{r}) = \nabla \wedge A(t, \mathbf{r})$ is the magnetic field. Since the external transverse electric field is

$$E(t, \mathbf{r}) = -\frac{1}{c} \frac{\partial A(t, \mathbf{r})}{\partial t},$$

the final expression of the transport equation at uniform temperature reads

$$I_{\mathbf{k}\sigma}(t, \mathbf{r}) = \frac{\partial n_{\mathbf{k}\sigma}(t, \mathbf{r})}{\partial t} + v_{\mathbf{k}\sigma}(t, \mathbf{r}) \cdot \nabla n_{\mathbf{k}\sigma}(t, \mathbf{r})$$
$$- \frac{\partial n_{\mathbf{k}\sigma}(t, \mathbf{r})}{\partial \mathbf{k}} \cdot \nabla \left(V_\sigma(t, \mathbf{r}) + \sum_{\mathbf{k}'\sigma'} f_{\mathbf{k}\sigma,\mathbf{k}'\sigma'}\, n_{\mathbf{k}'\sigma'}(t, \mathbf{r}) \right)$$
$$- e \frac{\partial n_{\mathbf{k}\sigma}(t, \mathbf{r})}{\partial \mathbf{k}} \cdot E(t, \mathbf{r}) - \frac{e}{c} \left(v_{\mathbf{k}\sigma}(t, \mathbf{r}) \times B(t, \mathbf{r}) \right) \cdot \frac{\partial n_{\mathbf{k}\sigma}(t, \mathbf{r})}{\partial \mathbf{k}}.$$
$$(5.143)$$

Let us now expand (5.143) at linear order in the deviation from equilibrium

$$n_{\mathbf{k}\sigma}(t, \mathbf{r}) = n_{\mathbf{k}\sigma}^0 + \delta n_{\mathbf{k}\sigma}(t, \mathbf{r}) .$$

First, we assume an ac electromagnetic field acting as a perturbation, in which case

$$\left(v_{\mathbf{k}\sigma}(t, \mathbf{r}) \times B(t, \mathbf{r}) \right) \cdot \frac{\partial n_{\mathbf{k}\sigma}(t, \mathbf{r})}{\partial \mathbf{k}} \simeq \frac{1}{V} \left(v_{\mathbf{k}} \wedge B(t, \mathbf{r}) \right) \cdot \frac{\partial f\left(\epsilon_*(\mathbf{k}) \right)}{\partial \epsilon_*(\mathbf{k})}\, v_{\mathbf{k}} = 0.$$

This shows that an ac magnetic field does not contribute to the linearised transport equation, that becomes

$$I_{\mathbf{k}\sigma}(t, \mathbf{r}) = \frac{\partial \delta n_{\mathbf{k}\sigma}(t, \mathbf{r})}{\partial t} + v_{\mathbf{k}}(t, \mathbf{r}) \cdot \nabla \delta n_{\mathbf{k}\sigma}(t, \mathbf{r}) - e \frac{1}{V} \frac{\partial f\left(\epsilon_*(\mathbf{k}) \right)}{\partial \epsilon_*(\mathbf{k})}\, v_{\mathbf{k}} \cdot E(t, \mathbf{r})$$
$$- \frac{1}{V} \frac{\partial f\left(\epsilon_*(\mathbf{k}) \right)}{\partial \epsilon_*(\mathbf{k})}\, v_{\mathbf{k}} \cdot \nabla \left(V_\sigma(t, \mathbf{r}) + \sum_{\mathbf{k}'\sigma'} f_{\mathbf{k}\sigma,\mathbf{k}'\sigma'}\, n_{\mathbf{k}'\sigma'}(t, \mathbf{r}) \right)$$
$$= \frac{\partial \delta n_{\mathbf{k}\sigma}(t, \mathbf{r})}{\partial t} + v_{\mathbf{k}} \cdot \nabla \overline{\delta n}_{\mathbf{k}\sigma}(t, \mathbf{r})$$
$$+ \frac{1}{V} \frac{\partial f\left(\epsilon_*(\mathbf{k}) \right)}{\partial \epsilon_*(\mathbf{k})}\, v_{\mathbf{k}} \cdot \left[-\nabla V_\sigma(t, \mathbf{r}) - e\, E(t, \mathbf{r}) \right].$$
$$(5.144)$$

In presence of a dc magnetic field $B(r)$, a subtle issue arises in connection with the Lorentz's force term

$$\text{Lorentz's force} = -\frac{e}{c}\left(v_{k\sigma}(t, r) \wedge B(r)\right) \cdot \frac{\partial n_{k\sigma}(t, r)}{\partial k}.$$

Indeed, a dc field, unlike an ac one, can be assumed as an integral part of the unperturbed Hamiltonian and, if large, can be taken as a zeroth order term. This requires to expand at linear order $v_{k\sigma}(r)$ and $\partial n_{k\sigma}(r)/\partial k$,

$$v_{k\sigma}(r) = \frac{\partial \bar{\epsilon}_{*k\sigma}(r)}{\partial k} \simeq v_k + \sum_{k'\sigma'} \frac{\partial f_{k\sigma,k'\sigma'}}{\partial k} \delta n_{k'\sigma'}(t, r),$$

$$\frac{\partial n_{k\sigma}(t, r)}{\partial k} \simeq \frac{1}{V}\frac{\partial f(\epsilon_*(k))}{\partial \epsilon_*(k)} v_k + \frac{\partial \delta n_{k\sigma}(t, r)}{\partial k},$$

leading to the linearised transport equation, including back the temperature gradient term,

$$I_{k\sigma}(t, r) = \frac{\partial \delta n_{k\sigma}(t, r)}{\partial t} + v_k \cdot \nabla \overline{\delta n}_{k\sigma}(t, r) - \frac{e}{c}\left(v_k \wedge B(r)\right) \cdot \frac{\partial \overline{\delta n}_{k\sigma}(t, r)}{\partial k}$$
$$+ \frac{1}{V}\frac{\partial f(\epsilon_*(k))}{\partial \epsilon_*(k)} v_k \cdot \left[-\nabla V_\sigma(t, r) - e\,E(t, r)\right]$$
$$+ \frac{1}{V}\frac{\partial f(\epsilon_*(k))}{\partial \epsilon_*(k)}\frac{\epsilon_*(k)}{T} v_k \cdot \nabla T(t, r).$$

$$(5.145)$$

The electric current is defined as, see (5.138) and (5.139),

$$J(t, r) = -e \sum_{k\sigma} v_k\, \overline{\delta n}_{k\sigma}(t, r) = -e \sum_{k\sigma} \bar{v}_k\, \delta n_{k\sigma}(t, r) \equiv \sum_{k\sigma} J_{k\sigma}(t, r),$$

whose component i, in absence of the collision term and for $V_\sigma = 0$, satisfies the equation of motion

$$\frac{\partial J_i(t, r)}{\partial t} = -e \sum_{k\sigma} \bar{v}_{ik} \frac{\partial \delta n_{k\sigma}(t, r)}{\partial t} = -\nabla \cdot \sum_{k\sigma} \bar{v}_{ik} J_{k\sigma}(t, r)$$
$$- \frac{e}{c} \sum_{k\sigma} \epsilon_{\ell j k} \frac{\partial \bar{v}_{ik}}{\partial k_\ell} J_{jk\sigma}(t, r) B_k - \frac{e^2}{V} \sum_{k\sigma} \frac{\partial f(\epsilon_*(k))}{\partial \epsilon_*(k)} \bar{v}_{ik}\, v_k \cdot E(t, r),$$

where $\epsilon_{\ell j k}$ is the antisymmetric tensor and repeated indices are summed. The above equation defines the cyclotron effective mass tensor $\hat{m}_*(k)$, whose inverse has components

$$\left(\hat{m}_*(\mathbf{k})^{-1}\right)_{ij} \equiv \frac{\partial \bar{v}_{i\mathbf{k}}}{\partial k_j} \,, \quad \mathrm{Tr}\left(\hat{m}_*(\mathbf{k})^{-1}\right) = \nabla_{\mathbf{k}} \cdot \bar{v}_{\mathbf{k}} \,.$$

For a uniform ac electric field, and assuming $\bar{v}_{\mathbf{k}} = \mathbf{k}/m_*$, the equation simplifies into the conventional equation of the electric current in presence of a uniform magnetic field,

$$\frac{\partial \boldsymbol{J}(t)}{\partial t} = -\frac{e}{m_* c} \boldsymbol{J}(t) \wedge \boldsymbol{B} - \frac{e^2}{V} \sum_{\mathbf{k}\sigma} \frac{\partial f\left(\epsilon_*(\mathbf{k})\right)}{\partial \epsilon_*(\mathbf{k})} \bar{v}_{\mathbf{k}} \, v_{\mathbf{k}} \cdot \boldsymbol{E}(t)$$

$$= -\frac{e}{m_* c} \boldsymbol{J}(t) \wedge \boldsymbol{B} + D_* \, \boldsymbol{E}(t) \,,$$

where D_* is the Drude weight (5.74), which can be written in $d = 2, 3$ dimensions and assuming space isotropy as

$$D_* = -\frac{2e^2}{d V} \sum_{\mathbf{k}} \frac{\partial f\left(\epsilon_*(\mathbf{k})\right)}{\partial \epsilon_*(\mathbf{k})} \bar{v}_{\mathbf{k}} \cdot \bar{v}_{\mathbf{k}} = -\frac{2e^2}{d V} \sum_{\mathbf{k}} \frac{\partial f\left(\epsilon_*(\mathbf{k})\right)}{\partial \mathbf{k}} \cdot \bar{v}_{\mathbf{k}}$$

$$= \frac{2e^2}{d V} \sum_{\mathbf{k}} f\left(\epsilon_*(\mathbf{k})\right) \nabla_{\mathbf{k}} \cdot \bar{v}_{\mathbf{k}} = \frac{2e^2}{d V} \sum_{\mathbf{k}} f\left(\epsilon_*(\mathbf{k})\right) \mathrm{Tr}\left(\hat{m}_*(\mathbf{k})^{-1}\right) \,,$$

yielding the conventional result $D_* = ne^2/m_*$ for $\bar{v}_{\mathbf{k}} = \mathbf{k}/m_*$, where n is the electron density. In the Mott insulator of Sect. 5.6, the Drude weight $D_* = 0$, which is equivalent to a vanishing trace of the inverse cyclotron mass tensor averaged over the volume enclosed by the Luttinger surface.

5.9 Application: Transport Coefficients with Rotational Symmetry

In this section, we consider the simple case where perturbation theory is valid and there is just a single Fermi surface. The aim is to derive from the linearised Boltzmann equation (5.137) the expression of the transport coefficients of a conventional Fermi liquid.

Equation (5.145) in presence of an internal electric field $\boldsymbol{E}(t, \mathbf{r})$ becomes, once summed over σ, thus for $\sum_\sigma \delta n_{\mathbf{k}\sigma}(t, \mathbf{r}) \equiv \delta n_{\mathbf{k}}(t, \mathbf{r})$,

$$\frac{\partial \delta n_{\mathbf{k}}(t, \mathbf{r})}{\partial t} + v_{\mathbf{k}} \cdot \nabla \overline{\delta n}_{\mathbf{k}}(t, \mathbf{r}) + \frac{2}{V} \frac{\partial f\left(\epsilon_*(\mathbf{k})\right)}{\partial \epsilon_*(\mathbf{k})} \frac{\epsilon_*(\mathbf{k})}{T} v_{\mathbf{k}} \cdot \nabla T(t, \mathbf{r})$$

$$- \frac{2e}{V} \frac{\partial f\left(\epsilon_*(\mathbf{k})\right)}{\partial \epsilon_*(\mathbf{k})} v_{\mathbf{k}} \cdot \boldsymbol{E}(t, \mathbf{r}) = I_{\mathbf{k}}(t, \mathbf{r}) \,.$$

$$(5.146)$$

We recall the definitions (5.138) for electric and heat current densities

$$\boldsymbol{J}(t, \mathbf{r}) = -e \sum_{\mathbf{k}\sigma} \boldsymbol{v}_{\mathbf{k}} \, \overline{\delta n}_{\mathbf{k}\sigma}(t, \mathbf{r}) \,, \quad \boldsymbol{J}_E(t, \mathbf{r}) = \sum_{\mathbf{k}\sigma} \epsilon_*(\mathbf{k}) \, \boldsymbol{v}_{\mathbf{k}} \, \overline{\delta n}_{\mathbf{k}\sigma}(t, \mathbf{r}) \,.$$

(5.147)

We want to find their expressions in presence of uniform internal electric field and temperature gradient. For that, we suppose that the model is, to a good approximation, $O(3)$ rotational invariant. Therefore $\epsilon_*(\mathbf{k}) = \epsilon_*(k)$ depends just on $k = |\mathbf{k}|$, and the Landau f-parameters are invariant under $O(3)$ rotations, so that, close to the quasiparticle Fermi surface,

$$\boldsymbol{v}_{\mathbf{k}} = \frac{\partial \epsilon_*(k)}{\partial \mathbf{k}} \simeq v_F \frac{\mathbf{k}}{k} \,, \quad f_{\mathbf{k},\mathbf{k}'}^{S/A} = \sum_{\ell} f_\ell^{S/A} \, P_\ell(\cos\theta_{\mathbf{k},\mathbf{k}'}) \,,$$

(5.148)

where $P_\ell(\cos\theta_{\mathbf{k},\mathbf{k}'})$ are Legendre polynomials, with $\theta_{\mathbf{k},\mathbf{k}'}$ the angle between $\mathbf{k}$ and $\mathbf{k}'$.

We recall , since it will be useful in what follows, that

$$\int \frac{d\Omega_{\mathbf{k}'}}{4\pi} \, P_\ell(\cos\theta_{\mathbf{k},\mathbf{k}'}) \, Y_{\ell'm'}(\Omega_{\mathbf{k}'}) = \delta_{\ell\ell'} \, \frac{1}{2\ell+1} \, Y_{\ell'm'}(\Omega_{\mathbf{k}}) \,,$$

(5.149)

where $\Omega_{\mathbf{k}}$ is the solid angle that represents a unit vector parallel to $\mathbf{k}$, and $Y_{\ell m}(\Omega) \equiv Y_{\ell m}(\theta, \phi)$ are the spherical harmonics. Moreover, one can easily demonstrate, since $\epsilon_*(\mathbf{k}) = \epsilon_*(k)$, that

$$\frac{1}{V} \sum_{\mathbf{k}} \left(-\frac{\partial f(\epsilon_*(k))}{\partial \epsilon_*(k)} \right) F(\epsilon_*(k)) \, G(\Omega_{\mathbf{k}}) = \int d\epsilon \, \rho_{qp}(\epsilon) \, F(\epsilon) \left(-\frac{\partial f(\epsilon)}{\partial \epsilon} \right) \int \frac{d\Omega_{\mathbf{k}}}{4\pi} \, G(\Omega_{\mathbf{k}})$$

$$\simeq \left(\rho_{qp} \, F(0) + \frac{\pi^2 \, T^2}{6} \frac{\partial^2 (\rho_{qp}(\epsilon) \, F(\epsilon))}{\partial \epsilon^2} \Big|_{\epsilon=0} \right) \int \frac{d\Omega_{\mathbf{k}}}{4\pi} \, G(\Omega_{\mathbf{k}}) \,,$$

(5.150)

where $\rho_{qp}(\epsilon)$ is the quasiparticle density of states, $\rho_{qp} = \rho_{qp}(0)$, and 4π is the total solid angle in three dimensions.

Exploiting the $O(3)$ symmetry and the fact that all physical quantities depend on momenta close to the quasiparticle Fermi surface, we write

$$\delta n_{\mathbf{k}}(t, \mathbf{r}) = -\frac{1}{V} \frac{\partial f(\epsilon_*(k))}{\partial \epsilon_*(k)} \sum_{\ell m} \delta n_{\ell m}(t, \mathbf{r}; \epsilon_*(k)) \, Y_{\ell m}(\Omega_{\mathbf{k}}) \,,$$

$$\overline{\delta n}_{\mathbf{k}}(t, \mathbf{r}) = -\frac{1}{V} \frac{\partial f(\epsilon_*(k))}{\partial \epsilon_*(k)} \sum_{\ell m} \overline{\delta n}_{\ell m}(t, \mathbf{r}; \epsilon_*(k)) \, Y_{\ell m}(\Omega_{\mathbf{k}}) \,.$$

We further assume that the collision integral is due to elastic scattering off non-magnetic impurities, thus, according to Fermi golden rule, it can be written as

$$I_{\mathbf{k}\sigma}(t, \mathbf{r}) = \frac{2\pi}{V} \sum_{\mathbf{k}'} W_{\mathbf{k}\mathbf{k}'} \left[n_{\mathbf{k}'\sigma}(t, \mathbf{r}) \left(1 - n_{\mathbf{k}\sigma}(t, \mathbf{r})\right) - n_{\mathbf{k}\sigma}(t, \mathbf{r}) \left(1 - n_{\mathbf{k}'\sigma}(t, \mathbf{r})\right) \right]$$

$$\delta\left(\bar{\epsilon}_{*\mathbf{k}\sigma}(t, \mathbf{r}) - \bar{\epsilon}_{*\mathbf{k}'\sigma}(t, \mathbf{r})\right)$$

$$= -\frac{2\pi}{V} \sum_{\mathbf{k}'} W_{\mathbf{k}\mathbf{k}'} \left(n_{\mathbf{k}\sigma}(t, \mathbf{r}) - n_{\mathbf{k}'\sigma}(t, \mathbf{r}) \right) \delta\left(\bar{\epsilon}_{*\mathbf{k}\sigma}(t, \mathbf{r}) - \bar{\epsilon}_{*\mathbf{k}'\sigma}(t, \mathbf{r})\right).$$

We remark that the energy conservation must involve the local quasiparticle energies. The reason is that the collision integral must vanish at local equilibrium $n_{\mathbf{k}\sigma}(t, \mathbf{r}) = \bar{n}^0_{\mathbf{k}\sigma}(t, \mathbf{r}) \equiv f\left(\bar{\epsilon}_{*\mathbf{k}\sigma}(t, \mathbf{r})\right)$, which indeed it does. Therefore

$$I_{\mathbf{k}\sigma}(t, \mathbf{r}) = -\frac{2\pi}{V} \sum_{\mathbf{k}'} W_{\mathbf{k}\mathbf{k}'} \left(\overline{\delta n}_{\mathbf{k}\sigma}(t, \mathbf{r}) - \overline{\delta n}_{\mathbf{k}'\sigma}(t, \mathbf{r}) \right) \delta\left(\bar{\epsilon}_{*\mathbf{k}\sigma}(t, \mathbf{r}) - \bar{\epsilon}_{*\mathbf{k}'\sigma}(t, \mathbf{r})\right)$$

$$- \frac{2\pi}{V} \sum_{\mathbf{k}'} W_{\mathbf{k}\mathbf{k}'} \left(\bar{n}^0_{\mathbf{k}\sigma}(t, \mathbf{r}) - \bar{n}^0_{\mathbf{k}'\sigma}(t, \mathbf{r}) \right) \delta\left(\bar{\epsilon}_{*\mathbf{k}\sigma}(t, \mathbf{r}) - \bar{\epsilon}_{*\mathbf{k}'\sigma}(t, \mathbf{r})\right)$$

$$= -\frac{2\pi}{V} \sum_{\mathbf{k}'} W_{\mathbf{k}\mathbf{k}'} \left(\overline{\delta n}_{\mathbf{k}\sigma}(t, \mathbf{r}) - \overline{\delta n}_{\mathbf{k}'\sigma}(t, \mathbf{r}) \right) \delta\left(\bar{\epsilon}_{*\mathbf{k}\sigma}(t, \mathbf{r}) - \bar{\epsilon}_{*\mathbf{k}'\sigma}(t, \mathbf{r})\right)$$

$$\simeq -\frac{2\pi}{V} \sum_{\mathbf{k}'} W_{\mathbf{k}\mathbf{k}'} \left(\overline{\delta n}_{\mathbf{k}\sigma}(t, \mathbf{r}) - \overline{\delta n}_{\mathbf{k}'\sigma}(t, \mathbf{r}) \right) \delta\left(\epsilon_*(\mathbf{k}) - \epsilon_*(\mathbf{k}')\right),$$

$$(5.151)$$

only depends on the deviations from local equilibrium, and the last equation is the linearised expression. We also write, consistently with the $O(3)$ symmetry,

$$W_{\mathbf{k}\mathbf{k}'} = \sum_{\ell} W_\ell\left(\epsilon_*(\mathbf{k})\right) P_\ell\left(\cos\theta_{\mathbf{k}\mathbf{k}'}\right), \tag{5.152}$$

so that, using (5.149), the collision integral (5.151) can be written as

$$\sum_{\sigma} I_{\mathbf{k}\sigma}(t, \mathbf{r}) = \frac{\partial f\left(\epsilon_*(\mathbf{k})\right)}{\partial \epsilon_*(\mathbf{k})} \sum_{\ell m} \frac{1}{\tau_\ell(\epsilon_*(\mathbf{k}))} \overline{\delta n}_{\ell m}(t, \mathbf{r}; \epsilon_*(\mathbf{k})) Y_{\ell m}(\Omega_{\mathbf{k}}),$$

$$(5.153)$$

where

$$\frac{1}{\tau_\ell\left(\epsilon_*(\mathbf{k})\right)} = 2\pi \rho_{\text{qp}}(\epsilon_*(\mathbf{k})) \left(W_0\left(\epsilon_*(\mathbf{k})\right) - \frac{W_\ell\left(\epsilon_*(\mathbf{k})\right)}{2\ell + 1} \right), \tag{5.154}$$

and vanishes at $\ell = 0$. We introduce the euclidean combinations of spherical harmonics

$$Y_{1z} = Y_{10} = \sqrt{\frac{3}{4\pi}} \cos\theta,$$

$$Y_{1x} = \frac{1}{\sqrt{2}}(Y_{1-1} - Y_{11}) = \sqrt{\frac{3}{4\pi}} \sin\theta\cos\phi,$$

$$Y_{1y} = \frac{i}{\sqrt{2}}(Y_{1-1} + Y_{11}) = \sqrt{\frac{3}{4\pi}} \sin\theta\sin\phi,$$

that transform like the components of a unit vector Y_1, so that (5.148) can be written as

$$\boldsymbol{v_k} = v_F \frac{\mathbf{k}}{k} \equiv \sqrt{\frac{4\pi}{3}}\, v_F \, \boldsymbol{Y}_1(\Omega_{\mathbf{k}}).$$

We are interested in the transport coefficients, which are calculated through the charge and heat currents in presence of uniform electric field $\boldsymbol{E}(t)$ and temperature gradient $\nabla T(t)$. In that case, $\delta n_{\mathbf{k}}(t, \mathbf{r}) = \delta n_{\mathbf{k}}(t)$ becomes independent of $\mathbf{r}$ and the Boltzmann equation (5.146) simplifies into

$$I_{\mathbf{k}}(t) = \frac{\partial \delta n_{\mathbf{k}}(t)}{\partial t} + \frac{2}{V}\frac{\epsilon_*(\mathbf{k})}{T}\frac{\partial f\big(\epsilon_*(\mathbf{k})\big)}{\partial \epsilon_*(\mathbf{k})}\sqrt{\frac{4\pi}{3}}\, v_F\, \boldsymbol{Y}_1(\Omega_{\mathbf{k}}) \cdot \nabla T(t)$$
$$- \frac{2e}{V}\frac{\partial f\big(\epsilon_*(\mathbf{k})\big)}{\partial \epsilon_*(\mathbf{k})}\sqrt{\frac{4\pi}{3}}\, v_F\, \boldsymbol{Y}_1(\Omega_{\mathbf{k}}) \cdot \boldsymbol{E}(t).$$

or, equivalently, after Fourier transform in frequency,

$$-\sum_{\ell m} \frac{1}{\tau_\ell\big(\epsilon_*(\mathbf{k})\big)}\, \overline{\delta n}_{\ell m}(\omega; \epsilon_*(\mathbf{k}))\, Y_{\ell m}(\Omega_{\mathbf{k}}) = -i\omega \sum_{\ell m} \delta n_{\ell m}(\omega; \epsilon_*(\mathbf{k}))\, Y_{\ell m}(\Omega_{\mathbf{k}})$$
$$- \frac{2e}{V}\sqrt{\frac{4\pi}{3}}\, v_F\, \boldsymbol{Y}_1(\Omega_{\mathbf{k}}) \cdot \boldsymbol{E}(\omega) + \frac{2}{V}\frac{\epsilon_*(\mathbf{k})}{T}\sqrt{\frac{4\pi}{3}}\, v_F\, \boldsymbol{Y}_1(\Omega_{\mathbf{k}}) \cdot \nabla T(\omega).$$

We note that only the $\ell = 1$ components are different from zero. Therefore, if we define two vectors $\delta\boldsymbol{n}_1 = \big(\delta n_{1x}, \delta n_{1y}, \delta n_{1z}\big)$ and $\overline{\delta\boldsymbol{n}}_1 = \big(\overline{\delta n}_{1x}, \overline{\delta n}_{1y}, \overline{\delta n}_{1z}\big)$ in real spherical harmonics, then the Boltzmann equation has the simple expression

$$i\, \frac{\overline{\delta\boldsymbol{n}}_1(\omega; \epsilon_*(\mathbf{k}))}{\tau_1\big(\epsilon_*(\mathbf{k})\big)} = -\omega\, \delta\boldsymbol{n}_1(\omega\epsilon_*(\mathbf{k})) + 2i\, \frac{1}{V}\frac{\epsilon_*(\mathbf{k})}{T}\sqrt{\frac{4\pi}{3}}\, v_F\, \nabla T(\omega) - 2i\, e\, \frac{1}{V}\sqrt{\frac{4\pi}{3}}\, v_F\, \boldsymbol{E}(\omega).$$

We now set $\omega = 0$, so that

$$\overline{\delta\boldsymbol{n}}_1\big(\epsilon_*(\mathbf{k})\big) = 2\, \tau_1\big(\epsilon_*(\mathbf{k})\big)\frac{1}{V}\frac{\epsilon_*(\mathbf{k})}{T}\sqrt{\frac{4\pi}{3}}\, v_F\, \nabla T - 2\, e\, \tau_1\big(\epsilon_*(\mathbf{k})\big)\frac{1}{V}\sqrt{\frac{4\pi}{3}}\, v_F\, \boldsymbol{E}.$$
$$\tag{5.155}$$

The current densities are, therefore, at linear order

$$
\begin{aligned}
\boldsymbol{J} &= -e \sum_{\mathbf{k}} \boldsymbol{v_k} \,\overline{\delta n_{\mathbf{k}}} = -e\, v_F \sqrt{\frac{4\pi}{3}} \sum_{\mathbf{k}} \boldsymbol{Y}_1\!\left(\Omega_{\mathbf{k}}\right) \overline{\delta n_{\mathbf{k}}} \\
&= -\frac{e\, v_F}{4\pi} \sqrt{\frac{4\pi}{3}} \sum_{\mathbf{k}} \left(-\frac{\partial f\!\left(\epsilon_*(\mathbf{k})\right)}{\partial \epsilon_*(\mathbf{k})} \right) \overline{\delta n}_1(\epsilon_*(\mathbf{k})) \\
&= \frac{2e\, v_F^2}{3} \int d\epsilon\, \rho_{\mathrm{qp}}(\epsilon)\, \tau_1(\epsilon) \left(-\frac{\partial f(\epsilon)}{\partial \epsilon} \right) \left(e\,\boldsymbol{E} - \frac{\epsilon}{T}\,\boldsymbol{\nabla} T \right) \\
&= \frac{2e\, v_F^2\, \rho_{\mathrm{qp}} \tau_1}{3} \left(e\,\boldsymbol{E} - \frac{\pi^2\, T\, C}{3}\,\boldsymbol{\nabla} T \right), \\
\boldsymbol{J}_E &= \sum_{\mathbf{k}} \epsilon_*(\mathbf{k})\, \boldsymbol{v_k}\,\overline{\delta n_{\mathbf{k}}} = -\frac{2e\, v_F^2}{3} \int d\epsilon\, \rho_{\mathrm{qp}}(\epsilon)\, \tau_1(\epsilon)\, \epsilon \left(-\frac{\partial f(\epsilon)}{\partial \epsilon} \right) \left(e\,\boldsymbol{E} - \frac{\epsilon}{T}\,\boldsymbol{\nabla} T \right) \\
&= \frac{2e\, v_F^2\, \rho_{\mathrm{qp}} \tau_1}{3}\, \frac{\pi^2\, T}{3} \left(\boldsymbol{\nabla} T - e\, C\, T\, \boldsymbol{E} \right),
\end{aligned}
\tag{5.156}
$$

where $\tau_1 = \tau_1(0)$ and

$$
C = \left. \frac{\partial \ln\left(\rho_{\mathrm{qp}}(\epsilon)\, \tau_1(\epsilon)\right)}{\partial \epsilon} \right|_{\epsilon=0}.
$$

The expression of the charge conductivity σ is defined as

$$
\sigma = \frac{2\, e^2\, v_F^2\, \rho_{\mathrm{qp}}\, \tau_1}{3},
\tag{5.157}
$$

through which we can write

$$
\begin{aligned}
\boldsymbol{J} &= \sigma\, \boldsymbol{E} - \sigma\, \frac{\pi^2 T C}{3e}\, \boldsymbol{\nabla} T, \\
\boldsymbol{J}_E &= -\sigma\, \frac{\pi^2 T^2 C}{3e}\, \boldsymbol{E} + \sigma\, \frac{\pi^2 T}{3e^2}\, \boldsymbol{\nabla} T,
\end{aligned}
\tag{5.158}
$$

that provide the expressions of the transport coefficients, namely the proportionality constants between the currents and the electric field or temperature gradient.

Through (5.158), we can now evaluate other measurable quantities besides electrical conductivity. However, as discussed at the end of Sect. 5.5.1, in common experiments the temperature drop exists at equilibrium, and that simply amounts to invert the sign of $\boldsymbol{\nabla} T$ in (5.158), thus

$$
\begin{aligned}
\boldsymbol{J} &= \sigma\, \boldsymbol{E} + \sigma\, \frac{\pi^2 T C}{3e}\, \boldsymbol{\nabla} T, \\
\boldsymbol{J}_E &= -\sigma\, \frac{\pi^2 T^2 C}{3e}\, \boldsymbol{E} - \sigma\, \frac{\pi^2 T}{3e^2}\, \boldsymbol{\nabla} T,
\end{aligned}
\tag{5.159}
$$

Therefore, if a temperature drop exists across a finite sample, electrons from the hot side will start flowing towards the cold one. This charge flow will stop once the opposite charge accumulated on the opposite sides generates an electric field that prevents further charge flowing. We can determine the value of the electric field generated as the one that cancels $\boldsymbol{J}$ in (5.159), namely

$$\boldsymbol{E} = -\frac{\pi^2 T \mathcal{C}}{3e} \nabla T. \tag{5.160}$$

This value substituted in (5.159) provides the expression of the heat current that keeps flowing at zero charge current

$$\boldsymbol{J}_E = -\sigma \frac{\pi^2 T}{3e^2} \left(1 - \frac{\pi^2 T^2 \mathcal{C}^2}{3}\right) \nabla T \equiv -K \nabla T,$$

where K is the thermal conductivity. We note that

$$\lim_{T \to 0} \frac{K}{T\sigma} = \frac{\pi^2}{3e^2},$$

is a universal constant, which is known as the Wiedemann-Franz law.

Another property that can be measured is the thermoelectric power. As we mentioned, in an open circuit of length ΔL subject to a temperature drop $\Delta L \nabla T$, a potential drop $-E \Delta L$ is generated to stop current flowing, see (5.160), which reads

$$Q = \frac{-E \Delta L}{\Delta L \nabla T} = \frac{-E}{\nabla T} = \frac{\pi^2 T \mathcal{C}}{3e}, \tag{5.161}$$

and is the thermopower or Seebeck coefficient.

Finally, suppose that the temperature gradient is zero. A constant charge current will induce a heat current which, solving (5.158) with $\nabla T = 0$ defines the so-called Peltier coefficient

$$\Pi = \frac{J_E}{J} = T Q. \tag{5.162}$$

Problems

5.1 Bosonization of quasiparticles' Fermi surface vibrations—Recall Landau's energy functional and transport equation without external field and with vanishing collision integral:

$$E = \sum_{k\sigma} \int d\boldsymbol{r} \, \epsilon_*(\mathbf{k}) \, \delta n_{k\sigma}(\boldsymbol{r}) + \frac{1}{2} \sum_{\mathbf{k}\mathbf{k}'} \sum_{\sigma\sigma'} \int d\boldsymbol{r} \, f_{\mathbf{k}\sigma,\mathbf{k}'\sigma'} \, \delta n_{k\sigma}(\boldsymbol{r}) \, \delta n_{\mathbf{k}'\sigma'}(\boldsymbol{r}),$$

$$\frac{\partial \delta n_{\mathbf{k}\sigma}(t, \boldsymbol{r})}{\partial t} = -\boldsymbol{v}_{\mathbf{k}} \cdot \nabla \delta n_{\mathbf{k}\sigma}(t, \boldsymbol{r}) + \frac{1}{V} \frac{\partial f(\epsilon_*(\mathbf{k}))}{\partial \epsilon_*(\mathbf{k})} \sum_{\mathbf{k}'\sigma'} f_{\mathbf{k}\sigma,\mathbf{k}'\sigma'} \, \boldsymbol{v}_{\mathbf{k}} \cdot \nabla \delta n_{\mathbf{k}'\sigma'}(t, \boldsymbol{r}),$$

and consider excitations of the form of a space-dependent transformation in momentum space

$$\mathbf{k} \rightarrow \mathbf{k}_\sigma(\mathbf{r}) = \mathbf{k} - \delta \mathbf{u}_{\mathbf{k}\sigma}(\mathbf{r}),$$

$$\epsilon_*(\mathbf{k}) \rightarrow \epsilon_*\big(\mathbf{k}_\sigma(\mathbf{r})\big) \equiv \epsilon_*(\mathbf{k}) + \delta\epsilon_{*\sigma}(\mathbf{k}, \mathbf{r}) \simeq \epsilon_*(\mathbf{k}) - \mathbf{v_k} \cdot \delta \mathbf{u}_{\mathbf{k}\sigma}(\mathbf{r}).$$

It follows that

$$V \, \delta n_{\mathbf{k}\sigma}(\mathbf{r}) = f\left(\epsilon_*\big(\mathbf{k}_\sigma(\mathbf{r})\big)\right) - f\big(\epsilon_*(\mathbf{k})\big)$$

$$\simeq \frac{\partial f\big(\epsilon_*(\mathbf{k})\big)}{\partial \epsilon_*(\mathbf{k})} \, \delta\epsilon_{*\sigma}(\mathbf{k}, \mathbf{r}) + \frac{1}{2} \frac{\partial^2 f\big(\epsilon_*(\mathbf{k})\big)}{\partial \epsilon_*(\mathbf{k})^2} \, \delta\epsilon_{*\sigma}(\mathbf{k}, \mathbf{r})^2.$$

After defining in generic dimensions d

$$\mathbf{v_k} \cdot \delta \mathbf{u}_{\mathbf{k}\sigma}(\mathbf{r}) \equiv (2\pi)^d \, |\mathbf{v_k}| \, u_{\mathbf{k}\sigma}(\mathbf{r}) \equiv (2\pi)^d \, v(\mathbf{k}) \, u_{\mathbf{k}\sigma}(\mathbf{r}), \qquad (5.163)$$

and recalling that

$$\frac{1}{V} \sum_{\mathbf{k}} \delta\big(\epsilon_*(\mathbf{k})\big) (\dots) = \iint_{S_F} \frac{dS}{(2\pi)^d} \frac{1}{|\mathbf{v_k}|} (\dots) = \iint_{S_F} \frac{dS}{(2\pi)^d} \frac{1}{v(\mathbf{k})} (\dots),$$

where $\iint_S dS \dots$ denotes a surface integral, and S_F is the quasiparticles' Fermi surface

- rewrite Landau's energy functional and transport equation in terms of the new variables $u_{\mathbf{k}\sigma}(\mathbf{r})$.

Assume to promote $u_{\mathbf{k}\sigma}(\mathbf{r})$ to a field operator defined on the quasiparticles' Fermi surface, the Landau's energy functional to its Hamiltonian $H\big[\{u_{\mathbf{k}\sigma}(\mathbf{r})\}\big]$ and the transport equation to its Heisenberg equation of motion

$$\frac{\partial u_{\mathbf{k}\sigma}(\mathbf{r})}{\partial t} \equiv -i \left[u_{\mathbf{k}\sigma}(\mathbf{r}), \, H\big[\{u_{\mathbf{k}'\sigma'}(\mathbf{r}')\}\big] \right].$$

- Show that such identification is consistent if

$$\left[u_{\mathbf{k}\sigma}(\mathbf{r}), \, u_{\mathbf{k}'\sigma'}(\mathbf{r}') \right] = -\frac{i}{(2\pi)^d} \, \delta_{\sigma\sigma'} \, \delta_{S_{FL}}(\mathbf{k} - \mathbf{k}') \, \frac{\mathbf{v_k}}{|\mathbf{v_k}|} \cdot \nabla_{\mathbf{r}} \, \delta(\mathbf{r} - \mathbf{r}'),$$

where $\delta_{S_{FL}}(\mathbf{k} - \mathbf{k}')$ is a δ-function on the Fermi or Luttinger surface.
- Define a new field $\phi_{\mathbf{k}\sigma}(\mathbf{r})$ through

$$u_{\mathbf{k}\sigma}(\mathbf{r}) \equiv \frac{1}{(2\pi)^d} \frac{\mathbf{v_k}}{|\mathbf{v_k}|} \cdot \nabla \phi_{\mathbf{k}\sigma}(\mathbf{r}) \implies \nabla \phi_{\mathbf{k}\sigma}(\mathbf{r}) = \delta \mathbf{u}_{\mathbf{k}\sigma}(\mathbf{r}),$$

and find the expression of its commutator with $u_{\mathbf{k}\sigma}(\mathbf{r})$.

The operator that yields the variation in the number of quasiparticles with respect to its equilibrium value N_* is defined as

$$\delta \hat{N}_* \equiv \int dr \sum_{\mathbf{k}\sigma} \delta n_{\mathbf{k}\sigma} \simeq - \int dr \frac{1}{V} \sum_{\mathbf{k}\sigma} \frac{\partial f\left(\epsilon_*(\mathbf{k})\right)}{\partial \epsilon_*(\mathbf{k})} \, \boldsymbol{v}_{\mathbf{k}} \cdot \delta \boldsymbol{u}_{\mathbf{k}\sigma}(\mathbf{r})$$

$$= \int dr \sum_{\sigma} \iint_{S_F} dS \, u_{\mathbf{k}\sigma}(\mathbf{r}) \, .$$

• Assume a state $\mid \delta N_* \rangle$ eigenstate of $\delta \hat{N}_*$ with eigenvalue δN_*, i.e., such that

$$\delta \hat{N}_* \mid \delta N_* \rangle = \delta N_* \mid \delta N_* \rangle \, .$$

Show that $\mathrm{e}^{-i \phi_{\mathbf{k}\sigma}(\mathbf{r})} \mid \delta N_* \rangle$ is still eigenstate of $\delta \hat{N}_*$ and calculate the eigenvalue.

The above quantum field operators correspond to the bosonization of the quasiparticle-quasihole excitations in the time-dependent Hartree-Fock approximation to the quasiparticle Hamiltonian, see Sect. 3.3.1. This representation, reminiscent of the abelian bosonization in one dimension that we discuss in Chap. 6, was introduced in [20], and further elaborated in [21,22].

5.2 Transport coefficients in presence of a magnetic field—Repeat the same calculations as in Sect. 5.9 but now in presence of a static and uniform magnetic field **B** and in a Hall bar geometry where **E** and ∇T are in plane and **B** out of plane. In this case, the electric and heat currents satisfy a matrix equation

$$\boldsymbol{J} = \hat{\sigma} \, \boldsymbol{E} + \hat{\alpha} \, \nabla T \, ,$$

$$\boldsymbol{J}_E = \hat{\alpha} \, \boldsymbol{E} + \hat{K} \, \nabla T \, ,$$

where $\hat{\sigma}$, $\hat{\alpha}$ and $\hat{K}$ are 2×2 matrices in the space of the two planar coordinates. Calculate those matrices.

References

1. L. Landau, Zh. Eskp. Teor. Fiz. **30**, 1058 (1956). [Sov. Phys. JETP **3**, 920 (1957)]
2. L. Landau, Zh. Eskp. Teor. Fiz. **32**, 59 (1957). [Sov. Phys. JETP **5**, 101 (1957)]
3. P. Noziéres, J.M. Luttinger, Phys. Rev. **127**, 1423 (1962). https://doi.org/10.1103/PhysRev.127.1423
4. J.M. Luttinger, P. Noziéres, Phys. Rev. **127**, 1431 (1962). https://doi.org/10.1103/PhysRev.127.1431
5. A. Abrikosov, L. Gorkov, I. Dzyaloshinskii, *Methods of Quantum Field Theory in Statistical Physics* (Dover, New York, 1975). See Sect. 19.4
6. P. Noziéres, *Theory of Interacting Fermi Systems* (CRC Press, Boca Raton, 1998)
7. D. Pines, P. Noziéres, *The Theory of Quantum Liquids* (CRC Press, Boca Raton, 1989). https://doi.org/10.4324/9780429492662

8. P. Anderson, Mater. Res. Bull. **8**(2), 153 (1973). https://doi.org/10.1016/0025-5408(73)90167-0
9. C.M. Varma, P.B. Littlewood, S. Schmitt-Rink, E. Abrahams, A.E. Ruckenstein, Phys. Rev. Lett. **63**, 1996 (1989). https://doi.org/10.1103/PhysRevLett.63.1996
10. X.G. Wen, Phys. Rev. B **65**, 165113 (2002). https://doi.org/10.1103/PhysRevB.65.165113
11. T. Senthil, S. Sachdev, M. Vojta, Phys. Rev. Lett. **90**, 216403 (2003). https://doi.org/10.1103/PhysRevLett.90.216403
12. P.A. Lee, N. Nagaosa, X.G. Wen, Rev. Mod. Phys. **78**, 17 (2006). https://doi.org/10.1103/RevModPhys.78.17
13. S. Hartnoll, A. Lucas, S. Sachdev, *Holographic Quantum Matter* (MIT Press, Cambridge, 2018)
14. T. Andrade, A. Krikun, K. Schalm, J. Zaanen, Nat. Phys. **14**(10), 1049 (2018). https://doi.org/10.1038/s41567-018-0217-6
15. M. Fabrizio, Phys. Rev. B **102**, 155122 (2020). https://doi.org/10.1103/PhysRevB.102.155122
16. M. Fabrizio, Nat. Commun. **13**(1), 1561 (2022). https://doi.org/10.1038/s41467-022-29190-y
17. I. Dzyaloshinskii, Phys. Rev. B **68**, 085113 (2003). https://doi.org/10.1103/PhysRevB.68.085113
18. A.V. Chubukov, D.L. Maslov, S. Gangadharaiah, L.I. Glazman, Phys. Rev. Lett. **95**, 026402 (2005). https://doi.org/10.1103/PhysRevLett.95.026402
19. M. Oshikawa, Phys. Rev. Lett. **84**, 3370 (2000). https://doi.org/10.1103/PhysRevLett.84.3370
20. F.D.M. Haldane, Luttinger's Theorem and Bosonization of the Fermi Surface (2005)
21. D.V. Else, R. Thorngren, T. Senthil, Phys. Rev. X **11**, 021005 (2021). https://doi.org/10.1103/PhysRevX.11.021005
22. X.G. Wen, Phys. Rev. B **103**, 165126 (2021). https://doi.org/10.1103/PhysRevB.103.165126

Brief Introduction to Luttinger Liquids

6

In Sect. 4.3.4 we showed that perturbatively $\text{Im}\Sigma(\epsilon \to 0, \mathbf{k}) \to |\epsilon|$ in one dimension, see (4.64), which implies through (4.65) that the quasiparticle residue vanishes as $\epsilon \to 0$, and thus that conventional Landau's Fermi liquid theory in not applicable whatever the interaction strength is. Nonetheless, we can still uncover in great detail the properties of interacting electrons in one dimension, commonly refereed to as Luttinger Liquids. That is actually the content of the present Chapter, where we briefly discuss Luttinger Liquids and abelian bosonization. More details can be found in the books *Bosonization and Strongly Correlated Systems*, authors A.O. Gogolin, A.A. Nersesyan and A.M. Tsvelik [1], and *Quantum Physics in One Dimension*, author T. Giamarchi [2].

6.1 What Is Special in One Dimension?

Let us imagine to apply the conserving Hartree-Fock approximation to calculate linear response functions, considering, for instance, the Hamiltonian on an L site chain

$$H = \sum_{k\sigma} \epsilon(k)\, c^\dagger_{k\sigma} c_{k\sigma} + \frac{U}{2L} \sum_{kq\sigma\sigma'} c^\dagger_{k\sigma} c^\dagger_{p+q\sigma'}\, c_{p\sigma'}\, c_{k+q\sigma}, \tag{6.1}$$

with an on-site interaction U, either positive or negative, and a generic band structure as in Fig. 6.1. At $U = 0$, the ground state is obtained by filling all energy states up to the Fermi energy, or, equivalently, occupying all momenta $|k| \leq k_F$, see Fig. 6.2, where

$$\sum_\sigma \sum_{k=-k_F}^{k_F} = 2L \int_{-k_F}^{k_F} \frac{dk}{2\pi} = 2L\, \frac{k_F}{\pi} = N \quad \Rightarrow \quad k_F = \pi\, \frac{N}{2L} \equiv \frac{\pi}{2} n,$$

© The Author(s), under exclusive license to Springer Nature Switzerland AG 2022
M. Fabrizio, *A Course in Quantum Many-Body Theory*, Graduate Texts in Physics,
https://doi.org/10.1007/978-3-031-16305-0_6

Fig. 6.1 Generic one-dimensional energy dispersion $\epsilon(k)$ in momentum space. The Brillouin zone is defined for $k \in [-\pi/a, \pi/a]$, where a is the lattice spacing

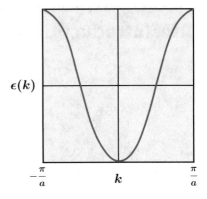

with n the electron density. If we force spin $SU(2)$ symmetry and translation invariance, the Hartree-Fock Green's function is the same as for non-interacting electrons, apart from a renormalisation of the chemical potential, which is irrelevant if we work at fixed density n. In other words,

$$G_0(i\epsilon, k) = \frac{1}{i\epsilon - \epsilon(k)} .$$

For weak interaction compared to the Fermi energy, the low energy physics is controlled by the excitations around the two Fermi points, $\pm k_F$, see Fig. 6.2, where $\epsilon(k \simeq k_F) \simeq v_F(k - k_F)$ and $\epsilon(k \simeq -k_F) \simeq -v_F(k + k_F)$, so that we can write

$$G_0(i\epsilon, k) \simeq \begin{cases} \dfrac{1}{i\epsilon - v_F(k - k_F)} \equiv G_{0R}(i\epsilon, k - k_F), \quad \left|k - k_F\right| \ll k_F , \\[4mm] \dfrac{1}{i\epsilon + v_F(k + k_F)} \equiv G_{0L}(i\epsilon, k + k_F), \quad \left|k + k_F\right| \ll k_F . \end{cases}$$

(6.2)

The labels R and L indicate electrons close to k_F and $-k_F$, respectively, conventionally named right (R) and left (L) moving electrons. With that notation, the particle-hole bubble at zero transferred frequency and momentum transferred $q = 2k_F$,

$$\chi_0(0, 2k_F) = \frac{1}{L} \sum_k T \sum_\epsilon G_0(i\epsilon, k + k_F) \, G_0(i\epsilon, k - k_F)$$

$$\simeq \frac{1}{L} \sum_{|k| \ll k_F} T \sum_\epsilon G_{0R}(i\epsilon, k) \, G_{0L}(i\epsilon, k) = -\frac{1}{L} \sum_{|k| \ll k_F} \frac{f(-v_F k) - f(v_F k)}{2v_F k}$$

$$\simeq -\int_{-k_F}^{k_F} \frac{dk}{2\pi} \tanh \frac{v_F k}{2T} \frac{1}{2v_F k} \simeq -\frac{1}{2\pi v_F} \ln \frac{v_F k_F}{T} ,$$

Fig. 6.2 One-dimensional
Fermi sea

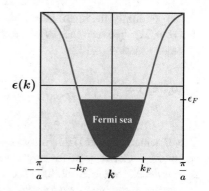

diverges as $T \to 0$. The reason of such singular behaviour is the nesting property of
the one-dimensional Fermi surface. In general, a d-dimensional Fermi surface has
nesting if there are parallel portions of it, thus connected by a single nesting vector $\mathbf{Q}$
defined apart from a reciprocal lattice vector. Whenever that occurs, the non interact-
ing particle-hole bubble at momentum transferred $\mathbf{Q} + \mathbf{q}$, with $|\mathbf{q}| \ll |\mathbf{Q}|$, diverges
at small temperature and frequency transferred ω at least as $\ln \max(T, v_F |\mathbf{q}|, \omega)$.

Similarly, we can consider the particle-particle bubble at zero total frequency and
momentum,

$$
\begin{aligned}
S_0(0,0) &= \frac{1}{L} \sum_k T \sum_\epsilon G_0(i\epsilon, k + k_F)\, G_0(-i\epsilon, -k - k_F) \\
&\simeq \frac{1}{L} \sum_{|k| \ll k_F} T \sum_\epsilon G_{0R}(i\epsilon, k)\, G_{0L}(-i\epsilon, -k) \\
&= -\frac{1}{L} \sum_{|k| \ll k_F} T \sum_\epsilon G_{0R}(i\epsilon, k)\, G_{0L}(i\epsilon, k) \simeq \frac{1}{2\pi v_F} \ln \frac{v_F k_F}{T} ,
\end{aligned}
$$

which also diverges logarithmically. That actually occurs in any dimensions and for
any Fermi surface provided the non-interacting local density of states is finite at zero
energy, and it is known as the Cooper singularity. In one dimension, the Cooper singularity and the nesting at $q = 2k_F$ coexist, and that has very peculiar consequences,
that are already clear within the conserving Hartree-Fock approximation.

In that approximation, the irreducible vertices in the particle-hole singlet, S, and
triplet, T, channels can be readily found through (4.99) and (4.100), and read

$$
\Gamma_{0S}^{\text{p-h}} = \frac{U}{2}, \quad \Gamma_{0T}^{\text{p-h}} = -\frac{U}{2},
$$

whereas those in the singlet and triplet particle-particle channels are, see (4.109) and
(4.110),

$$
\Gamma_{0S=0}^{\text{p-p}} = U, \quad \Gamma_{0S=1}^{\text{p-p}} = 0.
$$

Correspondingly, the interaction vertex that transfers frequency $i\omega = 0$ and momentum $q = 2k_F$ between the particle and the hole can be obtained through the Bethe-Salpeter equation (4.102)

$$\Gamma_S(0, 2k_F) = \frac{U}{2} \frac{1}{1 - U \chi_0(0, 2k_F)} \,, \qquad \Gamma_T(0, 2k_F) = -\frac{U}{2} \frac{1}{1 + U \chi_0(0, 2k_F)} \,. \tag{6.3}$$

Since $\chi_0(0, 2k_F) \sim \ln T$, the spin-triplet vertex at $2k_F$ becomes singular at any $U > 0$, which indicates instability towards a magnetic ordering modulated with wave vector $2k_F$. Conversely, for any $U < 0$, $\Gamma_S(0, 2k_F)$ is singular, now signal of a charge-density wave instability at $2k_F$.

Similarly, the interaction vertex where the two incoming lines in a spin-singlet configuration with zero total frequency and momentum is, still through the Bethe-Salpeter equation,

$$\Gamma_{S=0}(i\Omega, P) = \frac{U}{1 + U \, S_0(0, 0)} \,,$$

and is singular for any $U < 0$, reflecting the conventional superconducting instability for attractive interactions.

In conclusion, however small U is, and whatever its sign is, the system is unstable and would like to spontaneously break a continuous symmetry, either the translational symmetry if $2k_F$ is incommensurate, or spin $SU(2)$, or global charge $U(1)$. However, a continuous symmetry cannot be broken in one-dimension because of the Mermin-Wagner theorem, see Sect. 1.8.3, which implies that quantum fluctuations are strong and prevent symmetry breaking. As a consequence, we anticipate that long-range order is replaced by a quasi long-range one where the correlation functions decay power-law at long distances/times with anomalous interaction-dependent exponents, which is the first distinctive feature of one-dimensional interacting electrons.

Besides that, there is another peculiar property in one dimension. At generic transferred real frequency ω, with an infinitesimal positive imaginary part that we do not indicate unless necessary, and momentum q, the vertices in (6.3) are

$$\Gamma_S(\omega, q) = \frac{U}{2} \frac{1}{1 - U \chi_0(\omega, q)} \,, \qquad \Gamma_T(\omega, q) = -\frac{U}{2} \frac{1}{1 + U \chi_0(\omega, q)} \,,$$

where

$$\chi_0(\omega, q) = \int \frac{dk}{2\pi} \frac{f\big(\epsilon(k)\big) - f\big(\epsilon(k + q)\big)}{\omega - \epsilon(k + q) + \epsilon(k)} \,.$$

For small $|q| \ll k_F$ and $T = 0$,

$$\chi_0(\omega, q) \simeq \int \frac{dk}{2\pi} \delta\big(\epsilon(k)\big) \frac{v_k \, q}{\omega - v_k \, q} = \frac{1}{2\pi} \left(\frac{q}{\omega - v_F \, q} - \frac{q}{\omega + v_F \, q} \right) = \frac{1}{\pi} \frac{v_F \, q^2}{\omega^2 - v_F^2 \, q^2} \,, \tag{6.4}$$

so that

$$\Gamma_S(\omega, q) \simeq \frac{U}{2} \frac{\omega^2 - v_F^2 q^2}{\omega^2 - v_F^2 q^2 - U v_F q^2 / \pi} \,, \quad \Gamma_T(\omega, q) = -\frac{U}{2} \frac{\omega^2 - v_F^2 q^2}{\omega^2 - v_F^2 q^2 + U v_F q^2 / \pi} \,,$$

This result implies that, at small $|q|$, the particle-hole excitations are exhausted by two acoustic modes, a charge mode and a spin one with velocities, respectively,

$$v_\rho = v_F \sqrt{1 + \frac{U}{2\pi v_F}} \,, \quad v_\sigma = v_F \sqrt{1 - \frac{U}{2\pi v_F}} \,. \tag{6.5}$$

These two modes correspond to two linearly dispersive bosons, completely control the long wavelength behaviour, even beyond the weak coupling conserving Hartree-Fock approximation, as we shall see, and are actually the building blocks of Bosonization. We further note that these bosons carry separately charge and spin degrees of freedom, reflecting the so-called spin-charge separation in one dimension, which we will also discuss in what follows.

6.2   Interacting Spinless Fermions

Let us begin considering a one dimensional model of spinless fermions described by a non interacting Hamiltonian of the form

$$H_0 = \sum_k \epsilon(k) c_k^\dagger c_k \,, \tag{6.6}$$

with an energy dispersion $\epsilon(k)$ in momentum space of the form in Fig. 6.1, measured with respect to the chemical potential. The electrons mutually interact through

$$H_{\text{int}} = \frac{1}{2L} \sum_q V(q) \, \rho(q) \, \rho(-q) \,, \tag{6.7}$$

where L is the number of sites, $V(q) = V(-q)$ the interaction, and

$$\rho(q) = \sum_k c_k^\dagger c_{k+q} \,, \tag{6.8}$$

the Fourier transform of the density. In perturbation theory, as discussed earlier, one finds singularities that, in this case, indicate instabilities towards spontaneous breakdown of translational symmetry with wave vector $2k_F$ or towards superconductivity, which, as we mentioned, are prevented by quantum fluctuations. We will see in detail what that means.

For that, we adopt an approach substantially similar to the one we applied in Sects. 1.7 and 3.3.1. We start from the non interacting Hamiltonian (6.6). The ground state with N particles is just the Fermi sea (FS) obtained by occupying momentum eigenstates up to the Fermi momentum k_F, which for spinless fermions is simply $k_F = \pi n$, with n the average density, i.e., the number of electrons per site, thus the occupation number in momentum space

$$\langle c_k^\dagger c_k \rangle_{\text{FS}} = \theta(k_F - |k|) \equiv n^0(k) \, .$$

With Sect. 1.7 in mind, we shall regard the Fermi sea, see Fig. 6.2, like a classical ground state characterised by the macroscopic 'classical order parameter'

$$\Delta \equiv \sum_{|k| \le k_F} n^0(k) - \sum_{|k| > k_F} n^0(k) = N \, , \tag{6.9}$$

as well as the 'vacuum' of the quantum fluctuations brought about by the interaction (6.7). Such 'semiclassical' approximation is valid provided the order parameter (6.9) is only weakly affected by interaction. That implies, in the first place, that the effective strength of interaction, to be specified later, is much smaller that ϵ_F of Fig. 6.2, or, equivalently, that a macroscopic number of states deep inside, $|k| \ll k_F$, and outside, $|k| \gg k_F$, the Fermi sea are totally unaffected by interaction. We can therefore define new operators corresponding to the right-moving (R) and left-moving (L) fermions

$$c_{k \sim +k_F} \longrightarrow c_{Rk} \, , \quad \langle c_{Rk}^\dagger c_{Rk} \rangle_{\text{FS}} = n_R^0(k) = \theta(k_F - k) \, ,$$

$$c_{k \sim -k_F} \longrightarrow c_{Lk} \, , \quad \langle c_{Lk}^\dagger c_{Lk} \rangle_{\text{FS}} = n_L^0(k) = \theta(k + k_F) \, , \tag{6.10}$$

without running the risk of overcounting states. Correspondingly, we introduce the densities of right and left moving fermions,

$$\rho_R(q) \equiv \sum_k c_{Rk}^\dagger c_{Rk+q} \, , \quad \rho_L(q) \equiv \sum_k c_{Lk}^\dagger c_{Lk+q} \, , \tag{6.11}$$

which are unambiguously defined only for $|q| \ll k_F$. For that, whenever we shall need to sum over q, we will add a short-distance cutoff through the function $e^{-\alpha|q|}$ where $0 < \alpha \simeq 1/k_F \simeq a$, with a the lattice spacing.

The density operators (6.11) creates particle-hole excitation over the Fermi sea, or destroy them if they are already present. If we regard $\rho_R(q)$ and $\rho_L(q)$ as the spin-density operators in Sect. 1.7, then the counterpart of the spin-wave approximation is replacing commutators with their expectation values over the 'vacuum', i.e., the Fermi sea. Therefore, through (6.10) and since $q \ll k_F$,

$$\left[\rho_R(q), \rho_R(-q')\right] = \sum_k \left(c_{Rk}^\dagger c_{Rk+q-q'} - c_{Rk+q'}^\dagger c_{Rk+q}\right)$$

$$\simeq \sum_k \langle c_{Rk}^\dagger c_{Rk+q-q'} - c_{Rk+q'}^\dagger c_{Rk+q} \rangle_{FS}$$

$$= \delta_{q,q'} \sum_k \left(n_R^0(k) - n_R^0(k+q)\right) \simeq q\,\delta_{q,q'} \sum_k \left(-\frac{\partial n_R^0(k)}{\partial k}\right)$$

$$= q\,\delta_{q,q'}\, L \int \frac{dk}{2\pi} \delta(k - k_F) = \delta_{q,q'}\frac{qL}{2\pi}\,,$$

$$(6.12)$$

for the right-moving density,

$$\left[\rho_L(q), \rho_L(-q')\right] = \sum_k \left(c_{Lk}^\dagger c_{Lk+q-q'} - c_{Lk+q'}^\dagger c_{Lk+q}\right)$$

$$\simeq \sum_k \langle c_{Lk}^\dagger c_{Lk+q-q'} - c_{Lk+q'}^\dagger c_{Lk+q} \rangle_{FS}$$

$$= \delta_{q,q'} \sum_k \left(n_L^0(k) - n_L^0(k+q)\right) \simeq q\,\delta_{q,q'} \sum_k \left(-\frac{\partial n_R^0(k)}{\partial k}\right)$$

$$= -q\,\delta_{q,q'}\, L \int \frac{dk}{2\pi} \delta(k + k_F) = -\delta_{q,q'}\frac{qL}{2\pi}\,,$$

$$(6.13)$$

for the left-moving one, while

$$\left[\rho_R(q), \rho_L(-q')\right] = 0\,.$$

Therefore, in this approximation $\rho_R(q)$ and $\rho_L(q)$ behave as independent bosonic operator. We also note that the right hand sides of (6.12) and (6.13) imply that $\rho_{R(L)}(q) \sim \sqrt{L}$ while the 'classical order parameter' (6.9) is of order L at fixed density $n = N/L$, consistently with the semiclassical approximation.

6.2.1 Bosonized Expression of the Non-interacting Hamiltonian

For $q \ll k_F$, the density (6.8) is just

$$\rho(q) = \rho_R(q) + \rho_L(q)\,. \qquad (6.14)$$

The continuity equation provides the expression of the current, specifically, through (6.6) and (6.8) and still assuming $q \ll k_F$

$$i \dot{\rho}(q) = q \, J(q) = \left[\rho(q), \, H_0 \right] = \sum_k \left(\epsilon(k+q) - \epsilon(k) \right) c_k^\dagger c_{k+q} \simeq q \sum_k \frac{\partial \epsilon(k)}{\partial k} c_k^\dagger c_{k+q}$$

$$\simeq v_F \, q \left(\rho_R(q) - \rho_L(q) \right),$$

where $v_F > 0$ is the Fermi velocity at $+k_F$, while $-v_F$ that at $-k_F$. Since $\rho_R(q)$ and $\rho_L(q)$ commute with each other, it follows that

$$i \dot{\rho}_R(q) = \left[\rho_R(q), \, H_0 \right] = v_F \, q \, \rho_R(q), \quad i \dot{\rho}_L(q) = \left[\rho_L(q), \, H_0 \right] = -v_F \, q \, \rho_L(q),$$
$$(6.15)$$

which implies that $\rho_R(0) \equiv N_R$ and $\rho_L(0) \equiv N_L$ are separately conserved and their corresponding currents read

$$J_R(q) = v_F \, \rho_R(q), \quad J_L(q) = -v_F \, \rho_R(q). \tag{6.16}$$

If we consider an equilibrium state with a finite current, as shown in Fig. 6.3, then

$$\rho_R(0) = N_R = \int_0^{k_{RF}} \frac{dk}{2\pi} = \frac{L \, k_{RF}}{2\pi}, \quad \rho_L(0) = N_L = \int_0^{k_{LF}} \frac{dk}{2\pi} = \frac{L \, k_{LF}}{2\pi},$$
$$(6.17)$$

with $N = N_R + N_L$ number of fermions, and current

$$J = v_F \left(N_R - N_L \right). \tag{6.18}$$

Given the commutation relations (6.12) and (6.13), (6.15) can be reproduced by

$$H_0 = \frac{\pi v_F}{L} \sum_q \left(\rho_R(q) \rho_R(-q) + \rho_L(q) \rho_L(-q) \right) - \mu_R N_R - \mu_L N_L$$

$$= \frac{\pi v_F}{L} \left(N_R^2 + N_L^2 \right) + \frac{2\pi v_F}{L} \sum_{q>0} \left(\rho_R(q) \rho_R(-q) + \rho_L(q) \rho_L(-q) \right) - \mu_R N_R - \mu_L N_L$$

$$= \frac{\pi v_F}{2L} \left(N^2 + J^2 \right) + \frac{2\pi v_F}{L} \sum_{q>0} \left(\rho_R(q) \rho_R(-q) + \rho_L(q) \rho_L(-q) \right) - \mu N - \mu_J J,$$
$$(6.19)$$

where

$$\mu_R = v_F \, k_{RF}, \quad \mu_L = v_F \, k_{LF}, \tag{6.20}$$

guarantee through (6.17) that the ground state has the desired number of right and left moving fermions, while

$$\mu = \frac{\mu_R + \mu_L}{2}, \quad \mu_J = \frac{\mu_R - \mu_L}{2v_F}.$$

Fig. 6.3 Excited state with finite current due to $k_{RF} \neq k_{LF}$

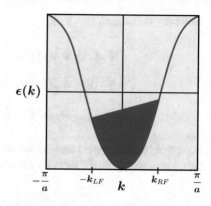

6.2.2 Bosonic Representation of the Fermi Fields

In real space the right- and left-moving densities are defined trough

$$\rho_{R(L)}(x) = \frac{1}{L} \sum_q e^{iqx} \, \rho_{R(L)}(q) \equiv \Psi^\dagger_{R(L)}(x) \, \Psi_{R(L)}(x) \,, \tag{6.21}$$

which also defines the corresponding Fermi fields. Equation (6.12) thus implies that

$$\begin{aligned}
\left[\rho_R(x), \, \rho_R(y) \right] &= \frac{1}{2\pi L} \sum_q q \, e^{iq(x-y)} \, e^{-\alpha|q|} = -\frac{i}{2\pi} \frac{\partial}{\partial x} \int \frac{dq}{2\pi} e^{iq(x-y)} \, e^{-\alpha|q|} \\
&= -\frac{i}{2\pi} \frac{\partial}{\partial x} \left(\frac{1}{\pi} \frac{\alpha}{(x-y)^2 + \alpha^2} \right) = -\frac{i}{2\pi} \frac{\partial}{\partial x} \delta_\alpha(x-y) \\
&= -\frac{i}{2\pi^2} \frac{\partial^2}{\partial x^2} \tan^{-1} \frac{x-y}{\alpha} \,,
\end{aligned} \tag{6.22}$$

where $\delta_\alpha(x-y)$ is, for $\alpha \to 0$, the representation of a δ-function through a Lorentzian. Similarly,

$$\left[\rho_L(x), \, \rho_L(y) \right] = \frac{i}{2\pi} \frac{\partial}{\partial x} \delta_\alpha(x-y) \,. \tag{6.23}$$

We further note, through (6.21), that

$$\begin{aligned}
\left[\Psi_R(x), \, \rho_R(y) \right] &= \delta(x-y) \, \Psi_R(x) \,, \quad \left[\Psi_L(x), \, \rho_L(y) \right] = \delta(x-y) \, \Psi_L(x) \,, \\
\left[\Psi_R(x), \, N_R \right] &= \Psi_R(x) \,, \qquad\qquad\quad \left[\Psi_L(x), \, N_L \right] = \Psi_L(x) \,,
\end{aligned} \tag{6.24}$$

while, by definition,

$$\left\{\Psi_R(x),\, \Psi_R^\dagger(y)\right\} = \left\{\Psi_L(x),\, \Psi_L^\dagger(y)\right\} = \delta(x-y)\,,$$

$$\left\{\Psi_R(x),\, \Psi_R(y)\right\} = \left\{\Psi_L(x),\, \Psi_L(y)\right\} = 0\,, \tag{6.25}$$

$$\left\{\Psi_R(x),\, \Psi_L^\dagger(y)\right\} = \left\{\Psi_R(x),\, \Psi_L(y)\right\} = 0\,.$$

Equation (6.24) suggest that the Fermi fields act on the densities like the unit translation operator $T = e^{ip}$ does on the coordinate x, i.e., through $[x, p] = i$,

$$e^{-ip}\, x\, e^{ip} = x - 1 \;\Rightarrow\; \left[e^{ip},\, x\right] = e^{ip}\,.$$

Therefore, let us assume

$$\Psi_R(x) = \mathcal{N}_R\, e^{i\,\phi_R(x)}\,, \quad \Psi_L(x) = \mathcal{N}_L\, e^{-i\,\phi_L(x)}\,, \tag{6.26}$$

with $\mathcal{N}_{R(L)}$ normalisation constants, and try to find the real $\phi_R(x)$ and $\phi_L(x)$ that allow fulfilling (6.24) and (6.25). The simplest condition to fulfil is that right and left moving fields anticommute with each other, namely,

$$e^{i\phi_R(x)}\, e^{\mp i\phi_L(y)} = -e^{\mp i\phi_L(y)}\, e^{i\phi_R(x)}\,, \quad \forall x, y\,. \tag{6.27}$$

We recall that, if two operators A and B commute with $[A, B]$, then the following equations hold

$$e^A\, e^B = e^B\, e^A\, e^{-[B,A]}\,, \quad e^{A+B} = e^A\, e^B\, e^{-[A,B]/2} = e^B\, e^A\, e^{-[B,A]/2}\,, \tag{6.28}$$

through which (6.27) can be satisfied if

$$\left[\phi_R(x),\, \phi_L(y)\right] = -i\,\pi\,, \quad \forall x, y\,. \tag{6.29}$$

It is easy to realise that (6.24) are instead fulfilled if

$$\left[\phi_R(x),\, \rho_R(y)\right] = -i\,\delta(x-y)\,, \quad \left[\phi_R(x),\, N_R\right] = -i\,,$$

$$\left[\phi_L(x),\, \rho_L(y)\right] = i\,\delta(x-y)\,, \quad \left[\phi_L(x),\, N_L\right] = i\,. \tag{6.30}$$

It follows, through (6.22) and (6.23), that

$$\rho_R(x) = \frac{1}{2\pi}\, \frac{\partial\phi_R(x)}{\partial x}\,, \quad \rho_L(x) = \frac{1}{2\pi}\, \frac{\partial\phi_L(x)}{\partial x}\,, \tag{6.31}$$

where, including also (6.29),

$$\left[\phi_R(x),\,\phi_R(y)\right] = 2i\,\tan^{-1}\frac{x-y}{\alpha} \simeq i\pi\,\mathrm{sign}(x-y)\,,$$

$$\left[\phi_L(x),\,\phi_L(y)\right] = -2i\,\tan^{-1}\frac{x-y}{\alpha} \simeq -i\pi\,\mathrm{sign}(x-y)\,, \qquad (6.32)$$

$$\left[\phi_R(x),\,\phi_L(y)\right] = -i\,\pi\,, \qquad \forall\,x,\,y\,,$$

are the commutation relations between the ϕ-fields. We note that

$$N_{R(L)} = \int dx\,\rho_{R(L)}(x) = \int \frac{dx}{2\pi}\,\frac{\partial\phi_{R(L)}(x)}{\partial x}\,,$$

is a conserved quantity. Therefore, the vacuum expectation value

$$\langle 0 \mid \phi_{R(L)}(x) \mid 0 \rangle = \frac{N_{R(L)}}{2\pi L}\,x = k_{R(L)F}\,x\,. \qquad (6.33)$$

We can now easily demonstrate that, for $x \neq y$,

$$\begin{aligned}
\left\{\Psi_R(x),\,\Psi_R(y)\right\} &= \mathcal{N}_R^2\left(e^{i\phi_R(x)}\,e^{i\phi_R(y)} + e^{i\phi_R(y)}\,e^{i\phi_R(x)}\right) \\
&= \mathcal{N}_R^2\,e^{i\phi_R(x)}\,e^{i\phi_R(y)}\left(1 + e^{-\left[i\phi_R(x),i\phi_R(y)\right]}\right) \\
&= \mathcal{N}_R^2\,e^{i\phi_R(x)}\,e^{i\phi_R(y)}\left(1 + e^{i\pi\,\mathrm{sign}(x-y)}\right) = 0\,,
\end{aligned}$$

and similarly for left movers.

However, we still need to prove that the representation (6.26) with fields satisfying (6.32) fulfils the first equation in (6.25). Let us write, recalling (6.33),

$$\phi_R(x) = k_{RF}\,x + \phi_R^{(+)}(x) + \phi_R^{(-)}(x)\,, \quad \phi_L(x) = k_{LF}\,x + \phi_L^{(+)}(x) + \phi_L^{(-)}(x)\,, \qquad (6.34)$$

where $\phi^{(+)}$ include creation operators and $\phi^{(-)}$ annihilation ones, and, since ϕ is hermitian, then $\phi^{(+)} = \phi^{(-)\,\dagger}$. Therefore, upon defining

$$\left[\phi_{R(L)}^{(+)}(x),\,\phi_{R(L)}^{(-)}(y)\right] \equiv F_{R(L)}(x-y)\,, \qquad (6.35)$$

where F only depends on $x - y$ because of translation symmetry, then,

$$\begin{aligned}
\left[\phi_R(x),\,\phi_R(y)\right] &= \left[\phi_R^{(+)}(x),\,\phi_R^{(-)}(y)\right] + \left[\phi_R^{(-)}(x),\,\phi_R^{(+)}(y)\right] \\
&= F_R(x-y) - F_R(y-x) = 2i\,\tan^{-1}\frac{x-y}{\alpha}\,, \\
\left[\phi_L(x),\,\phi_L(y)\right] &= \left[\phi_L^{(+)}(x),\,\phi_L^{(-)}(y)\right] + \left[\phi_L^{(-)}(x),\,\phi_L^{(+)}(y)\right] \\
&= F_L(x-y) - F_L(y-x) = -2i\,\tan^{-1}\frac{x-y}{\alpha}\,,
\end{aligned}$$

which are satisfied if

$$F_R(x - y) = \ln\left(\alpha + i(x - y)\right), \quad F_L(x - y) = \ln\left(\alpha - i(x - y)\right). \tag{6.36}$$

Let us continue and prove that $\left\{\Psi_R(x), \Psi_R^\dagger(y)\right\} = \delta(x - y)$. We observe, exploiting (6.28), that

$$\Psi_R(x) = \mathcal{N}_R\, e^{i\phi_R(x)} = \mathcal{N}_R\, e^{ik_{RF}x}\, e^{i\phi_R^{(+)}(x)}\, e^{i\phi_R^{(-)}(x)}\, e^{-[i\phi_R^{(+)}(x),\, i\phi_R^{(-)}(x)]/2}$$
$$= \mathcal{N}_R\, e^{ik_{RF}x}\, e^{i\phi_R^{(+)}(x)}\, e^{i\phi_R^{(-)}(x)}\, e^{F_R(0)/2} = \mathcal{N}_R\, e^{ik_{RF}x}\, \sqrt{\alpha}\, e^{i\phi_R^{(+)}(x)}\, e^{i\phi_R^{(-)}(x)}\,,$$

and thus

$$\Psi_R^\dagger(y) = \mathcal{N}_R\, e^{-ik_{RF}x}\, \sqrt{\alpha}\, e^{-i\phi_R^{(+)}(y)}\, e^{-i\phi_R^{(-)}(y)}\,.$$

It follows that, repeatedly using (6.28),

$$\Psi_R(x)\, \Psi_R^\dagger(y) = \mathcal{N}_R^2\, \alpha\, e^{ik_{RF}(x-y)}\, e^{i\phi_R^{(+)}(x)}\, e^{i\phi_R^{(-)}(x)}\, e^{-i\phi_R^{(+)}(y)}\, e^{-i\phi_R^{(-)}(y)}$$
$$= \mathcal{N}_R^2\, \alpha\, e^{ik_{RF}(x-y)}\, e^{i\phi_R^{(+)}(x)}\, e^{-i\phi_R^{(+)}(y)}\, e^{i\phi_R^{(-)}(x)}\, e^{-i\phi_R^{(-)}(y)}\, e^{-\left[-i\phi_R^{(+)}(y),\, i\phi_R^{(-)}(x)\right]}$$
$$= \mathcal{N}_R^2\, \alpha\, e^{ik_{RF}(x-y)}\, e^{-F_R(y-x)}\, e^{i\left(\phi_R^{(+)}(x)-\phi_R^{(+)}(y)\right)}\, e^{i\left(\phi_R^{(-)}(x)-\phi_R^{(-)}(y)\right)}$$
$$= \mathcal{N}_R^2\, \frac{\alpha}{\alpha - i(x-y)}\, :\, e^{i\left(\phi_R(x)-\phi_R(y)\right)}\, :\,, \tag{6.37}$$

$$\Psi_R^\dagger(y)\, \Psi_R(x) = \mathcal{N}_R^2\, \alpha\, e^{ik_{RF}(x-y)}\, e^{-i\phi_R^{(+)}(y)}\, e^{-i\phi_R^{(-)}(y)}\, e^{i\phi_R^{(+)}(x)}\, e^{i\phi_R^{(-)}(x)}$$
$$= \mathcal{N}_R^2\, \alpha\, e^{ik_{RF}(x-y)}\, e^{-F_R(x-y)}\, e^{i\left(\phi_R^{(+)}(x)-\phi_R^{(+)}(y)\right)}\, e^{i\left(\phi_R^{(-)}(x)-\phi_R^{(-)}(y)\right)}$$
$$= \mathcal{N}_R^2\, \frac{\alpha}{\alpha + i(x-y)}\, :\, e^{i\left(\phi_R(x)-\phi_R(y)\right)}\, :\,,$$

where $:(\dots):$ means the operators inside are normal ordered, i.e., creations operators are on the left of annihilation ones. For left moving fermions, the above equations read

$$\Psi_L(x)\, \Psi_L^\dagger(y) = \mathcal{N}_L^2\, \frac{\alpha}{\alpha + i(x-y)}\, :\, e^{-i\left(\phi_L(x)-\phi_L(y)\right)}\, :\,,$$
$$\Psi_L^\dagger(y)\, \Psi_L(x) = \mathcal{N}_L^2\, \frac{\alpha}{\alpha - i(x-y)}\, :\, e^{-i\left(\phi_L(x)-\phi_L(y)\right)}\, :\,. \tag{6.38}$$

Therefore,

$$\left\{\Psi_R(x), \Psi_R^\dagger(y)\right\} = \mathcal{N}_R^2\, :\, e^{i\left(\phi_R(x)-\phi_R(y)\right)}\, :\, \left(\frac{\alpha}{\alpha + i(x-y)} + \frac{\alpha}{\alpha - i(x-y)}\right)$$
$$= \mathcal{N}_R^2\, :\, e^{i\left(\phi_R(x)-\phi_R(y)\right)}\, :\, (-i\alpha)\left(\frac{1}{(x-y)-i\alpha} - \frac{1}{(x-y)+i\alpha}\right)$$

$$\simeq \mathcal{N}_R^2 \; : e^{i\left(\phi_R(x) - \phi_R(y)\right)} : \; 2\pi\alpha\,\delta(x - y) = \mathcal{N}_R^2 \; 2\pi\alpha\,\delta(x - y),$$

$$\left\{\Psi_L(x), \Psi_L^\dagger(y)\right\} = \mathcal{N}_L^2 \; 2\pi\alpha\,\delta(x - y),$$

since $: e^{i\left(\phi_R(x) - \phi_R(y)\right)} := 1$ and $: e^{-i\left(\phi_L(x) - \phi_L(y)\right)} := 1$ for $x = y$, which implies that $\mathcal{N}_R^2 = \mathcal{N}_L^2 = 1/(2\pi\alpha)$, and thus

$$\Psi_R(x) = \frac{1}{\sqrt{2\pi\alpha}}\, e^{i\phi_R(x)}, \quad \Psi_L(x) = \frac{1}{\sqrt{2\pi\alpha}}\, e^{-i\phi_L(x)}. \tag{6.39}$$

6.2.3 Operator Product Expansion

Equations (6.21) and (6.31) imply, e.g., for right moving electrons, that

$$\rho_R(x) = \frac{1}{2\pi}\frac{\partial\phi_R(x)}{\partial x} = \Psi_R^\dagger(x)\,\Psi_R(x) \stackrel{?}{=} \frac{1}{2\pi\alpha}\, e^{-i\phi_R(x)}\, e^{i\phi_R(x)}.$$

At first sight, the bosonized expressions of the Fermi fields should simply yield $1/2\pi\alpha$, which is a wrong result. The issue is that the exponential operators are delicate to dealt with. The proper way to proceed is through *point splitting*, which defines the so-called Operator Product Expansion (OPE). For that, we note that

$$\rho_R(x) = \frac{1}{2\pi}\frac{\partial\phi_R(x)}{\partial x} = \lim_{x \to y}\frac{1}{2}\left(\Psi_R^\dagger(x)\,\Psi_R(y) + \Psi_R^\dagger(y)\,\Psi_R(x)\right)$$

$$= \lim_{x \to y}\frac{1}{4\pi\alpha}\left(e^{-i\phi_R(x)}\, e^{i\phi_R(y)} + e^{-i\phi_R(y)}\, e^{i\phi_R(x)}\right)$$

$$= \lim_{x \to y}\frac{1}{4\pi}\left(\frac{: e^{-i\left(\phi_R(x) - \phi_R(y)\right)} :}{\alpha - i(x - y)} + \frac{: e^{i\left(\phi_R(x) - \phi_R(y)\right)} :}{\alpha + i(x - y)}\right).$$

The limit $x \to y$ cannot be straightly done. The reason is that in our approach α quantifies the indeterminacy of the coordinates. It follows that we are allowed to send x to y only after we send the indeterminacy $\alpha \to 0$. With such prescription

$$\rho_R(x) = \frac{1}{2\pi}\frac{\partial\phi_R(x)}{\partial x} = \lim_{x \to y}\lim_{\alpha \to 0}\frac{1}{4\pi}\left(\frac{: e^{-i\left(\phi_R(x) - \phi_R(y)\right)} :}{\alpha - i(x - y)} + \frac{: e^{i\left(\phi_R(x) - \phi_R(y)\right)} :}{\alpha + i(x - y)}\right)$$

$$= \frac{i}{4\pi}\lim_{x \to y}\left\{\frac{1}{x - y}\left(1 - i\frac{\partial\phi_R(x)}{\partial x}(x - y)\right) - \frac{1}{x - y}\left(1 + i\frac{\partial\phi_R(x)}{\partial x}(x - y)\right)\right\}$$

$$= \frac{1}{2\pi}\frac{\partial\phi_R(x)}{\partial x}, \tag{6.40}$$

which is the desired equivalence. Similarly, one can readily verify with that same prescription that

$$
-i \, \Psi_R^\dagger(x) \, \frac{\partial \Psi_R(x)}{\partial x} = \lim_{x \to y} \frac{i}{2} \left(\frac{\partial \Psi_R^\dagger(x)}{\partial x} \, \Psi_R(y) - \Psi_R^\dagger(y) \, \frac{\partial \Psi_R(x)}{\partial x} \right)
$$

$$
= -\lim_{x \to y} \frac{1}{8\pi} \frac{\partial}{\partial x} \left\{ \frac{: e^{-i\left(\phi_R(x)-\phi_R(y)\right)} :}{x-y} + \frac{: e^{i\left(\phi_R(x)-\phi_R(y)\right)} :}{x-y} \right\}
$$

$$
\simeq \frac{1}{4\pi} \left(\frac{\partial \phi_R(x)}{\partial x} \right)^2 , \tag{6.41}
$$

$$
i \, \Psi_L^\dagger(x) \, \frac{\partial \Psi_L(x)}{\partial x} \simeq \frac{1}{4\pi} \left(\frac{\partial \phi_L(x)}{\partial x} \right)^2 ,
$$

thus $\Psi_R(x)$ and $\Psi_L(x)$ have Dirac-like dispersion. The almost equivalence means we have dropped an infinite positive term that, strictly speaking, cancels the infinite negative energy of the linearised spectrum to yield a finite energy.

6.2.4   Non-interacting Green's Functions and Density-Density Response Functions

The non-interacting Hamiltonian (6.19) can be written, through (6.20), as

$$
H_0 = \pi \, v_F \int dx \left(\rho_R(x)^2 + \rho_L(x)^2 \right) - v_F \int dx \left(k_{RF} \, \rho_R(x) + k_{LF} \, \rho_L(x) \right)
$$

$$
= \frac{v_F}{4\pi} \int dx \left[\left(\frac{\partial \phi_R(x)}{\partial x} \right)^2 + \left(\frac{\partial \phi_L(x)}{\partial x} \right)^2 \right] - \frac{v_F}{2\pi} \int dx \left[k_{RF} \frac{\partial \phi_R(x)}{\partial x} + k_{LF} \frac{\partial \phi_L(x)}{\partial x} \right] ,
$$

so that, through (6.30),

$$
i \, \dot{\phi}_R(x, t) = \left[\phi_R(x, t) , \, H_0 \right] = -i \, 2\pi \, v_F \, \rho_R(x, t) + i \, v_F \, k_{RF}
$$

$$
= -i \, v_F \left(\frac{\partial \phi_R(x, t)}{\partial x} - k_{RF} \right) ,
$$

$$
i \, \dot{\phi}_L(x, t) = \left[\phi_L(x, t) , \, H_0 \right] = i \, 2\pi \, v_F \, \rho_L(x, t) - i \, v_F \, k_{LF}
$$

$$
= i \, v_F \left(\frac{\partial \phi_L(x, t)}{\partial x} - k_{LF} \right) ,
$$

namely,

$$
\left(\frac{\partial}{\partial t} + v_F \frac{\partial}{\partial x} \right) \phi_R(x, t) = v_F \, k_{RF} , \qquad \left(\frac{\partial}{\partial t} - v_F \frac{\partial}{\partial x} \right) \phi_L(x, t) = -v_F \, k_{LF} , \tag{6.42}
$$

with solution

$$\phi_R(x, t) = v_F k_{RF} t + \phi_R(x - v_F t), \quad \phi_L(x, t) = -v_F k_{LF} t + \phi_R(x + v_F t).$$
(6.43)

We observe that (6.33) implies that

$$\langle 0 \mid \phi_R(x, t) \mid 0 \rangle = v_F k_{RF} t + \langle 0 \mid \phi_R(x - v_F t) \mid 0 \rangle = k_{RF} x,$$
$$\langle 0 \mid \phi_L(x, t) \mid 0 \rangle = -v_F k_{LF} t + \langle 0 \mid \phi_L(x + v_F t) \mid 0 \rangle = k_{LF} x,$$

and therefore the expectation value is constant in time, as it should since it is associated to a conserved quantity.

6.2.4.1 Green's Functions

Using (6.37) and (6.38), we readily find the non-interacting Green's functions

$$G_R^0(x, t) = -i\,\theta(t)\,\langle \Psi_R(x, t)\,\Psi_R^\dagger(0, 0)\rangle + i\,\theta(-t)\,\langle \Psi_R^\dagger(0, 0)\,\Psi_R(x, t)\rangle$$
$$= \frac{1}{2\pi}\left[-i\,\theta(t)\,\frac{1}{\alpha - i(x - v_F t)} + i\,\theta(-t)\,\frac{1}{\alpha + i(x - v_F t)} \right] \langle\, :e^{i(\phi_R(x - v_F t) - \phi_R(0))}\, :\rangle$$
$$= \frac{1}{2\pi}\,\frac{e^{ik_{RF}x}}{x - v_F t + i\alpha\,\mathrm{sign}(t)},$$
(6.44)

and, similarly,

$$G_L^0(x, t) = -i\,\theta(t)\,\langle \Psi_L(x, t)\,\Psi_L^\dagger(0, 0)\rangle + i\,\theta(-t)\,\langle \Psi_L^\dagger(0, 0)\,\Psi_L(x, t)\rangle$$
$$= \frac{1}{2\pi}\left[-i\,\theta(t)\,\frac{1}{\alpha + i(x + v_F t)} + i\,\theta(-t)\,\frac{1}{\alpha - i(x + v_F t)} \right] \langle\, :e^{-i(\phi_L(x - v_F t) - \phi_L(0))}\, :\rangle$$
$$= \frac{1}{2\pi}\,\frac{e^{-ik_{LF}x}}{-x - v_F t + i\alpha\,\mathrm{sign}(t)}.$$
(6.45)

Through $G_R^0(x, t)$ we can calculate the non-interacting momentum distribution of right moving fermions, $n_R^0(k)$,

$$n_R^0(k) = -i \int dx\, e^{-ikx}\, G_R^0(x, t = 0^-) = \int \frac{dx}{2\pi i}\,\frac{e^{-i(k - k_{RF})x}}{x - i\alpha}.$$
(6.46)

We can turn the integral over x into a contour integral in the complex plane $z = x + iy$, where the contour runs along the real axis and closes in the upper half plane for $k < k_{RF}$ and in the lower one otherwise. Only in the former case the contour integral yields a finite result, since the function to be integrated has a single pole at $z = i\alpha$ with residue 1, thus in the upper half plane. Therefore $n_R^0(k) = \theta(k_{RF} - k)$, which is the correct result. An alternative way to get the same result, which will turn useful in the interacting case, is to calculated

$$
\begin{aligned}
\frac{\partial n_R^0(k)}{\partial k} &= -\int dx\, x\, e^{-ikx}\, G_R^0(x, t = 0^-) = -\int \frac{dx}{2\pi}\, \frac{x}{x - i\alpha}\, e^{-i(k-k_{RF})x} \\
&= -\int \frac{dx}{2\pi}\, e^{-i(k-k_{RF})x} - i\,\alpha \int \frac{dx}{2\pi}\, \frac{1}{x - i\alpha}\, e^{-i(k-k_{RF})x} \\
&\xrightarrow[\alpha \to 0]{} -\int \frac{dx}{2\pi}\, e^{-i(k-k_{RF})x} = -\delta\big(k - k_{RF}\big).
\end{aligned}
$$

$$(6.47)$$

For later convenience, we define the vacuum expectation value

$$
\Omega_{\beta_R, \beta_L}(x, t) \equiv \langle e^{i\beta_R\,\phi_R(x,t) + i\beta_L\,\phi_L(x,t)}\, e^{-i\beta_R\,\phi_R(0,0) - i\beta_L\,\phi_L(0,0)} \rangle . \tag{6.48}
$$

We observe that, upon repeatedly using (6.28) and (6.29)

$$
\begin{aligned}
&e^{i\beta_R\,\phi_R(x,t) + i\beta_L\,\phi_L(x,t)}\, e^{-i\beta_R\,\phi_R(0,0) - i\beta_L\,\phi_L(0,0)} \\
&= e^{-i\pi\,\beta_R\,\beta_L}\, e^{i\beta_R\,\phi_R(x,t)}\, e^{i\beta_L\,\phi_L(x,t)}\, e^{-i\beta_R\,\phi_R(0,0)}\, e^{-i\beta_L\,\phi_L(0,0)} \\
&= e^{i\beta_R\,\phi_R(x,t)}\, e^{-i\beta_R\,\phi_R(0,0)}\, e^{i\beta_L\,\phi_L(x,t)}\, e^{-i\beta_L\,\phi_L(0,0)} ,
\end{aligned}
$$

so that, through (6.37) and (6.38),

$$
\begin{aligned}
\Omega_{\beta_R, \beta_L}(x, t) &= \langle e^{i\beta_R\,\phi_R(x,t)}\, e^{-i\beta_R\,\phi_R(0,0)} \rangle \langle e^{i\beta_L\,\phi_L(x,t)}\, e^{-i\beta_L\,\phi_L(0,0)} \rangle \\
&= e^{ik_{RF}\,\beta_R\, x + ik_{LF}\,\beta_L\, x} \left(\frac{\alpha}{\alpha - i(x - v_F t)} \right)^{\beta_R^2} \left(\frac{\alpha}{\alpha + i(x + v_F t)} \right)^{\beta_L^2} .
\end{aligned}
$$

$$(6.49)$$

6.2.4.2 Density-Density Correlation Functions

The density-density correlation function that is simplest to calculate is the long wavelength one, $|q| \ll k_{R(L)F}$. We first note that

$$
\begin{aligned}
\langle 0 \mid \phi_R(x, t)\, \phi_R(y, 0) \mid 0 \rangle &= \langle 0 \mid \phi_R(x - v_F t)\, \phi_R(y) \mid 0 \rangle \\
&= k_{RF}^2\, x\, y + \langle 0 \mid \phi_R^{(-)}(x - v_F t)\, \phi_R^{(+)}(y) \mid 0 \rangle \\
&= k_{RF}^2\, x\, y - \langle 0 \mid \left[\phi_R^{(+)}(y),\, \phi_R^{(-)}(x - v_F t) \right] \mid 0 \rangle \\
&= k_{RF}^2\, x\, y - F_R\big(v_F t - x + y\big) = k_{RF}^2\, x\, y - \ln\big(\alpha + i\big(v_F t - x + y\big)\big) \\
&= k_{RF}^2\, x\, y + i\,\frac{\pi}{2} - \ln\big(x - y - v_F t + i\alpha\big), \\
\langle 0 \mid \phi_R(x, t)\, \phi_R(y, 0) \mid 0 \rangle &= k_{LF}^2\, x\, y - i\,\frac{\pi}{2} - \ln\big(x - y + v_F t - i\alpha\big),
\end{aligned}
$$

so that

$$\langle 0 \mid \rho_R(x,t)\,\rho_R(y,0) \mid 0\rangle = \frac{1}{4\pi^2} \frac{\partial^2}{\partial x \partial y} \langle 0 \mid \phi_R(x,t)\,\phi_R(y,0) \mid 0\rangle$$

$$= \frac{k_{RF}^2}{4\pi^2} - \frac{1}{4\pi^2} \left(\frac{1}{x-y-v_F t + i\alpha} \right)^2, \quad (6.50)$$

$$\langle 0 \mid \rho_L(x,t)\,\rho_L(y,0) \mid 0\rangle = \frac{k_{LF}^2}{4\pi^2} - \frac{1}{4\pi^2} \left(\frac{1}{x-y+v_F t - i\alpha} \right)^2.$$

It follows that the Fourier transforms at $q \neq 0$

$$\langle 0 \mid \rho_R(q,t)\,\rho_R(-q,0) \mid 0\rangle = -\frac{1}{4\pi^2} \int dx\, e^{-iqx} \left(\frac{1}{x - v_F t + i\alpha} \right)^2$$

$$= \theta(q) \frac{q}{2\pi} e^{-iv_F q t},$$

$$\langle 0 \mid \rho_R(-q,0)\,\rho_R(q,t) \mid 0\rangle = -\frac{1}{4\pi^2} \int dx\, e^{-iqx} \left(\frac{1}{x - v_F t - i\alpha} \right)^2$$

$$= \theta(-q) \frac{|q|}{2\pi} e^{-iv_F q t},$$

$$\langle 0 \mid \rho_L(q,t)\,\rho_L(-q,0) \mid 0\rangle = -\frac{1}{4\pi^2} \int dx\, e^{-iqx} \left(\frac{1}{x + v_F t - i\alpha} \right)^2$$

$$= \theta(-q) \frac{|q|}{2\pi} e^{iv_F q t},$$

$$\langle 0 \mid \rho_L(-q,0)\,\rho_L(q,t) \mid 0\rangle = -\frac{1}{4\pi^2} \int dx\, e^{-iqx} \left(\frac{1}{x + v_F t + i\alpha} \right)^2$$

$$= \theta(q) \frac{q}{2\pi} e^{iv_F q t},$$

so that

$$\chi_R(q,t) = -i\,\theta(t) \langle 0 \mid \left[\rho_R(q,t),\, \rho_R(-q,0) \right] \mid 0\rangle = -i\,\theta(t)\, e^{-iv_F q t} \frac{q}{2\pi},$$

$$\chi_L(q,t) = -i\,\theta(t) \langle 0 \mid \left[\rho_L(q,t),\, \rho_L(-q,0) \right] \mid 0\rangle = i\,\theta(t)\, e^{iv_F q t} \frac{q}{2\pi},$$

and thus

$$\chi_R(q,\omega) = \chi_L(q,\omega) = \frac{|q|}{2\pi} \left(\frac{1}{\omega - v_F|q| + i\eta} - \frac{1}{\omega + v_F|q| + i\eta} \right). \quad (6.51)$$

Therefore, the density-density response function in the long-wavelength limit is simply

$$\chi^0(q,\omega) = \chi_R(q,\omega) + \chi_L(q,\omega) = \frac{|q|}{\pi} \left(\frac{1}{\omega - v_F|q| + i\eta} - \frac{1}{\omega + v_F|q| + i\eta} \right), \quad (6.52)$$

which is the correct expression (6.4) one explicitly derives without passing through bosonization. The non-interacting compressibility is therefore

$$\kappa_0 = -\chi^0(q \to 0, 0) = \frac{1}{2\pi v_F} . \tag{6.53}$$

Let us now consider the density-density correlation function at $q \sim k_{RF} + k_{LF}$, namely

$$\chi^0_{k_{RF}+k_{LF}}(x, t) = -i\,\theta(t)\,\langle 0 \mid \left[\Psi_R^\dagger(x, t)\,\Psi_L(x, t),\ \Psi_L^\dagger(0, 0)\,\Psi_R(0, 0) \right] \mid 0 \rangle$$

$$= -i\,\theta(t)\,\frac{1}{4\pi^2\alpha^2}\,\Big(\Omega_{1,1}(x, t) - \Omega_{-1,-1}(-x, -t) \Big)$$

$$= -i\,\theta(t)\,\frac{e^{i(k_{RF}+k_{LF})x}}{4\pi^2}\,\left(\frac{1}{x - v_F t + i\alpha}\,\frac{1}{x + v_F t - i\alpha} - c.c. \right) . \tag{6.54}$$

The Fourier transform right at $q = k_{RF} + k_{LF}$ is

$$\chi^0(k_{RF} + k_{LF}, t) = -\theta(t)\,\frac{1}{4\pi}\,\left(\frac{1}{v_F t - i\alpha} + \frac{1}{v_F t + i\alpha} \right) .$$

One easily realises that $\chi(k_{RF} + k_{LF}, \omega)$ diverges logarithmically for $\omega \to 0$, as discussed in Sect. 6.1.

6.2.5 Interaction

Let us now take into account the interaction (6.7), namely

$$H_{\text{int}} = \frac{1}{2L} \sum_q V(q)\,\rho(q)\,\rho(-q) .$$

Consistently with the 'semiclassical' approximation, we will keep only the interaction scattering processes that correspond to low-energy particle-hole excitations, i.e., those close to right and left Fermi momenta. Those processes are shown in Fig. 6.4, and are associated to the coupling constants $g_4 \simeq V(0)$, $g_2 \simeq V(0) - V(k_{RF} + k_{LF})$ and $g_3 \simeq V(k_{RF} + k_{LF})$ in the g-ology jargon. We observe that the umklapp scattering g_4 is the only one that does not conserve separately N_R and N_L. However, it is allowed by momentum conservation only if $2k_{RF} \equiv -2k_{LF}$, where the equivalence holds apart from the primitive reciprocal lattice vector 2π, which requires $k_{RF} + k_{LF} = \pi$ or, equivalently, $N_R + N_L = L/2$, thus half-filled density.

We start assuming that the density is away from half-filling, so that only the scattering processes g_4 and g_2 are active at low energy. In this case N_R and N_L are still conserved quantity. Even though that is not a true symmetry property of H_{int}, it is asymptotically so at low enough energy and temperature. In other words, exactly like in a Fermi liquid, at low energy the system recovers the large symmetry of the non-interacting Hamiltonian, which includes independent $U_R(1)$ and $U_L(1)$ symmetries for R and L moving fermions, respectively. Here, we are actually assuming such dynamical symmetry recovery, while in Landau's Fermi liquids

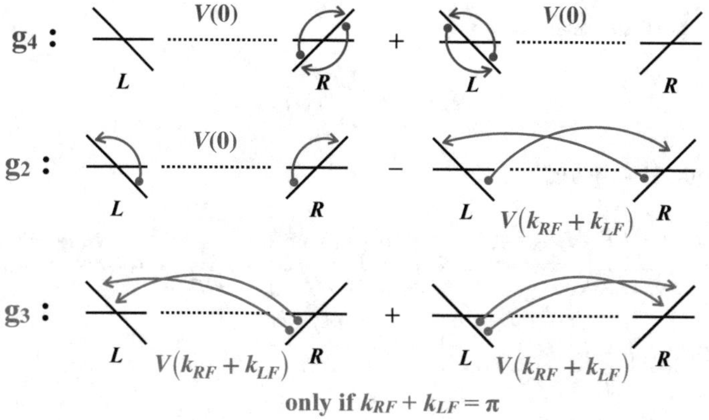

Fig. 6.4 Leading scattering processes yielded by interaction. By assumption, each excitation drawn as an arrow carries small momentum $|q| \ll k_{R(L)F}$, thus both initial and final states are close to a Fermi point

we proved that explicitly. In reality, also in Luttinger liquids one can convincingly demonstrate that away from half-filling the low-energy effective theory possesses the large $U_R(1) \times U_L(1)$ of the non-interacting Hamiltonian. In that case, we can assume that the effective interaction at low energy has the expression

$$H_{\text{int}} \simeq \frac{1}{2L} \sum_q V(q)\, \rho(q)\, \rho(-q)$$

$$\simeq \frac{1}{L} \sum_q \left[g_2\, \rho_R(q)\, \rho_L(-q) + \frac{g_4}{2} \left(\rho_R(q)\, \rho_R(-q) + \rho_L(q)\, \rho_L(-q) \right) \right],$$

$$(6.55)$$

where g_4 and g_2 might be different from their bare values, $g_4 \simeq V(0)$, $g_2 \simeq V(0) - V(k_{RF} + k_{LF})$, and just correspond to effective parameters like the f-ones in Landau's Fermi liquid theory. Comparing with (6.19), we note that g_4 simply yields a correction to the Fermi velocity $v_F \to v_F + g_4/2\pi$. If we redefine v_F accordingly, the full Hamiltonian reads

$$H = \frac{1}{L} \sum_q \left[\pi v_F \left(\rho_R(q)\, \rho_R(-q) + \rho_L(q)\, \rho_L(-q) \right) + g_2\, \rho_R(q)\, \rho_L(-q) \right] - \mu_R N_R - \mu_L N_L$$

$$= \frac{1}{L} \sum_{q \neq 0} \left[\pi v_F \left(\rho_R(q)\, \rho_R(-q) + \rho_L(q)\, \rho_L(-q) \right) + g_2\, \rho_R(q)\, \rho_L(-q) \right]$$

$$+ \frac{\pi v_F}{L} \left(N_R^2 + N_L^2 \right) + \frac{g_2}{L} N_R N_L - \mu_R N_R - \mu_L N_L,$$

$$(6.56)$$

which implies that now

$$\mu_R = v_F \left(k_{RF} + \frac{g_2}{2\pi v_F} k_{LF} \right), \qquad \mu_L = v_F \left(k_{LF} + \frac{g_2}{2\pi v_F} k_{RF} \right), \qquad (6.57)$$

in order to have the desired

$$N_{R(L)} = \frac{L}{2\pi} k_{R(L)F} .$$

We can equivalently write

$$H = \frac{v_F}{4\pi} \int dx \left[\left(\frac{\partial \phi_R(x)}{\partial x} \right)^2 + \left(\frac{\partial \phi_L(x)}{\partial x} \right)^2 \right] + \frac{g_2}{4\pi^2} \int dx \, \frac{\partial \phi_R(x)}{\partial x} \, \frac{\partial \phi_L(x)}{\partial x}$$
$$- \frac{1}{2\pi} \int dx \left[\mu_R \frac{\partial \phi_R(x)}{\partial x} + \mu_L \frac{\partial \phi_L(x)}{\partial x} \right].$$

$$(6.58)$$

We introduce the linear combinations

$$\Phi(x) \equiv \frac{1}{\sqrt{4\pi}} \left(\phi_R(x) + \phi_L(x) \right), \quad \Theta(x) \equiv \frac{1}{\sqrt{4\pi}} \left(\phi_L(x) - \phi_R(x) \right),$$

$$(6.59)$$

which, through (6.32), satisfy

$$\left[\Phi(x), \, \Theta(y) \right] = \frac{1}{4\pi} \left(-2i \, \pi \, \mathrm{sign}(x - y) - 2i \, \pi \right) = -i \, \theta(x - y) ,$$

so that, introducing the field,

$$\Pi(x) \equiv \frac{\partial \Theta(x)}{\partial x} ,$$

$$(6.60)$$

then

$$\left[\Phi(x), \, \Pi(y) \right] = i \, \delta(x - y) .$$

namely, $\Phi(x)$ and $\Pi(x)$ are conjugate fields. Correspondingly,

$$\phi_R(x) = \sqrt{\pi} \left(\Phi(x) - \Theta(x) \right), \quad \phi_L(x) = \sqrt{\pi} \left(\Phi(x) + \Theta(x) \right),$$

$$(6.61)$$

which, substituted in the Hamiltonian (6.58), lead to

$$H = \frac{v_F}{2} \int dx \left[\Pi(x)^2 + \left(\frac{\partial \Phi(x)}{\partial x} \right)^2 \right] + \frac{g_2}{4\pi} \int dx \left[\left(\frac{\partial \Phi(x)}{\partial x} \right)^2 - \Pi(x)^2 \right]$$
$$- \frac{1}{2\pi} \int dx \left[\mu_\Phi \frac{\partial \Phi(x)}{\partial x} + \mu_\Pi \, \Pi(x) \right],$$

$$(6.62)$$

where

$$\mu_\Phi = \sqrt{\pi} \left(\mu_R + \mu_L \right), \quad \mu_\Pi = -\sqrt{\pi} \left(\mu_R - \mu_L \right).$$

We apply the canonical transformation

$$\Phi(x) = \sqrt{K}\,\overline{\Phi}(x), \quad \Theta(x) = \frac{1}{\sqrt{K}}\,\overline{\Theta}(x), \quad K^2 = \frac{2\pi v_F - g_2}{2\pi v_F + g_2}, \qquad (6.63)$$

after which

$$H = \frac{v}{2}\int dx\left[\overline{\Pi}(x)^2 + \left(\frac{\partial\overline{\Phi}(x)}{\partial x}\right)^2\right] - \frac{1}{2\pi}\int dx\left[\mu_\Phi\,\sqrt{K}\,\frac{\partial\overline{\Phi}(x)}{\partial x} + \frac{\mu_\Pi}{\sqrt{K}}\,\overline{\Pi}(x)\right],$$

$$(6.64)$$

where

$$v = v_F\sqrt{1 - \frac{g_2^2}{4\pi^2 v_F^2}}. \qquad (6.65)$$

We now recall that the Fermi velocity is actually $v_F + g_4/2\pi$, so that

$$v^2 = \left(v_F + \frac{g_4}{2\pi}\right)^2 - \frac{g_2^2}{4\pi^2} = \left(v_F + \frac{g_4}{2\pi} + \frac{g_2}{2\pi}\right)\left(v_F + \frac{g_4}{2\pi} - \frac{g_2}{2\pi}\right)$$

$$\simeq \left(v_F + \frac{2V(0) - V(k_{RF} + k_{LF})}{2\pi}\right)\left(v_F + \frac{V(k_{RF} + k_{LF})}{2\pi}\right),$$

which is always finite if the interaction in real space $V(x)$ is repulsive, thus $K < 1$. If instead $V(x) < 0$ is attractive, thus $K > 1$, the effective velocity vanishes when

$$g_4 + g_2 \simeq 2V(0) - V(k_{RF} + k_{LF}) = -2\pi v_F,$$

which signals a thermodynamic instability towards phase separation.

Following (6.61), we define

$$\overline{\phi}_R(x) = \sqrt{\pi}\left(\overline{\Phi}(x) - \overline{\Theta}(x)\right), \quad \overline{\phi}_L(x) = \sqrt{\pi}\left(\overline{\Phi}(x) + \overline{\Theta}(x)\right), \qquad (6.66)$$

which satisfy the same commutation relations as $\phi_R(x)$ and $\phi_L(x)$ and allow rewriting (6.64) as

$$H = \frac{v}{4\pi}\int dx\left[\left(\frac{\partial\overline{\phi}_R(x)}{\partial x}\right)^2 + \left(\frac{\partial\overline{\phi}_L(x)}{\partial x}\right)^2\right]$$

$$- \frac{1}{2\pi}\int dx\left[\overline{\mu}_R\,\frac{\partial\overline{\phi}_R(x)}{\partial x} + \overline{\mu}_L\,\frac{\partial\overline{\phi}_L(x)}{\partial x}\right],$$

$$(6.67)$$

where

$$\overline{\mu}_R = \frac{1}{2}\left(\left(\sqrt{K} + \frac{1}{\sqrt{K}}\right)\mu_R + \left(\sqrt{K} - \frac{1}{\sqrt{K}}\right)\mu_L\right) \equiv v\,\overline{k}_{RF},$$

$$\overline{\mu}_L = \frac{1}{2}\left(\left(\sqrt{K} + \frac{1}{\sqrt{K}}\right)\mu_L + \left(\sqrt{K} - \frac{1}{\sqrt{K}}\right)\mu_R\right) \equiv v\,\overline{k}_{LF}.$$

$$(6.68)$$

The Hamiltonian (6.67) has the form of a non-interacting one for the new Fermi operators

$$\overline{\Psi}_R(x) = \frac{1}{\sqrt{2\pi\alpha}} \, e^{i\overline{\phi}_R(x)} \, , \quad \overline{\Psi}_L(x) = \frac{1}{\sqrt{2\pi\alpha}} \, e^{i\overline{\phi}_L(x)} \, ,$$

apart from a different Fermi velocity v.

6.2.6   Interacting Green's Functions and Correlation Functions

We note that

$$
\begin{aligned}
\phi_R(x) &= \sqrt{\pi} \left(\Phi(x) - \Theta(x) \right) = \sqrt{\pi} \left(\sqrt{K} \, \overline{\Phi}(x) - \frac{1}{\sqrt{K}} \overline{\Theta}(x) \right) \\
&= \frac{1}{2} \left(\left(\sqrt{K} + \frac{1}{\sqrt{K}} \right) \overline{\phi}_R(x) + \left(\sqrt{K} - \frac{1}{\sqrt{K}} \right) \overline{\phi}_L(x) \right), \qquad (6.69) \\
\phi_L(x) &= \frac{1}{2} \left(\left(\sqrt{K} + \frac{1}{\sqrt{K}} \right) \overline{\phi}_L(x) + \left(\sqrt{K} - \frac{1}{\sqrt{K}} \right) \overline{\phi}_R(x) \right),
\end{aligned}
$$

which also implies, through (6.68) and (6.57) that

$$
\begin{aligned}
\langle 0 \mid \phi_R(x) \mid 0 \rangle &= \frac{x}{2} \left(\left(\sqrt{K} + \frac{1}{\sqrt{K}} \right) \overline{k}_{RF} + \left(\sqrt{K} - \frac{1}{\sqrt{K}} \right) \overline{k}_{LF} \right) = k_{RF} \, x \, , \\
\langle 0 \mid \phi_L(x) \mid 0 \rangle &= \frac{x}{2} \left(\left(\sqrt{K} + \frac{1}{\sqrt{K}} \right) \overline{k}_{LF} + \left(\sqrt{K} - \frac{1}{\sqrt{K}} \right) \overline{k}_{RF} \right) = k_{LF} \, x \, ,
\end{aligned}
$$

as expected since both are conserved quantities.

6.2.6.1 Green's Functions
Through (6.48) and (6.49) we readily find that the interacting Green's functions are

$$
\begin{aligned}
G_R(x,t) &= \frac{1}{2\pi} \frac{e^{ik_{RF}x}}{x - vt + i\alpha \, \text{sign}(t)} \left[\frac{\alpha^2}{(x - vt + i\alpha \, \text{sign}(t))(x + vt - i\alpha \, \text{sign}(t))} \right]^{\beta^2}, \\
G_L(x,t) &= \frac{1}{2\pi} \frac{e^{-ik_{LF}x}}{-x - vt + i\alpha \, \text{sign}(t)} \left[\frac{\alpha^2}{(x - vt + i\alpha \, \text{sign}(t))(x + vt - i\alpha \, \text{sign}(t))} \right]^{\beta^2},
\end{aligned}
$$

$$(6.70)$$

where

$$\beta^2 = \frac{1}{4} \left(\sqrt{K} - \frac{1}{\sqrt{K}} \right)^2 .$$

Since the Green's functions at $t = 0^-$ are non-analytic on both sides of the complex plane $x \to z = x + iy$, the momentum distributions

$$n_{R(L)}(k) = -i \int dx\, e^{-ikx}\, G_{R(L)}(x, 0^-),$$

does not have anymore a jump at the Fermi momenta. However,

$$\frac{\partial n_{R(L)}(k)}{\partial k}\Big|_{k=k_{R(L)F}} = -\int dx\, e^{-ik_{R(L)F}x}\, x\, G_{R(L)}(x, 0^-),$$

is singular due to the large $|x|$ convergence of the integrand provided $2\beta^2 \le 1$. In that case, $n_{R(L)}(k)$ has no jump at $k_{R(L)F}$ but it has a diverging slope. If $n \le 2\beta^2 \le n + 1$ with integer $n \ge 1$, all derivatives of order less than $n + 1$ are finite, but those with order $\ge n + 1$ are singular. Therefore, even though $n_{R(L)}(k)$ is smooth, the singularity of its derivatives still allows identifying a Fermi surface.

6.2.6.2 Correlation Functions

The long-wavelength component of the local density is

$$\rho(x) = \rho_R(x) + \rho_L(x) = \sqrt{4\pi}\, \frac{\partial \Phi(x)}{\partial x} = \sqrt{K}\, \sqrt{4\pi}\, \frac{\partial \overline{\Phi}(x)}{\partial x} = \sqrt{K}\, \overline{\rho}(x)$$

so that the interacting density-density response function at small $|q| \ll k_F$ can be readily found through (6.52) and is

$$\chi(q, \omega) = K\, \frac{|q|}{\pi}\, \left(\frac{1}{\omega - v|q| + i\eta} - \frac{1}{\omega + v|q| + i\eta}\right), \tag{6.71}$$

which shows that the particle-hole excitations at small $|q|$ are still exhausted by an acoustic mode with renormalised velocity v. The interacting compressibility

$$\kappa = -\chi(q \to 0, 0) = K\, \frac{1}{2\pi v} = K\, \frac{v_F}{v}\, \kappa_0, \tag{6.72}$$

and is lower than the non-interacting one (6.53) for repulsive interaction, whereas it is bigger for attractive one diverging for $v \to 0$. That, as mentioned, signals the onset of phase separation.

Let us consider now the component of the local density at $q \sim -k_{RF} - k_{LF}$, namely,

$$\begin{aligned}
\rho_{-k_{RF}-k_{LF}}(x) &= \Psi_R^\dagger(x)\, \Psi_L(x) = \frac{1}{2\pi\alpha}\, e^{-i\phi_R(x)}\, e^{-i\phi_L(x)} \\
&= \frac{1}{2\pi\alpha}\, e^{-i(\phi_R(x)+\phi_L(x))}\, e^{+\left[-i\phi_R(x), -i\phi_L(x)\right]/2} = \frac{i}{2\pi\alpha}\, e^{-i(\phi_R(x)+\phi_L(x))} \\
&= \frac{i}{2\pi\alpha}\, e^{-i\sqrt{4\pi}\,\Phi(x)} = \frac{i}{2\pi\alpha}\, e^{-i\sqrt{4\pi K}\,\overline{\Phi}(x)} = \frac{i}{2\pi\alpha}\, e^{-i\sqrt{K}\,\left(\overline{\phi}_R(x)+\overline{\phi}_L(x)\right)},
\end{aligned} \tag{6.73}$$

and thus

$$\rho_{k_{RF}+k_{LF}}(x) = \Psi_L^\dagger(x)\,\Psi_R(x) = -\frac{i}{2\pi\alpha}\,e^{i\sqrt{K}\left(\bar{\phi}_R(x)+\bar{\phi}_L(x)\right)}\,. \tag{6.74}$$

It follows that

$$\langle\,\rho_{-k_{RF}-k_{LF}}(x,t)\,\rho_{k_{RF}k_{LF}}(0)\,\rangle = \frac{e^{i\left(k_{RF}+k_{LF}\right)x}}{4\pi^2\alpha^2}\left(\frac{\alpha^2}{(x-vt+i\alpha)(x+vt-i\alpha)}\right)^K,$$

which vanishes for $x \to \infty$ with an interaction dependent exponent K. In particular, it vanishes more slowly than in absence of interaction if the latter is repulsive, which, as mentioned, corresponds to a quasi long-range CDW order; the true long-range one prevented by Mermin-Wagner theorem.

Let us consider now the Cooper pair operator at zero total momentum, assuming zero current and thus $k_{RF} = k_{LF} = k_F$, which is

$$\begin{aligned}
\Delta^\dagger(x) &= \Psi_R^\dagger(x)\,\Psi_L^\dagger(x) = \frac{1}{2\pi\alpha}\,e^{-i\phi_R(x)}\,e^{i\phi_L(x)}\\
&= \frac{1}{2\pi\alpha}\,e^{i(\phi_L(x)-\phi_R(x))}\,e^{+\left[-i\phi_R(x),i\phi_L(x)\right]/2} = -\frac{i}{2\pi\alpha}\,e^{i(\phi_L(x)-\phi_R(x))}\\
&= -\frac{i}{2\pi\alpha}\,e^{i\sqrt{4\pi}\,\Theta(x)} = -\frac{i}{2\pi\alpha}\,e^{i\sqrt{4\pi/K}\,\Theta(x)} = -\frac{i}{2\pi\alpha}\,e^{-i\left(\bar{\phi}_R(x)-\bar{\phi}_L(x)\right)/\sqrt{K}}\,,
\end{aligned} \tag{6.75}$$

hence its correlation function

$$\langle\,\Delta^\dagger(x,t)\,\Delta(0)\,\rangle = \frac{1}{4\pi^2\alpha^2}\left(\frac{\alpha^2}{(x-vt+i\alpha)(x+vt-i\alpha)}\right)^{1/K},$$

decays more slowly than in absence of interaction when the latter is attractive, a quasi long-range superconducting order, whereas it decays faster when the interaction is repulsive.

It follows that the corresponding CWD and superconducting (SC) susceptibilities, both of which diverge as $\ln 1/T$ in absence of interaction, behave as

$$\kappa_{CDW} \sim T^{2K-2}, \quad \kappa_{SC} \sim T^{2/K-2}\,. \tag{6.76}$$

6.2.7   Umklapp Scattering

Let us now consider the case in which the umklapp scattering g_3 in Fig. 6.4 becomes allowed at low-energy. That occurs at half-filling when $k_{RF} + k_{LF} = \pi \equiv -k_{RF} - k_{LF}$, and corresponds to an interaction

$$\begin{aligned}
H_{umklapp} &\simeq g_3 \int dx \left(\Psi_R^\dagger(x+\epsilon)\,\Psi_R^\dagger(x)\,\Psi_L(x)\,\Psi_L(x+\epsilon) + H.c.\right)\\
&= -\frac{g_3}{2\pi^2\alpha^2}\int dx\,\cos 2(\phi_R(x)+\phi_L(x)) = -\frac{g_3}{2\pi^2\alpha^2}\int dx\,\cos\sqrt{16\pi}\,\Phi(x)\\
&= -\frac{g_3}{2\pi^2\alpha^2}\int dx\,\cos\sqrt{16\pi K}\,\Phi(x)\,,
\end{aligned} \tag{6.77}$$

where the bare value of g_3 is $V(q = \pi)$. For simplicity, we assume that $k_{RF} = k_{LF} \equiv k_F = \pi/2$. The fully interacting Hamiltonian at half-filling thus reads, dropping for simplicity the chemical potential terms and assuming zero current,

$$
H = \frac{v_F}{2} \int dx \left[\Pi(x)^2 + \left(\frac{\partial \Phi(x)}{\partial x}\right)^2 \right] + \frac{g_2}{4\pi} \int dx \left[\left(\frac{\partial \Phi(x)}{\partial x}\right)^2 - \Pi(x)^2 \right]
$$

$$
- \frac{g_3}{2\pi^2 \alpha^2} \int dx \, \cos \sqrt{16\pi} \, \Phi(x)
$$

$$
= \frac{v}{2} \int dx \left[\overline{\Pi}(x)^2 + \left(\frac{\partial \overline{\Phi}(x)}{\partial x}\right)^2 \right] - \frac{g_3}{2\pi^2 \alpha^2} \int dx \, \cos \sqrt{16\pi K} \, \overline{\Phi}(x).
$$

$$(6.78)$$

This Hamiltonian is known as sine-Gordon model, and its physics is equivalent to the model shown in Fig. 6.5. If we denote the position of the atom n in that figure as $r_n = na + u(na)$, where $u(na)$ is the displacement with conjugate variable $p(na)$, the Hamiltonian of the model in Fig. 6.5 is

$$
H_* = \sum_{n=1}^{L} \left(\frac{p(na)^2}{2M} + \frac{C}{2a^2} \left(u(na) - u((n+1)a) \right)^2 \right) - U \sum_{n=1}^{L} \cos u(na).
$$

If $a \to 0$ such that $na \to x$ with x a continuous variable $\in [0, La]$, where La goes to infinity in the thermodynamic limit, then

$$
H_* = \int \frac{dx}{a} \left[\frac{p(x)^2}{2M} + \frac{C}{2} \left(\frac{\partial u(x)}{\partial x}\right)^2 \right] - \frac{U}{a} \int dx \, \cos u(x)
$$

$$
= \frac{v_*}{2a} \int dx \left[P(x)^2 + \left(\frac{\partial Q(x)}{\partial x}\right)^2 \right] - \frac{U}{a} \int dx \, \cos \beta \, Q(x),
$$

$$(6.79)$$

where, taking as usual $\hbar = 1$,

$$
p(x) = \frac{1}{\beta} P(x), \quad u(x) = \beta Q(x), \quad \beta^4 = \frac{1}{MC}, \quad v_*^2 = \frac{C}{M}.
$$

The Hamiltonian (6.79) coincides with (6.78) upon renaming the parameters and defining a dimensionless variable $x \to x/a$. However, in the model of Fig. 6.5 the physics is more transparent. The acoustic mode with velocity v_* is just the acoustic phonon of the lattice of atoms, which is gapless since a uniform shift of all atomic positions does not cost elastic energy. However, that shift costs potential energy. If the mass M or the elastic constant C are very small, thus β is large, the gapless acoustic mode must survive. In the opposite case, the potential pins the atoms, which can only oscillate in each potential well without drifting. We thus expect a transition as function of β from a phase at large β where the model is gapless into a phase at small β where the atoms localise and the spectrum has a gap.

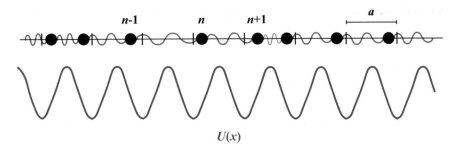

Fig. 6.5 A simple model whose Hamiltonian is equivalent to that in (6.78): L atoms of mass M on a one-dimensional rail are coupled to each other by springs whose equilibrium length and stiffness are a and C/a^2, respectively. Specifically, atom n is coupled to atoms $n - 1$ and $n + 1$. In addition, the atoms feel a periodic cosine potential with period a: $U(x) = -U \cos(x)$

The simplest way to identify such transition in the Hamiltonian (6.78) is recognising that in the gapless phase the precise value of the short-distance cutoff α, namely of the lattice spacing a in the model of Fig. 6.5, must not matter and therefore can be sent to zero provided the umklapp operator is properly normal-ordered, namely,

$$\frac{1}{\alpha^2} \cos \sqrt{16\pi}\ \Phi(x) = \frac{1}{\alpha^2} \cos \sqrt{16\pi K}\ \overline{\Phi}(x) = \frac{1}{\alpha^2} \cos 2\sqrt{K}\ \left(\overline{\phi}_R(x) + \overline{\phi}_L(x)\right)$$

$$= \alpha^{4K-2}\ :\cos \sqrt{16\pi K}\ \overline{\Phi}(x):\ .$$

$$(6.80)$$

In this way, we readily find that α can be safely sent to zero only if $K > 1/2$.

If, instead, $K < 1/2$, which may occur only if the interaction is sufficiently repulsive, the umklapp term grows in strength as $\alpha \to 0$, which implies that it is more appropriate first to minimise the potential, and next add quantum effects. Assuming a nearest neighbour repulsion, thus $V(q) = V \cos q$ and $g_3 \simeq V(\pi) = -V < 0$, the potential is minimum when

$$-\frac{g_3}{2\pi^2\alpha^2}\ \langle \cos \sqrt{16\pi}\ \Phi(x)\rangle \simeq \frac{V}{2\pi^2\alpha^2}\ \langle \cos \sqrt{16\pi}\ \Phi(x)\rangle = -1\,,$$

thus

$$\langle \sqrt{16\pi}\ \Phi(x)\rangle = \langle 2\big(\phi_R(x) + \phi_L(x)\big)\rangle = \big(2n + 1\big)\pi\,, \quad n \in \mathbb{Z}\,,$$

noticing that the uniform term $2\big(k_{RF} + k_{LF}\big)x = 2\pi x$ is also multiple of 2π because x is actually an integer ℓ in units of the lattice spacing.

Since $\phi_{R(L)}(x)$ is defined modulo 2π, there are actually only two inequivalent minima: $\sqrt{16\pi}\ \Phi(x) = \pm\pi$. That actually corresponds to the fact that, for $K < 1/2$, the fermions localise either on the even sites or on the odd ones, thus a double degenerate state. More specifically, the local density that includes both uniform component $1/2$ and staggered component at $2k_F = \pi$, is

$$
\begin{aligned}
\langle \rho(x = \ell) \rangle &= \frac{1}{2} + \langle \rho_{2k_F}(x) \rangle + \langle \rho_{-2k_F}(x) \rangle \\
&= \frac{1}{2} + \langle \frac{i}{2\pi\alpha} \, e^{-i\sqrt{4\pi}\,\Phi(x)} - \frac{i}{2\pi\alpha} \, e^{i\sqrt{4\pi}\,\Phi(x)} \rangle = \frac{1}{2} + \frac{1}{\pi\alpha} \langle \sin\left(\sqrt{4\pi}\,\Phi(x)\right) \rangle \\
&= \frac{1}{2} + \frac{e^{i\pi x}}{\pi\alpha} \, \sin\left(\pm\frac{\pi}{2}\right) = \frac{1}{2} \pm (-1)^{\ell} \, \delta\rho \,,
\end{aligned}
$$

(6.81)

thus a staggered modulation of the local density.

6.2.8 Behaviour Close to the $K = 1/2$ Marginal Case

We intentionally have not discussed the case of $K = 1/2$, where, according to (6.80) the umklapp term remains finite for $\alpha \to 0$, which implies it is a marginal operator. To assess whether it is marginally irrelevant or relevant, we write that

$$
K = \frac{1}{2} \sqrt{\frac{1 - g/2\pi v}{1 + g/2\pi v}} \,, \qquad |g| \ll 2\pi v \,,
$$

where g quantifies the deviation of K from $1/2$, and assume that also $g_3 \ll 2\pi v$. In (6.78) we perform the following canonical transformation

$$
\overline{\Phi}(x) = \sqrt{\frac{1}{2K}} \, \Phi'(x), \qquad \overline{\Pi}(x) = \sqrt{2K} \, \Pi'(x) \,,
$$

where

$$
2K \simeq 1 - \frac{g}{2\pi v} \,, \qquad \frac{1}{2K} \simeq 1 + \frac{g}{2\pi v} \,,
$$

so that the Hamiltonian becomes

$$
\begin{aligned}
H &= \frac{v}{2} \int dx \left[2K \, \Pi'(x)^2 + \frac{1}{2K} \left(\frac{\partial \Phi'(x)}{\partial x} \right)^2 \right] - \frac{g_3}{2\pi^2\alpha^2} \int dx \, \cos\sqrt{8\pi} \, \Phi'(x) \\
&\simeq \frac{v}{2} \int dx \left[\Pi'(x)^2 + \left(\frac{\partial \Phi'(x)}{\partial x} \right)^2 \right] \\
&\quad + \frac{g}{4\pi} \int dx \left[\left(\frac{\partial \Phi'(x)}{\partial x} \right)^2 - \Pi'(x)^2 \right] - \frac{g_3}{2\pi^2\alpha^2} \int dx \, \cos\sqrt{8\pi} \, \Phi'(x) \\
&\equiv H_0 + v \left\{ \frac{g}{4\pi v} \int dx \left[\left(\frac{\partial \Phi'(x)}{\partial x} \right)^2 - \Pi'(x)^2 \right] - \frac{g_3}{2\pi^2 v\alpha^2} \int dx \, \cos\sqrt{8\pi} \, \Phi'(x) \right\} \\
&\equiv H_0 + v \, \delta H \,.
\end{aligned}
$$

(6.82)

We shall treat δH in perturbation theory on the imaginary time-axis $ivt \to \tau$, with τ having dimension of a length. The S-operator is therefore

$$S_\alpha = T_\tau \left(e^{-\int_0^{v\beta} d\tau\, \delta H(\tau)} \right), \tag{6.83}$$

where $\delta H(\tau)$ is the imaginary time evolution with the unperturbed Hamiltonian H_0, and we explicitly indicate the short-distance cutoff α. The reason is that, in the 'semiclassical' approach we have adopted, α represents the indeterminacy of the position (x, τ), as we earlier mentioned. Therefore, when expanding the exponential in S_α, the multidimensional integral must be performed excluding regions where pair of coordinates (x, τ) and (x', τ') get closer than α. The procedure we shall follow will be to integrate out regions where the distance between any pair of coordinates (x, τ) and (x', τ') is greater than α but lower than $\lambda\alpha$, with $\lambda > 1$, making use of the operator product expansion of Sect. 6.2.3. The outcome of such partial integration will be interpreted in terms of a new effective $S_{\lambda\alpha}$ defined with a larger short-distance cutoff $\lambda\alpha$. Since $\tau \in [0, v\beta]$ and $x \in [0, L]$, should we iterate that procedure till $\lambda\alpha \sim v\beta \simeq L$ without encountering any singularity, we would end up in a theory where perturbation theory is well behaved. Put differently, we mentioned that perturbation theory in one dimension is ill defined because of the proliferation of logarithmic singularities like $\ln vk_F\beta \simeq \ln v\beta/\alpha$ at finite temperature, or $\ln L/\alpha$ on a finite size. Evidently, should $\alpha \sim v\beta \simeq L$, the logarithms would be vanishingly small and perturbation theory could be safely performed. This procedure essentially implements a Renormalisation Group (RG) scheme, here akin a real space decimation.

In imaginary time, $\phi'_R(x, \tau) + \phi'_L(x, \tau) = \phi'_R(x + i\tau) + \phi'_L(x - i\tau)$. We can therefore define new variables $z \equiv x + i\tau$ and $\bar{z} \equiv z^* = x - i\tau$, so that $\phi'_R(x + i\tau) \equiv \phi'_R(z)$ is holomorphic, while $\phi'_L(x - i\tau) \equiv \phi'_L(\bar{z})$ antiholomorphic. It follows that

$$\Phi'(z, \bar{z}) = \frac{1}{\sqrt{4\pi}} \left(\phi'_R(z) + \phi'_L(\bar{z}) \right),$$

so that

$$\frac{\partial \Phi'(z, \bar{z})}{\partial z} = \frac{1}{\sqrt{4\pi}} \frac{\partial \phi_R(z)}{\partial z}, \qquad \frac{\partial \Phi'(z, \bar{z})}{\partial \bar{z}} = \frac{1}{\sqrt{4\pi}} \frac{\partial \phi_L(\bar{z})}{\partial \bar{z}}.$$

Therefore

$$\left(\frac{\partial \Phi'(x, \tau)}{\partial x} \right)^2 - \Pi'(x, \tau)^2 = \frac{1}{\pi} \frac{\partial \phi_R(x, \tau)}{\partial x} \frac{\partial \phi_L(x, \tau)}{\partial x} = \frac{1}{\pi} \frac{\partial \phi_R(z)}{\partial z} \frac{\partial \phi_L(\bar{z})}{\partial \bar{z}}$$

$$= 4 \frac{\partial \Phi'(z, \bar{z})}{\partial z} \frac{\partial \Phi'(z, \bar{z})}{\partial \bar{z}},$$

and thus the first order correction of the S-operator can be simply written as

$$
S_\alpha^{(1)} = -\int_0^{\upsilon\beta} d\tau \, \delta H(\tau) \equiv -\frac{i}{2} \int dz \, d\bar{z} \, \delta H(z, \bar{z})
$$

$$
= -\frac{i}{2} \int dz \, d\bar{z} \left\{ \frac{g}{\pi\upsilon} \frac{\partial \Phi'(z, \bar{z})}{\partial z} \frac{\partial \Phi'(z, \bar{z})}{\partial \bar{z}} - \frac{g_3}{2\pi^2\upsilon\alpha^2} \cos\sqrt{8\pi} \, \Phi'(z, \bar{z}) \right\}.
$$

$$(6.84)$$

We observe that $S_\alpha^{(1)}$ is in reality independent of α, since $\cos\sqrt{8\pi} \; \Phi'(z, \bar{z})/\alpha^2 =:$ $\cos\sqrt{8\pi} \; \Phi'(z, \bar{z})$:, and the normal ordered product is what really enters the perturbative calculations.

To simplify notations, we shall drop the apex and redefine $\Phi' \to \Phi$ and $\Pi' \to \Pi$, as well as

$$
\frac{g}{2\pi\upsilon} \to g, \qquad \frac{g_3}{2\pi\upsilon} \to g_3,
$$

and finally write $\delta H(z, \bar{z}) = \delta H_2(z, \bar{z}) + \delta H_3(z, \bar{z})$ where

$$
\delta H_2(z, \bar{z}) = 2g \frac{\partial \Phi(z, \bar{z})}{\partial z} \frac{\partial \Phi(z, \bar{z})}{\partial \bar{z}}, \qquad \delta H_3(z, \bar{z}) = -\frac{g_3}{\pi\alpha^2} \cos\sqrt{8\pi} \; \Phi(z, \bar{z}).
$$

The second order term, fixing a specific time order, is instead

$$
S_\alpha^{(2)} = -\frac{1}{4} \int dz \, d\bar{z} \, dz' \, d\bar{z}' \; \theta(\tau - \tau') \, \delta H(z, \bar{z}) \, \delta H(z', \bar{z}'). \qquad (6.85)
$$

As discussed above, we divide $S_\alpha^{(2)}$ into a component $S_{\lambda\alpha}^{(2)}$ where $|z - z'| = |\bar{z} - \bar{z}'| > \lambda\alpha > \alpha$, and another $\delta S^{(2)}$ where $\alpha < |z - z'| = |\bar{z} - \bar{z}'| \le \lambda\alpha$, in which case we make use of the operator product expansion. Specifically, in $\delta S^{(2)}$ we set $z' = z + \epsilon e^{i\theta}$ and $\bar{z}' = \bar{z} + \epsilon e^{-i\theta}$ where $\theta \in [-\pi, 0]$ since $\tau > \tau'$ in the time-ordered equation (6.85), and $\epsilon \in [\alpha, \lambda\alpha]$. That implies a phase space vanishing as ϵ^2, so that only singular terms diverging as least as $1/\epsilon^2$ can contribute to $\delta S^{(2)}$.

To accomplish our goal, we first derive some preliminary results. Since

$$
\left[\phi_R^{(+)}(z), \phi_R^{(-)}(z') \right] = \ln\left(\alpha + i(z - z') \right), \qquad \left[\phi_L^{(+)}(\bar{z}), \phi_L^{(-)}(\bar{z}') \right] = \ln\left(\alpha - i(\bar{z} - \bar{z}') \right),
$$

the following expressions readily follow for $z \sim z'$

$$
e^{\pm i\sqrt{2} \, \phi_R(z)} \, e^{\mp i\sqrt{2} \, \phi_R(z')} = \frac{\alpha^2}{\left(\alpha - i(z - z') \right)^2} \; : e^{\pm i\sqrt{2} \left(\phi_R(z) - \phi_R(z') \right)} :
$$

$$
\simeq \frac{\alpha^2}{\left(\alpha - i(z - z') \right)^2} \left(1 \pm i\sqrt{2} \, \frac{\partial \phi_R(z)}{\partial z} (z - z') \right)
$$

$$
= \frac{\alpha^2}{\left(\alpha - i(z - z') \right)^2} \left(1 \pm i\sqrt{8\pi} \, \frac{\partial \Phi(z, \bar{z})}{\partial z} (z - z') \right),
$$

$$\frac{\partial \phi_R(z)}{\partial z} e^{\pm i\sqrt{2}\, \phi_R(z')} = \; : \frac{\partial \phi_R(z)}{\partial z} e^{\pm i\sqrt{2}\, \phi_R(z')} : \; \mp \frac{\sqrt{2}}{\alpha - i(z - z')} e^{\pm i\sqrt{2}\, \phi_R(z')} \, ,$$

$$e^{\pm i\sqrt{2}\, \phi_L(\bar{z})} e^{\mp i\sqrt{2}\, \phi_L(\bar{z}')} = \frac{\alpha^2}{\left(\alpha + i(\bar{z} - \bar{z}')\right)^2} \; : e^{\pm i\sqrt{2}\, \left(\phi_L(\bar{z}) - \phi_L(\bar{z}')\right)} :$$

$$\simeq \frac{\alpha^2}{\left(\alpha + i(\bar{z} - \bar{z}')\right)^2} \left(1 \pm i\sqrt{2}\, \frac{\partial \phi_L(\bar{z})}{\partial \bar{z}} (\bar{z} - \bar{z}')\right)$$

$$= \frac{\alpha^2}{\left(\alpha + i(\bar{z} - \bar{z}')\right)^2} \left(1 \pm i\sqrt{8\pi}\, \frac{\partial \Phi(z, \bar{z})}{\partial \bar{z}} (\bar{z} - \bar{z}')\right) \, ,$$

$$\frac{\partial \phi_L(\bar{z})}{\partial \bar{z}} e^{\pm i\sqrt{2}\, \phi_L(\bar{z}')} = \; : \frac{\partial \phi_L(\bar{z})}{\partial \bar{z}} e^{\pm i\sqrt{2}\, \phi_L(\bar{z}')} : \; \pm \frac{\sqrt{2}}{\alpha + i(\bar{z} - \bar{z}')} e^{\pm i\sqrt{2}\, \phi_L(\bar{z}')} \, .$$

$$(6.86)$$

By means of the above results, we readily find that for $z = z' + \epsilon e^{i\theta}$, with the OPE prescription and dropping constant terms as well as terms non-singular as $\epsilon \to 0$,

$$\delta H_3(z, \bar{z}) \, \delta H_3(z', \bar{z}') \simeq \frac{g_3^2}{2\pi^2 \alpha^4} \cos\sqrt{32\pi}\, \Phi(z, \bar{z}) + \frac{g_3^2}{4\pi^2} \frac{1}{\left(z - z'\right)^2 \left(\bar{z} - \bar{z}'\right)^2}$$

$$\left\{ \left(1 + i\sqrt{8\pi}\, \frac{\partial \Phi(z, \bar{z})}{\partial z} (z - z')\right) \left(1 + i\sqrt{8\pi}\, \frac{\partial \Phi(z, \bar{z})}{\partial \bar{z}} (\bar{z} - \bar{z}')\right) \right.$$

$$\left. + \left(1 - i\sqrt{8\pi}\, \frac{\partial \Phi(z, \bar{z})}{\partial z} (z - z')\right) \left(1 - i\sqrt{8\pi}\, \frac{\partial \Phi(z, \bar{z})}{\partial \bar{z}} (\bar{z} - \bar{z}')\right) \right\}$$

$$\simeq \frac{g_3^2}{2\pi^2 \alpha^4} \cos\sqrt{32\pi}\, \Phi(z, \bar{z}) - \frac{4g_3^2}{\pi} \frac{1}{\epsilon^2} \frac{\partial \Phi(z, \bar{z})}{\partial z} \frac{\partial \Phi(z, \bar{z})}{\partial \bar{z}} \simeq -\frac{4g_3^2}{\pi} \frac{1}{\epsilon^2} \frac{\partial \Phi(z, \bar{z})}{\partial z} \frac{\partial \Phi(z, \bar{z})}{\partial \bar{z}} \, .$$

$$(6.87)$$

Similarly,

$$\delta H_2(z, \bar{z}) \, \delta H_2(z', \bar{z}') \sim O(\epsilon^0) \simeq 0 \, ,$$

$$\delta H_2(z, \bar{z}) \, \delta H_3(z', \bar{z}') + \delta H_3(z, \bar{z}) \, \delta H_2(z', \bar{z}') = \frac{2\, g g_3}{\pi^2 \alpha^2} \frac{1}{\epsilon^2} \cos\sqrt{8\pi}\, \Phi(z, \bar{z})$$

We still have to integrate over ϵ and θ:

$$\int dz' d\bar{z}' \frac{1}{\epsilon^2} = -2i \int_\alpha^{\lambda\alpha} \epsilon\, d\epsilon \frac{1}{\epsilon^2} \int_{-\pi}^0 d\theta = -2\pi i\, \ln\lambda \, ,$$

so that, finally,

$$\delta S^{(2)} = -\frac{i}{2}\, \ln\lambda \int dz\, d\bar{z} \left(4g_3^2\, \frac{\partial \Phi(z, \bar{z})}{\partial z} \frac{\partial \Phi(z, \bar{z})}{\partial \bar{z}} - \frac{2\, g g_3}{\pi\alpha^2} \cos\sqrt{8\pi}\, \Phi(z, \bar{z})\right) .$$

$$(6.88)$$

In conclusion, we have found that

$$S_\alpha = S_\alpha^{(1)} + S_\alpha^{(2)} + \cdots = S_\alpha^{(1)} + \delta S^{(2)} + S_{\lambda\alpha}^{(2)} + \cdots ,$$

which allows defining $S_{\lambda\alpha}^{(1)} \equiv S_\alpha^{(1)} + \delta S^{(2)}$. Comparing (6.88) with (6.84) we can formally define up to second order the new coupling constants that refer to the larger short-distance cutoff $\lambda\alpha$,

$$g + \delta g = g + 2g_3^2 \ln \lambda , \quad g_3 + \delta g_3 = g_3 + 2g \, g_3 \ln \lambda .$$

Similarly, writing $\lambda = e^s$ and denoting as $g(s)$ and $g_3(s)$ the coupling constants with short-distance cutoff $e^s \alpha$, we could repeat the same procedure of integrating over the regions $e^s \alpha < |z - z'| < e^{s+\delta s} \alpha$, and thus find

$$g(s + \delta s) = g(s) + 2g_3(s)^2 \, \delta s , \quad g_3(s + \delta s) = g_3(s) + 2g(s) \, g_3(s) \, \delta s ,$$

which can be recast as the following differential equations

$$\frac{\partial g(s)}{\partial s} = 2g_3(s)^2 , \quad \frac{\partial g_3(s)}{\partial s} = 2g(s) \, g_3(s) , \tag{6.89}$$

with boundary conditions the bare coupling constants $g(0) = g$ and $g_3(0) = g_3$.[1] We observe that (6.89) is invariant under $g_3 \to -g_3$. That is not surprising since

---

[1] We emphasise that the validity of the equations (6.89) is not guaranteed by the second order calculation we performed. Indeed, suppose that $s \ll 1$, and we write

$$g(s) \simeq \sum_{n=0}^{\infty} g^{(n)} s^n , \quad g_3(s) \simeq \sum_{n=0}^{\infty} g_3^{(n)} s^n ,$$

where $g^{(n)}$ and $g_3^{(n)}$ derive from $S_\alpha^{(n+1)}$, then, if (6.89) were exact, for $n \geq 0$

$$(n+1) \, g^{(n+1)} = 2 \sum_{m=0}^{n} g_3^{(n-m)} \, g_3^{(m)} , \quad (n+1) \, g_3^{(n+1)} = 2 \sum_{m=0}^{n} g^{(n-m)} \, g_3^{(m)} ,$$

implying that higher order terms are all determined from the first order correction we have calculated. For instance, the third order $S_\alpha^{(3)}$ yields corrections $\propto \ln^2 \lambda$ which must correspond to

$$\delta g = 4 \, g(0) \, g_3(0)^2 \ln^2 \lambda , \quad \delta g_3 = 2 \, g_3(0)^3 \ln^2 \lambda + 2 \, g(0)^2 \, g_3(0) \ln^2 \lambda .$$

If that indeed happens at all orders, we say that the theory is renormalisable. In the present case, renormalisability can be indeed proven at few orders in perturbation theory. We shall assume it does hold at any order. We also note that the right hand sides of the equations (6.89) are just the leading order terms. In general,

$$\frac{\partial g(s)}{\partial s} = \beta\big(g(s), g_3(s)\big) , \quad \frac{\partial g_3(s)}{\partial s} = \beta_3\big(g(s), g_3(s)\big) ,$$

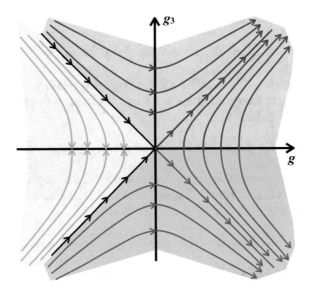

Fig. 6.6 Renormalisation group flow corresponding to the equations (6.89). Starting from the initial values of $g(0) = g$ and $g_3(0) = g_3$, the solutions $g(s)$ and $g_3(s)$ move along the paths that are shown as s grows. Any initial point within the blue coloured region flows at $g(\infty) = g_3(\infty) = \infty$, while within the red coloured one at $g(\infty) = g_3(\infty) = -\infty$. All initial points within the green coloured region flow to $g_3(\infty) = 0$ and $g(\infty)$ finite. We remark that the horizontal axis $g_3 = 0$ corresponds to a gapless system, and is stable for $g < 0$ and unstable otherwise

$g_3 \rightarrow -g_3$ under the unitary transformation $\Psi_L(x) \rightarrow i\,\Psi_L(x)$. Therefore the RG flow is independent of the sign of g_3, though the physical properties do depend on it, as we discuss later. The equations in (6.89) can be readily solved noticing that

$$\frac{\partial^2 g(s)}{\partial s^2} = 4\,g_3(s)^2\,g(s) = 2\,g(s)\,\frac{\partial g(s)}{\partial s} = \frac{\partial g(s)}{\partial s}^2\,.$$

We shall not explicitly solve the above simple equation, but just show the renormalisation group flow diagram in Fig. 6.6. For positive g, thus $K < 1/2$, both coupling constants flows to strong coupling, as expected from (6.80). However, the flow is still towards strong coupling even if $g < 0$ provided $|g_3| > -g$.

We already discussed the fixed point at $g_3 \rightarrow -\infty$, which corresponds to a pinned CDW where the fermions occupy either the even or the odd sites, see top panels in Fig. 6.7. The fixed point at $g_3 \rightarrow +\infty$ implies instead that

$$-\frac{|g_3|}{2\pi^2\alpha^2}\,\langle\cos\sqrt{16\pi}\,\Phi(x)\rangle = -1\,,$$

---

where β and β_3 can be expanded in powers of the coupling constants. For instance, $S_\alpha^{(3)}$ contains terms $\sim g^3 \ln \lambda$ that contribute to the higher order corrections of β. Equation (6.89) is therefore valid provided $g, g_3 \ll 1$.

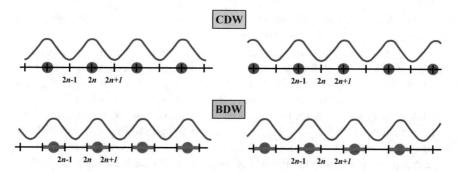

Fig. 6.7 Charge density wave (CDW), top panels, versus bond density wave (BDW), bottom panels. The density profile is the blue curve, and is peaked when there is excess charge, red dots. The BDW is associate to spontaneous dimerisation where the bond strengths alternate along the chain, strong bonds in green and weak one in black

thus

$$\langle \sqrt{16\pi}\, \Phi(x) \rangle = \langle 2\big(\phi_R(x) + \phi_L(x)\big) \rangle = 2n\pi, \quad n \in \mathbb{Z},$$

and still admits only two inequivalent minima at $n = 0, 1$. This state actually corresponds to a spontaneously dimerised chain, or bond density wave (BDW), where the density is peaked in the middle of a bond, and minimum in the nearest neighbour ones, see bottom panels in Fig. 6.7. The two minima at $n = 0, 1$ correspond to the two dimerisation patterns shown in that figure. To better understand the difference between CDW and BDW from the bosonization viewpoint, let us redefine $\phi_{R(L)}(x) \to k_F x + \phi_{R(L)}(x)$ so that $\phi_{R(L)}(x)$ are now slowly varying fields with vanishing space average. It follows that the Fermi fields become

$$\Psi_R(x) \sim e^{ik_F x}\, e^{i\phi_R(x)} = e^{i\frac{\pi}{2}x}\, e^{i\phi_R(x)}, \quad \Psi_L(x) \sim e^{-ik_F x}\, e^{-i\phi_L(x)} = e^{-i\frac{\pi}{2}x}\, e^{-i\phi_L(x)}.$$

The CDW, $\Delta_{CDW}(x)$, and BDW, $\Delta_{BDW}(x)$, order parameters are the slowly varying (s-v) components of $\cos(\pi x)\, \rho(x)$ and $\sin(\pi x)\, \rho(x)$, as one can easily realise from Fig. 6.7. Therefore

$$\Delta_{CDW}(x) = \Big[\cos(\pi x)\, \rho(x)\Big]_{s\text{-}v} \simeq \Big[\cos(\pi x)\,\big(\Psi_R^\dagger(x)\,\Psi_L(x) + \Psi_L^\dagger(x)\,\Psi_R(x)\big)\Big]_{s\text{-}v}$$
$$= \frac{1}{2\pi\alpha}\Big(e^{-i\phi_R(x)}\, e^{-i\phi_L(x)} + e^{i\phi_L(x)}\, e^{i\phi_R(x)}\Big) = \frac{1}{\pi\alpha}\,\sin\sqrt{4\pi}\,\Phi(x),$$

$$\Delta_{BDW}(x) = \Big[\sin(\pi x)\, \rho(x)\Big]_{s\text{-}v} \simeq \Big[\sin(\pi x)\,\big(\Psi_R^\dagger(x)\,\Psi_L(x) + \Psi_L^\dagger(x)\,\Psi_R(x)\big)\Big]_{s\text{-}v} \tag{6.90}$$
$$= \frac{1}{2\pi\alpha}\Big(-i\, e^{-i\phi_R(x)}\, e^{-i\phi_L(x)} + i\, e^{i\phi_L(x)}\, e^{i\phi_R(x)}\Big) = \frac{1}{\pi\alpha}\,\cos\sqrt{4\pi}\,\Phi(x).$$

We recall that the CDW state is characterised by $\sqrt{16\pi}\,\Phi(x) = 2\sqrt{4\pi}\,\Phi(x) = \pm\pi$, and indeed $\Delta_{CDW}(x) \sim \sin\sqrt{4\pi}\,\Phi(x) = \pm 1$ is finite, while $\Delta_{BDW}(x) \sim \cos\sqrt{4\pi}\,\Phi(x) = 0$. On the contrary, the BDW state is characterised by $\sqrt{16\pi}\,\Phi(x)$

$= 2\sqrt{4\pi}\ \Phi(x) = 0, 2\pi$, and indeed $\Delta_{CDW}(x) \sim \sin\sqrt{4\pi}\ \Phi(x) = 0$ is zero, while $\Delta_{BDW}(x) \sim \cos\sqrt{4\pi}\ \Phi(x) = \pm 1$ is finite.

Let us end discussing the transition from the green to either the red or blue regions in the RG flow diagram, Fig. 6.6. The transition point corresponds to any initial g and g_3 along the black lines in Fig. 6.6, i.e., $g_3 = g$ or $g_3 = -g$ with $g < 0$. Along these lines, the flow tends towards $g(\infty) = g_3(\infty) = 0$, which, in the original model, corresponds to $K(\infty) = 1/2$. On the contrary, beyond the transition, the system develops a single particle gap, so that the compressibility vanishes and thus $K(\infty) = 0$, see (6.72). The jump of $K(\infty)$ is one of the features of such transition, which actually belongs to the Berezinskii-Kosterlitz-Thouless universality class.

For simplicity, let us consider an initial point $g_3 = -g$, in which case such relationship persists along the flow so that the RG equation (6.89) simplifies into

$$\frac{\partial g(s)}{\partial s} = 2g(s)^2 \Rightarrow \int_0^s \frac{dg}{g^2} = 2s \Rightarrow g(s) = \frac{g}{1 - 2gs} = \frac{g}{1 - 2g\ln\lambda}.$$

Indeed, if $g < 0$, $g(s) \to 0$ for $s \to \infty$. On the contrary, if $0 \lesssim g \ll 1$, $g(s)$ diverges at λ corresponding to a length scale $\xi = \alpha\,e^{1/2g}$, which is actually the correlation length in the CDW phase.

6.2.8.1 $SU(2)$ Invariant Line

The generic spinless fermion Hamiltonian (6.6) plus (6.7) is invariant under the global charge $U(1)$ symmetry. Here we show that the line $g_3 = g$ in the flow diagram of Fig. 6.6 actually corresponds to models with enlarged $SU(2)$ symmetry.

Let us consider again (6.86) but now at equal times $\tau = \tau'$, and $x \sim y$

$$\frac{1}{4\pi^2\alpha^2}\left[e^{i\sqrt{2}\,\phi_R(x)}, e^{-i\sqrt{2}\,\phi_R(y)}\right] = \frac{1}{4\pi^2}\left[\left(\frac{\alpha + i(x-y)}{(x-y)^2 + \alpha^2}\right)^2 - \left(\frac{\alpha - i(x-y)}{(x-y)^2 + \alpha^2}\right)^2\right]$$
$$\left(1 + i\sqrt{2}\,\frac{\partial\phi_R(x)}{\partial x}(x-y)\right)$$
$$= i\frac{1}{2\pi}\frac{x-y}{(x-y)^2 + \alpha^2}\,\delta_\alpha(x-y)\left(1 + i\sqrt{2}\,\frac{\partial\phi_R(x)}{\partial x}(x-y)\right)$$
$$\simeq -\delta(x-y)\,2\,\frac{\sqrt{2}}{4\pi}\,\frac{\partial\phi_R(x)}{\partial x},$$
$$\left[\frac{\sqrt{2}}{4\pi}\frac{\partial\phi_R(x)}{\partial x}, \frac{e^{i\sqrt{2}\,\phi_R(x)}}{2\pi\alpha}\right] = \frac{1}{4\pi^2\alpha}\left\{\frac{-1}{\alpha - i(x-y)} + \frac{1}{\alpha + i(x-y)}\right\}e^{i\sqrt{2}\,\phi_R(y)}$$
$$= -\delta(x-y)\,\frac{e^{i\sqrt{2}\,\phi_R(x)}}{2\pi\alpha}.$$

We note that the operators

$$J_{zR}(x) \equiv \frac{\sqrt{2}}{4\pi}\frac{\partial\phi_R(x)}{\partial x}, \quad J_R^+(x) \equiv i\,\frac{e^{-i\sqrt{2}\,\phi_R(x)}}{2\pi\alpha}, \quad J_R^-(x) \equiv -i\,\frac{e^{i\sqrt{2}\,\phi_R(x)}}{2\pi\alpha},$$

(6.91)

realise an $SU(2)$ algebra of right-moving fields. Similarly,

$$J_{zL}(x) \equiv \frac{\sqrt{2}}{4\pi} \frac{\partial \phi_L(x)}{\partial x} , \quad J_L^+(x) \equiv i \frac{e^{i\sqrt{2}\,\phi_L(x)}}{2\pi\alpha} , \quad J_L^-(x) \equiv -i \frac{e^{-i\sqrt{2}\,\phi_L(x)}}{2\pi\alpha} ,$$

(6.92)

do the same but for left-moving ones. We also note that, using the OPE,

$$\frac{1}{2}\Big(J_{R(L)}^+(x)\, J_{R(L)}^-(x) + J_{R(L)}^-(x)\, J_{R(L)}^+(x) \Big) = \frac{1}{4\pi^2}\left(\frac{\partial \phi_{R(L)}(x)}{\partial x}\right)^2 = 2\,J_{zR(L)}(x)^2 ,$$

so that

$$\frac{v}{4\pi}\left(\frac{\partial \phi_{R(L)}(x)}{\partial x}\right)^2 = 2\pi\,\frac{v}{3}\, \boldsymbol{J}_{R(L)}(x) \cdot \boldsymbol{J}_{R(L)}(x) ,$$

where $\boldsymbol{J}_{R(L)}(x) = \big(J_{xR(L)}(x), J_{yR(L)}(x), J_{zR(L)}(x) \big)$, and, as usual, $J_{R(L)}^{\pm}(x) = J_{xR(L)}(x) \pm i\, J_{yR(L)}(x)$. Therefore, H_0 in (6.82) can be simply written as

$$H_0 = 2\pi\,\frac{v}{3}\int dx\,\Big(\boldsymbol{J}_R(x) \cdot \boldsymbol{J}_R(x) + \boldsymbol{J}_L(x) \cdot \boldsymbol{J}_L(x) \Big),$$

(6.93)

which makes the $SU(2)$ symmetry explicit. In addition,[2]

$$\frac{g}{4\pi^2}\frac{\partial \phi_R(x)}{\partial x}\frac{\partial \phi_L(x)}{\partial x} = 2g\,J_{Rz}(x)\,J_{Lz}(x) ,$$

$$-\frac{g_3}{2\pi^2\alpha^2}\cos\sqrt{2}\big(\phi_R(x) + \phi_L(x)\big) = g_3\left(J_R^+(x)\, J_L^-(x) + J_R^-(x)\, J^+L(x) \right)$$

$$= 2g_3\left(J_{Rx}(x)\, J_{Lx}(x) + J_{Ry}(x)\, J_{Ly}(x) \right),$$

(6.94)

so that the Hamiltonian (6.82) can be written as

$$H = 2\pi\,\frac{v}{3}\int dx\,\Big(\boldsymbol{J}_R(x) \cdot \boldsymbol{J}_R(x) + \boldsymbol{J}_L(x) \cdot \boldsymbol{J}_L(x) \Big)$$

$$+ 2\int dx\,\Big[g\,J_{Rz}(x)\,J_{Lz}(x) + g_3\big(J_{Rx}(x)\, J_{Lx}(x) + J_{Ry}(x)\, J_{Ly}(x) \big) \Big],$$

(6.95)

and is $SU(2)$ symmetric if $g_3 = g$. Moreover, if $g = g_3 < 0$, the black line in the lower half-plane of Fig. 6.6, the model asymptotically flows to $g(\infty) = g_3(\infty) = 0$, which thus describes a gapless $SU(2)$ invariant phase right at $K(\infty) = 1/2$. If, on the

---

[2] Note that

$$e^{-i\sqrt{2}\,\phi_R(x)}\, e^{-i\sqrt{2}\,\phi_L(x)} = e^{-i\sqrt{2}\big(\phi_R(x)+\phi_L(x)\big)}\, e^{-2\big[\phi_R(x),\phi_L(x)\big]/2} = -e^{-i\sqrt{2}\big(\phi_R(x)+\phi_L(x)\big)} .$$

contrary, $g = g_3 > 0$, the model flows to $g(\infty) = g_3(\infty) = \infty$. As we discussed, this point corresponds to a BDW which, in spin language, is commonly known as a spin Peierls state where the strong bonds in Fig. 6.7 represent pairs of spins coupled into singlets.

The line $g_3 = -g$, though behaves under RG as that with $g = g_3$, corresponds instead to an anisotropic system that still flows to the gapless $SU(2)$ point if $g < 0$, but to $g(\infty) = -g_3(\infty) = -\infty$ for $g > 0$.

6.2.8.2 Luther-Emery Point

For initial g and g_3 within the red and blue regions in Fig. 6.6, the RG equations (6.89) predict that g flows to infinity for any $g_3 \neq 0$. Therefore, g will unavoidably reach along its evolution a value $g_*/2\pi v = 3/5$ such that

$$K_* = \frac{1}{2}\sqrt{\frac{1 - g_*/2\pi v}{1 + g_*/2\pi v}} = \frac{1}{4} .$$

At this point, known as the Luther-Emery point, the Hamiltonian becomes, see (6.78),

$$H_* = \frac{v_*}{2}\int dx \left[\overline{\Pi}(x)^2 + \left(\frac{\partial\overline{\Phi}(x)}{\partial x}\right)^2 \right] - \frac{g_{3*}}{2\pi^2\alpha^2}\int dx \cos\sqrt{4\pi}\,\overline{\Phi}(x)$$

$$= \frac{v_*}{4\pi}\int dx \left[\left(\frac{\partial\overline{\phi}_R(x)}{\partial x}\right)^2 + \left(\frac{\partial\overline{\phi}_L(x)}{\partial x}\right)^2 \right] - \frac{g_{3*}}{2\pi^2\alpha^2}\int dx \cos\left(\overline{\phi}_R(x) + \overline{\phi}_L(x)\right),$$
$$(6.96)$$

where v_* and g_{3*} are the renormalised velocity and umklapp strength at g_*. We observe that $\overline{\phi}_R(x) + \overline{\phi}_L(x) = 2(\phi_R(x) + \phi_L(x))$, which implies that $\overline{\phi}_{R(L)}(x)$ has a Fermi momentum $\overline{k}_{R(L)F} = 2k_{R(L)F} = \pi$. If we define new Fermi fields

$$\overline{\Psi}_R(x) = \frac{1}{\sqrt{2\pi\alpha}}\,e^{i\overline{\phi}_R(x)} , \quad \overline{\Psi}_L(x) = \frac{1}{\sqrt{2\pi\alpha}}\,e^{-i\overline{\phi}_L(x)} , \qquad (6.97)$$

assuming that $\left[\overline{\phi}_R(x), \overline{\phi}_L(y)\right] = -i\pi$, then

$$\frac{1}{2\pi^2\alpha^2}\cos\left(\overline{\phi}_R(x) + \overline{\phi}_L(x)\right) = -i\,\overline{\Psi}_R^\dagger(x)\,\overline{\Psi}_L(x) + i\,\overline{\Psi}_L^\dagger(x)\,\overline{\Psi}_R(x).$$

Therefore, through (6.41), the Hamiltonian can be written as

$$H_* = -i\,v_*\int dx \left(\overline{\Psi}_R^\dagger(x)\,\frac{\partial\overline{\Psi}_R(x)}{\partial x} - \overline{\Psi}_L^\dagger(x)\,\frac{\partial\overline{\Psi}_L(x)}{\partial x}\right)$$

$$+ i\,g_3\int dx \left(\overline{\Psi}_R^\dagger(x)\,\overline{\Psi}_L(x) - \overline{\Psi}_L^\dagger(x)\,\overline{\Psi}_R(x)\right),$$

which reads in momentum space

$$H_* = \sum_k \left[v_* k \left(d^\dagger_{Rk+\bar{k}_{RF}} d_{Rk+\bar{k}_{RF}} - d^\dagger_{Lk-\bar{k}_{LF}} d_{Lk-\bar{k}_{LF}} \right) \right.$$
$$\left. + i \, g_3 \left(d^\dagger_{Rk+\bar{k}_{RF}} d_{Lk-\bar{k}_{LF}} - d^\dagger_{Lk-\bar{k}_{LF}} d_{Rk+\bar{k}_{RF}} \right) \right]. \tag{6.98}$$

The Hamiltonian (6.98) describes an insulator with conduction and valence band dispersions, respectively,

$$E_c(k) = \sqrt{v_*^2 k^2 + g_3^2} \,, \quad E_v(k) = -\sqrt{v_*^2 k^2 + g_3^2} \,.$$

In the spirit of the renormalisation group, the Hamiltonian (6.98) should be representatives of all Hamiltonians with parameters inside the red and blue regions of Fig. 6.6.

6.3   Spin-1/2 Heisenberg Chain

Let us now consider an anisotropic spin-1/2 Heisenberg model in one dimension,

$$H = J \sum_{i=1}^{L-1} \left(S_{ix} S_{i+1x} + S_{iy} S_{i+1y} + \Delta S_{iz} S_{i+1z} \right), \tag{6.99}$$

where Δ quantifies the anisotropy. The Hamiltonian for $\Delta \neq \pm 1$ has only $U(1)$ symmetry of rotations around the spin z-axis. On the contrary, at $\Delta = \pm 1$ the symmetry enlarges to $SU(2)$, and the model becomes the isotropic Heisenberg antiferromagnet for $\Delta = 1$ and ferromagnet for $\Delta = -1$.[3]

We can map the Hamiltonian (6.99) onto a spinless fermion model via the so called Jordan-Wigner transformation. Specifically, we write

$$S_{iz} = \frac{1}{2} - c_i^\dagger c_i = \frac{1}{2} - n_i \,, \tag{6.100}$$

so that $S_{iz} = 1/2$ and $S_{iz} = -1/2$ correspond, respectively, to an empty and occupied site. We note that

$$S_i^+ = (-1)^i \, \xi_i \, c_i \,, \quad S_i^- = (-1)^i \, c_i^\dagger \, \xi_i \,, \tag{6.101}$$

---

[3] We observe that it is always possible to change the sign of $S_{ix} S_{i+1x} + S_{iy} S_{i+1y}$ by a staggered rotation around z such that

$$S_{ix} \to (-1)^i \, S_{ix} \,, \quad S_{iy} \to (-1)^i \, S_{iy} \,.$$

are good representation for spin operators, where the 'string' operator is defined as

$$\xi_i \equiv e^{i\pi \sum_{j=1}^{i-1} n_j} = \prod_{j=1}^{i-1} (1 - 2n_j) \tag{6.102}$$

Indeed, since $\xi_i^2 = 1$ and $\left[\xi_i, c_i^\dagger c_i\right] = \left[\xi_i, c_i c_i^\dagger\right] = 0$, then

$$\left[S_i^+, S_i^-\right] = c_i c_i^\dagger - c_i^\dagger c_i = 1 - 2n_i = 2S_{iz}, \quad \left[S_i^+, S_{iz}\right] = -(-1)^i \xi_i \left[c_i, n_i\right] = -S_i^\dagger.$$

Moreover, noticing that, for $i > j, \xi_i c_j = -c_j \xi_i$ while, for $i \le j, \xi_i c_j = c_j \xi_i$, one can readily verify that the operators (6.101) commute at different sites. Let us rewrite the Hamiltonian (6.99) in terms of the above representation of the spin operators. We note that

$$S_{ix} S_{i+1x} + S_{iy} S_{i+1y} = \frac{1}{2} \left(S_i^+ S_{i+1}^- + S_i^- S_{i+1}^+\right) = -\frac{1}{2} \left(\xi_i c_i \, c_{i+1}^\dagger \xi_{i+1} + c_i^\dagger \xi_i \, \xi_{i+1} c_{i+1}\right)$$

$$= -\frac{1}{2} \left(c_i \, (1 - 2n_i) c_{i+1}^\dagger + c_i^\dagger \, (1 - 2n_i) c_{i+1}\right) = -\frac{1}{2} \left(c_i^\dagger c_{i+1} + c_{i+1}^\dagger c_i\right),$$

$$S_{iz} S_{i+1z} = \left(\frac{1}{2} - n_i\right)\left(\frac{1}{2} - n_{i+1}\right) = \frac{1}{4} + n_i n_{i+1} - \frac{1}{2}n_i - \frac{1}{2}n_{i+1}.$$

Therefore, the Hamiltonian (6.99) can be also written as

$$H = -\frac{J}{2} \sum_i \left(c_i^\dagger c_{i+1} + c_{i+1}^\dagger c_i\right) + J \Delta \sum_i \left(\frac{1}{4} + n_i n_{i+1} - \frac{1}{2}n_i - \frac{1}{2}n_{i+1}\right)$$

$$\equiv -\frac{J}{2} \sum_i \left(c_i^\dagger c_{i+1} + c_{i+1}^\dagger c_i\right) + J \Delta \sum_i n_i n_{i+1}$$

$$= \sum_k \epsilon(k) c_k^\dagger c_k + \sum_q V(q) \rho(q) \rho(-q), \tag{6.103}$$

where we have dropped a constant as well as a term proportional to the total number N of fermions, which is conserved, and $\epsilon(k) = J \cos k$ while $V(q) = J \Delta \cos(q)$. The Hamiltonian (6.103) is therefore a particular case of the spinless fermion Hamiltonian we have earlier discussed, so that we can simply borrow all previous results.

Specifically, in the subspace of vanishing total magnetisation, i.e.,

$$0 = S_z = \frac{1}{2} \sum_i (1 - 2n_i) = \frac{L}{2} - N,$$

the model has half-filled density of spinless fermions, and therefore the bosonized Hamiltonian reads

$$H = \frac{v_F}{4\pi} \int dx \left[\left(\frac{\partial \phi_R(x)}{\partial x} \right)^2 + \left(\frac{\partial \phi_L(x)}{\partial x} \right)^2 \right] + \frac{g_2}{4\pi^2} \int dx \frac{\partial \phi_R(x)}{\partial x} \frac{\partial \phi_L(x)}{\partial x}$$

$$- \frac{g_3}{2\pi^2 \alpha^2} \int dx \cos 2 \big(\phi_R(x) + \phi_L(x) \big)$$

$$\equiv H_0 + \int dx \left\{ \frac{g_2}{4\pi^2} \frac{\partial \phi_R(x)}{\partial x} \frac{\partial \phi_L(x)}{\partial x} - \frac{g_3}{2\pi^2 \alpha^2} \cos 2 \big(\phi_R(x) + \phi_L(x) \big) \right\},$$

(6.104)

where now the bare values of the coupling constants are $g_2 = 2J \Delta$ and $g_3 = -J \Delta$. If we absorb g_2 through the canonical transformation (6.63), we arrive at the same conclusion as before that the umklapp scattering is irrelevant as long as $K > 1/2$. On the ferromagnetic side, $\Delta < 0$, $K > 1$ and the umklapp is always irrelevant. There, the worst can happen is phase separation when the effective velocity (6.65) vanishes. The only region where we physically expect phase separation is for $\Delta \leq -1$, where the easy axis anisotropy would favour a ferromagnetic state with all spin polarised parallel or anti parallel to z. Since we force to have equal number of up and down spins, all equal spins have to cluster together, thus the phase separation. Therefore, we expect a gapless phase for any $-1 < \Delta \leq 0$.

Considering instead the antiferromagnetic side, $\Delta > 0$ and $K < 1$, the only region where we expect a gap opening is for $\Delta > 1$, which corresponds to an easy axis anti-ferromagnet. That is essentially an Ising antiferromagnet that can order at zero temperature even in one dimension, since the ordered phase breaks a discrete symmetry. Therefore $\Delta \lesssim 1$ corresponds to $K \gtrsim 1/2$, and its gapless nature implies that the model is within the green region in Fig. 6.6, which also implies that, as $\Delta \to 1$ from below, the model flows to the origin $g(\infty) = g_3(\infty) = 0$ in Fig. 6.6. As discussed in Sect. 6.2.8.1, $g(\infty) = g_3(\infty) = 0$ with $K(\infty) = 1/2$ is an $SU(2)$ invariant point, consistently with the fact that at $\Delta = 1$ the Hamiltonian (6.99) describes an isotropic antiferromagnetic Heisenberg model.

6.4 The One-Dimensional Hubbard Model

We now consider the spinful Hubbard model

$$H = -t \sum_{i\sigma} \left(c_{i\sigma}^\dagger c_{i+1\sigma} + c_{i+1\sigma}^\dagger c_{i\sigma} \right) + U \sum_i n_{i\uparrow} n_{i\downarrow},$$

(6.105)

with $U > 0$, as representative of generic models of electrons with short-range inter-action and two Fermi points, $\pm k_F$, and apply the same bosonization scheme as above. For that, we need to introduce two copies of right and moving electrons, one for each spin, thus $\phi_{R\uparrow}(x)$, $\phi_{R\downarrow}(x)$, $\phi_{L\uparrow}(x)$ and $\phi_{L\downarrow}(x)$, as well as the fields $\Phi_\uparrow(x)$, $\Theta_\uparrow(x)$, $\Phi_\downarrow(x)$ and $\Theta_\downarrow(x)$. Each of them have the same properties of $\phi_R(x)$, $\phi_L(x)$, $\Phi(x)$

and $\Theta(x)$ previously discussed. In addition, in order to make different Fermi fields anticommute, we assume that

$$
\left[\phi_{R\uparrow}(x),\ \phi_{L\uparrow}(y)\right] = \left[\phi_{R\uparrow}(x),\ \phi_{R\downarrow}(y)\right] = \left[\phi_{R\uparrow}(x),\ \phi_{L\downarrow}(y)\right] = -i\pi,
$$
$$
\left[\phi_{L\uparrow}(x),\ \phi_{R\downarrow}(y)\right] = \left[\phi_{L\uparrow}(x),\ \phi_{L\downarrow}(y)\right] = -i\pi,
$$
$$
\left[\phi_{R\downarrow}(x),\ \phi_{L\downarrow}(y)\right] = -i\pi.
$$

$$(6.106)$$

We further introduce the combinations

$$
\phi_{R(L)\rho}(x) = \frac{1}{\sqrt{2}}\left(\phi_{R(L)\uparrow}(x) + \phi_{R(L)\downarrow}(x)\right),
$$
$$
\phi_{R(L)\sigma}(x) = \frac{1}{\sqrt{2}}\left(\phi_{R(L)\uparrow}(x) - \phi_{R(L)\downarrow}(x)\right),
$$

$$(6.107)$$

referring to charge, ρ, and spin, σ, degrees of freedom, as well as the fields

$$
\Phi_{\rho(\sigma)}(x) = \frac{1}{\sqrt{4\pi}}\left(\phi_{R\rho(\sigma)}(x) + \phi_{L\rho(\sigma)}(x)\right),
$$
$$
\Theta_{\rho(\sigma)}(x) = \frac{1}{\sqrt{4\pi}}\left(\phi_{L\rho(\sigma)}(x) - \phi_{R\rho(\sigma)}(x)\right),
$$

$$(6.108)$$

or, equivalently,

$$
\Phi_\rho(x) = \frac{1}{\sqrt{2}}\left(\Phi_\uparrow(x-)55 + 5\Phi_\downarrow(x)\right), \quad \Theta_\rho(x) = \frac{1}{\sqrt{2}}\left(\Theta_\uparrow(x) + \Theta_\downarrow(x)\right),
$$
$$
\Phi_\sigma(x) = \frac{1}{\sqrt{2}}\left(\Phi_\uparrow(x) - \Phi_\downarrow(x)\right), \qquad \Theta_\sigma(x) = \frac{1}{\sqrt{2}}\left(\Theta_\uparrow(x) - \Theta_\downarrow(x)\right).
$$

We note that the total electron number

$$
N = \frac{1}{2\pi}\int dx \sum_{a=R\uparrow,L\uparrow,R\downarrow,L\downarrow} \frac{\partial\phi_a(x)}{\partial x} = \sqrt{\frac{2}{\pi}}\int dx\, \frac{\partial\Phi_\rho(x)}{\partial x},
$$

$$(6.109)$$

so that the non-interacting Hamiltonian, assuming

$$
N_{R\uparrow} = N_{L\uparrow} = N_{R\downarrow} = N_{L\downarrow} = \frac{N}{4} = \frac{k_F}{2\pi}\,L,
$$

is simply

$$
H_0 = \frac{v_F}{2}\int dx\left[\Pi_\rho(x)^2 + \left(\frac{\partial\Phi_\rho(x)}{\partial x}\right)^2 + \Pi_\sigma(x)^2 + \left(\frac{\partial\Phi_\sigma(x)}{\partial x}\right)^2\right]
$$
$$
- \mu\sqrt{\frac{2}{\pi}}\int dx\, \frac{\partial\Phi_\rho(x)}{\partial x},
$$

where μ must be fixed by imposing the desired electron number.

In order to bosonize the interaction, we follow a simpler approach than we did before for spinless fermions. In the long wavelength limit the interaction can be written as

$$U \sum_i n_{i\uparrow} n_{i\downarrow} \rightarrow U \int dx\, \rho_\uparrow(x)\, \rho_\downarrow(x),$$

where

$$\begin{aligned}
\rho_\uparrow(x) &= \rho_{R\uparrow}(x) + \rho_{L\uparrow}(x) + \Psi_{R\uparrow}^\dagger(x)\Psi_{L\uparrow}(x) + \Psi_{L\uparrow}^\dagger(x)\Psi_{R\uparrow}(x) \\
&= \frac{1}{2\pi}\frac{\partial\phi_{R\uparrow}(x)}{\partial x} + \frac{1}{2\pi}\frac{\partial\phi_{L\uparrow}(x)}{\partial x} + \frac{i}{2\pi\alpha}e^{-i\left(\phi_{R\uparrow}(x)+\phi_{L\uparrow}(x)\right)} - \frac{i}{2\pi\alpha}e^{i\left(\phi_{R\uparrow}(x)+\phi_{L\uparrow}(x)\right)} \\
&= \frac{1}{\sqrt{\pi}}\frac{\partial\Phi_\uparrow(x)}{\partial x} + \frac{i}{2\pi\alpha}e^{-i\sqrt{4\pi}\,\Phi_\uparrow(x)} - \frac{i}{2\pi\alpha}e^{i\sqrt{4\pi}\,\Phi_\uparrow(x)} \\
&\equiv \frac{1}{\sqrt{\pi}}\frac{\partial\Phi_\uparrow(x)}{\partial x} + \frac{1}{\pi\alpha}\sin\sqrt{4\pi}\,\Phi_\uparrow(x),
\end{aligned}$$

(6.110)

and similarly for $\rho_\downarrow(x)$. It follows that

$$\begin{aligned}
U\int dx\, \rho_\uparrow(x)\,\rho_\downarrow(x) &\simeq U\int dx\left\{\frac{1}{\pi}\frac{\partial\Phi_\uparrow(x)}{\partial x}\frac{\partial\Phi_\downarrow(x)}{\partial x}\right. \\
&\qquad\qquad \left. + \frac{1}{\pi^2\alpha^2}\sin\sqrt{4\pi}\,\Phi_\uparrow(x)\,\sin\sqrt{4\pi}\,\Phi_\downarrow(x)\right\} \\
&= U\int dx\left\{\frac{1}{2\pi}\left(\frac{\partial\Phi_\rho(x)}{\partial x}\right)^2 - \frac{1}{2\pi}\left(\frac{\partial\Phi_\sigma(x)}{\partial x}\right)^2\right. \\
&\qquad\qquad \left. + \frac{1}{2\pi^2\alpha^2}\cos\sqrt{8\pi}\,\Phi_\sigma(x) - \frac{1}{2\pi^2\alpha^2}\cos\sqrt{8\pi}\,\Phi_\rho(x)\right\} \\
&\equiv \int dx\left\{\frac{g_\rho}{2\pi}\left(\frac{\partial\Phi_\rho(x)}{\partial x}\right)^2 + \frac{g_\sigma}{2\pi}\left(\frac{\partial\Phi_\sigma(x)}{\partial x}\right)^2\right. \\
&\qquad\qquad \left. - \frac{g_{3\sigma}}{2\pi^2\alpha^2}\cos\sqrt{8\pi}\,\Phi_\sigma(x) - \frac{g_{3\rho}}{2\pi^2\alpha^2}\cos\sqrt{8\pi}\,\Phi_\rho(x)\right\},
\end{aligned}$$

where $g_\rho = U$, $g_\sigma = -U$, $g_{3\rho} = U$ and $g_{3\sigma} = -U$. Therefore, the fully interacting Hamiltonian can be written as

$$H = H_\rho + H_\sigma,$$

where the charge component is

$$\begin{aligned}
H_\rho &= \frac{v_F}{2}\int dx\left[\Pi_\rho(x)^2 + \left(\frac{\partial\Phi_\rho(x)}{\partial x}\right)^2\right] - \mu\sqrt{\frac{2}{\pi}}\int dx\,\frac{\partial\Phi_\rho(x)}{\partial x} \\
&\quad + \int dx\left\{\frac{g_\rho}{2\pi}\left(\frac{\partial\Phi_\rho(x)}{\partial x}\right)^2 - \frac{g_{3\rho}}{2\pi^2\alpha^2}\cos\sqrt{8\pi}\,\Phi_\rho(x)\right\},
\end{aligned}$$

(6.111)

with

$$\mu = v_F \left(1 + \frac{g_\rho}{\pi v_F}\right) k_F,$$

so as to have the desired electron number, while the spin component reads

$$H_\sigma = \frac{v_F}{2} \int dx \left[\Pi_\sigma(x)^2 + \left(\frac{\partial \Phi_\sigma(x)}{\partial x}\right)^2 \right]$$

$$+ \int dx \left\{ \frac{g_\sigma}{2\pi} \left(\frac{\partial \Phi_\sigma(x)}{\partial x}\right)^2 - \frac{g_{3\sigma}}{2\pi^2 \alpha^2} \cos \sqrt{8\pi} \, \Phi_\sigma(x) \right\}. \tag{6.112}$$

We note that both H_ρ and H_σ look like the Hamiltonian of interacting spinless fermions around $K = 1/2$ that we studied in Sect. 6.2.8. This observation allows us to draw several conclusions.

First, H_σ, with $g_\sigma = g_{3\sigma}$, refers to a model endowed with an $SU(2)$ symmetry, as discussed in Sects. 6.2.8.1 and 6.3.

In addition, since $g_\sigma < 0$, H_σ lies on the black line in the upper half-plane of Fig. 6.6. Therefore, under RG it will flow to $g_\sigma(\infty) = g_{3\sigma}(\infty) = 0$, thus to an effective non-interacting spinless fermion model with $K_\sigma(\infty) = 1$.

Let us now discuss more closely the charge Hamiltonian (6.111). The chemical potential term can be eliminated by the canonical transformation

$$\Phi_\rho(x) \to \Phi_\rho(x) + \sqrt{\frac{2}{\pi}} \, k_F x, \quad \Pi_\rho(x) \to \Pi_\rho(x),$$

after which

$$H_\rho = \frac{v_F}{2} \int dx \left[\Pi_\rho(x)^2 + \left(\frac{\partial \Phi_\rho(x)}{\partial x}\right)^2 \right]$$

$$+ \int dx \left\{ \frac{g_\rho}{2\pi} \left(\frac{\partial \Phi_\rho(x)}{\partial x}\right)^2 - \frac{g_{3\rho}}{2\pi^2 \alpha^2} \cos \left(4k_F x + \sqrt{8\pi} \, \Phi_\rho(x)\right) \right\}, \tag{6.113}$$

If $4k_F \neq 2\pi$, i.e., away from half-filling, the umklapp oscillates fast and thus can be neglected in the asymptotic long-wavelength limit. In that case, H_ρ describes a gapless system with

$$K_\rho = \sqrt{\frac{1}{1 + g_\rho/\pi v_F}}, \quad v_\rho = v_F \sqrt{1 + \frac{v_\rho}{\pi v_F}}. \tag{6.114}$$

On the contrary, at half-filling $4k_F = 2\pi$, the umklapp cannot be neglected. Since $g_\rho > 0$ and $g_{3\rho} > 0$, the system will flow to the strong-coupling fixed point $g_\rho(\infty) = g_{3\rho}(\infty) = \infty$. This implies that we must fix $\Phi_\rho(x)$ to the value that minimises the umklapp, thus

$$\langle \sqrt{8\pi} \, \Phi_\rho(x) \rangle = \sqrt{4\pi} \, \langle \Phi_\uparrow(x) + \Phi_\downarrow(x) \rangle = 2\pi n.$$

In this case there is only one independent value $n = 0$, which implies that $\langle \Phi_\uparrow(x) \rangle = -\langle \Phi_\downarrow(x) \rangle$ and therefore

$$\langle \rho(x) \rangle = 1 + \frac{1}{\pi\alpha} \langle \sin \sqrt{4\pi} \, \Phi_\uparrow(x) + \sin \sqrt{4\pi} \, \Phi_\downarrow(x) \rangle = 1 . \tag{6.115}$$

This phase thus corresponds to a charge insulator, the electrons localised one at each site, with gapless spin excitations, which behaves asymptotically just alike the isotropic antiferromagnetic Heisenberg model of Sect. 6.3. As we know, the two models exactly maps onto each other at large $U \gg t$, hence in that limit the correspondence holds not only asymptotically at low energy and long distances.

If $N/L = 1 + \delta$, thus $4k_F = 2\pi + 2\pi\delta \equiv 2\pi + 2\bar{k}_F$, with $\delta \ll 1$, the model asymptotically approaches a doped Luther-Emery model, see Sect. 6.2.8.2, at $K_\rho = 1/2$ with Hamiltonian

$$H_\rho \simeq \frac{v_*}{2} \int dx \left[\overline{\Pi}_\rho(x)^2 + \left(\frac{\partial \overline{\Phi}_\rho(x)}{\partial x} \right)^2 \right] - \frac{g_*}{2\pi^2\alpha^2} \int dx \, \cos \left(2\bar{k}_F x + \sqrt{4\pi} \, \overline{\Phi}_\rho(x) \right),$$

which describes an insulator with a doping of $\delta > 0$ electrons in the conduction band or $-\delta > 0$ holes in the valence one.

6.4.1 Luttinger Versus Fermi Liquids

Away from half-filling, the one-dimensional Hubbard model has gapless charge and spin modes that behave as acoustic waves with different velocities $v_\rho > v_F > v_\sigma > 0$, and with $K_\rho < 1$ and $K_\sigma = 1$. The single-particle Green's function can be readily calculated as we previously did for spinless fermions,

$$G_R(x,t) = \frac{1}{2\pi} \frac{e^{ik_{RF}x}}{\sqrt{x - v_\rho t + i\alpha \, \text{sign}(t)} \sqrt{x - v_\sigma t + i\alpha \, \text{sign}(t)}}$$
$$\left[\frac{\alpha^2}{(x - v_\rho t + i\alpha \, \text{sign}(t))(x + v_\rho t - i\alpha \, \text{sign}(t))} \right]^{\beta^2},$$

$$G_L(x,t) = \frac{1}{2\pi} \frac{e^{-ik_{LF}x}}{\sqrt{-x - v_\rho t + i\alpha \, \text{sign}(t)} \sqrt{-x - v_\sigma t + i\alpha \, \text{sign}(t)}}$$
$$\left[\frac{\alpha^2}{(x - v_\rho t + i\alpha \, \text{sign}(t))(x + v_\rho t - i\alpha \, \text{sign}(t))} \right]^{\beta^2}, \tag{6.116}$$

where now

$$\beta^2 = \frac{1}{8} \left(\sqrt{K_\rho} - \frac{1}{\sqrt{K_\rho}} \right)^2 .$$

These expressions clearly highlight the spin-charge separation; the electron decouples into independent charge and spin components that move with different velocities. In addition, the Green's functions at $t = 0^-$ are non analytic on both sides of the complex $z = x + iy$ plane, and thus the momentum distribution does not jump at the Fermi momenta.

Although the Fourier transform $G_{R(L)}(\epsilon + i\eta, k)$ of those heavily non-analytic Green's functions is rather cumbersome and does not allow easily extracting the self-energies $\Sigma_{R(L)}(\epsilon + i\eta, k)$, spin-charge separation makes hardly conceivable that filtering $G_{R(L)}(\epsilon + i\eta, k)$ by

$$Z_{R(L)}(\epsilon, k)^{-1} = 1 - \frac{\partial \Sigma_{R(L)}(\epsilon + i\eta, k)}{\partial \epsilon} , \tag{6.117}$$

may yield well defined quasiparticles. Therefore, the Landau-Fermi liquid theory discussed in Chap. 5 cannot be applied in one dimensional interacting electron systems, all the more since all correlation functions at momentum transferred $\sim 2k_F$ decay with anomalous interaction-dependent exponents, which, as discussed in Sect. 6.1, signals the unavoidable quasi-long range order that characterises interacting electrons in one-dimension.

Nonetheless, through $\overline{\Phi}_{\rho(\sigma)}(x)$ and $\overline{\Theta}_{\rho(\sigma)}(x)$ we can define new Fermi fields $\Psi_{R\rho(\sigma)}(x)$ and $\Psi_{L\rho(\sigma)}(x)$ whose low-energy long-wavelength dynamics is determined by free Hamiltonians. Those fields thus play the role of the 'quasiparticles' in Luttinger liquids. As a consequence, the specific heat receives linear in temperature contributions from both charge and spin fermions, each having its own density of states $1/2\pi v_{\rho(\sigma)}$. Moreover, the charge, χ_ρ, and spin, χ_σ, density-density response function at long wavelengths also look like those of non-interacting electrons, specifically,

$$\chi_{\rho(\sigma)}(q, \omega) = 2K_{\rho(\sigma)} \frac{|q|}{\pi} \left(\frac{1}{\omega - v_{\rho(\sigma)} |q| + i\eta} - \frac{1}{\omega + v_{\rho(\sigma)} |q| + i\eta} \right),$$

yielding charge compressibility

$$\kappa = \frac{2K_\rho}{2\pi v_\rho} = K_\rho \frac{v_F}{v_\rho} \kappa_0 < \kappa_0 ,$$

and spin susceptibility, recalling that $K_\sigma = 1$,

$$\chi = \frac{2}{2\pi v_\sigma} = \frac{v_F}{v_\sigma} \kappa_0 > \kappa_0 .$$

All those long-wavelength properties, as well as their temperature dependence, behave as in ordinary Fermi liquids, with K_ρ, v_ρ and v_σ playing the role of Landau's parameters.

Therefore, Luttinger and Fermi Liquids, in spite of their differences, share a very unique feature: both of them recovers at low energy the larger symmetry of the non-interacting system. However, there is an important caveat. Landau's Fermi liquids are asymptotically described by quasiparticles whose Hamiltonian conserves their occupation numbers in momentum space, $n_{\mathbf{k}\uparrow}$ and $n_{\mathbf{k}\downarrow}$, separately for each momentum $\mathbf{k}$ on the quasiparticle Fermi surface and for each spin. On the contrary, 'quasiparticles' in Luttinger liquids are spin-charge separated. Therefore, only the charge $n_{k\uparrow} + n_{k\downarrow}$ and the spin $n_{k\uparrow} - n_{k\downarrow}$ are separately conserved at each Fermi point, k_{RF} and $-k_{LF}$, while each spin component is not.

Problems

6.1 Spin-charge separation—Consider the following one-dimensional Hamiltonian:

$$
\begin{aligned}
H_0 = v_F \sum_{k\sigma} & \left[(k - k_F) c_{Rk\sigma}^{\dagger} c_{Rk\sigma} - (k + k_F) c_{Lk\sigma}^{\dagger} c_{Lk\sigma} \right] \\
& + \frac{g_4}{2L} \sum_{q} \sum_{p=R,L} \left(\rho_{p\uparrow}(q) + \rho_{p\downarrow}(q) \right) \left(\rho_{p\uparrow}(-q) + \rho_{p\downarrow}(-q) \right),
\end{aligned}
\tag{6.118}
$$

which contains only a g_4 charge scattering process, see Fig. 6.4, and where the dispersion $\epsilon(k)$ of Fig. 6.2 has been linearised around the two Fermi points $\pm k_F$.

- Calculate the single-particle Green's function.

Now assume to have two chains, each described by the Hamiltonian (6.118), so that

$$
\begin{aligned}
H_0 = \sum_{n=1,2} & \left\{ v_F \sum_{k\sigma} \left[(k - k_F) c_{nRk\sigma}^{\dagger} c_{nRk\sigma} - (k + k_F) c_{nLk\sigma}^{\dagger} c_{nLk\sigma} \right] \right. \\
& \left. + \frac{g_4}{2L} \sum_{q} \sum_{p=R,L} \left(\rho_{np\uparrow}(q) + \rho_{np\downarrow}(q) \right) \left(\rho_{np\uparrow}(-q) + \rho_{np\downarrow}(-q) \right) \right\},
\end{aligned}
\tag{6.119}
$$

where $n = 1, 2$ is the chain label. The two chains are mutually coupled by an inter-chain hopping

$$
H_\perp = -t \sum_{\sigma} \sum_{p=R,L} \int dx \left(\Psi_{1p\sigma}^{\dagger}(x) \Psi_{2p\sigma}(x) + \Psi_{2p\sigma}^{\dagger}(x) \Psi_{1p\sigma}(x) \right),
\tag{6.120}
$$

where $\Psi_{1R(L)\sigma}^{\dagger}(x)$ creates a right(left) moving fermion with spin σ on chain 1, while $\Psi_{2R(L)\sigma}^{\dagger}(x)$ does the same but on chain 2.

- Rewrite the interchain hopping (6.120) using the bosonized expressions of right and left moving fields.

- Introduce the charge (c) and spin (s) symmetric (S) and antisymmetric (A) combinations of the Bose fields $\phi_{np\sigma}(x)$, $n = 1, 2$, $p = R, L$ and $\sigma = \uparrow, \downarrow$:

$$\phi_{c\,S\,p}(x) = \frac{1}{2}\left(\phi_{1p\uparrow}(x) + \phi_{2p\uparrow}(x) + \phi_{1p\downarrow}(x) + \phi_{2p\downarrow}(x)\right),$$

$$\phi_{s\,S\,p}(x) = \frac{1}{2}\left(\phi_{1p\uparrow}(x) + \phi_{2p\uparrow}(x) - \phi_{1p\downarrow}(x) - \phi_{2p\downarrow}(x)\right),$$

$$\phi_{c\,A\,p}(x) = \frac{1}{2}\left(\phi_{1p\uparrow}(x) - \phi_{2p\uparrow}(x) + \phi_{1p\downarrow}(x) - \phi_{2p\downarrow}(x)\right), \tag{6.121}$$

$$\phi_{s\,A\,p}(x) = \frac{1}{2}\left(\phi_{1p\uparrow}(x) - \phi_{2p\uparrow}(x) - \phi_{1p\downarrow}(x) + \phi_{2p\downarrow}(x)\right),$$

and through them the new Fermi fields

$$\Psi_{c\,S\,R}(x) = \frac{1}{\sqrt{2\pi\,\alpha}}\,e^{i\phi_{c\,S\,R}(x)}, \qquad \Psi_{s\,S\,R}(x) = \frac{1}{\sqrt{2\pi\,\alpha}}\,e^{i\phi_{s\,S\,R}(x)},$$

$$\Psi_{c\,A\,R}(x) = \frac{1}{\sqrt{2\pi\,\alpha}}\,e^{i\phi_{c\,A\,R}(x)}, \qquad \Psi_{s\,A\,R}(x) = \frac{1}{\sqrt{2\pi\,\alpha}}\,e^{i\phi_{c\,A\,R}(x)},$$

$$\Psi_{c\,S\,L}(x) = \frac{1}{\sqrt{2\pi\,\alpha}}\,e^{-i\phi_{c\,S\,L}(x)}, \qquad \Psi_{s\,S\,L}(x) = \frac{1}{\sqrt{2\pi\,\alpha}}\,e^{-i\phi_{s\,S\,L}(x)},$$

$$\Psi_{c\,A\,L}(x) = \frac{1}{\sqrt{2\pi\,\alpha}}\,e^{-i\phi_{c\,A\,L}(x)}, \qquad \Psi_{s\,A\,L}(x) = \frac{1}{\sqrt{2\pi\,\alpha}}\,e^{-i\phi_{c\,A\,L}(x)}.$$

Rewrite the unperturbed Hamiltonian H_0 (6.119) plus the interchain hopping (6.120) in terms of those Fermi fields, similarly to Sect. 6.2.8.2 but here for generic k_F, and show that $H = H_0 + H_\perp$ can be easily diagonalised in this representation.
- Calculate the intrachain and interchain Green's functions.

References

1. A. Gogolin, A. Nersesyan, A. Tsvelik, *Bosonization and Strongly Correlated Systems* (Cambridge University Press, 1998)
2. T. Giamarchi, *Quantum Physics in One Dimension* (Oxford University Press, 2004)

Kondo Effect and the Physics of the Anderson Impurity Model

The behaviour of magnetic impurities (Fe, Mn, Cr) diluted into non-magnetic metals, e.g., Cu, Ag, Au, Al, is the simplest manifestation of strong electron-electron correlations that escapes a description in terms of independent particles. Moreover, that same phenomenology has progressively become the paradigm of strongly correlated metals close to a Mott transition as unveiled by Dynamical Mean Field Theory.

We know that the behaviour of a good metal with a large Fermi temperature, $T_F \sim 10^4$ K, is dominated at low temperature $T \ll T_F$ by Pauli principle. For instance, the magnetic susceptibility is roughly constant, $\chi \sim 1/T_F$, and one should in principle heat the sample to very high temperatures $T \gg T_F$, where the metal likely melts, to release the spin entropy and recover a Curie-Weiss behaviour $\chi \sim 1/T$. Moreover, the resistivity is an increasing function of temperature, since the channels that may dissipate current, the coupling to phonons and the electron decay into particle-hole excitations brought by interaction, become available only upon heating.

At odds with that expectation, if one introduces a very diluted (few parts per million) concentration of magnetic impurities the above behaviour changes drastically. We just mention three distinct features.

(1) The magnetic susceptibility shows a Curie-Weiss behaviour well below T_F, proportional to the impurity concentration n_i and roughly with the same g-factor of the isolated magnetic impurity, apart from corrections due to the crystal field. Around a very low temperature, called Kondo-temperature T_K, the Curie-Weiss behaviour turns into a logarithmic one, and finally the susceptibility saturates at low temperature to a value $\chi(0) \sim n_i/T_K$, with $\chi(T) - \chi(0) \sim -T^2$.

(2) The resistivity $R(T)$ displays a minimum around T_K, followed at $T < T_K$ by a logarithmic increase. At very low temperatures, $R(T)$ approaches a constant value $R(0) \propto n_i$ with $R(T) - R(0) \sim -T^2$. The value of the residual resistivity $R(0)$ suggests very strong scattering potential, near the so-called unitary limit.

© The Author(s), under exclusive license to Springer Nature Switzerland AG 2022
M. Fabrizio, *A Course in Quantum Many-Body Theory*, Graduate Texts in Physics,
https://doi.org/10.1007/978-3-031-16305-0_7

(3) The entropy which is released above the Kondo temperature, which can be extracted from the specific heat, includes the spin degrees of freedom of the isolated impurity but not the charge ones, which indicates that the magnetic impurities behave as local moments above T_K.

Explaining this behaviour amounts actually to understand three different problems.

- The first concerns the region $T_K \ll T \ll T_F$ and can be formulated as follows: How is it possible to sustain a local moment inside a metal? We shall see that to answer this question one is obliged to abandon conventional independent particle schemes.
- The second issue regards what happens around the Kondo temperature, namely to understand the logarithmic crossover.
- Finally, the last question concerns the low temperature behaviour, the way local moments get screened and why resistivity is a decreasing function of temperature.

The last two questions turn out to represent a complicated many-body problem which we shall here discuss using the tools we have presented so far.

7.1   Brief Introduction to Scattering Theory

We start by showing why the existence of local moments in a metal for $T \ll T_F$ is so puzzling.

Let us consider an impurity imbedded in a wide-band metal. We only consider the valence band, which can be safely described by a non-interacting Hamiltonian

$$H_0 = \sum_{\mathbf{k},\sigma} \epsilon_{\mathbf{k}} \, c^{\dagger}_{\mathbf{k}\sigma} c_{\mathbf{k}\sigma} . \tag{7.1}$$

The scattering potential provided by the impurity has the general form

$$V = \sum_{\sigma} \sum_{\mathbf{k},\mathbf{p}} V_{\mathbf{k}\mathbf{p}} \, c^{\dagger}_{\mathbf{k}\sigma} c_{\mathbf{p}\sigma} , \tag{7.2}$$

where $V_{\mathbf{k}\mathbf{p}}$ are the matrix elements of the impurity potential $\hat{V}$ between the valence band Block waves. The single-particle Green's function in complex frequency for the full Hamiltonian $H = H_0 + V$ can be formally written as the matrix

$$\hat{G}(z) = \frac{1}{z - \hat{H}} , \tag{7.3}$$

where $\hat{H} = \hat{H}_0 + \hat{V}$, with $\hat{H}_0$ diagonal with elements $\epsilon_{\mathbf{k}}$. Similarly, the unperturbed Green's function reads

$$\hat{G}_0(z) = \frac{1}{z - \hat{H}_0} . \tag{7.4}$$

The above operators have the following interpretation. If we consider for instance the inverse of the Green's function $\hat{G}$, namely

$$\hat{G}(z)^{-1} = z - \hat{H},$$

in the basis set of the valence band Block waves, this is the matrix

$$\left(\hat{G}(z)^{-1} \right)_{\mathbf{k}\sigma, \mathbf{p}\sigma'} = \delta_{\sigma\sigma'} \left[\delta_{\mathbf{kp}} z - \delta_{\mathbf{kp}} \epsilon_{\mathbf{k}} - V_{\mathbf{kp}} \right], \tag{7.5}$$

hence the Green's function is the inverse of the above matrix. Suppose we have diagonalised the full Hamiltonian and obtained the eigenvalues ϵ_a. In the diagonal basis the Green's function has matrix elements

$$G(z)_{ab} = \delta_{ab} \frac{1}{z - \epsilon_a} .$$

Therefore, we readily find that for $\eta > 0$ infinitesimal

$$-\frac{1}{\pi} \operatorname{Im} \operatorname{Tr} \left(\hat{G}(z = \omega + i\eta) \right) = \sum_{a\sigma} \delta(\omega - \epsilon_a) \equiv \rho(\omega),$$

where ω is a real frequency and $\rho(\omega)$ is the density of states (DOS). Since the trace is invariant under unitary transformations, hence also under the transformation that diagonalises the Hamiltonian, it is generally true that

$$-\frac{1}{\pi} \operatorname{Im} \operatorname{Tr} \left[\hat{G}(z = \omega + i\eta) \right] = \rho(\omega). \tag{7.6}$$

Similarly,

$$-\frac{1}{\pi} \operatorname{Im} \operatorname{Tr} \left[\hat{G}_0(z = \omega + i\delta) \right] = \rho_0(\omega) = \sum_{\mathbf{k}\sigma} \delta(\epsilon_{\mathbf{k}} - \omega), \tag{7.7}$$

gives the DOS of the host metal.

Let us formally write the full Green's function as

$$\hat{G} = \frac{1}{z - \hat{H}_0 - \hat{V}} = \frac{1}{\hat{G}_0^{-1} - \hat{V}} = \frac{1}{\hat{I} - \hat{G}_0 \hat{V}} \hat{G}_0$$

$$= \hat{G}_0 + \hat{G}_0 \hat{V} \left[\hat{I} - \hat{G}_0 \hat{V} \right]^{-1} \hat{G}_0 \equiv \hat{G}_0 + \hat{G}_0 \hat{T} \hat{G}_0, \tag{7.8}$$

which provides the definition of the so-called T-matrix, namely

$$\hat{T}(z) = \hat{V} \left[\hat{I} - \hat{G}_0 \, \hat{V} \right]^{-1}. \tag{7.9}$$

Hereafter, we shall assume $z = \omega + i\eta$ with $\eta > 0$ infinitesimal. We note that

$$\frac{\partial}{\partial z} \ln \hat{G}(z) = -\hat{G}(z),$$

so that

$$\rho(\omega) = \frac{1}{\pi} \frac{\partial}{\partial z} \left\{ \mathrm{Im}\, \mathrm{Tr} \Big(\ln \hat{G}(z) \Big) \right\}. \tag{7.10}$$

We are interested in the variation of the DOS induced by the impurity, which, by making use of (7.8) and (7.9), is given by

$$\Delta\rho(\omega) = \rho(\omega) - \rho_0(\omega) = \frac{1}{\pi} \frac{\partial}{\partial z} \left\{ \mathrm{Im}\, \mathrm{Tr} \Big(\ln \hat{G}(z) \, \hat{G}_0(z)^{-1} \Big) \right\}$$

$$= \frac{1}{\pi} \frac{\partial}{\partial z} \left\{ \mathrm{Im}\, \mathrm{Tr} \Big(\ln \frac{1}{\hat{I} - \hat{G}_0(z) \, \hat{V}} \Big) \right\} = \frac{1}{\pi} \frac{\partial}{\partial z} \left[\mathrm{Im}\, \mathrm{Tr} \Big(\ln \hat{V}^{-1} \, \hat{T}(z) \Big) \right]. \tag{7.11}$$

We define the matrix of the scattering phase shifts

$$\hat{\delta}(z) = \mathrm{Im}\, \ln \hat{V}^{-1} \, \hat{T}(z) = \mathrm{Arg} \left(\hat{V}^{-1} \, \hat{T}(z) \right), \tag{7.12}$$

through which

$$\Delta\rho(\omega) = \frac{1}{\pi} \frac{\partial}{\partial \omega} \mathrm{Tr} \Big(\hat{\delta}(\omega) \Big). \tag{7.13}$$

We note that for $|z| \to \infty$, $\hat{G}_0(z) \to 1/z \to 0$, hence $\hat{T}(z) \to \hat{V}$ and $\hat{\delta}(z) \to 0$.

The variation of the total number of electrons, ΔN_{els}, induced by the impurity at fixed chemical potential μ is therefore

$$\Delta N_{els} = \int_{-\infty}^{\mu} d\omega \, \Delta\rho(\omega) = \frac{1}{\pi} \mathrm{Tr} \Big(\hat{\delta}(\mu) \Big), \tag{7.14}$$

the so-called Friedel sum rule.

Let us go back to the T-matrix defined in (7.9). One readily finds its inverse

$$\hat{T}(z)^{-1} = \left[\hat{V}^{-1} - \hat{G}_0(z) \right].$$

Since the Hamiltonian is hermitean, then

$$\left[\hat{T}(z)^{-1} \right]^{\dagger} = \left[\hat{V}^{-1} - \hat{G}_0(z^*) \right],$$

so that

$$\left[\hat{T}(z)^{-1}\right]^{\dagger} - \hat{T}(z)^{-1} = \left[\hat{G}_0(z) - \hat{G}_0(z^*)\right].$$

Therefore

$$\hat{T}(z)^{\dagger}\left\{\left[\hat{T}(z)^{-1}\right]^{\dagger} - \hat{T}(z)^{-1}\right\} \hat{T}(z) = \hat{T}(z) - \hat{T}(z)^{\dagger} = \hat{T}(z)^{\dagger}\left[\hat{G}_0(z) - \hat{G}_0(z^*)\right]\hat{T}(z).$$
(7.15)

Since

$$\hat{G}_0(z) - \hat{G}_0(z^*) = -2\pi i \; \delta\left(\omega - \hat{H}_0\right),$$

Equation (7.15) implies the following identity

$$T_{\mathbf{kp}}(\omega) - T_{\mathbf{kp}}^{\dagger}(\omega) = -2\pi i \sum_{\mathbf{q}} T_{\mathbf{kq}}^{\dagger}(\omega) \, \delta(\omega - \epsilon_{\mathbf{q}}) \, T_{\mathbf{qp}}(\omega), \qquad (7.16)$$

which is the so-called optical theorem. It also shows that the imaginary part of the T-matrix is finite only within the conduction band.

Let us now analyse the on-shell T-matrix $T_{\mathbf{kp}}(\epsilon)$, where $\epsilon_{\mathbf{k}} = \epsilon_{\mathbf{p}} = \epsilon$. We can rewrite the on-shell optical theorem as follows

$$-2\pi i \; \delta\left(\epsilon_{\mathbf{k}} - \epsilon_{\mathbf{p}}\right)\left[T_{\mathbf{kp}}(\epsilon) - T_{\mathbf{kp}}^{\dagger}(\epsilon)\right] = (-2\pi i)^2 \; \delta\left(\epsilon_{\mathbf{k}} - \epsilon_{\mathbf{p}}\right) \sum_{\mathbf{q}} T_{\mathbf{kq}}^{\dagger}(\epsilon) \, \delta(\epsilon_{\mathbf{k}} - \epsilon_{\mathbf{q}}) \, T_{\mathbf{qp}}(\epsilon).$$

The above equation implies that, if we introduce the so-called on-shell S-matrix through

$$S_{\mathbf{kp}}(\epsilon) = \delta_{\mathbf{kp}} - 2\pi i \; \delta\left(\epsilon_{\mathbf{k}} - \epsilon_{\mathbf{p}}\right) T_{\mathbf{kp}}(\epsilon), \qquad (7.17)$$

where, as before, $\epsilon_{\mathbf{k}} = \epsilon_{\mathbf{p}} = \epsilon$, then it follows that the S-matrix is unitary, i.e., $\hat{S} \, \hat{S}^{\dagger} = \hat{I}$. $S_{\mathbf{kp}}(\epsilon)$ is the transition probability that an electron in state $\mathbf{k}$ scatters elastically into state $\mathbf{p}$. Since it is unitary, it follows that only elastic scattering survives.

Let us consider the simpler case of a spherical Fermi surface, i.e. $\epsilon_{\mathbf{k}} = \epsilon_k$ depending only on the modulous of the wavevector. In addition we assume that the matrix elements $V_{\mathbf{kp}}$ only depend on the angle between the two wavevectors, $\theta_{\mathbf{kp}}$, hence can be expanded in Legendre polynomials

$$V_{\mathbf{kp}} = \sum_{l} V_l \, (2l + 1) \, P_l \left(\cos\theta_{\mathbf{kp}}\right), \qquad (7.18)$$

as well as the T-matrix

$$T_{\mathbf{kp}}(z) = \sum_{l} t_l(z) \, (2l + 1) \, P_l \left(\cos\theta_{\mathbf{kp}}\right). \qquad (7.19)$$

Since

$$\int \frac{d\Omega_q}{4\pi} \, P_l\big(\cos\theta_{kq}\big) \, P_{l'}\big(\cos\theta_{qp}\big) = \frac{1}{2l+1} \, \delta_{ll'} \, P_l\big(\cos\theta_{kp}\big),$$

the optical theorem transforms into

$$t_l\,(\omega) - t_l^*(\omega) = 2i \, \mathrm{Im}\, t_l\,(\omega) = -2\pi i \, \rho_0(\omega) \, \big|t_l\,(\omega)\big|^2, \tag{7.20}$$

where $\rho_0(\omega)$ is the bare conduction electron density of states per spin. Since

$$t_l(\omega) = \big|t_l(\omega)\big| \, e^{i\delta_l(\omega)},$$

it follows that

$$t_l(\omega) = -\frac{1}{\pi\rho_0(\omega)} \, \sin\delta_l(\omega) \, e^{i\delta_l(\omega)}. \tag{7.21}$$

If the concentration of impurities is very low, then the scattering rate suffered by the electrons is given by the Fermi-golden rule with the T-matrix playing the role of the operator driving electron scattering, namely

$$\frac{1}{\tau_k} = 2\pi \, n_i \sum_q T_{kq}^\dagger(\omega) \, \delta(\omega - \epsilon_q) \, T_{qk}(\omega) \left(1 - \cos\theta_{kq}\right),$$

where n_i is the impurity concentration and the last term is a geometric factor which guarantees that forward scattering, $\theta = 0$, does not contribute to the current dissipation. Since

$$(2l+1) \, z \, P_l(z) = (l+1) \, P_{l+1}(z) + l \, P_{l-1}(z),$$

we obtain

$$\frac{1}{n_i \, \tau_k(\omega)} = 2\pi \, \rho(\omega) \sum_{ll'} t_l^*(\omega) \, t_{l'}(\omega) \int \frac{d\Omega_q}{4\pi} \, (2l+1) \, P_l(\cos\theta_{kq})$$

$$\big[(2l'+1) \, P_{l'}(\cos\theta_{kq}) + (l'+1) \, P_{l'+1}(\cos\theta_{kq}) + l' \, P_{l'-1}(\cos\theta_{kq})\big]$$

$$= 2\pi \, \rho(\omega) \sum_l \Big[(2l+1) \, \big|t_l(\omega)\big|^2 + l \, t_l^*(\omega) \, t_{l-1}(\omega) + (l+1) \, t_l^*(\omega) \, t_{l+1}(\omega)\Big],$$

$$\tag{7.22}$$

which as expected gives a scattering time $\tau_k(\omega) \equiv \tau(\omega)$ independent of momentum, whose value at zero energy determines the residual resistivity

$$R(T=0) \propto \frac{1}{\tau(0)}. \tag{7.23}$$

7.1.1 General Analysis of the Phase-Shifts

Let us keep assuming a spherically symmetric case. In addition we assume that the scattering components V_l are non-zero only for a well defined $l = L$. For instance, in the case of magnetic impurities with partially filled d-shell, $L = 2$. Then the only non-zero phase shift is

$$\delta_L(\omega) = \tan^{-1} \frac{\operatorname{Im} t_L(\omega)}{\operatorname{Re} t_L(\omega)}.$$

An important role is played by the frequencies at which the real part vanishes. The first possibility is that occurs outside the conduction band, either below or above. In this case we already know that the imaginary part is zero. This implies that the phase-shift jumps by π at these energies so that the variation of the DOS is δ-like. In this case, one speaks of bound states that appear outside the conduction band. Clearly this possibility can not explain a Curie-Weiss behaviour for $T \ll T_F$. Indeed, at very low temperature the bound state is either doubly occupied, if below the conduction band, or empty, if above, and one needs a temperature larger than the conduction bandwidth, hence larger that T_F, to release its spin entropy.

The other possibility is that the real part vanishes at a frequency ω_* within the conduction band, where the imaginary part is therefore finite. Around ω_*, assuming a real part linearly vanishing and an imaginary part roughly constant, we get

$$\delta_L(\omega) \simeq \tan^{-1} \frac{\Gamma}{\omega - \omega_*},$$

leading to a Lorentzian DOS variation yielded by the impurity

$$\Delta\rho(\omega) \simeq \frac{1}{\pi} \frac{\Gamma}{(\omega - \omega_*)^2 + \Gamma^2}, \tag{7.24}$$

which is called a resonance. Clearly, in other for this resonance to contribute at $T \ll T_F$, ω_* should be very close to the chemical potential, in particular $|\omega_* - \mu| \leq T_K$, and, in addition, $\Gamma \simeq T_K$.

Let us therefore assume $\Gamma = T_K$ and, for simplicity, $\omega_* = 0$. The contribution of the resonant state to the magnetic susceptibility is then given by

$$\Delta\chi(T) = -\mu_B g (2L + 1) \int d\epsilon \, \frac{\partial f(\epsilon)}{\partial \epsilon} \frac{1}{\pi} \frac{T_K}{\epsilon^2 + T_K^2}, \tag{7.25}$$

where $f(\epsilon)$ is the Fermi distribution function. We readily find that for $T \gg T_K$,

$$\Delta\chi \simeq \mu_B g \frac{2L + 1}{T},$$

while for $T \ll T_K$

$$\Delta\chi \simeq \mu_B g \frac{2L + 1}{\pi T_K},$$

consistent with the observed behaviour. In addition, the resistivity would be given through (7.22) and (7.23) by

$$R(0) \propto \frac{1}{\tau(0)} = n_i \frac{2(2L+1)}{\pi \rho_0} \sin^2 \delta_L(0) = n_i \frac{2(2L+1)}{\pi \rho_0}, \qquad (7.26)$$

where $\rho_0 = \rho_0(0)$, again compatible with the almost unitary limit, $\delta(0) = \pi/2$, observed experimentally. Therefore, the existence of a narrow resonance near the chemical potential with width exactly given by T_K seems the natural explanation to what experiments find.

However, this conclusion is rather strange, since the Kondo behaviour is observed in many different host metals and for different magnetic impurities, hence it would be really surprising that in all those cases a resonance appears and it is always pinned near the chemical potential.

Moreover, this simple single-particle scenario fails to explain the entropy released above T_K. Indeed, for $T \gg T_K$, the resonance is effectively like an isolated level which can be empty, singly occupied with a spin up or down, or doubly occupied. Therefore its entropy per impurity should be $S_{\text{imp}} = 2(2L+1)\ln 2$. On the other hand the experiments tells us that only the spin degrees of freedom are released above T_K, which amounts in the above simplified model to an entropy $S_{\text{imp}} = \ln S$, where S is the spin of the isolated atom. In other words, a resonance at the chemical potential and at temperatures higher than its width is not at all the same as a local moment, since the former does have valence fluctuations which are absent in the latter. Therefore, although the resonance scenario is suggestive, it is not at all the solution to the puzzle.

7.2   The Anderson Impurity Model

Something which is obviously missing in the above analysis are the electron-electron correlations at the magnetic ions. A narrow resonance induced by an impurity is quite localised around it, and is essentially akin an atomic level slightly broadened by the hybridizsation with the conduction electrons of the host metal. Therefore, like an atomic level, such resonance can accomodate only a fixed number of electrons, say N, paying a finite amount of energy by adding or removing electrons with respect to the reference valency N. As usual, it is convenient to define a so-called Hubbard repulsion U through

$$U = E(N+1) + E(N-1) - 2E(N), \qquad (7.27)$$

where $E(M)$ is the total energy when the resonance is forced to have M electrons, N being the equilibrium value. In principle one might include this additional ingredient by adding to the Hamiltonian a term

$$\frac{U}{2}\left(\hat{n} - N\right)^2, \qquad (7.28)$$

with $\hat{n}$ the occupation number of the resonance. If it were possible to define such an operator, then (7.28) would actually solve one of the puzzles. Indeed, if $U \gg T_K$ and for temperatures $T_K \ll T \ll U$, valence fluctuations with respect to the equilibrium value N would be suppressed and the only degrees of freedom contributing to the entropy would be those related to the degeneracy of the N-electron configurations at the resonance. Assuming the angular momentum quenched by the crystal field, those residual degrees of freedom are just the spin ones, now consistent with experiments.

Unfortunately the resonance occupation number operator is an ill defined object. To overcome such difficulty, Anderson had the idea to represent the resonance as an additional level inside the conduction band, even if that would be, rigorously speaking, an overcomplete basis. This leads to the so-called Anderson Impurity Model (AIM) [1] defined, in the simplest case of a single-orbital impurity, as

$$H = \sum_{\mathbf{k}\sigma} \epsilon_{\mathbf{k}} c^{\dagger}_{\mathbf{k}\sigma} c_{\mathbf{k}\sigma} + \sum_{\mathbf{k}\sigma} V_{\mathbf{k}} \left(c^{\dagger}_{\mathbf{k}\sigma} d_{\sigma} + d^{\dagger}_{\sigma} c_{\mathbf{k}\sigma} \right) + \frac{U}{2}(n-1)^2 + \epsilon_d\, n.$$

(7.29)

This model describes a band of conduction electrons with energy dispersion $\epsilon_{\mathbf{k}}$, hybridised with a single level, d_{σ}, with energy ϵ_d, both $\epsilon_{\mathbf{k}}$ and ϵ_d being measured with respect to the chemical potential, and $n = n_{\uparrow} + n_{\downarrow}$ with $n_{\sigma} = d^{\dagger}_{\sigma} d_{\sigma}$. Because of $U > 0$, this level tends to accomodate no more no less than a single electron.

Before switching the hybridisation $V_{\mathbf{k}}$ and the interaction U, the conduction electron Green's function in complex frequency z is

$$G^{(0)}(z, \mathbf{k}) = \frac{1}{z - \epsilon_{\mathbf{k}}},$$

(7.30)

and is represented as a dashed tiny line in Fig. 7.1, while the impurity one, solid tiny line in Fig. 7.1, is

$$\mathcal{G}^{(0)}(z) = \frac{1}{z - \epsilon_d}.$$

(7.31)

The impurity Green's function at finite hybridisation, the solid black line in Fig. 7.1, can be readily calculated through Dyson's equation,

$$\mathcal{G}(z) = \frac{1}{z - \epsilon_d - \Delta(z)},$$

(7.32)

where the so-called hybridisation function is defined through

$$\Delta(z) = \sum_{\mathbf{k}} \frac{V^2_{\mathbf{k}}}{z - \epsilon_{\mathbf{k}}}.$$

(7.33)

A finite U yields an impurity self-energy $\Sigma(z)$, so that the fully interacting impurity Green's function $\mathcal{G}(z)$, blue bold line in Fig. 7.1, reads read

$$\mathcal{G}(z) = \frac{1}{\mathcal{G}_0(z)^{-1} - \Sigma(z)} = \frac{1}{z - \epsilon_d - \Delta(z) - \Sigma(z)}.$$

(7.34)

Correspondingly, the interacting conduction electron Green's function, $G(z, \mathbf{k}, \mathbf{k}')$, not anymore diagonal in momentum due to the impurity breaking translational symmetry, and the mixed conduction electron-impurity Green's function, $\overline{G}(z, \mathbf{k})$, are, see Fig. 7.1

$$
\begin{aligned}
G(z, \mathbf{k}, \mathbf{k}') &= \delta_{\mathbf{k}, \mathbf{k}'}\, G^{(0)}(z, \mathbf{k}) + G^{(0)}(z, \mathbf{k})\, V_{\mathbf{k}}\, \mathcal{G}(z)\, V_{\mathbf{k}'}\, G^{(0)}(z, \mathbf{k}') \\
&\equiv \delta_{\mathbf{k}, \mathbf{k}'}\, G^{(0)}(z, \mathbf{k}) + G^{(0)}(z, \mathbf{k})\, T_{\mathbf{k}\mathbf{k}'}(z)\, G^{(0)}(z, \mathbf{k}') ,
\end{aligned}
\tag{7.35}
$$
$$
\overline{G}(z, \mathbf{k}) = G^{(0)}(z, \mathbf{k})\, V_{\mathbf{k}}\, \mathcal{G}(z) .
$$

The hybridisation function has a branch cut on the real axis,

$$
\Delta(z \to \epsilon \pm i0^+) = \mp i\pi \sum_{\mathbf{k}} V_{\mathbf{k}}^2\, \delta(\epsilon - \epsilon_{\mathbf{k}}) \equiv \mp i\Gamma(\epsilon),
$$

with $\Gamma(\epsilon) > 0$, so that

$$
\Delta(z) = \int \frac{d\epsilon}{\pi}\, \frac{\Gamma(\epsilon)}{z - \epsilon} .
$$

Since the physics of interest occurs on energy scales $\ll T_F$, we can safely assume that $\Gamma(\epsilon) = \Gamma$ for $|\epsilon| < W$ and $\Gamma(\epsilon) = 0$ otherwise, with W the conduction bandwidth. In other words, we assume that both conduction electron DOS, $\rho_0(\epsilon)$, and $\Gamma(\epsilon)$ are particle-hole symmetric, i.e., are even function of ϵ, which is measured with respect to the chemical potential. Therefore, if $|z| \ll W$,

$$
\Delta(z) \simeq -i\Gamma\, \mathrm{sgn}\Big(\mathrm{Im}\, z\Big),
\tag{7.36}
$$

and thus

$$
\mathcal{G}(z) = \frac{1}{z - \epsilon_d - \Sigma(z) + i\Gamma\, \mathrm{sgn}\Big(\mathrm{Im}\, z\Big)} .
\tag{7.37}
$$

7.2.1 Non Interacting Impurity

If $U = 0$, thus $\Sigma(z) = 0$, the impurity Green's function (7.37) is simply

$$
\mathcal{G}_0(z) = \frac{1}{z - \epsilon_d + i\Gamma\, \mathrm{sgn}\Big(\mathrm{Im}\, z\Big)} ,
\tag{7.38}
$$

from which the impurity DOS readily follows

$$
\begin{aligned}
A_0(\epsilon) &= -\frac{1}{\pi}\, \mathrm{Im}\, \mathcal{G}_0(\epsilon + i0^+) = \frac{1}{\pi}\, \frac{\Gamma}{(\epsilon - \epsilon_d)^2 + \Gamma^2} \\
&= \frac{1}{\pi}\, \frac{\partial}{\partial\epsilon}\, \tan^{-1} \frac{-\Gamma}{\epsilon - \epsilon_d} = \frac{1}{\pi}\, \frac{\partial\, \mathrm{Im}\, \ln \mathcal{G}_0(\epsilon + i0^+)}{\partial\epsilon} ,
\end{aligned}
\tag{7.39}
$$

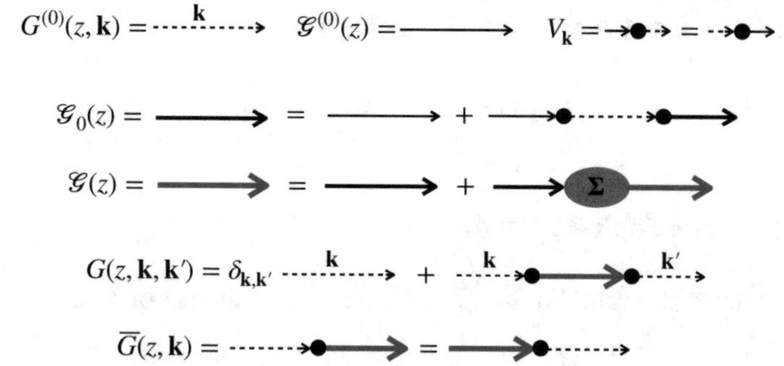

Fig. 7.1 Green's functions of the Hamiltonian (7.29). $G^{(0)}(z, \mathbf{k})$ (dashed line) and $\mathcal{G}^{(0)}(z)$ (solid line) are, respectively, the Green's functions of conduction electrons and impurity in the absence of hybridisation, the black dot vertex with dashed and solid external legs, and for $U = 0$. $\mathcal{G}(z)$ is the impurity Green's function that includes the hybridisation but still at $U = 0$, which is obtained by the shown Dyson equation. A finite U yields an impurity self-energy $\Sigma(z)$, in red, that defines the fully interacting impurity Green's function $\mathcal{G}(z)$, blue bold line, through Dyson's equation. In turn, $\mathcal{G}(z)$ determines the interacting electron Green's function $G(z, \mathbf{k}, \mathbf{k}')$, not anymore diagonal in $\mathbf{k}$, and the mixed impurity-conduction electron Green's function $\overline{G}(z, \mathbf{k})$

which is just a resonance within the conduction band for $|\epsilon_d| < W$. Therefore, at $U = 0$, the Anderson impurity model (7.29) does yield the desired physical behaviour.

Correspondingly, the non interacting conduction electron and mixed Green's functions are

$$G_0(z, \mathbf{k}, \mathbf{k}') = \delta_{\mathbf{k}, \mathbf{k}'} \, G^{(0)}(z, \mathbf{k}) + G^{(0)}(z, \mathbf{k}) \, V_{\mathbf{k}} \, \mathcal{G}_0(z) \, V_{\mathbf{k}'} \, G^{(0)}(z, \mathbf{k}') \,,$$
$$\overline{G}_0(z, \mathbf{k}) = G^{(0)}(z, \mathbf{k}) \, V_{\mathbf{k}} \, \mathcal{G}_0(z) \,.$$

The variation of electron number at fixed chemical is therefore

$$\Delta N = \sum_\sigma T \sum_\epsilon e^{i\epsilon 0^+} \left\{ \mathcal{G}_0(i\epsilon) + \sum_\mathbf{k} G_0(i\epsilon, \mathbf{k}, \mathbf{k}) \right\}$$

$$= 2T \sum_\epsilon e^{i\epsilon 0^+} \mathcal{G}_0(i\epsilon) \left\{ 1 + \sum_\mathbf{k} \frac{V_\mathbf{k}^2}{(i\epsilon - \epsilon_\mathbf{k})^2} \right\} = 2T \sum_\epsilon e^{i\epsilon 0^+} \mathcal{G}_0(i\epsilon) \left\{ 1 - \frac{\partial \Delta(i\epsilon)}{\partial i\epsilon} \right\} \qquad (7.40)$$

$$= -2T \sum_\epsilon e^{i\epsilon 0^+} \frac{\partial \ln \mathcal{G}_0(i\epsilon)}{\partial i\epsilon} = 2 \int \frac{d\epsilon}{\pi} f(\epsilon) \frac{\partial \text{Im} \ln \mathcal{G}_0(\epsilon + i0^+)}{\partial \epsilon} = 2 \int d\epsilon \, f(\epsilon) \, A_0(\epsilon) \,,$$

and is just the expectation value $\langle n \rangle$ of the number of electrons n occupying the impurity. That is a consequence of our assumption (7.36). More generally, ΔN is equal to $\langle n \rangle$ plus corrections of order Γ/W, which we assume vanishingly small.

At zero temperature, (7.40) becomes

$$\Delta N = \langle n \rangle = \langle n_\uparrow + n_\downarrow \rangle = 2 \int_{-\infty}^0 \frac{d\epsilon}{\pi} \frac{\partial}{\partial \epsilon} \tan^{-1} \frac{-\Gamma}{\epsilon - \epsilon_d} = \frac{2}{\pi} \left(\tan^{-1} \frac{-\Gamma}{-\epsilon_d} + \pi \right)$$

$$= \frac{2}{\pi} \tan^{-1} \frac{\Gamma}{\epsilon_d} = 1 - \frac{2}{\pi} \tan^{-1} \frac{\epsilon_d}{\Gamma} \,. \qquad (7.41)$$

If a magnetic field is present that splits $\epsilon_{d\uparrow}$ from $\epsilon_{d\downarrow}$, then

$$\langle n_\sigma \rangle = \frac{1}{2} - \frac{1}{\pi} \tan^{-1} \frac{\epsilon_{d\sigma}}{\Gamma} \,. \qquad (7.42)$$

7.2.2 Hartree-Fock Approximation

When U is finite, the Hamiltonian (7.29) includes the interaction term

$$\frac{U}{2} (n-1)^2 = U n_\uparrow n_\downarrow - \frac{U}{2} n \,.$$

The second term renormalises $\epsilon_d \to \epsilon_d - U/2$. Let us treat the interaction $U n_\uparrow n_\downarrow$ within the Hartree-Fock approximation [1], allowing for spontaneous spin-polarisation of the impurity, thus $\mathcal{G}_\uparrow(z) \neq \mathcal{G}_\downarrow(z)$. The Hartree-Fock self-energy in Matsubara frequencies and for spin-σ impurity electrons contains in our case only the Hartree term

$$\Sigma_\sigma(i\epsilon) = U T \sum_\epsilon e^{i\epsilon 0^+} \mathcal{G}_{-\sigma}(i\epsilon) = U \langle n_{-\sigma} \rangle \,, \qquad (7.43)$$

where $\langle n_\sigma \rangle$ must be determined self-consistently. Therefore the impurity Green's function is

$$\mathcal{G}_\sigma(\epsilon + i0^+) = \frac{1}{\epsilon - \epsilon_d + U/2 - U \langle n_{-\sigma} \rangle + i \Gamma} \equiv \frac{1}{\epsilon - \epsilon_{d\sigma} + i \Gamma} \,, \qquad (7.44)$$

from which it readily follows through (7.42) that

$$\langle n_\sigma \rangle = \frac{1}{2} - \frac{1}{\pi} \tan^{-1} \frac{\epsilon_{d\sigma}}{\Gamma} = \frac{1}{2} - \frac{1}{\pi} \tan^{-1} \frac{\epsilon_d - U/2 + U \langle n_{-\sigma} \rangle}{\Gamma} \,. \qquad (7.45)$$

We define $\langle n_\uparrow \rangle = (n+m)/2$ and $\langle n_\downarrow \rangle = (n-m)/2$, with $n \in [0, 2]$ and $m = [-1, 1]$, so that the self-consistency conditions are

$$n = 1 - \frac{1}{\pi} \tan^{-1} \frac{2\epsilon_d - U + Un - Um}{2\Gamma} - \frac{1}{\pi} \tan^{-1} \frac{2\epsilon_d - U + Un + Um}{2\Gamma} \,,$$

$$m = \frac{1}{\pi} \tan^{-1} \frac{2\epsilon_d - U + Un + Um}{2\Gamma} - \frac{1}{\pi} \tan^{-1} \frac{2\epsilon_d - U + Un - Um}{2\Gamma} \equiv F(m) \,. \qquad (7.46)$$

In order to assess under which conditions a spontaneous magnetisation emerges, i.e., a solution with $m \neq 0$ exists, we note that the function $F(m)$ on the right hand side of the second equation increases monotonically in m, and, in addition, $F(0) = 0$ while $F(\pm\infty) = \pm 1$. Therefore a solution $m = \pm m_*$, with $m_* > 0$, exists if and only if $F'(0) \geq 1$, namely

$$\frac{1}{\pi} \frac{\Gamma}{\left(\epsilon_d - U/2 + Un/2\right)^2 + \Gamma^2} U = A_{m=0}(0) U \geq 1 \,, \qquad (7.47)$$

Fig. 7.2 Impurity density of states of the Anderson impurity model (7.29) within the Hartree-Fock approximation for $U \gg \Gamma, |\epsilon_d|$

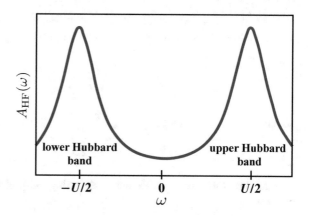

where $A_{m=0}(0)$ is the Hartree-Fock impurity DOS at the chemical potential in absence of spontaneous magnetisation. Specifically, if $U \gg \Gamma, |\epsilon_d|$, (7.46) has solution $n = 1$ and $m = \pm 1$, i.e., the impurity is half-filled and spin polarised. Since the Hamiltonian (7.29) has spin $SU(2)$ symmetry, we can freely rotate the axis of spin polarisation and get always the same result. Therefore, if U is large the impurity effectively behaves as $S = 1/2$ local moment, in accordance with the experimental observation at $T_K \ll T \ll T_F$. In other words, the Curie-like behaviour is consistent with the Anderson impurity model (7.29) if $U \gg \Gamma, |\epsilon_d|$, which we assume hereafter.

Since, as we mentioned, the direction of the impurity spin is arbitrary, averaging over that direction yields a spin-independent impurity Green's function

$$\mathcal{G}(\epsilon + i0^+) \simeq \frac{1}{2} \left(\frac{1}{\epsilon + U/2 + i\Gamma} + \frac{1}{\epsilon + U/2 + i\Gamma} \right), \qquad (7.48)$$

and thus the impurity DOS is composed by two Lorentzian functions, one centred at $-U/2$ below the chemical potential, and the other at $U/2$ above, which are called the Mott-Hubbard bands, see Fig. 7.2. In conclusion, the large U Hartree-Fock solution of the Anderson impurity model (7.29) seems to account for the behaviour above the Kondo temperature, and satisfactorily explains how local moments appear. The key ingredient is a large Coulomb repulsion at the magnetic impurity as compared to the resonance broadening Γ, which suppresses valence fluctuations. However, we still have to understand what happens as temperature decreases.

7.3   From the Anderson Impurity Model to the Kondo Model

The Hartree-Fock solution of the Anderson impurity model (7.29) does explain in the large U/Γ-limit the existence of free local moments in a metal well below its Fermi temperature. Nonetheless, the $SU(2)$ symmetry breaking of the mean-field solution is an artefact, since $SU(2)$ symmetry cannot be locally broken by the Mermin-Wagner theorem. The issue in that the Hamiltonian still allows for a matrix element coupling two solutions corresponding to different orientations of the impurity spin.

In order to understand the role of such coupling, it is more convenient to introduce the conduction wave-function at the impurity site through

$$c_{0\sigma} = \sqrt{\frac{1}{t}} \sum_{\mathbf{k}} V_{\mathbf{k}} \, c_{\mathbf{k}\sigma} \,, \qquad (7.49)$$

where

$$t^2 = \sum_{\mathbf{k}} V_{\mathbf{k}}^2 \,,$$

so that the impurity Hamiltonian, neglecting $|\epsilon_d| \ll U$, plus the hybridisation becomes

$$t \sum_{\sigma} \left(c_{0\sigma}^{\dagger} \, d_{\sigma} + d_{\sigma}^{\dagger} \, c_{0\sigma} \right) + \frac{U}{2} \, (n-1)^2 \,. \qquad (7.50)$$

In the large U limit, the impurity traps a single electron, either with spin up or down, and we can treat the hybridisation as a small perturbation lifting the degeneracy between the two spin direction. Second order perturbation theory through intermediate states in which the impurity is either empty or doubly occupied, with energy difference $U/2$ for large U, yields the following effective Hamiltonian

$$H_K = \sum_{\mathbf{k}\sigma} \epsilon_{\mathbf{k}} \, c_{\mathbf{k}\sigma}^{\dagger} \, c_{\mathbf{k}\sigma} + J_K \, \mathbf{S}_0 \cdot \mathbf{S} \,, \qquad (7.51)$$

where

$$J_K = \frac{8 V^2}{U} \,,$$

and

$$\mathbf{S}_0 = \frac{1}{2} \sum_{\alpha\beta} c_{0\alpha}^{\dagger} \, \boldsymbol{\sigma}_{\alpha\beta} \, c_{0\beta} = \frac{1}{2t^2} \sum_{\mathbf{kp}\alpha\beta} V_{\mathbf{k}} \, V_{\mathbf{p}} \, c_{\mathbf{k}\alpha}^{\dagger} \, \boldsymbol{\sigma}_{\alpha\beta} \, c_{\mathbf{p}\beta} \,, \qquad \mathbf{S} = \frac{1}{2} \sum_{\alpha\beta} d_{\alpha}^{\dagger} \, \boldsymbol{\sigma}_{\alpha\beta} \, d_{\beta} \,,$$

$$(7.52)$$

are the conduction electron spin-density operator at the impurity site, $\mathbf{S}_0$, and the impurity spin-1/2 operator $\mathbf{S}$. The Hamiltonian (7.51) is known as the Kondo model [2], and describes conduction electrons antiferromagnetically coupled to a local moment.

We note that H_K has built-in a local moment, hence by construction correctly describes the regime $T_K \ll T \ll T_F$. That local moment provides a scattering potential for the conduction electrons. The major difference with respect to a conventional scalar potential is that the Kondo exchange has a non-trivial structure yielded by the commutation relations between spin operators, whose consequences we now investigate.

Fig. 7.3 First order corrections to the Kondo exchange. Solid and dashed lines indicate impurity and conduction electron Green's functions. Also indicated are the spin labels. All external lines are assumed to be at zero frequency $\epsilon = 0$. The internal frequency, which is going to be summed over, is also indicated

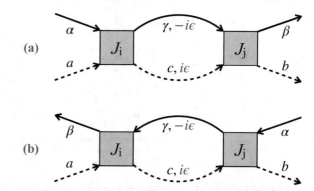

7.3.1 The Emergence of Logarithmic Singularities and the Kondo Temperature

Let us analyse the role of the Kondo exchange

$$J_K \, \mathbf{S}_0 \cdot \mathbf{S} \,, \tag{7.53}$$

in perturbation theory [2]. Since (7.53) conserves independently the number of conduction and d-electrons, we treat the impurity spin in terms of d-electrons with an unperturbed Green's function

$$\mathcal{G}^{(0)}(z) = \frac{1}{z} \,,$$

namely as a localised state right at the chemical potential, which thus contains just a single electron, number that is not going to be changed by (7.53). Since a single electron has a well defined spin, it indeed acts like a local spin-1/2 moment. Moreover, the Green's function for the conduction electron c_0 is

$$G_0^{(0)}(z) = \frac{1}{t^2} \sum_{\mathbf{k}} \frac{V_{\mathbf{k}}^2}{z - \epsilon_{\mathbf{k}}} = \frac{1}{t^2} \Delta(z) \,. \tag{7.54}$$

For convenience let us rewrite (7.53) in a spin-asymmetric form as

$$\sum_{i=x,y,z} J_i \, S_0^i \, S^i \,, \tag{7.55}$$

and calculate the first order corrections to the exchange as given by the diagrams in Fig. 7.3. For simplicity we assume that all external lines are at zero frequencies. Each diagram is multiplied by (-1) because of first order perturbation theory. The explicit expression of diagram (a) is

$$
\begin{aligned}
\text{(a)} &= - \sum_{i,j=x,y,z} J_i \, J_j \sum_{\gamma c} S^i_{\alpha\gamma} \, S^i_{ac} \, S^j_{\gamma\beta} \, S^j_{cb} \, T \sum_{\epsilon} G^{(0)}_0(i\epsilon) \, \mathcal{G}^{(0)}(-i\epsilon) \\
&= - \sum_{i,j=x,y,z} \frac{J_i \, J_j}{t^2} \sum_{\gamma c} S^i_{\alpha\gamma} \, S^i_{ac} \, S^j_{\gamma\beta} \, S^j_{cb} \, T \sum_{\epsilon} \frac{\Delta(i\epsilon)}{-i\epsilon} \\
&= -I(T) \sum_{i,j=x,y,z} \frac{J_i \, J_j}{t^2} \sum_{\gamma c} S^i_{\alpha\gamma} \, S^j_{\gamma\beta} \, S^i_{ac} \, S^j_{cb},
\end{aligned}
$$

where the function of temperature $I(T)$ is defined through

$$
\begin{aligned}
I(T) &= -T \sum_{\epsilon} \frac{\Delta(i\epsilon)}{i\epsilon} = \int \frac{d\epsilon}{\pi} f(\epsilon) \, \mathrm{Im}\left(\frac{\Delta(\epsilon + i0^+)}{\epsilon} \right) \\
&= -\Gamma \int_{-W}^{W} \frac{d\epsilon}{\pi} f(\epsilon) \frac{1}{\epsilon} = -\Gamma \int_{0}^{W} \frac{d\epsilon}{\pi} \frac{1}{\epsilon} \big(f(\epsilon) - f(-\epsilon) \big) \qquad (7.56) \\
&= \Gamma \int_{0}^{W} \frac{d\epsilon}{\pi} \frac{1}{\epsilon} \tanh \frac{\epsilon}{2T} \simeq \Gamma \int_{T}^{W} \frac{d\epsilon}{\pi} \frac{1}{\epsilon} = \frac{\Gamma}{\pi} \ln \frac{W}{T} .
\end{aligned}
$$

Similarly, diagram (b) is

$$
\begin{aligned}
\text{(b)} &= - \sum_{i,j=x,y,z} J_i \, J_j \sum_{\gamma c} S^i_{\alpha\gamma} \, S^i_{cb} \, S^j_{\gamma\beta} \, S^j_{ac} \, T \sum_{\epsilon} G^{(0)}_0(i\epsilon) \, \mathcal{G}^{(0)}(i\epsilon) \\
&= I(T) \sum_{i,j=x,y,z} \frac{J_i \, J_j}{t^2} \sum_{\gamma c} S^i_{\alpha\gamma} \, S^j_{\gamma\beta} \, S^j_{ac} \, S^i_{cb},
\end{aligned}
$$

hence the sum is

$$
\begin{aligned}
\text{(a)} + \text{(b)} &= I(T) \sum_{i,j=x,y,z} \frac{J_i \, J_j}{t^2} \sum_{\gamma} S^i_{\alpha\gamma} \, S^j_{\gamma\beta} \sum_{c} \left(S^j_{ac} \, S^i_{cb} - S^i_{ac} \, S^j_{cb} \right) \\
&= I(T) \sum_{i,j,k=x,y,z} \frac{J_i \, J_j}{t^2} \sum_{\gamma} S^i_{\alpha\gamma} \, S^j_{\gamma\beta} \, i \, \epsilon_{jik} \, S^k_{ab},
\end{aligned}
$$

where we used the commutation relations between the spin operators

$$
\left[S^i, S^j \right] = i \, \epsilon_{ijk} \, S^k,
$$

where ϵ_{ijk} is the antisymmetric tensor. Since we sum over i and j, we can also write

$$(a) + (b) = \frac{1}{2} I(T) \sum_{i,j,k=x,y,z} \frac{J_i J_j}{t^2} \sum_\gamma \left[S_{\alpha\gamma}^i S_{\gamma\beta}^j \, i \, \epsilon_{jik} S_{ab}^k + (i \leftrightarrow j) \right]$$

$$= \frac{1}{2} I(T) \sum_{i,j,k=x,y,z} \frac{J_i J_j}{t^2} \, i \, \epsilon_{jik} S_{ab}^k \sum_\gamma \left(S_{\alpha\gamma}^i S_{\gamma\beta}^j - S_{\alpha\gamma}^j S_{\gamma\beta}^i \right)$$

$$= \sum_{k=x,y,z} S_{ab}^k S_{\alpha\beta}^k \left[\frac{1}{2} I(T) \sum_{i,j} \frac{J_i J_j}{t^2} \left| \epsilon_{ijk} \right|^2 \right] ,$$

which provides the first order correction to the exchange constants according to

$$\begin{aligned} J_k + \delta J_k &= J_k + \frac{1}{2} I(T) \sum_{i,j} \frac{J_i J_j}{t^2} \left| \epsilon_{ijk} \right|^2 \\ &= J_k + \frac{1}{2\pi} \frac{\Gamma}{t^2} \ln \frac{W}{T} \sum_{i,j} J_i J_j \left| \epsilon_{ijk} \right|^2 , \end{aligned} \tag{7.57}$$

namely,

$$\delta J_x = \frac{\Gamma}{\pi t^2} \ln \frac{W}{T} J_y J_z , \quad \delta J_y = \frac{\Gamma}{\pi t^2} \ln \frac{W}{T} J_z J_x , \quad \delta J_z = \frac{\Gamma}{\pi t^2} \ln \frac{W}{T} J_x J_y . \tag{7.58}$$

Therefore, perturbation theory generates logarithmic singularities [2] that become visible roughly around a temperature, which has to be identified with the Kondo temperature T_K, at which the correction exceeds the bare exchange. In the isotropic case, $J_x = J_y = J_z = J$, that implies

$$J = \frac{\Gamma}{\pi t^2} J^2 \ln \frac{W}{T_K} ,$$

thus

$$T_K = W \exp\left(-\frac{\pi t^2}{\Gamma J} \right) = W \exp\left(-\frac{\pi U}{8\Gamma} \right) \ll W \sim T_F , \tag{7.59}$$

since $U \gg \Gamma$. For instance, the resistivity calculated through the Fermi golden rule, see (7.22) and (7.23), at first order in perturbation theory would looks like

$$R(T) - R_0(T) \propto n_i \, \rho_0 \left(J + \frac{\Gamma}{\pi t^2} J^2 \ln \frac{W}{T_K} \right)^2 ,$$

where $R_0(T)$ is the value in absence of magnetic impurities and decreases monotonically lowering T. It follows a minimum metallic resistivity around T_K, followed by a logarithmic raise, just like in experiments. In agreement with the latter, we do find that the Kondo temperature is much smaller than the host-metal Fermi temperature. Since perturbation theory becomes meaningless below T_K, the next obvious question is how to proceed further.

7.3.2   Anderson's Poor Man's Scaling

If $J_x = J_y \equiv J_\perp \neq J_z$, through (7.58) we can write

$$
\begin{aligned}
J_z[W] + \delta J_z[W] &= J_z[W] + \frac{\Gamma}{\pi\, t^2}\, J_\perp[W]^2\, \ln\frac{W}{T}\,, \\
J_\perp[W] + \delta J_\perp[W] &= J_\perp[W] + \frac{\Gamma}{\pi\, t^2}\, J_\perp[W]\, J_z[W]\, \ln\frac{W}{T}\,.
\end{aligned}
\tag{7.60}
$$

In (7.60) we explicitly indicate the dependence upon the high-energy cutoff W. Now suppose we have another model with a smaller conduction electron bandwidth cutoff $W(\lambda) = W/\lambda$ with $\lambda > 1$, different J's but equal Γ/t^2. In this case, we would get the first order corrections

$$
\begin{aligned}
J_z[W(\lambda)] + \delta J_z[W(\lambda)] &= J_z[W(\lambda)] + \frac{\Gamma}{\pi\, t^2}\, J_\perp[W(\lambda)]^2\, \ln\frac{W}{\lambda T}\,, \\
J_\perp[W(\lambda)] + \delta J_\perp[W(\lambda)] &= J_\perp[W(\lambda)] + \frac{\Gamma}{\pi\, t^2}\, J_\perp[W(\lambda)]\, J_z[W(\lambda)]\, \ln\frac{W}{\lambda T}\,.
\end{aligned}
$$

On the other hand, it makes not really a big difference for the low energy behaviour whether those electrons close to the chemical potential derive from a bandwidth W or $W(\lambda)$ provided they suffer the same scattering off the impurity. Therefore we can ask the following question [3]: What should $J[W(\lambda)]$ be in order for the effective exchange up to first order to be the same as that with bandwidth W?

The answer is quite simple, since we can, for instance, write

$$
\begin{aligned}
J_z[W] + \delta J_z[W] &= J_z[W] + \frac{\Gamma}{\pi\, t^2}\, J_\perp[W]^2\, \ln\lambda + \frac{\Gamma}{\pi\, t^2}\, J_\perp[W]^2\, \ln\frac{W}{\lambda T} \\
&\simeq \left(J_z[W] + \frac{\Gamma}{\pi\, t^2}\, J_\perp[W]^2\, \ln\lambda \right) \\
&\quad + \frac{\Gamma}{\pi\, t^2}\left(J_z[W] + \frac{\Gamma}{\pi\, t^2}\, J_\perp[W]^2\, \ln\lambda \right)^2 \ln\frac{W}{\lambda T} + O\left(J^3\right),
\end{aligned}
$$

the equality being valid up to the order at which we stop the expansion. Therefore the two models with W and W/λ have the same spin exchange up to first order in perturbation theory provided the bare exchange constants satisfy

$$
\begin{aligned}
J_z[W(\lambda)] &= J_z[W] + \frac{\Gamma}{\pi\, t^2}\, J_\perp[W]^2\, \ln\lambda\,, \\
J_\perp[W(\lambda)] &= J_\perp[W] + \frac{\Gamma}{\pi\, t^2}\, J_\perp[W]\, J_z[W]\, \ln\lambda\,,
\end{aligned}
$$

which can be cast in a differential form

$$
\frac{dj_z(s)}{ds} = 2\, j_\perp(s)^2\,, \qquad \frac{dj_\perp(s)}{ds} = 2\, j_\perp(s)\, j_z(s)\,,
\tag{7.61}
$$

having defined $\lambda = e^s$ with $s \geq 0$, and

$$j_i(s) \equiv J_i \left[W(\lambda) \right] \frac{\Gamma}{2\pi t^2} .$$

The equations (7.61) are exactly the renormalisation group equations (6.89) in Sect. 6.2.8, with g replaced by j_z and g_3 by $j_\perp$.[1] These equations describe how the bare exchange constants have to be modified in order for the model with a reduced bandwidth $W e^{-s}$ to have the low-energy scattering amplitudes equal to those of the original model [3]. The idea behind is the same as in Sect. 6.2.8: if we are able to follow the evolution of $j_i(s)$ until $W e^{-s} \sim T$, at this point we can rely on perturbation theory since $\ln(W/\lambda T) \simeq 0$. This is more or less how Anderson formulated his *poor man's scaling* theory for the Kondo problem in 1970, in essence the first implementation of a Renormalisation Group transformation.

The results can be simply borrowed from Fig. 6.6 recalling that $g = j_z$ and $g_3 = j_\perp$, which we can take positive. In particular, if $j_\perp < -j_z > 0$, the case of an easy-axis ferromagnetic Kondo exchange, $j_\perp(\infty) = 0$ and $j_z(\infty) = j_{z*} < 0$. In this case, the impurity polarises along z and provides an effective local magnetic field for the conduction electrons. This model can be readily solved. In the particular case $j_\perp = -j_z > 0$, both exchange constants flows to zero; the magnetic impurity asymptotically decouples from the conduction electrons.

Everywhere else in the flow diagram 6.6, the exchange constants flows to $+\infty$. As discussed in Sect. 6.2.8, in this case we must first diagonalise the Kondo exchange $J_K \mathbf{S}_0 \cdot \mathbf{S}$, and next treat what is left in perturbation theory. The Kondo exchange is minimised by locking the conduction electron $c_{0\sigma}$ in (7.49) with the impurity spin into a spin-singlet state. After that, the impurity site becomes inaccessible to other conduction electrons, which would break the Kondo singlet, in that equivalent to a scattering potential yielding a $\pi/2$ phase shift. More precisely, the conduction electron local Green's function at the impurity site is, see (7.54),

$$G_0^{(0)}(\epsilon + i0^+) = -i \frac{\Gamma}{t^2} .$$

If the magnetic impurity asymptotically behaves as a conventional scalar potential V_*, the local Green's function changes into, see (7.8),[2]

$$G_0(\epsilon + i0^+) = G_0^{(0)}(\epsilon + i0^+) + G_0^{(0)}(\epsilon + i0^+) \, T(\epsilon + i0^+) \, G_0^{(0)}(\epsilon + i0^+),$$

where now

$$T(\epsilon + i0^+) = \frac{V_*}{1 - V_* \, G_0^{(0)}(\epsilon + i0^+)} .$$

---

[1] We note that the sign of $j_\perp$ is unimportant as it can be changed by an 180° rotation around the z-axis.

[2] The conduction electron combination $c_{0\sigma}$ in (7.49) is the only coupled to the impurity, and thus the T-matrix is diagonal with the only finite element referring to $c_{0\sigma}$.

Since the impurity site in inaccessible after the Kondo singlet is formed, then $G_0(\epsilon + i0^+)$ must vanish, which implies $V_* = \infty$ and thus

$$T(\epsilon + i0^+) = -\frac{1}{G_0^{(0)}(\epsilon + i0^+)} = -i\,\frac{t^2}{\Gamma}\,,$$

hence a $\pi/2$ phase shift, consistent with experiments, as well as the low-temperature screening of the magnetic impurity yielding a Pauli-like magnetic susceptibility.

7.4   Noziéres's Local Fermi Liquid Theory

A better description of the low temperature behaviour can be obtained through the diagrammatic many-body perturbation theory presented in Chap. 4, especially in connection with the microscopic derivation of Landau's Fermi liquid theory in Chap. 5. Indeed, a similar Fermi liquid picture can be also derived for Anderson impurity models, in this case known as Noziéres local Fermi liquid theory [4].

Let us consider again the Anderson impurity model (7.29). The non-interacting and fully-interacting impurity Green's functions are

$$\begin{aligned}
\mathcal{G}_0(i\epsilon) &= \frac{1}{i\epsilon - \epsilon_d + i\,\Gamma\,\mathrm{sign}(\epsilon)} = \int d\omega\,\frac{A_0(\omega)}{i\epsilon - \omega}\,, \\
\mathcal{G}(i\epsilon) &= \frac{1}{i\epsilon - \epsilon_d - \Sigma(i\epsilon) + i\,\Gamma\,\mathrm{sign}(\epsilon)} = \int d\omega\,\frac{A(\omega)}{i\epsilon - \omega}\,,
\end{aligned} \tag{7.62}$$

where $A_0(\omega)$ and $A(\omega)$ are the corresponding densities of states, and we correctly assume unbroken $SU(2)$ symmetry.

We start calculating the contribution to the retarded self-energy given by the second order diagram in Fig. 7.4, exactly as we did in Sect. 4.3.4. We can simply borrow the result of (4.58) and adapt it to the present case, thus finding

$$\begin{aligned}
\Sigma(i\epsilon) = U^2 \int d\epsilon_0\,d\epsilon_1\,d\omega_1\,A_0(\epsilon_0)\,A_0(\epsilon_1)\,A_0(\omega_1)\,\frac{1}{i\epsilon + \omega_1 - \epsilon_0 - \epsilon_1} \\
\left[\,f(\omega_1)\left(1 - f(\epsilon_0)\right)\left(1 - f(\epsilon_1)\right) + \left(1 - f(\omega_1)\right)f(\epsilon_0)\,f(\epsilon_1)\,\right].
\end{aligned} \tag{7.63}$$

Fig. 7.4 Second order self-energy diagram of the Anderson impurity model (7.29)

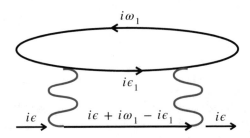

The $\text{Im} \, \Sigma(\epsilon + i0^+)$ for very small $|\epsilon|$ and temperature can be readily obtained as in Sect. 4.3.4,

$$
\begin{aligned}
\text{Im} \, \Sigma(\epsilon + i0^+) &= -\pi \, U^2 \int d\epsilon_0 \, d\epsilon_1 \, d\omega_1 \, A_0(\epsilon_0) \, A_0(\epsilon_1) \, A_0(\omega_1) \, \delta(\epsilon + \omega_1 - \epsilon_0 - \epsilon_1) \\
&\qquad \left[f(\omega_1) \left(1 - f(\epsilon_0)\right) \left(1 - f(\epsilon_1)\right) + \left(1 - f(\omega_1)\right) f(\epsilon_0) \, f(\epsilon_1) \right] \\
&\simeq -\pi \, U^2 \, A_0(0)^3 \int d\epsilon_0 \, d\epsilon_1 \, d\omega_1 \, \delta(\epsilon + \omega_1 - \epsilon_0 - \epsilon_1) \\
&\qquad \left[f(\omega_1) \left(1 - f(\epsilon_0)\right) \left(1 - f(\epsilon_1)\right) + \left(1 - f(\omega_1)\right) f(\epsilon_0) \, f(\epsilon_1) \right] \\
&\simeq -\frac{\pi}{2} \, U^2 \, A_0(0)^3 \left(\epsilon^2 + \pi^2 \, T^2 \right).
\end{aligned}
$$

(7.64)

Therefore, exactly like in the case of bulk interacting electrons, we arrive at the conclusion that, at any order in perturbation theory, $\text{Im} \, \Sigma(\epsilon + i0^+)$ vanishes at least quadratically in ϵ at $T = 0$. Since at $U = 0$ as well as at large U the Hamiltonian (7.29) supposedly describes a resonant level model with interaction dependent width, we can safely assume that perturbation theory never breaks down, and thus that the result $\text{Im} \, \Sigma(\epsilon + i0^+) \simeq -\gamma \, \epsilon^2$ holds true beyond the perturbative regime. It follows that, for very small $|\epsilon|$ at $T = 0$, and upon defining the conventional quasiparticle residue

$$
Z(\epsilon) \equiv \left(1 - \frac{\partial \text{Re} \, \Sigma(\epsilon + i0^+)}{\partial \epsilon}\right)^{-1},
$$

(7.65)

the inverse of the retarded impurity Green's function

$$
\begin{aligned}
\mathcal{G}(\epsilon)^{-1} &= \epsilon - \epsilon_d - \Sigma(\epsilon + i0^+) + i\Gamma \simeq Z^{-1}(0) \, \epsilon - \epsilon_d - \Sigma(0) + i\Gamma \\
&= Z^{-1}(0) \left[\epsilon - Z(0)\left(\epsilon_d + \Sigma(0)\right) + i \, Z(0) \, \Gamma \right] \equiv Z^{-1}(0) \left(\epsilon - \epsilon_{d*} + i\Gamma_* \right),
\end{aligned}
$$

where we have neglected $\gamma \, \epsilon^2$ which is much smaller than Γ, and we recall that $\Sigma(0) \in \mathbb{R}$. In conclusion,

$$
\mathcal{G}(\epsilon) \simeq \frac{Z(0)}{\epsilon - \epsilon_{d*} + i\Gamma_*} + \mathcal{G}_{\text{inc}}(\epsilon) \equiv Z(0) \, \mathcal{G}_{\text{qp}}(\epsilon) + \mathcal{G}_{\text{inc}}(\epsilon) ,
$$

(7.66)

where $\mathcal{G}_{\text{inc}}(\epsilon)$ is the high-energy incoherent component that we know must describe lower and upper Hubbard bands, on top of which a narrow resonance of weight $Z(0)$ appears near zero energy, see Fig. 7.5. Filtering out $Z(\epsilon)$ yields the quasiparticle Green's function, $\mathcal{G}_{\text{qp}}(\epsilon)$, that describes a resonant level with width $Z(0) \, \Gamma$ that must be identified with the Kondo temperature T_K, and centred at

$$
\epsilon_{d*} = Z(0) \left(\epsilon_d + \Sigma(0) \right) = \frac{T_K}{\Gamma} \left(\epsilon_d + \Sigma(0) \right).
$$

(7.67)

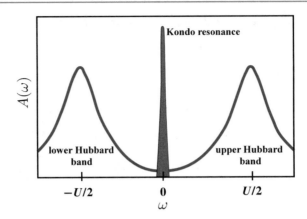

Fig. 7.5 Sketch of the interacting impurity density of states at low temperature. Besides the high energy Hubbard bands, a narrow resonance emerges at low temperature that is pinned at zero energy and whose width is the Kondo temperature T_K. This behaviour is believed to be the paradigm of the local density of states of correlated metals close to a Mott transition, where the resonance correspond to a very narrow band of quasiparticles

To get an order of magnitude of $\epsilon_d + \Sigma(0)$, let us recall the Hartree-Fock result (7.46) at $m = 0$, i.e., when $\Sigma(0) = \Sigma_{HF} = -U(1-n)/2 \equiv -U\,\delta n/2$, and assuming $\delta n \simeq 0$,

$$\delta n = \frac{2}{\pi} \tan^{-1} \frac{2\epsilon_d - U\,\delta n}{2\Gamma} \;\Rightarrow\; \epsilon_d - \frac{U}{2}\,\delta n = \Gamma \tan \frac{\pi\,\delta n}{2} \simeq \Gamma\,\frac{\pi\,\delta n}{2}\;.$$

In this case

$$\epsilon_d + \Sigma_{HF} = \epsilon_d - \frac{U}{2}\,\delta n \simeq \frac{\epsilon_d}{1 + U/\pi\Gamma} \simeq \frac{\pi\Gamma}{U}\,\epsilon_d\;.$$

Therefore, if we approximate $\Sigma(0) \simeq \Sigma_{HF}$, then

$$\epsilon_{d*} \simeq \frac{\pi\,T_K}{U}\,\epsilon_d \;\Rightarrow\; |\epsilon_{d*}| \ll T_K\;, \tag{7.68}$$

namely, the resonance is essentially pinned at the chemical potential, consistently with the experimental observations.

After the above preamble, let us apply the whole diagrammatic technique to better analyse the Anderson impurity model, which we here rewrite for convenience,

$$
\begin{aligned}
H &= \sum_{\mathbf{k}\sigma} \epsilon_{\mathbf{k}}\, c^\dagger_{\mathbf{k}\sigma} c_{\mathbf{k}\sigma} + \sum_{\mathbf{k}\sigma} V_{\mathbf{k}} \left(c^\dagger_{\mathbf{k}\sigma} d_\sigma + d^\dagger_\sigma c_{\mathbf{k}\sigma} \right) + \epsilon_d\, n + \frac{U}{2}\,(n-1)^2 \\
&= \sum_{\mathbf{k}\sigma} \epsilon_{\mathbf{k}}\, c^\dagger_{\mathbf{k}\sigma} c_{\mathbf{k}\sigma} + t \sum_\sigma \left(c^\dagger_{0\sigma} d_\sigma + d^\dagger_\sigma c_{0\sigma} \right) + \epsilon_d\, n + \frac{U}{2}\,(n-1)^2\;,
\end{aligned}
\tag{7.69}
$$

with $c_{0\sigma}$ defined in (7.49), and $n = \sum_\sigma d_\sigma^\dagger d_\sigma$. Even in the present case, we can introduce a Luttinger-Ward functional, here functional of the impurity Green's function only, i.e., $\Phi[\mathcal{G}]$, so that

$$\delta\Phi[\mathcal{G}] = T \sum_{i\epsilon} e^{i\epsilon\eta} \, \Sigma(i\epsilon) \, \delta\mathcal{G}(i\epsilon) \,,$$

and, as usual, $\eta > 0$ is infinitesimal. Similarly, in the generic case in which, because of external fields, $\mathcal{G}_{\sigma_1\sigma_2}(i\epsilon_1, i\epsilon_2) \equiv \mathcal{G}(1, 2)$ becomes non diagonal in frequency and spin, as well as $\Sigma(1, 2)$,

$$\frac{\delta\Sigma(1, 2)}{\delta\mathcal{G}(3, 2)} = \Gamma_0(1, 2; 3, 4) \equiv \Gamma_{0\,\sigma_1,\sigma_2;\sigma_3,\sigma_4}\left(i\epsilon_1, i\epsilon_2; i\epsilon_3, i\epsilon_4\right), \tag{7.70}$$

with $\epsilon_1 + \epsilon_2 = \epsilon_3 + \epsilon_4$ and $\sigma_1 + \sigma_2 = \sigma_3 + \sigma_4$, is by definition the irreducible interaction vertex in the particle-hole channel, which, through the Bethe-Salpeter equation in Fig. 4.32, where now the internal lines are the fully interacting impurity Green's functions, yields the reducible interaction vertex $\Gamma_{\sigma_1,\sigma_2;\sigma_3,\sigma_4}\left(i\epsilon_1, i\epsilon_2; i\epsilon_3, i\epsilon_4\right)$.

For later convenience, we recall that

$$\begin{aligned}
-\langle T_\tau\left(d_\sigma(\tau)\, c_{0\sigma}^\dagger(\tau')\right)\rangle &= \frac{1}{t} \int d\tau_1 \, \mathcal{G}(\tau - \tau_1) \, \Delta(\tau_1 - \tau') \,, \\
-\langle T_\tau\left(c_{0\sigma}(\tau)\, d_\sigma^\dagger(\tau')\right)\rangle &= \frac{1}{t} \int d\tau_1 \, \Delta(\tau - \tau_1) \, \mathcal{G}(\tau_1 - \tau') \,,
\end{aligned} \tag{7.71}$$

where

$$\Delta(\tau) = T \sum_\epsilon e^{-i\epsilon\tau} \, \Delta(i\epsilon) \,,$$

is the hybridisation function in imaginary time.

7.4.1 Ward-Takahashi Identity

The Hamiltonian (7.69) admits as conserved quantities the total charge and any component of the total spin. For simplicity we shall consider the z-component, so that we can use the following two conserved quantities

$$N_\sigma = \sum_\mathbf{k} c_{\mathbf{k}\sigma}^\dagger c_{\mathbf{k}\sigma} + d_\sigma^\dagger d_\sigma = \sum_\mathbf{k} c_{\mathbf{k}\sigma}^\dagger c_{\mathbf{k}\sigma} + n_\sigma \,, \quad \sigma = \uparrow, \downarrow \,. \tag{7.72}$$

We observe that

$$-\frac{\partial n_\sigma}{\partial \tau} = \left[n_\sigma, H\right] = -t\left(c_{0\sigma}^\dagger d_\sigma - d_\sigma^\dagger c_{0\sigma}\right) \equiv J_\sigma \,. \tag{7.73}$$

Fig. 7.6 Diagrammatic representation of the vertex function Λ in (7.74), and of the corresponding Ward-Takahashi identity (7.75), using (7.71). The dashed line between the two black circles represents the hybridisation function $\Delta(\tau)$

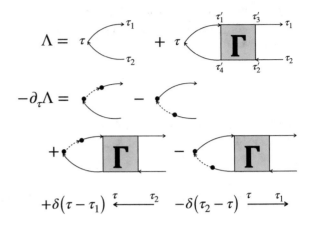

Following what we did in Sect. 4.9, we define, see Fig. 7.6,

$$\Lambda(\tau\,\sigma;\tau_1,\tau_2,\sigma') \equiv -\langle T_\tau\Big(n_\sigma(\tau)\,d^\dagger_{\sigma'}(\tau_1)\,d_{\sigma'}(\tau_2)\Big)\rangle = \delta_{\sigma,\sigma'}\,\mathcal{G}(\tau_2-\tau)\,\mathcal{G}(\tau-\tau_1)$$
$$+ \int d\tau'_1\,d\tau'_2\,d\tau'_3\,d\tau'_4\,\mathcal{G}(\tau'_4-\tau)\,\mathcal{G}(\tau-\tau'_1)$$
$$\Gamma_{\sigma,\sigma';\sigma',\sigma}\big(\tau'_1,\tau'_2;\tau'_3,\tau'_4\big)\,\mathcal{G}(\tau_2-\tau'_2)\,\mathcal{G}(\tau'_1-\tau_1),$$
$$(7.74)$$

and find that [5,6]

$$-\frac{\partial}{\partial\tau}\,\Lambda(\tau\,\sigma;\tau_1,\tau_2,\sigma') = -\langle T_\tau\Big(J_\sigma(\tau)\,d^\dagger_{\sigma'}(\tau_1)\,d_{\sigma'}(\tau_2)\Big)\rangle$$
$$+ \delta_{\sigma,\sigma'}\Big(\delta(\tau-\tau_1)\,\mathcal{G}(\tau_2-\tau) - \delta(\tau-\tau_2)\,\mathcal{G}(\tau-\tau_1)\Big),$$
$$(7.75)$$

shown diagrammatically in Fig. 7.6. If we now take the Fourier transform, we find

$$\delta_{\sigma,\sigma'}\Big(i\omega - \Delta(i\epsilon+i\omega) + \Delta(i\epsilon)\Big)\,\mathcal{G}(i\epsilon+i\omega)\,\mathcal{G}(i\epsilon)$$
$$+ T\sum_{\epsilon'}\Big(i\omega - \Delta(i\epsilon'+i\omega) + \Delta(i\epsilon')\Big)\,\mathcal{G}(i\epsilon'+i\omega)\,\mathcal{G}(i\epsilon')$$
$$\Gamma_{\sigma,\sigma';\sigma',\sigma}(i\epsilon'+i\omega,i\epsilon;i\epsilon+i\omega,i\epsilon')\,\mathcal{G}(i\epsilon+i\omega)\,\mathcal{G}(i\epsilon)$$
$$= \delta_{\sigma,\sigma'}\Big(\mathcal{G}(i\epsilon) - \mathcal{G}(i\epsilon+i\omega)\Big),$$

so that, dividing by $\mathcal{G}(i\epsilon + i\omega)\,\mathcal{G}(i\epsilon)$,

$$\delta_{\sigma,\sigma'} \Big(i\omega - \Delta(i\epsilon + i\omega) + \Delta(i\epsilon) \Big)$$
$$+ T \sum_{\epsilon'} \Big(i\omega - \Delta(i\epsilon' + i\omega) + \Delta(i\epsilon') \Big) \mathcal{G}(i\epsilon' + i\omega)\,\mathcal{G}(i\epsilon')$$
$$\Gamma_{\sigma,\sigma';\sigma',\sigma}(i\epsilon' + i\omega, i\epsilon; i\epsilon + i\omega, i\epsilon')$$
$$= \delta_{\sigma,\sigma'} \Big(\mathcal{G}(i\epsilon + i\omega)^{-1} - \mathcal{G}(i\epsilon)^{-1} \Big)$$
$$= \delta_{\sigma,\sigma'} \Big(i\omega - \Delta(i\epsilon + i\omega) + \Delta(i\epsilon) - \Sigma(i\epsilon + i\omega) + \Sigma(i\epsilon) \Big),$$

thus the final expression of the Ward-Takahashi identity [5,6]

$$\delta_{\sigma,\sigma'} \Big(\Sigma(i\epsilon + i\omega) - \Sigma(i\epsilon) \Big) = -T \sum_{\epsilon'} \Big(i\omega - \Delta(i\epsilon' + i\omega) + \Delta(i\epsilon') \Big) \mathcal{G}(i\epsilon' + i\omega)\,\mathcal{G}(i\epsilon')$$
$$\Gamma_{\sigma,\sigma';\sigma',\sigma}(i\epsilon' + i\omega, i\epsilon; i\epsilon + i\omega, i\epsilon'),$$
$$(7.76)$$

which can also be written as

$$\delta_{\sigma,\sigma'} \frac{\partial \Sigma(i\epsilon)}{\partial i\epsilon} = -\lim_{\omega \to 0^+} \Bigg\{ T \sum_{\epsilon'} \frac{i\omega - \Delta(i\epsilon' + i\omega) + \Delta(i\epsilon')}{i\omega} \mathcal{G}(i\epsilon' + i\omega)\,\mathcal{G}(i\epsilon')$$
$$\Gamma_{\sigma,\sigma';\sigma',\sigma}(i\epsilon' + i\omega, i\epsilon; i\epsilon + i\omega, i\epsilon') \Bigg\}.$$
$$(7.77)$$

The limit $\omega \to 0^+$ of the right hand side requires some care, similar to that we take in Sect. 5.2.1. The main issue is the term

$$\frac{i\omega - \Delta(i\epsilon' + i\omega) + \Delta(i\epsilon')}{i\omega},$$

which is simply one if $\epsilon'(\epsilon' + \omega) > 0$, the semi infinite lines of the imaginary axis within the red regions in Fig. 7.7, while it is

$$\frac{i\omega + 2i\Gamma}{i\omega} \simeq \frac{2i\Gamma}{i\omega},$$

for $-\omega < \epsilon' < 0$, the segment of the imaginary axis within the light green region in Fig. 7.7, thus also leading to a finite result since the sum over ϵ' is proportional to ω. The sum over the over the Matsubara frequency ϵ' in (7.77) can be transformed as usual in the contour integrals shown in Fig. 7.7. The two contours C_1 and C_2 involve singularities of the Green's functions but also of the reducible interaction vertex. However, their sum is simply the right hand side of (7.77) calculated right at $\omega = 0$, thus

$$C_1 + C_2 = -T \sum_{\epsilon'} \mathcal{G}(i\epsilon')^2\, \Gamma_{\sigma,\sigma';\sigma',\sigma}(i\epsilon', i\epsilon; i\epsilon, i\epsilon').$$

On the contrary,

$$
C_3 = -\lim_{\omega \to 0^+} \int \frac{d\epsilon'}{2\pi i} \, f(\epsilon') \, \frac{2i\,\Gamma}{i\omega} \Big\{ \mathcal{G}(\epsilon' + i\omega)\, \mathcal{G}(\epsilon' - i0^+)
$$
$$
\Gamma_{\sigma,\sigma';\sigma',\sigma}(\epsilon' + i\omega, i\epsilon; i\epsilon + i\omega, \epsilon' - i0^+)
$$
$$
- \mathcal{G}(\epsilon' + i0^+)\,\mathcal{G}(\epsilon' - i\omega)\, \Gamma_{\sigma,\sigma';\sigma',\sigma}(\epsilon' + i0^+, i\epsilon; i\epsilon + i\omega, \epsilon' - i\omega) \Big\}.
$$

We take the analytic continuation $i\omega \to \omega + i0^+$, after which

$$
C_3 = -\lim_{\omega \to 0} \int \frac{d\epsilon'}{2\pi i} \, f(\epsilon') \, \frac{2i\,\Gamma}{\omega} \Big\{ \mathcal{G}(\epsilon' + \omega + i0^+)\, \mathcal{G}(\epsilon' - i0^+)
$$
$$
\Gamma_{\sigma,\sigma';\sigma',\sigma}(\epsilon' + \omega + i0^+, i\epsilon; i\epsilon + \omega, \epsilon' - i0^+)
$$
$$
- \mathcal{G}(\epsilon' + i0^+)\,\mathcal{G}(\epsilon' - \omega - i0^+)\, \Gamma_{\sigma,\sigma';\sigma',\sigma}(\epsilon' + i0^+, i\epsilon; i\epsilon + \omega, \epsilon' - \omega - i0^+) \Big\}
$$
$$
= -\Gamma \lim_{\omega \to 0} \int \frac{d\epsilon'}{\pi} \, \frac{f(\epsilon') - f(\epsilon' + \omega)}{\omega} \, \mathcal{G}(\epsilon' + \omega + i0^+)\, \mathcal{G}(\epsilon' - i0^+)
$$
$$
\Gamma_{\sigma,\sigma';\sigma',\sigma}(\epsilon' + \omega + i0^+, i\epsilon; i\epsilon + \omega + i0^+, \epsilon' - i0^+)
$$
$$
\simeq \Gamma \int \frac{d\epsilon'}{\pi} \, \frac{\partial f(\epsilon')}{\partial \epsilon'} \, \mathcal{G}(\epsilon' + i0^+)\, \mathcal{G}(\epsilon' - i0^+)\, \Gamma_{\sigma,\sigma';\sigma',\sigma}(\epsilon', i\epsilon; i\epsilon, \epsilon'),
$$

where we assumed that the interaction vertex is independent of the infinitesimal imaginary part of the external frequencies. In conclusion, if we analytically continue $i\epsilon \to \epsilon + i0^+$ with small ϵ and recalling that $\mathrm{Im}\,\Sigma(\epsilon + i0^+) \sim -\epsilon^2$, we find

$$
\delta_{\sigma,\sigma'} \frac{\partial \Sigma(\epsilon + i0^+)}{\partial \epsilon} \simeq \delta_{\sigma,\sigma'} \left(1 - Z(\epsilon)^{-1} \right) = -T \sum_{\epsilon'} \mathcal{G}(i\epsilon')^2 \, \Gamma_{\sigma,\sigma';\sigma',\sigma}(i\epsilon', \epsilon; \epsilon, i\epsilon')
$$
$$
+ \Gamma \int \frac{d\epsilon'}{\pi} \, \frac{\partial f(\epsilon')}{\partial \epsilon'} \, \mathcal{G}(\epsilon' + i0^+)\, \mathcal{G}(\epsilon' - i0^+)\, \Gamma_{\sigma,\sigma';\sigma',\sigma}(\epsilon', \epsilon; \epsilon, \epsilon').
$$

Because of the Fermi distribution function derivative, $\epsilon' \simeq 0$, so that

$$
\mathcal{G}(\epsilon' + i0^+)\,\mathcal{G}(\epsilon' - i0^+) \simeq \frac{1}{\epsilon' - \epsilon_d - \mathrm{Re}\,\Sigma(\epsilon) + i\Gamma} \, \frac{1}{\epsilon' - \epsilon_d - \mathrm{Re}\,\Sigma(\epsilon) - i\Gamma}
$$
$$
= \frac{1}{2i\Gamma} \left(\mathcal{G}(\epsilon' - i0^+) - \mathcal{G}(\epsilon' + i0^+) \right) = \frac{\pi}{\Gamma}\, A(\epsilon'),
$$

and thus [5,6]

$$
\delta_{\sigma,\sigma'} \frac{\partial \Sigma(\epsilon + i0^+)}{\partial \epsilon} \simeq \delta_{\sigma,\sigma'} \left(1 - Z(\epsilon)^{-1} \right) = -T \sum_{\epsilon'} \mathcal{G}(i\epsilon')^2 \, \Gamma_{\sigma,\sigma';\sigma',\sigma}(i\epsilon', \epsilon; \epsilon, i\epsilon')
$$
$$
+ \int d\epsilon' \, \frac{\partial f(\epsilon')}{\partial \epsilon'} \, A(\epsilon')\, \Gamma_{\sigma,\sigma';\sigma',\sigma}(\epsilon', \epsilon; \epsilon, \epsilon').
$$

$$
\tag{7.78}
$$

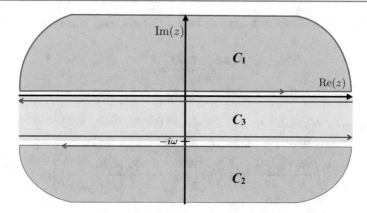

Fig. 7.7 Contour integral used to evaluate the sum over the Matsubara frequency ϵ' in (7.77). The contour actually comprises three different ones, C_1, C_2 and C_3, running anticlockwise and avoiding the imaginary axis, not explicitly shown

7.4.2 Luttinger's Theorem and Thermodynamic Susceptibilities

Since there is no breakdown of perturbation theory whatever the value of U is, Luttinger's theorem must be valid, and thus

$$N_\sigma = T \sum_\epsilon e^{i\epsilon 0^+} \left\{ \sum_{\mathbf{k}} \left(G^{(0)}(i\epsilon, \mathbf{k}) + V_{\mathbf{k}}^2 \, G^{(0)}(i\epsilon, \mathbf{k})^2 \, \mathcal{G}(i\epsilon) \right) + \mathcal{G}(i\epsilon) \right\}$$

$$= N_{0\sigma} + T \sum_\epsilon e^{i\epsilon 0^+} \left(1 - \frac{\partial \Delta(i\epsilon)}{\partial i\epsilon} \right) \mathcal{G}(i\epsilon) \equiv N_{0\sigma} - T \sum_\epsilon e^{i\epsilon 0^+} \frac{\partial \ln \mathcal{G}(i\epsilon)}{\partial i\epsilon} \, ,$$

where $N_{0\sigma}$ is the value in absence of the impurity. Therefore

$$\Delta N_\sigma \equiv N_\sigma - N_{0\sigma} = -T \sum_\epsilon e^{i\epsilon 0^+} \frac{\partial \ln \mathcal{G}(i\epsilon)}{\partial i\epsilon} = \int \frac{d\epsilon}{\pi} \, f(\epsilon) \, \frac{\partial \mathrm{Im} \ln \mathcal{G}(\epsilon)}{\partial \epsilon}$$

$$= -\int \frac{d\epsilon}{\pi} \, \frac{\partial f(\epsilon)}{\partial \epsilon} \, \mathrm{Im} \ln \mathcal{G}(\epsilon) \, ,$$

$$(7.79)$$

where, by definition, $\mathcal{G}(\epsilon) \equiv \mathcal{G}(\epsilon + i0^+)$. Suppose we add to the Hamiltonian the perturbation $\delta H = -\sum_\sigma h_\sigma N_\sigma$, in which case all Green's functions as well as the impurity self-energy become dependent on $h \equiv (h_\uparrow, h_\downarrow)$ and different for both spin directions. Specifically,

$$\mathcal{G}_\sigma(i\epsilon, h) = \frac{1}{i\epsilon - \epsilon_d + h_\sigma - \Delta_\sigma(i\epsilon, h) - \Sigma_\sigma(i\epsilon, h)} = \frac{1}{\mathcal{G}_{0\sigma}(i\epsilon, h) - \Sigma_\sigma(i\epsilon, h)} \, ,$$
$$(7.80)$$

where

$$\Delta_\sigma(i\epsilon) = \Delta(i\epsilon) \sum_{\mathbf{k}} \frac{V_{\mathbf{k}}^2}{i\epsilon - \epsilon_{\mathbf{k}}} \rightarrow \Delta_\sigma(i\epsilon, h) = \sum_{\mathbf{k}} \frac{V_{\mathbf{k}}^2}{i\epsilon - \epsilon_{\mathbf{k}} + h_\sigma} \, ,$$

which, in the large bandwidth limit we have so far assumed, is actually independent of h_σ, thus

$$\frac{\partial \Delta_\sigma (i\epsilon, h)}{\partial h_\sigma} = 0 \,.$$

By definition, the impurity contribution to the susceptibility is

$$
\begin{aligned}
\chi_{\sigma\sigma'}^{\text{imp}} &\equiv \frac{\partial \Delta N_\sigma (h)}{\partial h_{\sigma'}} \Big|_{h=0} = -\int \frac{d\epsilon}{\pi} \frac{\partial f(\epsilon)}{\partial \epsilon} \frac{\partial \text{Im} \ln \mathcal{G}_\sigma (\epsilon, h)}{\partial h_{\sigma'}} \Big|_{h=0} \\
&= \int \frac{d\epsilon}{\pi} \frac{\partial f(\epsilon)}{\partial \epsilon} \, \text{Im} \left[\mathcal{G}(\epsilon) \left(\delta_{\sigma,\sigma'} - \frac{\partial \Sigma_\sigma (\epsilon, h)}{\partial h_{\sigma'}} \Big|_{h=0} \right) \right] .
\end{aligned}
\tag{7.81}
$$

On the other hand[3]

$$\delta \Sigma_\sigma (i\epsilon, h) = -T \sum_{\epsilon'} \sum_{\sigma'} \Gamma_{\sigma,\sigma';\sigma',\sigma} (i\epsilon, i\epsilon'; i\epsilon', i\epsilon) \, \mathcal{G}_{\sigma'} (i\epsilon', h)^2 \, \delta \mathcal{G}_{0\sigma'}^{-1} (i\epsilon', h) \,,$$

and thus

$$
\begin{aligned}
\frac{\partial \Sigma_\sigma (i\epsilon, h)}{\partial h_{\sigma'}} \Big|_{h=0} &= -T \sum_{\epsilon'} \sum_{\sigma''} \Gamma_{\sigma,\sigma'';\sigma'',\sigma} (i\epsilon, i\epsilon'; i\epsilon', i\epsilon) \, \mathcal{G}_{\sigma''} (i\epsilon')^2 \, \frac{\partial \mathcal{G}_{0\sigma''}^{-1} (i\epsilon', h)}{\partial h_{\sigma'}} \Big|_{h=0} \\
&= -T \sum_{\epsilon'} \Gamma_{\sigma,\sigma';\sigma',\sigma} (i\epsilon, i\epsilon'; i\epsilon', i\epsilon) \, \mathcal{G}_{\sigma'} (i\epsilon')^2 \,,
\end{aligned}
$$

which, using (7.78), and after analytic continuation $i\epsilon \to \epsilon + i0^+$, can be also written as

$$\frac{\partial \Sigma_\sigma (\epsilon, h)}{\partial h_{\sigma'}} \Big|_{h=0} = \delta_{\sigma,\sigma'} \left(1 - Z(\epsilon)^{-1} \right) - \int d\epsilon' \, \frac{\partial f(\epsilon')}{\partial \epsilon'} \, A(\epsilon') \, \Gamma_{\sigma,\sigma';\sigma',\sigma} (\epsilon', \epsilon; \epsilon, \epsilon') \,.
\tag{7.82}$$

---

[3] In compact notations,

$$\delta \Sigma = \Gamma_0 \odot \delta \mathcal{G} = -\Gamma_0 \odot \mathcal{G}^2 \odot \left(\delta \mathcal{G}_0^{-1} - \delta \Sigma \right),$$

thus

$$\left(1 - \Gamma_0 \odot \mathcal{G}^2 \right) \odot \delta \Sigma = -\Gamma_0 \odot \mathcal{G}^2 \odot \delta \mathcal{G}_0^{-1} \implies \delta \Sigma = -\left(1 - \Gamma_0 \odot \mathcal{G}^2 \right)^{-1} \odot \Gamma_0 \odot \mathcal{G}^2 \odot \delta \mathcal{G}_0^{-1} \,.$$

Since the Bethe-Salpeter equation reads

$$\Gamma = \Gamma_0 + \Gamma_0 \odot \mathcal{G}^2 \odot \Gamma \implies \left(1 - \Gamma_0 \odot \mathcal{G}^2 \right)^{-1} \odot \Gamma = \Gamma_0 \,,$$

substituting in the previous equation we get the desired result

$$\delta \Sigma = -\Gamma \odot \mathcal{G}^2 \odot \delta \mathcal{G}_0^{-1} \,.$$

Substituting (7.82) in (7.81) we find

$$\chi_{\sigma\sigma'}^{\text{imp}} = \int \frac{d\epsilon}{\pi} \frac{\partial f(\epsilon)}{\partial \epsilon} \text{Im} \left[\mathcal{G}(\epsilon + i0^+) \left(\delta_{\sigma,\sigma'} Z(\epsilon)^{-1} \right. \right.$$
$$\left. \left. + \int d\epsilon' \frac{\partial f(\epsilon')}{\partial \epsilon'} A(\epsilon') \Gamma_{\sigma,\sigma';\sigma',\sigma}(\epsilon', \epsilon; \epsilon, \epsilon') \right) \right].$$
(7.83)

Since both ϵ and ϵ' are small due to the derivatives of the Fermi distribution functions, we expect the interaction vertex to be real and thus the imaginary part is only contributed by $\text{Im}\,\mathcal{G}(\epsilon + i0^+) = -\pi\,A(\epsilon)$, so that

$$\chi_{\sigma\sigma'}^{\text{imp}} = -\int d\epsilon \frac{\partial f(\epsilon)}{\partial \epsilon} A(\epsilon) \left[\delta_{\sigma,\sigma'} Z(\epsilon)^{-1} + \int d\epsilon' \frac{\partial f(\epsilon')}{\partial \epsilon'} A(\epsilon') \Gamma_{\sigma,\sigma';\sigma',\sigma}(\epsilon', \epsilon; \epsilon, \epsilon') \right].$$
(7.84)

For small ϵ, by definition

$$A(\epsilon) \equiv Z(\epsilon)\,A_{\text{qp}}(\epsilon),$$

where the quasiparticle DOS $A_{\text{qp}}(\epsilon)$ is normalised to one and represents a narrow resonance pinned close to zero energy. Therefore, upon defining the quasiparticle scattering amplitudes

$$A_{\sigma,\sigma';\sigma',\sigma}(\epsilon', \epsilon; \epsilon, \epsilon') \equiv Z(\epsilon) Z(\epsilon') \Gamma_{\sigma,\sigma';\sigma',\sigma}(\epsilon', \epsilon; \epsilon, \epsilon'),$$
(7.85)

Equation (7.84) can be written as

$$\chi_{\sigma\sigma'}^{\text{imp}} = -\int d\epsilon \frac{\partial f(\epsilon)}{\partial \epsilon} A_{\text{qp}}(\epsilon) \left[\delta_{\sigma,\sigma'} + \int d\epsilon' \frac{\partial f(\epsilon')}{\partial \epsilon'} A_{\text{qp}}(\epsilon') A_{\sigma,\sigma';\sigma',\sigma}(\epsilon', \epsilon; \epsilon, \epsilon') \right],$$
(7.86)

where the analogy with Landau's Fermi liquid theory is self-evident, see (5.57). If we assume that $h_\uparrow = h_c + h_s$ and $h_\downarrow = h_c - h_s$, where h_c refers to the charge compressibility and h_s to the magnetic susceptibility, we readily find [5,6]

$$\kappa^{\text{imp}} = \sum_{\sigma\sigma'} \chi_{\sigma\sigma'}^{\text{imp}} = -2 \int d\epsilon \frac{\partial f(\epsilon)}{\partial \epsilon} A_{\text{qp}}(\epsilon) \left[1 + \int d\epsilon' \frac{\partial f(\epsilon')}{\partial \epsilon'} A_{\text{qp}}(\epsilon') A^S(\epsilon', \epsilon; \epsilon, \epsilon') \right]$$
$$\simeq 2 A_{\text{qp}}(0) \left[1 - A_S \right],$$

$$\chi^{\text{imp}} = \sum_{\sigma\sigma'} \sigma \sigma' \chi_{\sigma\sigma'}^{\text{imp}} = -2 \int d\epsilon \frac{\partial f(\epsilon)}{\partial \epsilon} A_{\text{qp}}(\epsilon) \left[1 + \int d\epsilon' \frac{\partial f(\epsilon')}{\partial \epsilon'} A_{\text{qp}}(\epsilon') A^A(\epsilon', \epsilon; \epsilon, \epsilon') \right]$$
$$\simeq 2 A_{\text{qp}}(0) \left[1 - A_A \right],$$
(7.87)

where

$$A^S(\epsilon', \epsilon; \epsilon, \epsilon') \equiv \frac{1}{2} \left(A_{\uparrow,\uparrow;\uparrow,\uparrow}(\epsilon', \epsilon; \epsilon, \epsilon') + A_{\uparrow,\downarrow;\downarrow,\uparrow}(\epsilon', \epsilon; \epsilon, \epsilon') \right),$$

$$A^A(\epsilon', \epsilon; \epsilon, \epsilon') \equiv \frac{1}{2} \left(A_{\uparrow,\uparrow;\uparrow,\uparrow}(\epsilon', \epsilon; \epsilon, \epsilon') - A_{\uparrow,\downarrow;\downarrow,\uparrow}(\epsilon', \epsilon; \epsilon, \epsilon') \right),$$

and

$$
A_{qp}(0)\, A_{S(A)} \equiv \int d\epsilon\, d\epsilon'\, \frac{\partial f(\epsilon)}{\partial \epsilon}\, \frac{\partial f(\epsilon')}{\partial \epsilon'}\, A_{qp}(\epsilon)\, A_{qp}(\epsilon')\, A^{S(A)}(\epsilon', \epsilon; \epsilon, \epsilon')\,.
$$

We could readily follow the derivation of the specific heat in a Fermi liquid, see Sect. 5.4.3, through the Ward-Takahashi identity for the heat density, and find in the present case that

$$
c_V^{imp} \simeq -\frac{2}{T} \int d\epsilon\, \epsilon^2\, \frac{\partial f(\epsilon)}{\partial \epsilon}\, A_{qp}(\epsilon) = \frac{2\pi^2}{3}\, T\, A_{qp}(0)\,, \tag{7.88}
$$

namely, the impurity contributes to the specific heat with a linear in temperature term. Similarly, in the Kondo regime $U \gg \Gamma$, valence fluctuations of the impurity are suppressed thus

$$
\begin{aligned}
A_S &= \frac{1}{2}\Big(A_{\uparrow,\uparrow;\uparrow,\uparrow} + A_{\uparrow,\downarrow;\downarrow,\uparrow}\Big) \simeq \frac{A_{\uparrow,\downarrow;\downarrow,\uparrow}}{2} \simeq 1\,, \\
A_A &= \frac{1}{2}\Big(A_{\uparrow,\uparrow;\uparrow,\uparrow} - A_{\uparrow,\downarrow;\downarrow,\uparrow}\Big) \simeq -1\,,
\end{aligned} \tag{7.89}
$$

yielding a Wilson ratio (5.68) $R_W \simeq 2$. We mention that the above procedure can be readily extended to multi-orbital Anderson impurity model that better describe realistic magnetic ions with partially filled d or f shells.

We end remarking that a necessary condition for the local Fermi liquid description to apply is

$$
\begin{aligned}
0 \neq A_{qp}(0) &= \int d\epsilon \left(-\frac{\partial f(\epsilon)}{\partial \epsilon}\right) A_{qp}(\epsilon) = \int d\epsilon \left(-\frac{\partial f(\epsilon)}{\partial \epsilon}\right) Z(\epsilon)^{-1} A(\epsilon) \\
&= \int \frac{d\epsilon}{\pi} \left(-\frac{\partial f(\epsilon)}{\partial \epsilon}\right) \left(1 - \frac{\partial \mathrm{Re}\Sigma(\epsilon)}{\partial \epsilon}\right) \frac{\Gamma - \mathrm{Im}\Sigma(\epsilon)}{\Big(\epsilon - \epsilon_d - \mathrm{Re}\Sigma(\epsilon)\Big)^2 + \Big(\Gamma - \mathrm{Im}\Sigma(\epsilon)\Big)^2}\,,
\end{aligned} \tag{7.90}
$$

and is clearly verified in the single-orbital Anderson impurity model, where $Z(0)$ and $A(\epsilon)$ are both finite, but remains valid also when $\mathrm{Re}\Sigma(\epsilon) \sim 1/\epsilon$, in which case $Z(\epsilon)^{-1} \sim 1/\epsilon^2$ compensates the vanishing $A(\epsilon) \sim \epsilon^2$, which represents the local counterpart of the quasiparticles at Luttinger surfaces discussed in Sect. (5.1). Since the conduction electron T-matrix is $t^2\, \mathcal{G}(\epsilon)$, a $\mathrm{Re}\Sigma(\epsilon) \sim 1/\epsilon$ corresponds to vanishing phase shift.

Problems

7.1 Mean-field approximation of two coupled Anderson impurities—Consider two Anderson impurities with Hamiltonian

$$
H = \sum_{\ell=1}^{2} \left\{ \sum_{\mathbf{k}\sigma} \epsilon_{\mathbf{k}} c^{\dagger}_{\ell\mathbf{k}\sigma} c_{\ell\mathbf{k}\sigma} + \sum_{\mathbf{k}\sigma} V_{\mathbf{k}} \left(c^{\dagger}_{\ell\mathbf{k}\sigma} d_{\ell\sigma} + d^{\dagger}_{\ell\sigma} c_{\ell\mathbf{k}\sigma} \right) \right.
$$
$$
\left. + \frac{U}{2} (n_{\ell} - 1)^{2} \right\} - t_{\perp} \sum_{\sigma} \left(d^{\dagger}_{1\sigma} d_{2\sigma} + d^{\dagger}_{2\sigma} d_{1\sigma} \right),
$$

(7.91)

where $\ell = 1, 2$ labels the two impurities.

- Find under which condition the Hartree-Fock approximation yields spin-polarised impurities.
- Calculate the effective Kondo model in the limit of large U.

References

1. P.W. Anderson, Phys. Rev. **124**, 41 (1961). https://doi.org/10.1103/PhysRev.124.41
2. J. Kondo, Prog. Theor. Phys. **32**, 37 (1964)
3. P.W. Anderson, J. Phys. C: Solid State Phys. **3**(12), 2436 (1970)
4. P. Nozières, J. Low Temp. Phys. **17**(1), 31 (1974). https://doi.org/10.1007/BF00654541
5. A. Yoshimori, A. Zawadowski, J. Phys. C: Solid State Phys. **15**(25), 5241 (1982)
6. L. Mihaly, A. Zawadowski, J. Phys. Lett. (France) **39**, L483 (1978)

Printed in the United States
by Baker & Taylor Publisher Services